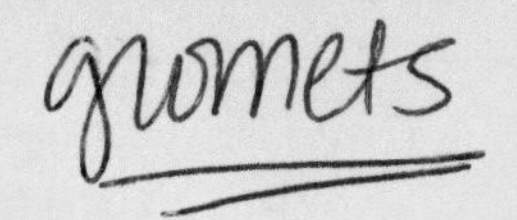

Independent Groups t Test *Continued*

$$[10.5] \quad \hat{s}_{\bar{X}_1-\bar{X}_2} = \sqrt{\left[\frac{(n_1-1)\hat{s}_1^2 + (n_2-1)\hat{s}_2^2}{n_1+n_2-2}\right]\left(\frac{1}{n_1}+\frac{1}{n_2}\right)}$$

$$[10.6] \quad \hat{s}_{\bar{X}_1-\bar{X}_2} = \sqrt{\left(\frac{SS_1 + SS_2}{n_1+n_2-2}\right)\left(\frac{1}{n_1}+\frac{1}{n_2}\right)}$$

$$[10.7] \quad t = \frac{(\bar{X}_1-\bar{X}_2)-(\mu_1-\mu_2)}{\hat{s}_{\bar{X}_1-\bar{X}_2}} \qquad [10.10] \quad \text{eta}^2 = \frac{SS_{\text{EXPLAINED}}}{SS_{\text{TOTAL}}} \qquad [10.11] \quad \text{eta}^2 = \frac{t^2}{t^2+\text{df}}$$

Correlated Groups t Test

$$[11.1] \quad \hat{s}_{\bar{D}} = \frac{\hat{s}_D}{\sqrt{N}} \qquad [11.2] \quad t = \frac{\bar{D}-\mu_{\bar{D}}}{\hat{s}_{\bar{D}}} \qquad [11.4] \quad \text{eta}^2 = \frac{t^2}{t^2+\text{df}}$$

One-Way Between-Subjects Analysis of Variance

$$[12.10] \quad F = \frac{MS_{\text{BETWEEN}}}{MS_{\text{WITHIN}}} \qquad [12.15] \quad \text{eta}^2 = \frac{SS_{\text{BETWEEN}}}{SS_{\text{TOTAL}}}$$

$$[12.16] \quad \text{eta}^2 = \frac{(\text{df}_{\text{BETWEEN}})F}{(\text{df}_{\text{BETWEEN}})F + \text{df}_{\text{WITHIN}}} \qquad [12.17] \quad CD = q\sqrt{\frac{MS_{\text{WITHIN}}}{n}}$$

One-Way Repeated Measures Analysis of Variance

$$[13.13) \quad F = \frac{MS_{\text{TREATMENTS}}}{MS_{\text{ERROR}}} \qquad [13.14] \quad \text{eta}^2 = \frac{SS_{\text{TREATMENTS}}}{SS_{\text{TREATMENTS}} + SS_{\text{ERROR}}}$$

$$[13.15] \quad \text{eta}^2 = \frac{(\text{df}_{\text{TREATMENTS}})F}{(\text{df}_{\text{TREATMENTS}})F + \text{df}_{\text{ERROR}}} \qquad [13.16] \quad CD = q\sqrt{\frac{MS_{\text{ERROR}}}{N}}$$

Chi-Square Test

$$[15.1] \quad E_j = \left(\frac{CMF_j}{N}\right)(RMF_j) \qquad [15.2] \quad \chi^2 = \sum\frac{(O_j-E_j)^2}{E_j} \qquad [15.5] \quad V = \sqrt{\frac{\chi^2}{N(L-1)}}$$

Two-Way Between-Subjects Analysis of Variance

$$[17.19] \quad F_A = \frac{MS_A}{MS_{\text{WITHIN}}} \qquad [17.20] \quad F_B = \frac{MS_B}{MS_{\text{WITHIN}}}$$

$$[17.21] \quad F_{A\times B} = \frac{MS_{A\times B}}{MS_{\text{WITHIN}}} \qquad [17.22] \quad \text{eta}_A^2 = \frac{SS_A}{SS_{\text{TOTAL}}}$$

$$[17.23] \quad \text{eta}_B^2 = \frac{SS_B}{SS_{\text{TOTAL}}} \qquad [17.24] \quad \text{eta}_{A\times B}^2 = \frac{SS_{A\times B}}{SS_{\text{TOTAL}}}$$

$$[17.25] \quad CD = q\sqrt{\frac{MS_{\text{WITHIN}}}{nb}} \qquad [17.26] \quad CD = q\sqrt{\frac{MS_{\text{WITHIN}}}{na}}$$

STATISTICS FOR THE BEHAVIORAL SCIENCES

SECOND EDITION

JAMES JACCARD & MICHAEL A. BECKER

STATE UNIVERSITY OF NEW YORK AT ALBANY

THE PENNSYLVANIA STATE UNIVERSITY AT HARRISBURG

Wadsworth Publishing Company
Belmont, California
A Division of Wadsworth, Inc.

To Liliana and Sara (JJ)
To my family, most especially Rhonda, Emily, and Janna (MAB)

Credits

P. 23, from *How to Lie with Statistics* by Darrell Huff, pictures by Irving Geis, by permission of W. W. Norton & Company, Inc., and A. Watkins, Inc. Copyright 1954 by Darrell Huff and Irving Geis. Copyright renewed 1982 by Darrell Huff and Irving Geis. ▪ Table 7.5, p. 159, and Figure 14.1, p. 333, from E. W. Minium, *Statistical Reasoning in Psychology and Education,* 1978. Used by permission of John Wiley & Sons. ▪ Figures 7.1, p. 161, and 17.1, p. 421, adapted from M. K. Johnson and R. M. Liebert, *Statistics: Tool of the Behavioral Sciences* © 1977. Reprinted by permission of Prentice-Hall, Inc., Englewood Cliffs, New Jersey. ▪ Figure 8.1, p. 173, adapted from R. B. McCall, *Fundamental Statistics for Psychology,* Third Edition. Harcourt Brace Jovanovich, 1980. ▪ Figure 8.2, p. 177, from *Statistics* by R. S. Witte. Copyright © 1980 by Holt, Rinehart & Winston, Inc. Reprinted by permission of the publisher. ▪ P. 203, "Mind Your Ms & Qs," from *Rival Hypotheses: Alternative Interpretations of Data-Based Conclusions* by Schuyler W. Huck and Howard M. Sandler. Copyright © 1979 by Schuyler W. Huck and Howard M. Sandler. Reprinted by permission of Harper & Row, Publishers, Inc.

Psychology Editor: Kenneth King
Editorial Assistant: Michelle Palacio
Production Editor: Vicki Friedberg
Designer: James Chadwick
Print Buyer: Randy Hurst
Permissions Editor: Robert M. Kauser
Copy Editor: Susan Reiland
Technical Illustrator: Judith Ogus/Random Arts
Compositor: G&S Typesetters
Cover Designer: James Chadwick
Cover Photograph: Copyright © Randy Green

Printed in the United States of America 34
4 5 6 7 8 9 10—94 93 92

Library of Congress Cataloging-in-Publication Data
Jaccard, James.
Statistics for the behavioral sciences.
Bibliography: p.
Includes index.
1. Psychometrics. I. Becker, Michael A., 1953– .
II. Title.
BF39.J28 1990 519.5′0243 89-16417
ISBN 0-534-10326-X

BRIEF CONTENTS

CONTENTS

PREFACE

While developing the outline for this text, we were haunted by our own reactions to new statistics books: "Oh no, not *another* introductory statistics text." There are dozens of introductory statistics books available, some of which are excellent. Several have been on the market for a decade and have had the benefit of two or more revisions. Introductory statistics, unlike content areas in the behavioral sciences, does not become dated quickly. Many of the concepts taught 10 years ago are still relevant today. So why another text?

Despite the existence of some excellent books, we have been unable to locate any that accomplish our personal goals in teaching introductory statistics at the undergraduate level. What follows is an elaboration of some of these goals and of how this text differs from those currently available.

Application and Integration

In our opinion, most introductory statistics texts fail to integrate sufficiently the subject matter of statistics with what students will encounter in behavioral science journals. A statistics course should not only teach students basic skills for analyzing data but also help make them intelligent consumers of scientific information. Students need to know how to make sense of research reports and journal articles, and a firm grasp of statistics is a major step in this direction.

In our view, there is a large gap between what students learn in statistics classes and the way statistics are presented in research reports. Accordingly, one goal of the present book is to teach the reader how to present the results of statistical analyses when writing research reports. By definition, this also conveys to the reader the form in which he or she will encounter statistical information when *reading* research reports. Most statistics texts emphasize the importance of stating null and alternative hypotheses, critical values, and formal decision rules for drawing inferences with respect to the null and alternative hypotheses. Yet these are rarely explicitly stated in research reports, and students often find this omission confusing. In fact, phrases like "the means were found to be significantly different, $t(18) = 2.89$, $p < .01$" have usually never been encountered by students who have had introductory statistics. The present book attempts to confront this discrepancy. A special section, "Method of Presentation," will be found at the end of most chapters. This section provides examples of how statistical analyses are typically presented in research reports (using the format of the American Psychological Association) and discusses the rationale underlying the format of such presentation.

Learning When to Use Statistical Tests

Because of the way chapters and exercises are organized in most texts, students are essentially told which statistical procedure to use on a given set of data. This state of affairs is simply unrealistic. It is just as important to teach students *when* to use a particular statistic and *why* it should be used as it is to teach them how to *compute* and *interpret* the statistic. Given a set of data, many students cannot determine where to begin in answering relevant research questions. We have attempted to address this problem. To be sure, it is impossible to state any hard-and-fast rules for data analysis. It is always possible to find an exception or alternative procedures that might give better insights into the data. But some rough guidelines can be given, and the issues involved in selecting a statistical test can be explicated. In the present text, each statistical technique is introduced by giving instances where the test is most typically applied, and an interesting research example is then given. Chapter 18 develops in detail issues to consider when selecting a statistical test to analyze one's data.

Relevance

A common complaint among students who take statistics courses is that statistics is irrelevant and boring. This view is fostered, in part, by the tendency of statistics texts to use examples and exercises that *are* irrelevant and boring. Sometimes mundane examples best illustrate the concepts. However, in many instances, it is possible to provide students with interesting applications of statistics. We have accomplished this in two ways: First, in most chapters, we have included a section ("Examples from the Literature") that presents interesting published research findings that are based on the statistical method being developed. Second, where relevant, chapter exercises present problems in which students are asked to take representative data from research that has been reported in the behavioral science literature and to perform statistical analyses using what they have learned. Supplementing these are other exercises to help students learn and understand the material covered in the chapter.

Unifying Theme

During the years we have taught statistics, we have found that students do not readily understand the common focus of the various statistical tests. Most students simply cannot appreciate the conceptual relationship between the t test, analysis of variance, Pearson correlation, and the chi-square test. As a result, no unifying theme is developed and students lose sight of the purpose of statistical analysis. In the present book, a unifying structure is provided. The fact that each of the major statistical tests concerns the relationship between variables is made explicit: The t test and analysis of variance usually (but not always) are applied when analyzing the relationship between a qualitative independent variable and a quantitative dependent variable; Pearson correlation is applied when analyzing the relationship between two quantitative variables; and the chi-square test is applied when analyzing the relationship between two qualitative variables. Three questions serve as the organizing framework for each test: (1) Given sample data, can we infer that a relationship exists between two variables in the population? (2) If so, what is the strength of the relationship? And, (3) if so, what is the nature of the relationship? As an example, in analysis of variance, the first question can be addressed by the test of the null hypothesis in the form of the test of the *F* ratio, the second by eta-squared, and the third by the Tukey HSD test. By relating these three questions to each of the major tests, a unified framework emerges.

Conceptual Versus Computational Emphasis

This book emphasizes a conceptual understanding of statistics. With rapid progress in the computer field and with the widespread use of hand calculators geared toward statistics, it seems unwise to spend considerable time on computational formulas and methods of calculation. Very few students who take introductory statistics ever find it necessary to calculate statistics by hand. Rather, they read about them in research reports or learn to program a computer to do the calculations. Because of this, we emphasize both conceptual *and* computational formulas in this book—computational formulas are introduced after the conceptual formulas have been introduced and reinforced.

Research Design

Another unique characteristic of this text is a chapter on research methods. We have always believed that statistics and research design are intertwined and that statistics should be placed in the context of design concerns. For example, how can students really grasp the meaning of error variance without some elementary understanding of disturbance variables? Chapter 9 is intended to provide an appropriate research context. In addition, each research example used to develop a statistical technique is discussed in the context of its methodological constraints. We hope this will encourage the student to consider the results of statistical analysis in a broader sense than most statistics books convey.

Advanced Students

We have included a special feature for advanced students. Appendixes to several chapters explain certain advanced concepts referred to in the body of the text in more detail; these appendixes are generally written at a higher level and can be readily excluded from class presentation.

Material Covered

At first glance, the table of contents might suggest that this text is more advanced than the typical introductory statistics book. This is not the case. We recognize that different instructors emphasize different material. An introductory class could not even begin to cover all 18 chapters of the present book. The chapters included are intended to provide the instructor with a useful set of topics from which to choose. The order of chapters is flexible, except for natural progressions (for example, no one would cover *t* tests before they covered means and standard deviations). The material not covered in class will be available as reference for students who pursue graduate work or advanced undergraduate research projects.

In talking with statistics instructors, we have found that one of the main differences in teaching statistics is in the treatment of probability. Some instructors prefer to cover it in some detail (as we have done in Chapter 6), while others prefer to give it less emphasis. *All* instructors recognize that probability is a key concept in statistics. However, some feel that topics such as conditional probabilities, joint probabilities, and sampling with versus without replacement have little practical relevance for statistical applications (for example, for computing *t* tests or analysis of variance). Thus, we have written Chapter 6 so that it can be omitted without disrupting succeeding chapters. The concept of probability is discussed in Chapters 1–4 in sufficient detail to give students the necessary appreciation for later statistical tests.

A second major difference among instructors is in the introduction of correlation and regression. Some instructors introduce these concepts early in the course in the context of descriptive statistics, whereas others introduce the material in the context of inferential statistics. Our own preference is for the latter, because it is rare that correlation and regression are used in a purely descriptive manner. We therefore omit Chapter 5 in the early part of the course and then assign it along with Chapter 14 after discussing analysis of variance. Instructors who prefer otherwise can present the chapters in sequence. We should note that Chapter 5 uses purely descriptive formulas (e.g., for the standard error of estimate), whereas some texts use inferential formulas when discussing correlation/regression in descriptive contexts. The inferential formulas are presented in Chapter 14 in the present book.

In our own one-semester courses (which are *very* introductory), we omit Chapters 6, 16, and 17. We try to give students a brief, one- or two-lecture overview of the remaining chapters and encourage them to read what we could not cover. Also, within certain chapters, we skip selected sections (for example, percentiles) so that we can emphasize material we think is more appropriate *for our particular students*. We have tried to structure sections within chapters so that instructors who want to skip a topic can easily do so.

We have also focused discussion on the most common techniques in the behavioral science literature. This is not to say that the omitted concepts are not important. The decision to exclude these reflects space demands and a cost-benefit analysis of what students need from the course more than anything else.

Chapter Structure

Chapters 10 to 17 develop the major statistical tests typically introduced in beginning statistics courses. We have imposed a common structure on these chapters in order to underscore the common focus of the tests.

Each chapter begins with a discussion of the conditions under which the test is typically applied. Attention then turns to inferring whether a relationship exists between the variables under study and, if so, to determining the strength and nature of the relationship. These issues are developed in the context of a research example, and critical computational stages are highlighted with study exercises that occur within the chapter. The test is then placed in context via a section on methodological considerations. This section underscores the importance of interpreting statistics relative to research design considerations. Then a numerical example takes the student through an application of the statistical test from start to finish. A discussion of planning an investigation using the test is then presented, with explicit consideration of power and sample size selection. The Method of Presentation section then discusses how the statistical results will typically be reported in journal articles, and several examples from the literature follow. The exercises for each chapter are of two types: (1) exercises designed to review and reinforce concepts the students have learned in the chapter and (2) exercises that require students to apply these concepts to real research situations.

Changes from First Edition

One major change from the first edition is that the variance extraction approach has been replaced as a major emphasis by more traditional approaches and formulas when discussing the correlated groups t test and repeated measures analysis of variance. At the same time, recognizing the advantages of a variance extraction approach, much of this material has been retained in a subsidiary role.

A second major change is that the material on Pearson correlation and regression has been reorganized such that regression now receives greater attention than it did previously, and descriptive and inferential aspects of these techniques are discussed in separate chapters (Chapters 5 and 14, respectively). Within each of these chapters, there are now separate, though interrelated, sections for correlation and regression.

Other new material includes the graphing of cumulative frequencies in Chapter 2, discussion of sphericity in Chapter 13, and an introduction to some advanced statistical techniques in Chapter 18. We have also adopted the use of X notation instead of Y notation in the formulas. Material deemed unnecessary for beginning statistics students has been deleted to streamline the presentation. In other instances, material has been retained but moved to more suitable locations. Lastly, the book has been made more "user friendly," which is manifested in numerous ways. For instance, we now refer to specific columns within a table rather than just to "Table X," and as a means of jarring readers' memories, we frequently refer to earlier chapters.

Acknowledgments

Many individuals have helped in the development of this book. First, we would like to acknowledge the excellent work of the staff at Wadsworth. Vicki Friedberg kept things moving along in an orderly and efficient manner and was a constant source of information and encouragement. The copy editing of Susan Reiland was superb. The rest of the production team, most particularly the designer, James Chadwick, went above and beyond the call of duty to make this book a better product. We would also like to thank our editor, Ken King, for his efforts.

We would like to thank the following reviewers of the second edition for their helpful comments: Raymond P. Carlson, Bemidji State University; John C. Jahnke, Miami University; William E. Jaynes, Oklahoma State University; Jack Kirschenbaum, Fullerton College; Mary E. Kite, Ball State University; George O. Rogers, Oak Ridge National Laboratory; Lanna Ruddy, State University of New York at Geneseo, who also wrote the Study Guide; Kirk H. Smith, Bowling Green State University; and Robert B. Stewart, Oakland University. We would also like to thank Doug Mandra for checking the accuracy of all the statistical calculations and for writing the Instructor's Manual.

We would also like to acknowledge those colleagues who reviewed the manuscript, either in part or in its entirety, during preparation of the first edition: Teresa Amabile, Brandeis University; Roger Baumgarte, Winthrop College; David Brinberg, State University of New York at Albany; Stephen Edgell, University of Louisville; Scott E. Graham, Allentown College; Alfred Hall, College of Wooster; Stephen W. Hinkle, Miami University, Ohio; John M. Knight, Central State University, Oklahoma; Scott E. Maxwell, University of Notre Dame; Ervin M. Segal, State University of New York at Buffalo; Karyl Swartz, Herbert Lehman College, CUNY; Jerry W. Thornton, Angelo State University; Brian A. Wandell, Stanford University; and Arnold Well, University of Massachusetts, Amherst.

If there is any merit in this book, it may well be the result of input from the above individuals. Of course, any shortcomings are our responsibility alone.

Typing of various portions of the manuscript was performed by Judy Dashnaw, Marilyn Bazzett, and Kathy Ritter, all of whom extended themselves to help us meet ever-looming deadlines. Additional help with indexing, proofreading, and various other aspects of manuscript preparation was provided by Flossie Wolf and Rhonda Becker. Finally, we would like to acknowledge the support and patience of our families, who were most understanding throughout this project.

TO THE STUDENT

This is an introductory text designed for a first course in statistics. We have written the text assuming the student possesses a minimum of mathematical background (basic algebra). For most of you, much of the material in the book will be new. Our experience in teaching statistics has led us to conclude that one of the most difficult things for students is the amount of new material they must assimilate and use. Statistics courses are unique in several respects. Later material relies heavily on a clear understanding of previous material. There is a continual building process, and you *must* keep up with the pace your instructor sets. Statistics is not the kind of material you can put off until the night before an exam and then cram for a test on the next day.

We have developed a number of features in the text that should help you in your study of statistics. First, we have included examples (called "Study Exercises") that present problems based upon the material just covered. Working through these examples will help you to acquire many important statistical concepts. Second, key terms are **boldfaced**. These terms should be reviewed after reading each chapter. Make sure you understand and can define each one. Third, extensive exercises have been provided. We strongly recommend that you work through *all* of these exercises, as they reinforce much of what you read. Some of the exercises are worked out step-by-step in the Study Guide. Fourth, we have included examples of interesting research that uses the concepts developed earlier in the chapter. If you read these carefully, not only will you learn a good deal about behavioral science research, but you will also be able to appreciate more fully the role of statistics. Fifth, we have included Method of Presentation sections that describe how statistical tests are reported in professional journals and reports. These sections should help you to understand more fully the material in the book and also the material you read in your study of the behavioral sciences.

1

STATISTICAL PRELIMINARIES

ONE

INTRODUCTION AND MATHEMATICAL PRELIMINARIES

1.1

The Study of Statistics

It has become common for courses in statistics to be required of students majoring in the behavioral sciences. Many such students question why statistical training is necessary. There are several reasons. Statistics is an integral part of research activity. Important questions and issues are addressed in behavioral science research, and statistics can be a valuable tool in developing answers to these questions. For the student who makes a career of conducting research, statistical analysis should prove to be a useful aid in the acquisition of knowledge.

But the fact is, many students who take statistics courses will not develop careers that require an active part in research. Although these students may not actually conduct research, they may be required to read, interpret, and use research reports. These reports will usually rely on statistical analyses to draw conclusions and suggest courses of action. Knowledge of statistics is therefore important to help one understand and interpret these reports.

Research that uses statistical analysis is clearly having a great impact on society, both in our everyday lives and in more abstract situations. On television we see commercials that report research "demonstrating" that brand A is 3 times as effective as brand X. In national magazines and newspapers we read the results of surveys of public opinion and attitudes toward politicians. Many magazines include special sections designed to disseminate to the public at large the results of research in the physical and behavioral sciences. As our society becomes more technologically complex, greater demands will be placed on professionals to understand and use results of research designed to answer applied problems. This will generally require a working understanding of statistical methods.

A knowledge of statistical analysis may also help to foster new and creative ways of thinking about problems. Several colleagues have remarked on the new insights they developed when they approached a problem from the perspective of statistical analysis. Statistical "thinking" can be a useful aid in suggesting alternative answers to questions and posing new ones. In addition, statistics helps to develop one's skills in critical thinking, in terms of both inductive and deductive logic. These skills can be applied to any area of inquiry, and hence are extremely useful.

1.2 Research in the Behavioral Sciences

The major concern with statistics in this book is how they are used in behavioral science research. As such, it will be useful to consider briefly the research process as it commonly occurs in the behavioral sciences.

Most people do not view scientific research as a process but rather as a product. Reference is made to a "body of facts" that is known about some phenomenon. Scientific research is better characterized as an ongoing process consisting of five stages. The first stage is the formulation of a question about some phenomenon or phenomena. Why do people smoke marijuana? Why do some children do better in school than others? Why do some people fail to help another person who is in need of help? The second stage is forming a **hypothesis** concerning the question. A hypothesis is a statement proposing that something is true about a given phenomenon. One might hypothesize that people smoke marijuana because of pressures from their peers to do so. Or one might hypothesize that children's school performance is influenced by the value placed on education in the home. The third stage involves designing an investigation to test the validity of the hypothesis. In such an investigation one makes systematic observations of individuals or groups of individuals in settings that are conducive to testing the hypothesis. The fourth stage is analyzing the data collected in the investigation in order to help the researcher draw the appropriate conclusions. This is generally done with the aid of statistics. The final stage is drawing a conclusion and thinking about the implications of the investigation for future research.

1.3 Variables

Most behavioral science research is concerned with relationships between **variables.** A variable is a phenomenon that takes on different values, or *levels*. For example, gender is a variable that takes on two values, male and female.

Weight is a variable that takes on values such as 101 pounds, 78 pounds, and so on. In contrast, a **constant** does not vary within given constraints. For instance, the value to four decimal places for the mathematical quantity π (pi) is always 3.1416. Since it takes on only one value that never changes, π is a constant. If an investigation is conducted only with females, then in the context of that investigation, gender is a constant because it takes on one and only one value (female).

Researchers distinguish between variables in many ways. One such distinction in the behavioral sciences is between an **independent variable** and a **dependent variable.** Suppose an investigator is interested in the relationship between two variables, the effect of information about the gender of a job applicant on hiring decisions by personnel managers. An experiment might be designed in which 50 personnel managers are provided with descriptions of job applicants and asked whether they would hire that applicant. The applicant is described in the same way on several pertinent dimensions to all 50 managers. The only difference is that 25 of the managers are told that the applicant is female, and the other 25 managers are told that the applicant is male. Each manager then indicates his or her hiring decision. In this experiment the information about the gender of the applicant is the independent variable and the hiring decision is the dependent variable. The hiring decision is termed the *dependent variable* because it is thought to "depend on" the information about the gender of the applicant. In other words, the independent variable is presumed to influence the dependent variable.

A useful tool for identifying independent and dependent variables is the phrase "The effects of __________ on __________." The variable that fits into the first blank is the independent variable, and the variable that fits into the second blank is the dependent variable. For example, in a study on the effects of sugar on the taste of coffee, the independent variable is the amount of sugar that is present and the dependent variable is the taste of the coffee. Similarly, if the effects of child-rearing practices on intelligence are being studied, the independent variable is the type of child-rearing practice and the dependent variable is the child's intelligence level.

The term *independent variable* has assumed different meanings in various areas of the behavioral sciences. Some investigators restrict the definition of an independent variable to a variable that is explicitly manipulated in the context of an experiment (such as the information about the gender of the applicant in the hiring example). We will adopt the more general definition of an independent variable as any variable that is presumed to influence a second variable (the dependent variable). According to this definition, it is not necessary for a variable to be experimentally manipulated in order to be conceptualized as an independent variable. Note that just because a researcher presumes that one variable influences another does not necessarily mean that it does. This is only a presumption made for purposes of the investigation.

1.4

Measurement

A major feature of behavioral science research is **measurement.** Measurement involves the assignment of phenomena such as people, objects, and events to numerical categories according to a set of rules. With respect to the variable of gender, for example, there are two categories into which we classify people: male and female. When a teacher assigns grades, five categories are typically used: A, B, C, D, and F. Categories used in assigning typing speed include 86 words per minute, 27 words per minute, 40 words per minute, and so forth. Obviously, there are many different types of rules one may use when making numerical assignments and, hence, many different levels or types of measurement. We will discuss the four levels of measurement typically used in the behavioral sciences: nominal, ordinal, interval, and ratio.

Nominal measurement involves using numbers merely as labels. An investigator might classify people according to their religion—Catholic, Protestant, Jewish, and all others—and use the numbers 1, 2, 3, and 4 to refer to these categories. In this case, the numbers have no special quality about them; they are merely used as labels. If we wished, we could have used any other set of numbers instead (for instance, 13, 48, 7, and 101). In behavioral science research, the basic statistics of interest for variables that involve nominal measurement are usually frequencies (for example, how many people are Democrats, how many are Republicans), proportions, and percentages.

Another level of measurement is **ordinal measurement.** A variable is said to be measured on an ordinal level when the categories can be *ordered* on some continuum or dimension. Suppose we take four individuals who differ in height and assign the number 1 to the shortest individual, the number 2 to the next shortest individual, the number 3 to the next shortest individual, and the number 4 to the tallest individual. In this case, we have measured height on an ordinal level, as it allows us to order the individuals from shortest to tallest. Thus, ordinal measurement allows the researcher to classify individuals into different categories that, in turn, are ordered along a dimension of interest.

Note in the preceding example that height was *not* measured in terms of feet or inches. The shortest individual had a score of 1 on the measurement scale, the next shortest individual had a score of 2, and so on. This set of measures (that is, the rank order from shortest to tallest) exhibits ordinal characteristics. As we will subsequently illustrate, height can be measured in other ways whereby the measures have more than just ordinal characteristics.

A third level of measurement is **interval measurement.** Interval measures have all the properties of ordinal measures but allow us to do more than order objects on a dimension. They have the additional property that numerically equal distances on the scale represent equal distances on the dimension being measured. For example, when temperature is measured in degrees Fahrenheit, the difference between 68 and 70 degrees is the same as the difference between 101 and 103 degrees. In both instances, the difference is 2 degrees. If we were to separately add each of these 2 degrees to the air temperature, the level of heat would increase by the identical amount, since any 2 degrees represent the same amount of heat as any other 2 degrees. Thus, interval measures not only provide us with information about the ordering of individuals on a dimension, but also with information about the *magnitude of differences* between scale units. Note that this was not true in the previous example using an ordinal measure of height. It was not necessarily true that the difference in height between individuals 3 and 4 was the same as the difference in height between individuals 1 and 2. It was only true that individual 4 was taller than individual 3, who, in turn, was taller than individual 2, who, in turn, was taller than individual 1.

A fourth level of measurement is **ratio measurement.** Ratio measures have all the properties of interval measures (and, hence, ordinal measures as well), but in addition possess a natural or *absolute zero point.* This allows the investigator to specify the exact amount of the property being measured. In measuring height, for example, 0 inches implies no height whatsoever. For this measurement scale, there is an absolute zero point. It follows that an individual who is 80 inches tall is twice as tall as an individual who is 40 inches tall. In contrast, interval measures do not have an absolute zero point. For instance, a temperature of 0 degrees Fahrenheit does not imply a complete absence of heat. Rather, the zero point on the Fahrenheit scale, as with all interval measures, is *arbitrary.* Without an absolute zero point, ratio statements such as the preceding are not possible.

We can illustrate the differences between ordinal, interval, and ratio levels of measurement by examining three different ways of measuring the height of buildings. Suppose an architect who specializes in determining the capacity of buildings to survive an earthquake is hired to study all buildings that are taller than 100 feet in a particular city. Figure 1.1a shows graphically the height of four such buildings and indicates how tall each one is. The first way of measuring the height of these buildings is to assign the number 1 to the shortest building, the number 2 to the next shortest building, the number 3 to the next shortest building, and the number 4 to the tallest building (Figure 1.1b). This assignment represents ordinal measurement. It allows us to order the buildings on a dimension of height, but does not tell us anything about the magnitude of the heights. A second method is to measure by how many feet each building exceeds the 100-foot criterion. In this case, we find that building D is 2 feet taller than the criterion, building B is 4 feet taller than the criterion, building C is 80 feet taller than the criterion, and building A is 104 feet taller than the criterion (Figure 1.1c). In contrast to the previous (ordinal) measurement, not only can we order the buildings on a dimension of height, but also we have information about the relative magnitude of the heights: Building B is 2 feet taller than building D ($4 - 2 = 2$), building C is 76 feet taller than building B ($80 - 4 = 76$), and so on. We have measured height on an interval scale. Note

FIGURE 1.1 Three Different Ways of Measuring the Height of Four Buildings

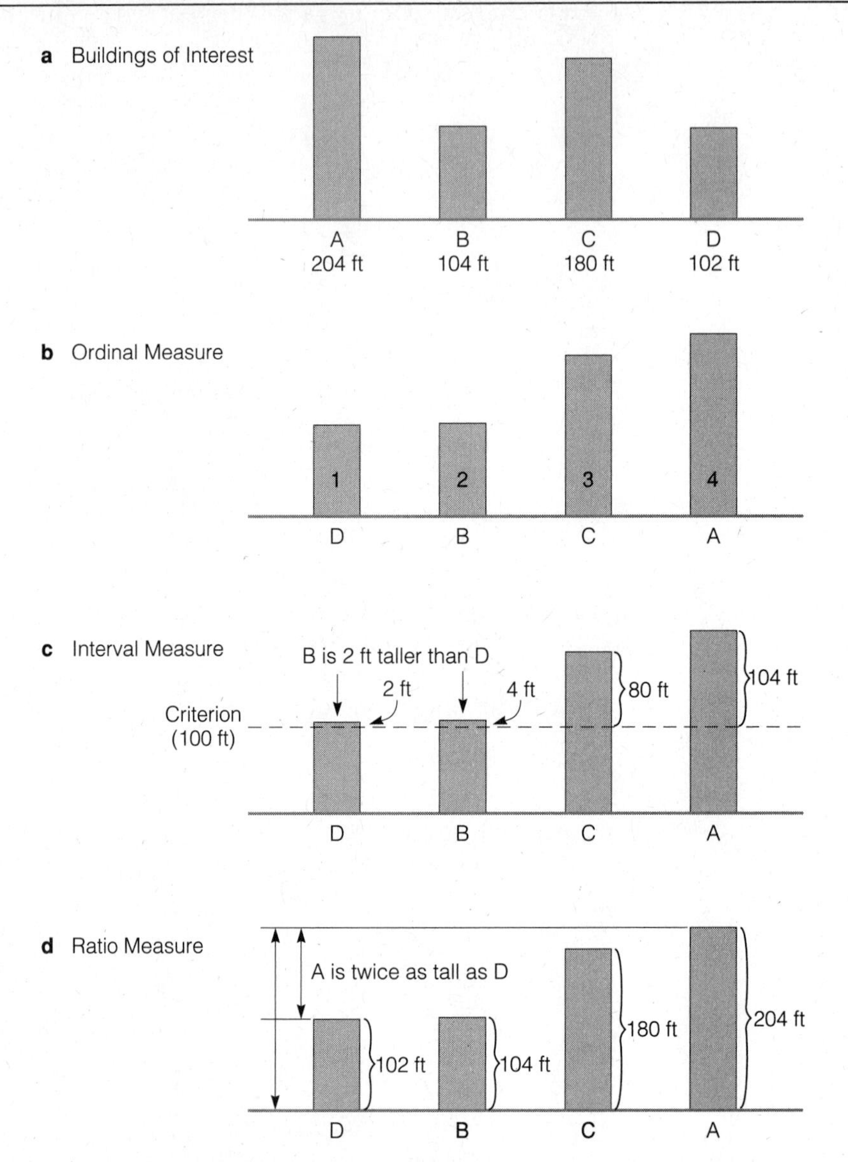

that on this scale even though building B has a score of 4 (that is, it is 4 feet above the criterion) and building D has a score of 2 (it is 2 feet above the criterion), it is not the case that building B is twice as tall as building D. We cannot make a ratio statement because we do not have an absolute zero point; all measures were taken relative to an arbitrary criterion (100 feet). Finally, we can measure each building from the ground, yielding an absolute zero point. Building D is 102 feet, building B is 104 feet, building C is 180 feet, and building A is 204 feet. We can now state with confidence that building A is twice as tall as building D (Figure 1.1d). We have measured height on a ratio scale.

The four types of measurement just outlined can be thought of as a hierarchy. At the lowest level, nominal measurement allows us to categorize phenomena into different groups. The second level, ordinal measurement, not only allows us to classify phenomena into different groups, but also indicates the relative ordering of the groups on a dimension of interest. Interval measurement, the next level, possesses the same properties as ordinal measurement but, in addition, is sensitive to the magnitude of the differences in the groups on the dimension. However, ratio statements are not possible at this level. It is only at the final level, ratio measurement, that such statements are possible. Ratio measures have all of the properties of nominal, ordinal, and interval measures, but also permit ratio judgments to be made.

An important distinction can be made between nominal measurement, on the one hand, and ordinal, interval, and ratio measurement, on the other. A variable measured at one of the latter levels takes on an ordered set of values along some dimension. Scores can thus be ordered on the dimension in question depending on their values. In contrast, scores measured on a nominal level cannot be ordered. Rather, they merely distinguish among categories.

Variables measured on ordinal, interval, or ratio levels are known as **quantitative variables,** while variables measured on a nominal level are called **qualitative variables.** Any variable can be classified as either quantitative or qualitative. As we will see later, the distinction between quantitative and qualitative variables is crucial in statistics.

When reading a research report, you may encounter references to interval scales, ordinal scales, and so on. Technically, the use of the word *scales* is somewhat misleading. Nominal, ordinal, interval, and ratio properties are characteristics of a set of measures, not just the scales used to generate those measures. A measure has as its referent not only a particular scale (for example, inches), but an individual on whom the measure is taken, a time at which the measure is taken, and a setting in which the measure is taken. All these must be considered when evaluating the properties of a set of measures. We can illustrate this idea using height as an example. Consider four individuals whose heights are 54, 53, 52, and 51 inches, respectively. We can rank order these individuals from shortest to tallest:

Individual	*Height in inches* (X)	*Rank order of height* (Y)
1	54	4
2	53	3
3	52	2
4	51	1

The rank-order measure (Y) has ordinal properties, as does any rank-order index. However, for this set of measures (that is, for these four individuals at this point in time), the measures on Y also have interval level properties. The difference between scores for any two individuals with adjacent scores corresponds to the same height difference as any other two individuals with adjacent scores $(54 - 53 = 53 - 52 = 52 - 51 = 1)$. For this set of measures, Y has interval properties. Note that if a 58-inch-tall individual were added to this set, receiving a rank of 5, Y would no longer exhibit interval properties. It would instead represent a set of measures with only ordinal properties. The point is that the concepts of nominal, ordinal, interval, and ratio properties are inherent in measures, not scales.

The determination of whether a variable is measured on a nominal or some other level is usually a straightforward matter in the behavioral sciences. This is not necessarily true for the other measures. For example, there is controversy as to whether intelligence test scores (such as the Wechsler Adult Intelligence Scale) reflect only an ordinal measure of intelligence or an interval measure of intelligence. The critical question is whether test score differences of a given magnitude *always* represent equivalent differences in intellectual ability; for instance, is the difference in *intelligence* of individuals with intelligence test scores of 110 and 120 the same as the difference in intelligence of individuals with intelligence test scores of 90 and 100? If so, scores on the Wechsler Adult Intelligence Scale represent an interval measure; if not, they represent an ordinal measure. What is clear is that intelligence test scores do not reflect a ratio measure—there is no evidence that an intelligence test score of 0 represents a complete absence of intellectual capacity.

Techniques for testing the measurement assumptions of different measures have been developed in psychophysical scaling (for example, Anderson, 1970). The majority of statistical techniques considered in this text assume that the dependent variables are measured on at least an ordinal level that approximates interval characteristics. We will consider the issue of the level of measurement of variables in more detail in later chapters.

STUDY EXERCISE 1.1

For each of the following experiments specify the independent variable and the dependent variable. Also identify any variables that are explicitly held constant by the experimenter. For both the independent variable and the dependent variable indicate the level of measurement and whether a quantitative or a qualitative variable is represented.

Experiment I

Goldberg (1968) was interested in investigating gender bias among females. One hundred female college students were asked to rate an article in terms of its persuasiveness. The article was on the topic of education. The subjects were assigned to one of two groups. One group of 50 women read the article and were told that it was authored by a woman named Joan McKay. The other group of 50 women read the same article but were told it was authored by a man named John McKay. After reading the article, each subject rated the article on a 7-point scale as follows:

not at all persuasive	1	2	3	4	5	6	7	very persuasive

The average rating scores were compared for the two groups—that is, the group with a male author vs. the group with a female author. Results indicated that the average rating was much higher when the article was attributed to a male author than to a female author.

Answer The independent variable is the gender of the author of the article, male or female. It is a nominal measure and, hence, constitutes a qualitative variable. The dependent variable is the persuasiveness rating. This measure constitutes at least an ordinal measure because the higher the rating, the more persuasive the article was perceived to be. Because the dependent measure has at least ordinal characteristics, it represents a quantitative variable. Numerous variables have been held constant. One of the most obvious ones is that the study was conducted only with females. Also, the content of the articles was held constant.

Experiment II

Research on extrasensory perception (ESP) has taken many different directions. Recently, attention has been given to the possibility that hypnosis may be helpful in fostering ESP in people. One standard ESP task involves the use of Zener cards. These are special cards that have only five denominations and look like this:

The standard task is to take a deck of 200 cards and have a "sender" shuffle them. The sender looks at the first card, thinks of the denomination of that card, and then the subject guesses what the card is. This process is repeated throughout the entire deck. ESP is measured by the number of correct guesses on the part of the subject (the receiver).

Casler (1964) used this task with 100 female college students. Two conditions were used. In the first condition, 50 women were hypnotized and then given the task described above. In the second condition, an additional 50 women completed the task without being hypnotized. The average number of correct predictions was

Study Exercise 1.1 continued

computed for the two groups. These averages turned out to be roughly equivalent to each other, and it was concluded that hypnosis does not affect ESP.

Answer The independent variable is whether or not the subject was hypnotized. This represents a nominal measure and, hence, is a qualitative variable. The dependent variable is the number of correct answers on the 200 trials. This represents at least an ordinal measure of ESP and, thus, it is a quantitative variable. You might be inclined to view this measure as one with ratio characteristics, since it appears that there is an absolute zero point (that is, none correct) and that ratio-type statements are possible (for instance, 10 correct is twice as many as 5 correct). Actually, it is unclear whether this is the case. One can conceptualize the number of correct trials as an *index* of ESP ability. Ten correct trials relative to 5 correct trials may not really reflect *twice* as much ESP ability. Numerous variables have been held constant. Again, the most obvious one is that all subjects were females. Try to specify some others.

1.5

Discrete and Continuous Variables

The Concept of Discrete and Continuous Variables

Another distinction made in statistics is between **discrete** and **continuous variables.** A discrete variable is one in which there are a finite number of values that can occur between any two points with respect to the variable. An example is the number of cars in a parking lot. For this variable, there can be only one value between the value of 1 car and 3 cars (namely, 2 cars). We do not think of there being 1.5 or 2.7 parked cars. In contrast, a continuous variable can theoretically include an infinite number of values between any two points. Time is an example of a continuous variable. Even between the values of 1 and 2 seconds, there are an infinite number of values that could occur (1.001 seconds, 1.873 seconds, 1.874 seconds, and so on).

It should be emphasized that whether a variable is classified as discrete or continuous depends on the nature of the underlying theoretical dimension and not on the scale used to measure that dimension. Tests used to measure intelligence, for example, yield scores that are whole numbers (101, 102, and so on). Nevertheless, intelligence is still continuous in nature because it involves a dimension that permits an infinite number of values to occur, even though existing measuring devices are not sensitive enough to make such fine distinctions.

When reading research reports, you may encounter instances where a statistic for a discrete variable, such as the number of children in families, is re-

FIGURE 1.2 Real Limits of 10

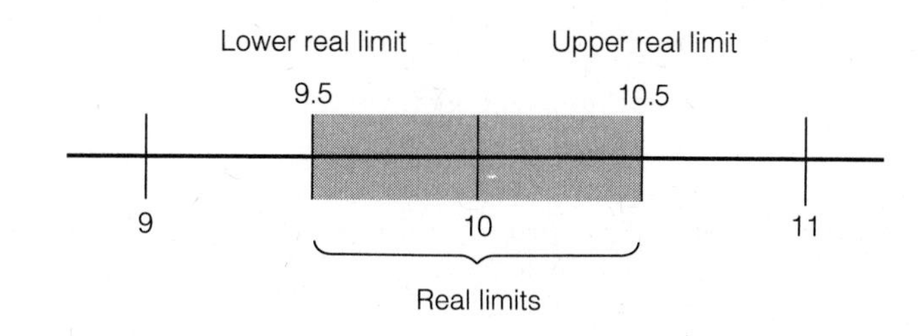

ported in a form representative of a continuous variable. For example, you might be told that the average number of children for married couples in the United States is 2.11. In this instance, the inclusion of decimal places is useful since it provides a more sensitive index of the number of children that couples in the United States, considered as a group, tend to have.

Real Limits of a Number

If a variable is continuous in nature, then it follows that the measurements taken on that variable must be approximate in nature. When we say a person reacted to a stimulus in 10 seconds, we do not usually mean exactly 10 seconds, but only approximately 10 seconds because more refined measures are always possible, such as 10.02 seconds or 10.093 seconds. When we say the reaction was in 10 seconds, we actually mean it was somewhere between 9.5 seconds and 10.5 seconds since any number less than 9.5 would be rounded to 9 and any number greater than 10.5 would be rounded to 11. Thus, the **real limits** of the number 10 are 9.5 to 10.5. The quantity 9.5 is called the *lower real limit* and the quantity 10.5 is called the *upper real limit. The real limits of a number are those points falling one-half a measurement unit above that number and one-half a measurement unit below that number.* Figure 1.2 graphically presents the concept of real limits.

The real limits of a number can be stated not only with respect to whole numbers, but also for numbers expressed as decimals. For example, consider the number 10.6. Since it is expressed in tenths, the unit of measurement is one-tenth, or .1. One-half a measurement unit is therefore $.1/2 = .05$. The lower real limit is thus $10.6 - .05 = 10.55$ and the upper real limit is thus $10.6 + .05 = 10.65$. Similarly, the lower and upper real limits of 10.63 are $10.63 - .005 = 10.625$ and $10.63 + .005 = 10.635$, respectively.

STUDY EXERCISE 1.2

State the real limits of the following numbers, assuming they are measured in the units reported:

(a) 20 (b) 8.4 (c) 12.23 (d) 16.0478

Answers

(a) 19.5 to 20.5
(b) 8.35 to 8.45
(c) 12.225 to 12.235
(d) 16.04775 to 16.04785

1.6

Mathematical Preliminaries: A Review

The purpose of this section is to review a number of mathematical symbols and concepts that will be used in this book.

Summation Notation

Suppose we have a measure of the number of months worked in the past year for each of five individuals. The scores on this variable are as follows:

Individual	*Number of months worked (X)*
1	3
2	2
3	3
4	3
5	2

In statistical notation, the capital letter X is used as a general name for a variable. In the present case, X stands for the variable "number of months worked." Sometimes the X will be subscripted with a number to indicate that a particular individual's score is being represented. For instance, X_1 is the first individual's score on X, which in this case is 3; X_2 is the second individual's score on X, which in this case is 2; and so on.

On some occasions, we will want to consider simultaneously two variables, such as the number of months worked and monthly income:

Individual	*Number of months worked (X)*	*Monthly income in dollars (Y)*
1	3	200
2	2	300
3	3	200
4	3	300
5	2	300

If we let X represent the number of months worked and Y represent monthly income, then we can refer to individual scores on X and Y using subscripts: $X_1 = 3$, $Y_1 = 200$, $X_2 = 2$, $Y_2 = 300$, and so on.

Suppose we want to sum the five scores on variable X to determine the total number of months worked by the five individuals. In statistics, we have a shorthand way of writing an instruction to sum a set of scores, called **summation notation.** The operation is written as follows:

$$\sum_{i=1}^{5} X_i$$

The summation operation is signaled by Σ (uppercase Greek S, called "sigma"). The notation below the sigma tells us to start with individual number 1, and the number above the sigma tells us to add through to individual number 5. The X_i to the right of the sigma is a general term that stands for the individual X scores. In this case,

$$\sum_{i=1}^{5} X_i = X_1 + X_2 + X_3 + X_4 + X_5 = 3 + 2 + 3 + 3 + 2 = 13$$

If the summation were written as

$$\sum_{i=2}^{4} X_i$$

it would mean to sum the scores of individuals 2 through 4 on variable X:

$$\sum_{i=2}^{4} X_i = X_2 + X_3 + X_4 = 2 + 3 + 3 = 8$$

Often we let the letter N represent the total number of individuals. You might thus encounter summation notation as follows:

$$\sum_{i=1}^{N} X_i \quad [1.1]$$

which is the same as having the number that represents the total number of individuals above the sigma. In our example, $N = 5$ since there are five individuals; thus,

$$\sum_{i=1}^{N} X_i = \sum_{i=1}^{5} X_i = 13$$

as calculated above. Similar terminology applies to the Y variable:

$$\begin{aligned}\sum_{i=1}^{N} Y_i = \sum_{i=1}^{5} Y_i &= Y_1 + Y_2 + Y_3 + Y_4 + Y_5 \\ &= 200 + 300 + 200 + 300 + 300 \\ &= 1{,}300\end{aligned}$$

In addition to the expression in Equation 1.1, we will encounter other summation terms in this book. We will briefly review some of these. One such expression is

$$\sum_{i=1}^{N} X_i^2 \quad [1.2]$$

This means that each X score should first be squared and then summed:

$$\begin{aligned}\sum_{i=1}^{N} X_i^2 &= X_1^2 + X_2^2 + X_3^2 + X_4^2 + X_5^2 \\ &= 3^2 + 2^2 + 3^2 + 3^2 + 2^2 \\ &= 35\end{aligned}$$

A third summation expression is

$$\left(\sum_{i=1}^{N} X_i\right)^2 \quad [1.3]$$

This is *not* the same as the expression in Equation 1.2. A general rule that we will follow throughout this book is to *perform any mathematical operations within parentheses before performing the operations outside the parentheses.* In Equation 1.3, the parentheses signal that the summation operation should be executed first (that is, the X scores should be summed) and then this sum should be squared:

$$\begin{aligned}\left(\sum_{i=1}^{N} X_i\right)^2 &= (X_1 + X_2 + X_3 + X_4 + X_5)^2 \\ &= (3 + 2 + 3 + 3 + 2)^2 = 13^2 = 169\end{aligned}$$

In short, then, *the expression in Equation 1.2 means to sum the squared X scores, whereas the expression in Equation 1.3 means to square the summed X scores.*

A fourth summation expression is

$$\sum_{i=1}^{N} X_i Y_i \quad [1.4]$$

This means that for each pair of scores, each X score should first be multiplied by its corresponding Y score, and then these products should be summed:

$$\begin{aligned}\sum_{i=1}^{N} X_i Y_i &= X_1 Y_1 + X_2 Y_2 + X_3 Y_3 + X_4 Y_4 + X_5 Y_5 \\ &= (3)(200) + (2)(300) + (3)(200) + (3)(300) + (2)(300) \\ &= 600 + 600 + 600 + 900 + 600 = 3{,}300\end{aligned}$$

Another summation term we will encounter is

$$\sum_{i=1}^{N}(X_i - c)^2 \qquad [1.5]$$

where c represents a constant. Suppose that $c = 2$. Then this expression indicates that we should subtract 2 from each X score, square each difference, and, lastly, sum these squared differences:

$$\begin{aligned}\sum_{i=1}^{N}(X_i - c)^2 &= (X_1 - 2)^2 + (X_2 - 2)^2 + (X_3 - 2)^2 + (X_4 - 2)^2 \\ &\quad + (X_5 - 2)^2 \\ &= (3 - 2)^2 + (2 - 2)^2 + (3 - 2)^2 + (3 - 2)^2 \\ &\quad + (2 - 2)^2 \\ &= 1^2 + 0^2 + 1^2 + 1^2 + 0^2 = 3\end{aligned}$$

The final summation expression that we will review is

$$\sum_{i=1}^{N}(X_i - c)(Y_i - k) \qquad [1.6]$$

where both c and k represent constants. This means for each individual, multiply the difference between X and c by the difference between Y and k, and then sum the resulting products. For instance, if $c = 2$ and $k = 100$, then

$$\begin{aligned}\sum_{i=1}^{N}(X_i - c)(Y_i - k) &= (X_1 - c)(Y_1 - k) + (X_2 - c)(Y_2 - k) \\ &\quad + (X_3 - c)(Y_3 - k) + (X_4 - c)(Y_4 - k) \\ &\quad + (X_5 - c)(Y_5 - k) \\ &= (3 - 2)(200 - 100) + (2 - 2)(300 - 100) + \\ &\quad (3 - 2)(200 - 100) + (3 - 2)(300 - 100) + \\ &\quad (2 - 2)(300 - 100) \\ &= (1)(100) + (0)(200) + (1)(100) + (1)(200) \\ &\quad + (0)(200) = 400\end{aligned}$$

It is very important to understand these six summation expressions, as we will refer to them constantly throughout this book. Investigators will frequently use a shorthand version of these terms. For instance, the first expression,

$$\sum_{i=1}^{N} X_i$$

may also be written

$$\Sigma X$$

Note that there is no subscript for X, no instruction beneath the sigma, and no symbol above the sigma. It is understood that X is subscripted with i, the instruction $i = 1$ is below the sigma, and the letter N is above it. Thus,

$$\sum_{i=1}^{N} X_i = \Sigma X$$

$$\sum_{i=1}^{N} X_i^2 = \Sigma X^2$$

$$\left(\sum_{i=1}^{N} X_i\right)^2 = (\Sigma X)^2$$

$$\sum_{i=1}^{N} X_i Y_i = \Sigma XY$$

$$\sum_{i=1}^{N} (X_i - c)^2 = \Sigma (X - c)^2$$

$$\sum_{i=1}^{N} (X_i - c)(Y_i - k) = \Sigma (X - c)(Y - k)$$

We will use the shorthand version of these terms wherever possible in this book.

STUDY EXERCISE 1.3

Given the following values for X and Y, complete the requested operations.

Individual	*X*	*Y*
1	3	3
2	4	3
3	2	3
4	7	5

(a) ΣX (c) $(\Sigma X)^2$ (e) ΣY (g) $(\Sigma Y)^2$ (i) ΣXY
(b) ΣX^2 (d) $\Sigma (X - 4)^2$ (f) ΣY^2 (h) $\Sigma (Y - 3)^2$ (j) $\Sigma (X - 1)(Y - 2)$

Answers

(a) $\Sigma X = 3 + 4 + 2 + 7 = 16$
(b) $\Sigma X^2 = 3^2 + 4^2 + 2^2 + 7^2 = 78$
(c) $(\Sigma X)^2 = 16^2 = 256$
(d) $\Sigma (X - 4)^2 = (3 - 4)^2 + (4 - 4)^2 + (2 - 4)^2 + (7 - 4)^2 = 14$
(e) $\Sigma Y = 3 + 3 + 3 + 5 = 14$
(f) $\Sigma Y^2 = 3^2 + 3^2 + 3^2 + 5^2 = 52$
(g) $(\Sigma Y)^2 = 14^2 = 196$
(h) $\Sigma (Y - 3)^2 = (3 - 3)^2 + (3 - 3)^2 + (3 - 3)^2 + (5 - 3)^2 = 4$
(i) $\Sigma XY = (3)(3) + (4)(3) + (2)(3) + (7)(5) = 62$
(j) $\Sigma (X - 1)(Y - 2) = (3 - 1)(3 - 2) + (4 - 1)(3 - 2) + (2 - 1)(3 - 2) + (7 - 1)(5 - 2) = 24$

Rounding It will often be desirable to round numbers to a certain number of decimal places. For instance, the fraction 7/3 in decimal notation is equivalent to 2. followed by an infinite number of 3s (that is, 2.33333 . . .). In a case like this, we will obviously have to round off. There are commonly accepted mathematical rules for rounding that will be adopted in this book. These rules are as follows:

1. If the remainder to the right of the decimal place you wish to round to is greater than one-half a measurement unit, increase the last digit kept by one.

   Suppose we wish to round to two decimal places. In this case, the unit of measurement would be one-hundredth, or .01, and one-half a measurement unit would be .01/2 = .005. Thus, according to this rule, 5.338 would be rounded to 5.34 because the remainder after the second decimal place, .008, is greater than .005. The quantity 5.335001 would also be rounded to 5.34 because .005001 is greater than .005.

2. If the remainder to the right of the decimal place you wish to round to is less than one-half a measurement unit, leave the last digit kept as it is.

   Thus, 7/3 would be represented in decimal notation as 2.33 because the remainder after the second decimal place, .00333 . . . , is less than .005. Similarly, 2.3348 would be rounded to 2.33 according to this rule.

3. If the remainder to the right of the decimal place you wish to round to is exactly one-half a measurement unit, leave the last digit kept as it is if it is an even number, but increase it by one if it is an odd number. *Note that when this rule is used, the last digit of the answer will always be an even number.*

   According to this rule, 10.345 would be rounded to 10.34 because the 4 in the hundredths column is an even number, and 10.335 would also be rounded to 10.34 because the 3 in the hundredths column is an odd number. The purpose of the rule is to avoid a bias in rounding up or down across a large set of numbers, since with the rule, approximately half the time you will round up and half the time you will round down.

The number of decimal places that are used in reporting a statistic in a research report will differ depending on the nature of the variable being reported. The average annual income of a group of individuals might be rounded to the nearest whole number (for example, $10,030), whereas the average number of seconds it takes a group of rats to run a maze might be reported to two decimal places (for example, 5.32 seconds). The number of decimal places you should report will depend on how precise you need to be in order to make your point. In practice, statistics in the behavioral sciences are most commonly reported to two decimal places.

When computing a statistic such as an average, it might be necessary to do intermediate calculations before you can arrive at a final answer. Again, the exact number of decimal places you should use in your calculations will depend on the variable you are studying and the nature of the calculations performed. No hard-and-fast rules can be given to reduce rounding error. As a rule of thumb, *intermediate calculations should be done using at least one decimal place beyond the number of decimal places you plan to report in your final answer.* If you are performing your computations on a calculator, rounding error can be substantially reduced by keeping all digits shown until rounding the final result.

In this book, calculations will generally be rounded to two decimal places. For clarity of presentation, we will follow this strategy even when reporting intermediate values.

STUDY EXERCISE 1.4

Round the following numbers to two decimal places:

(a) 8.337 (c) 7.555001 (e) 13.63500
(b) 7.443 (d) 10.54500

Answers

(a) 8.34
(b) 7.44
(c) 7.56
(d) 10.54
(e) 13.64

The Concept of Probability

The concept of probability is an essential aspect of statistics. All of us are somewhat familiar with this concept in our everyday life. A weather forecaster tells us that the chances of rain tomorrow are 70%. A bettor at the racetrack knows that the odds on a given horse are 3 to 1. A student thinks it is "likely" he or she will obtain an A in a course.

In statistics, the concept of probability has a precise meaning. For a given task, there may be a number of different possible outcomes. If you roll a die, there are six possible outcomes that could occur: 1, 2, 3, 4, 5, or 6. If you draw a card from a standard deck of playing cards, there are 52 different possible outcomes. The **probability** of some outcome, A, can be defined as the ratio

$$p(A) = \frac{\text{number of observations favoring outcome } A}{\text{total number of possible observations}}$$

On a die, what is the probability of rolling a 2? On a given roll, the total number of possible observations is six (that is, 1, 2, 3, 4, 5, or 6). There is only one 2 on a die and hence only one possible observation favoring the event "2." The probability is 1/6 or .17. If one draws a card at random from a standard deck, the probability of drawing an ace is 4/52 or .08. There are 52 possible observations, of which 4 favor an ace.

A given probability must always range from 0 to 1.00. It can never be less than 0 or greater than 1.00. The probability of an impossible event is always 0. The probability of rolling an 8 on a single die is 0 because the number of observations representing the event "8" is 0. It follows that 0/6 = 0. The probability of a completely certain event is always 1.00. For example, if an individual rolls a die, what is the probability that the individual will roll a 1, 2, 3, 4, 5, or 6? In this case there are six outcomes of which all six will satisfy the event 1, 2, 3, 4, 5, or 6. The probability is therefore 6/6 = 1.00.

Sometimes probabilities can be derived on logical grounds, as in the examples above. Other times, they are estimated empirically based on data an investigator has collected. A good example is that of a baseball player's batting average. If the player has been at bat 400 times and had 100 base hits, then his batting average is 100/400 = .250. All things being equal, the probability that the player will obtain a hit his next time at bat can be estimated to be .250.

The probability of an event can also be interpreted in terms of a "long run" perspective. If we flip a coin 10 times, it is unlikely that the coin will come up heads exactly 5 times and tails the other 5 times. As we continue to flip the coin, then across a large number of flips (that is, over the long run), the number of heads relative to the total number of flips will approach 1/2, or .50.

1.7 Populations and Samples

In scientific research, we are often interested in making descriptive statements about a group of individuals or objects. For example, one might state that the average number of children desired by married males in the United States is 2.3 or that the average weight of males who are 6 feet tall is 180 pounds. Such statements are made with reference to a **population.** A population is the aggregate of all cases to which one wishes to generalize statements. In the first example, the population consists of all married males in the United States. In the second example, the population consists of all males who are 6 feet tall.

It may be the case that an investigator is unable to make observations on every member of the population about which he or she wishes to make a descriptive statement. In this instance, the investigator will resort to a **sample** of the population. A sample is simply a subset of the population. On the basis of observing the sample, the researcher makes generalizations to the population.

When we select a sample for purposes of making a statement about a population on a given dimension (for example, the average number of children), we want to ensure that we are using a **representative sample** of the population. If the population has 60% males and 40% females, we want our sample to reflect this. We do not want a *biased* sample that will lead us to make erroneous statements about the population. In selecting a sample, we want to use procedures that will yield a representative sample. The box on page 23 presents an interesting example of biased sampling as described by Darrell Huff in his excellent book *How to Lie with Statistics* (1954).

One procedure for approximating representative samples is through the use of **random sampling.** The term *random* has a very precise meaning in scientific discourse. As applied to sampling problems, the essential characteristic of random sampling is that every member of the population has an equal chance of being selected for the sample.

Scientists use random sampling in a variety of ways. In order to obtain a random sample of a population, a survey researcher will make use of a very useful resource known as a *random number table.* A random number table is a list of numbers generated by a computer that has been programmed to yield a set of truly random numbers. Computers are used to construct such tables since the typical human is not capable of generating random numbers. For example, a person might have a tendency to list mostly even numbers, or those ending only in 5 or 0.

Appendix A presents a random number table. Suppose an investigator wanted to select a random sample of people from a population. This would involve obtaining a list of all members of the population and then arbitrarily assigning a number to each member. If the population consisted of 500 individuals, a list would be made of the names of these people and numbered from 1 to 500. The investigator would then consult a random number table, such as the one presented in Appendix A. Using the directions provided, the investigator would draw a sample of, say, 50 individuals. The use of random number tables ensures that the selection will be random and not influenced by any unknown selection bias the investigator may have.

It should be emphasized that random sampling is an ideal that is seldom achieved in practice. This is not surprising given the difficulty of compiling a complete listing of a population and ensuring that all selected individuals will agree to participate. Furthermore, the use of random sampling procedures does *not* guarantee that a sample will be representative of the population. Random sampling will tend to yield representative samples, but sometimes nonrepresentative samples will result even when random sampling is used. Nevertheless, random sampling is an important concept in statistical theory. We will discuss the problem of sampling in greater detail in later chapters.

Throughout this text we will refer to various numerical indices based on data from either populations or samples. When the indices are based on data from an entire population, they will be referred to as **parameters.** When they are based on data from a sample, they will be referred to as **statistics.**

BOX BIASED SAMPLING

"The average Yaleman, Class of '24," *Time* magazine noted once, commenting on something in the *New York Sun,* "makes $25,111 a year." Well, good for him! But wait a minute. What does this impressive [at the time of the report] figure mean? Is it, as it appears to be, evidence that if you send your boy to Yale you won't have to work in your old age and neither will he? Two things about the figure stand out at first suspicious glance. It is surprisingly precise. It is quite improbably salubrious. . . .

Let us put our finger on a likely source of error, a source that can produce $25,111 as the "average income" of some men whose actual average may well be nearer half that amount. This is the sampling procedure, which is the heart of the greater part of the statistics you meet on all sorts of subjects. Its basis is simple enough, although its refinements in practice have led into all sorts of by-ways, some less than respectable. If you have a barrel of beans, some red and some white, there is only one way to find out exactly how many of each color you have: Count 'em. However, you can find out approximately how many are red in much easier fashion by pulling out a handful of beans and counting just those, figuring that the proportion will be the same all through the barrel. If your sample is large enough and selected properly, it will represent the whole well enough for most purposes. If it is not, it may be far less accurate than an intelligent guess and have nothing to recommend it but a spurious air of scientific precision. It is sad truth that conclusions from such samples, biased or too small or both, lie behind much of what we read or think we know.

The report on the Yale men comes from a sample. We can be pretty sure of that because reason tells us that no one can get hold of all the living members of that class of '24. There are bound to be many whose addresses are unknown twenty-five years later.

And, of those whose addresses are known, many will not reply to a questionnaire, particularly a rather personal one. With some kinds of mail questionnaire, a five or ten per cent response is quite high. This one should have done better than that, but nothing like one hundred per cent.

So we find that the income figure is based on a sample composed of all class members whose addresses are known and who replied to the questionnaire. Is this a representative sample? That is, can this group be assumed to be equal in income to the unrepresented group, those who cannot be reached or who do not reply?

Who are the little lost sheep down in the Yale rolls as "address unknown"? Are they the big-income earners—the Wall Street men, the corporation directors, the manufacturing and utility executives? No; the addresses of the rich will not be hard to come by. Many of the most prosperous members of the class can be found through *Who's Who in America* and other reference volumes even if they have neglected to keep in touch with the alumni office. It is a good guess that the lost names are those of the men who, twenty-five years or so after becoming Yale bachelors of arts, have not fulfilled any shining promise. They are clerks, mechanics, tramps, unemployed alcoholics, barely surviving writers and artists . . . people of whom it would take half a dozen or more to add up to an income of $25,111. These men do not so often register at class reunions, if only because they cannot afford the trip.

Who are those who chucked the questionnaire into the nearest wastebasket? We cannot be so sure about these, but it is at least a fair guess that many of them are just not making enough money to brag about.

It becomes pretty clear that the sample has omitted two groups most likely to depress the average. The $25,111 figure is beginning to explain itself. If it is a true figure for anything it is one merely for that special group of the class of '24 whose addresses are known and who are willing to stand up and tell how much they earn. Even that requires an assumption that the gentlemen are telling the truth.

1.8

Descriptive and Inferential Statistics

The discipline of statistical analysis has traditionally been divided into two major subfields, descriptive statistics and inferential statistics. The two are highly related to one another, and in some respects, the distinction is arbitrary. **Descriptive statistics** involves the use of numerical indices to describe either a population (when measurements have been taken on all members of that population) or a sample.* In either case, the goal is to *describe* a group of scores in a clear and precise manner. **Inferential statistics,** in contrast, involves taking measurements on a sample and then from the observations, inferring something about a population. In this instance, we are again attempting to describe a population. However, we do so not by taking measures on all cases in the population, but rather by selecting a sample, observing scores on the variable of interest for that sample, and then *inferring* something with respect to that variable for the entire population. As we will see, this involves the use of sample statistics to estimate population parameters.

1.9

Summary

Scientific research can be conceptualized as a five-step process involving the analysis of data collected to test the validity of hypotheses. Most hypotheses in the behavioral sciences concern relationships between variables. A variable is a phenomenon that takes on different values. Important distinctions can be made between independent and dependent variables, quantitative (ordinal, interval, or ratio levels of measurement) and qualitative (nominal level of measurement) variables, and discrete and continuous variables. A distinction can also be made between parameters, which are numerical indices based on data from an entire population, and statistics, which are numerical indices based on data from a sample.

The use of numerical indices to describe either a sample or a population is referred to as descriptive statistics. The use of numerical indices to infer something about a population from observation of a sample is referred to as inferential statistics. Samples should be representative of the populations from which

* As stated above, numerical indices based on data from entire populations are called *parameters*. Nevertheless, the term *descriptive statistics* has historically been used to encompass the description of both sample and population data.

they were selected. One procedure for approximating representative samples is through the use of random sampling.

Exercises

Answers to asterisked (*) exercises appear at the back of the book.

1. What are the five stages of the scientific research process?

***2.** Identify each of the following as a variable or a constant. Explain the reasons for your choices.

- **a.** the number of hours in a day
- **b.** people's attitudes toward abortion
- **c.** the country of birth of presidents of the United States
- **d.** the value of a number divided by itself
- **e.** the total number of points scored in a football game
- **f.** the number of days in a month

***3.** Identify each of the following as a qualitative variable or a quantitative variable. Explain the reasons for your choices.

- **a.** weight
- **b.** religion
- **c.** income
- **d.** age
- **e.** gender
- **f.** eye color

For each of the studies described in Exercises 4–7, identify the independent variable and the dependent variable. Indicate whether each is a quantitative variable or a qualitative variable. Justify your answers.

***4.** Eron (1963) reported an investigation in which he examined the possible existence of a relationship between the exposure of young children to violent television shows and the amount of aggression they exhibited toward peers. Eron gathered information concerning the aggressive behavior and television viewing habits of 875 third-grade children. By questioning parents about their child's viewing habits, Eron developed a 4-point scale measuring preference on the part of the child for aggressive TV shows. The scale had the following categories: very low preference for aggressive TV shows, low preference for aggressive TV shows, moderately high preference for aggressive TV shows, and high preference for aggressive TV shows. Aggression was measured by peer ratings of each child by at least two other children. These ratings could range from 0 to 32, with higher scores indicating higher amounts of aggression.

***5.** Touhey (1974) was interested in studying the relationship between various types of occupations and the prestige people associated with them. In this study, five different occupations were studied: architect, professor, lawyer, physician, and research scientist. A large number of individuals were asked to rate each of these occupations on a 60-point scale measuring perceived prestige. Low scores indicated low levels of prestige and higher scores indicated increasingly higher levels of prestige.

6. Steiner (1972) discussed a series of experiments that studied the relationship between group size and how quickly a group could solve problems. In one experiment, six different group sizes were created: two members, three members, four members, five members, six members, or seven members. Each group was then given a series of problems to solve, and the time until solution was measured for each group and compared.

7. Rubovits and Maehr (1973) were interested in the effects of teacher expectancies on their behavior toward students. Female undergraduates who were enrolled in a teacher training class were asked to prepare a lesson for four seventh- and eighth-grade students. Just before meeting with the students, each teacher was told that two of the students were "gifted" and had high IQs, while the other two were "not gifted" and possessed average intelligence. In reality, all children were about equal in ability, and these labels were assigned in an arbitrary manner. The teachers were then observed during a 40-minute period while they interacted with the four students. Rubovits and Maehr measured the number of times the teacher interacted with each student. The average number of interactions was then compared for students who were labeled as "gifted" versus those who were labeled as "not gifted."

8. Indicate whether each measure listed below is a nominal measure, ordinal measure, interval measure, or ratio measure. Explain the reasons for your choices.

- **a.** inches on a yardstick
- **b.** Social Security numbers
- **c.** dollars as a measure of income
- **d.** order of finish in a car race
- **e.** intelligence test scores

9. What is the difference between ordinal, interval, and ratio measures? Give an example of each.

***10.** Indicate whether each of the following variables is discrete or continuous. Explain the reasons for your choices.

- **a.** grains of sand on a beach
- **b.** height
- **c.** the annual federal budget
- **d.** shyness

***11.** State the real limits of the following numbers, assuming they are measured in the units reported:

- **a.** 21,384.11
- **b.** .689
- **c.** 13
- **d.** 13.0
- **e.** 13.00

12. Consider the following data for eight individuals:

Individual	*X*	*Y*
1	4	5
2	1	7
3	2	3
4	8	2
5	8	1
6	2	1
7	5	4
8	7	7

Perform the following calculations:

- **a.** ΣX
- **b.** ΣY
- **c.** $\sum_{i=1}^{4} X_i$
- **d.** $\sum_{i=4}^{8} Y_i$
- **e.** ΣXY
- **f.** $(\Sigma X)/N$
- **g.** $(\Sigma Y)/N$
- **h.** $(\Sigma X)(\Sigma Y)$
- **i.** ΣX^2
- **j.** ΣY^2
- **k.** $\Sigma (X - 3)^2$
- **l.** $\Sigma (Y - 2)^2$
- **m.** $(\Sigma X)^2$
- **n.** $(\Sigma Y)^2$
- **o.** $\Sigma (X - 3)(Y - 2)$
- **p.** $\Sigma (X - 2)(Y - 3)$

***13.** Express the following statements in summation notation for $N = 5$:

- **a.** $X_1 + X_2 + X_3 + X_4 + X_5$
- **b.** $X_1^2 + X_2^2 + X_3^2 + X_4^2 + X_5^2$
- **c.** $(X_1 - 5) + (X_2 - 5) + (X_3 - 5) + (X_4 - 5) + (X_5 - 5)$
- **d.** $(X_1^2 + X_2^2 + X_3^2 + X_4^2 + X_5^2)/N$
- **e.** $(Y_1 + Y_2 + Y_3 + Y_4 + Y_5)^2$
- **f.** $Y_1 + Y_2 + Y_3$
- **g.** $(X_1 - 1)(Y_1 - 6) + (X_2 - 1)(Y_2 - 6) + (X_3 - 1)(Y_3 - 6) + (X_4 - 1)(Y_4 - 6) + (X_5 - 1)(Y_5 - 6)$
- **h.** $X_1Y_1 + X_2Y_2 + X_3Y_3 + X_4Y_4 + X_5Y_5$

***14.** Consider the following data for five individuals:

Individual	*X*
1	4
2	6
3	2
4	2
5	4

Let k be a constant, with $k = 2$.

- **a.** Compute ΣXk.
- **b.** Compute $k\Sigma X$.
- **c.** Compare your answer in **a** to your answer in **b**. What equation describes this relationship?
- **d.** Compute $\Sigma (X/k)$.
- **e.** Compute $(\Sigma X)/k$.
- **f.** Compare your answer in **d** to your answer in **e**. What equation describes this relationship?

***15.** Round each of the following numbers to three decimal places:

- **a.** 4.8932
- **b.** 8.9749
- **c.** 1.4153
- **d.** 4.1450
- **e.** 6.245002
- **f.** 2.615501
- **g.** 6.3155
- **h.** .39572
- **i.** .9999
- **j.** 3.6666
- **k.** 12.2538
- **l.** 9.724001
- **m.** 1.9950
- **n.** 2.0050

16. Round the numbers in Exercise 15 to two decimal places. Compare your answers with those you derived previously.

***17.** Consider the following data for five individuals:

Individual	X
1	3.8753
2	4.2660
3	4.1156
4	3.4954
5	4.2061

Perform the following calculations on the given scores, keeping all digits shown until rounding the final answers to two decimal places. Then repeat the calculations, rounding the scores and all intermediate values to two decimal places. Compare the two sets of results. What accounts for the difference between them?

a. ΣX **c.** $(\Sigma X)^2$
b. $(\Sigma X)/N$ **d.** ΣX^2

18. Consider the following data for five individuals:

Individual	X
1	5.4749
2	4.8348
3	4.2947
4	5.3650
5	4.7749

Perform the following calculations on the given scores, keeping all digits shown until rounding the final answers to two decimal places. Then repeat the calculations, rounding the scores and all intermediate values to two decimal places. Compare the two sets of results. What accounts for the difference between them?

a. ΣX **c.** $(\Sigma X)^2$
b. $(\Sigma X)/N$ **d.** ΣX^2

19. Compute the probability of each of the following events:

a. Drawing an ace, king, or queen from a standard deck of 52 cards
b. Throwing a 2 or a 3 on a single die
c. For a pair of dice, rolling a combination that totals 7
d. Drawing a face card from a standard deck of 52 cards

***20.** Suppose you are considering whether to have an operation that is important to your health. A total of 420 operations of this nature have been performed in the past, of which 21 have been successful. Given only this information, what is your best guess as to the probability of success of the operation? What would be your decision? Why?

21. What is the difference between a sample and a population? Give three examples of each.

***22.** A newspaper conducted a survey in which readers were asked to indicate their preference for either of two mayoral candidates in an election, John Doe or Jane Smith. People were asked to cut out a ballot provided in the paper that day and send it to the newspaper with their preference indicated. One week later the newspaper reported it received 1,000 ballots, of which 800 favored Jane Smith. It stated that the "spirit of the community lies with Jane Smith" and predicted her victory in the upcoming election. A total of 100,000 people live in the community. Is the newspaper's sample a representative sample of the community in general? Why or why not? What implications does this have for the newspaper's conclusion about the election?

23. What is the essential characteristic of random sampling?

24. Suppose you have a list of all 1,000 members of a population of interest to you. Using the random number table in Appendix A, select a random sample of 25 individuals to participate in your investigation.

***25.** Repeat the random sampling process described in Exercise 24. How many of the same individuals were selected to participate in *both* samples? What does this say about the use of random sampling to approximate representative samples?

26. How are descriptive and inferential statistics different?

TWO

FREQUENCY AND PROBABILITY DISTRIBUTIONS

2.1 Frequency Distributions for Quantitative Variables: Ungrouped Scores

Suppose an investigator administered to 15 individuals a test designed to measure aggressiveness. Scores on this test can range from 0 to 12, with higher scores indicating greater aggressiveness. The scores for the 15 individuals were as follows:

8	8	10	10	7
4	9	4	7	7
8	8	9	7	9

How can we best describe the scores on this test? One procedure is to list each of the 15 scores. By examination, we can then obtain an intuitive feel about what the scores tend to be like. But suppose that instead of 15 scores, there were 500 scores. It now becomes impractical to list all the scores individually. A useful tool for summarizing a large set of data is a **frequency distribution**. A frequency distribution is a table that lists scores on a variable and shows the number of individuals who obtained each value. We begin by listing the obtained score values from highest to lowest. We then derive *frequencies* by counting the number of individuals who received each score and indicate these frequencies next to the corresponding score values. For the 15 scores listed above, we obtain the following frequency distribution:

Score	*f*
10	2
9	3
8	4
7	4
4	2

The letter f is used to represent the word *frequency*. We can see from the frequency distribution that two people had a score of 10, three people had a score of 9, four people had a score of 8, and so on. Adding the individual frequencies, we find that there are a total of 15 scores, or $N = 15$.

If only a relatively small number of different scores are possible, researchers sometimes include all possible score values, even those that were not actually obtained, in the frequency table. When this is done, a frequency of 0 is indicated where appropriate. If we followed this approach, our frequency distribution would appear as follows:

Score	*f*
12	0
11	0
10	2
9	3
8	4
7	4
6	0
5	0
4	2
3	0
2	0
1	0
0	0

This illustrates an important point: Unlike most of the other statistical procedures we will discuss, there are few hard-and-fast rules for presenting frequency information. Rather, there are typically several possible approaches to a given problem. The guidelines that we have presented are those that we find most useful. They are, however, only guidelines, and as such, might have to be modified depending on the specific characteristics of the data.

Considered alone, an index of frequency is not easily interpreted. Suppose you are told the results of a study showing that 200 people in a given town are prejudiced. This tells you little unless you also know the size of the town. With the information that the town population is 400, however, the frequency of 200 takes on more meaning: Half (200/400) of the town is prejudiced! This illustrates a more informative statistic used by researchers, called a **relative frequency.** A relative frequency is the number of scores of a given value (for example, scores of 10) divided by the total number of scores. In the example on aggressiveness, the relative frequencies would be computed as follows:

Score	*f*	*rf*
10	2	2/15 = .133
9	3	3/15 = .200
8	4	4/15 = .267
7	4	4/15 = .267
4	2	2/15 = .133

The symbol *rf* is used to represent *relative frequency.* A relative frequency is simply the **proportion** of times that a score occurred. Thus, the relative frequencies in a given distribution will always sum to 1.00. When a proportion is multiplied by 100, it reflects the **percentage** of times the score occurred. In our example, 13.3% of the individuals had a score of 10, 20.0% had a score of 9, and so on.

Relative frequency bears an important relationship with probability. Recall from Chapter 1 that the probability of an outcome, *A*, is the ratio of the number of observations favoring outcome *A* to the total number of possible observations. This is exactly what a relative frequency reflects in a distribution: the number of individuals who obtain a particular score divided by the total number of scores. Thus, for instance, the probability of randomly selecting a score of 7 from the above distribution is .267.

In addition to frequencies and relative frequencies, we will sometimes want to compute **cumulative frequencies** (represented by *cf*) and **cumulative relative frequencies** (represented by *crf*). For the aggressiveness example, these would appear as follows:

Score	*f*	*rf*	*cf*	*crf*
10	2	.133	15	1.000
9	3	.200	13	.867
8	4	.267	10	.667
7	4	.267	6	.400
4	2	.133	2	.133

The entries in the cumulative frequency column are obtained by a process of successive addition of the entries in the frequency column. Specifically, for any given score (for example, 7) the cumulative frequency is the frequency associated with that score (for the score of 7 the frequency is 4) plus the sum of all frequencies below that score. For the score of 7, the cumulative frequency is 4 + 2 = 6. For the score of 9, the cumulative frequency is 3 + 4 + 4 + 2 = 13. Cumulative relative frequencies are computed in the same manner but use the column of relative frequencies instead of the column of frequencies. For the score of 7, the cumulative relative frequency is .267 + .133 = .400. For the score of 9, the cumulative relative frequency is .200 + .267 + .267 + .133 = .867.

The advantage of a cumulative frequency is that it allows us to tell at a glance the number of scores that are equal to or less than a given score value. We can readily see that 10 individuals had scores of 8 or less and that 13 individuals had scores of 9 or less. By looking at the cumulative relative frequency column, we can see that the proportion of people who had scores of 8 or less was .667 and that the proportion of people who had scores of 9 or less was .867.

When we are concerned with a continuous variable such as aggressiveness, frequencies and relative frequencies should be thought of in terms of the real limits of the scores. In the present example, although four individuals had scores of 8, this is more properly conceptualized as four individuals having scores between 7.5 and 8.5. Similarly, cumulative frequencies and cumulative relative frequencies are conceptualized with respect to the upper real limit of a score. The cumulative frequency in the aggressiveness data for a score of 8 is 10. Technically, this means that 10 individuals had scores of 8.5 or less.

STUDY EXERCISE 2.1

Compute the frequencies, relative frequencies, cumulative frequencies, and cumulative relative frequencies for the following set of scores:

87	75	87	83	93
72	77	70	91	90
91	83	74	75	74
75	87	91	75	83

Answer

Score	*f*	*rf*	*cf*	*crf*
93	1	.050	20	1.000
91	3	.150	19	.950
90	1	.050	16	.800
87	3	.150	15	.750
83	3	.150	12	.600
77	1	.050	9	.450
75	4	.200	8	.400
74	2	.100	4	.200
72	1	.050	2	.100
70	1	.050	1	.050

2.2

Frequency Distributions for Quantitative Variables: Grouped Scores

The preceding analysis of frequency distributions examined the case in which a quantitative variable took on relatively few different values. Specifically, there were five: 10, 9, 8, 7, and 4. Often, however, a quantitative variable takes on many different values. For example, we might have a sample of 100 people, each with a different income. In constructing a frequency table, it would be neither practical nor informative to list 100 different values, each with a frequency of 1. Rather, we would want to group the data before reporting it, perhaps as follows:

Income ($)	*f*	*rf*	*cf*	*crf*
30,000–34,999	14	.140	100	1.000
25,000–29,999	21	.210	86	.860
20,000–24,999	30	.300	65	.650
15,000–19,999	19	.190	35	.350
10,000–14,999	16	.160	16	.160

Since scores are grouped together into intervals, tables of this type are referred to as **grouped frequency distributions.** Note that as with the *ungrouped frequency distributions* discussed in Section 2.1, the scores in the above table are listed from high to low. Also note that the lower bound for each group of scores appears on the left and the upper bound on the right.

An important consideration in presenting grouped data is how to form the groups. Three questions are central: (1) How many groups should be reported? (2) What should the interval size be for each group? and (3) What should be the lowest value at which the first interval starts? There are no standard rules that govern these issues. In large part, the nature of the grouping will depend on the particular characteristics of the data. Nevertheless, useful guidelines are available for each of the above questions.

Number of Groups

In deciding how many groups to report, a balance must be struck between having so many groups that the data are incomprehensible and having so few groups that the table is imprecise. The problem of too many groups would occur in its extreme in our previous example on income, should each of the 100 incomes be listed individually. The problem of too few groups is illustrated in the following table reporting scores of 65 individuals who took a test having a possible score range of 0 to 100:

Score	*f*	*rf*	*cf*	*crf*
50–100	57	.877	65	1.000
0–49	8	.123	8	.123

This table is not very informative, as it provides little insight into how the individuals performed on the test.

In general, if the number of possible score values is small, fewer groups can be used, whereas if the number of possible score values is large, more groups will be required. As a rule of thumb, the use of 5 to 15 groups tends to strike the appropriate balance between imprecision and incomprehensibility in most instances.

Size of Interval

Once we have an idea of how many groups we wish to present, the question of the size of the interval arises (for example, should the interval for the income problem be $10,000, $5,000, or $1,000?). Typically, interval sizes of 2, 3, or multiples of 5 (5, 10, 15, and so on) are used. To determine the interval size for a particular set of data, first subtract the lowest score from the highest score. This difference should then be divided by the desired number of groups and the result rounded to the nearest of the commonly used interval-size values.

Returning to the test score example, suppose that the lowest obtained score was 47 and the highest obtained score was 99. Further suppose that the decision was made to present the frequency analysis in five groups. In this case, $99 - 47 = 52$ and $52/5 = 10.40$. Since 10 is a multiple of 5, we drop the .40 and adopt 10 as the interval size.

Beginning of Lowest Interval

We now know that we will present five groups with an interval of 10 units per group. The final question is where to begin the lowest interval (at 40? at 45? at 47?). The conventional starting point is the closest number evenly divisible by the interval size that is equal to or less than the lowest score. In our example, the lowest score is 47 and the interval size is 10. The closest number equal to or less than 47 that is evenly divisible by 10 is 40. This should be the starting point for the lowest interval. The distribution, according to these general rules, is:

Score	*f*	*rf*	*cf*	*crf*
90–99	3	.046	65	1.000
80–89	6	.092	62	.954
70–79	15	.231	56	.862
60–69	21	.323	41	.631
50–59	12	.185	20	.308
40–49	8	.123	8	.123

Note that the lower bound of each interval is a multiple of the interval size of 10. Also note that it was necessary to use six groups instead of five because of the determined interval size and lowest value. Again, the above rules are only guidelines that might be useful in presenting grouped data.

STUDY EXERCISE 2.2

Construct a grouped frequency distribution containing frequencies, relative frequencies, cumulative frequencies, and cumulative relative frequencies for the following set of scores:

9	29	36	15	37	12	38	19	37	47
29	14	8	24	13	6	48	19	2	25
39	28	39	49	27	31	9	26	18	1
5	22	16	23	6	46	38	4	49	39
34	15	21	35	11	32	2	19	29	11
43	26	7	44	49	28	20	41	1	41
39	25	24	19	17	5	33	32	27	23
26	44	17	25	31	42	43	12	3	33
22	24	23	45	27	28	3	18	42	22
4	16	21	34	14	35	13	36	21	46

Answer We will use 10 groups to characterize these data. The first step involves defining the interval size. The highest score is 49 and the lowest is 1. The difference between these is 49 − 1 = 48. This number divided by the number of groups is 48/10 = 4.80. We will thus define the interval size as 5. The next step is to define the beginning of the lowest interval. The lowest score is 1. The closest number equal to or less than 1 that is evenly divisible by 5 is 0. We will therefore use 10 groups beginning with the number 0, with intervals of size 5. The analysis is as follows:

Score	*f*	*rf*	*cf*	*crf*
45–49	8	.080	100	1.000
40–44	8	.080	92	.920
35–39	12	.120	84	.840
30–34	8	.080	72	.720
25–29	15	.150	64	.640
20–24	13	.130	49	.490
15–19	12	.120	36	.360
10–14	8	.080	24	.240
5–9	8	.080	16	.160
0–4	8	.080	8	.080

2.3

Frequency Distributions for Qualitative Variables

Frequency distributions for qualitative variables begin with a listing of the variable categories in the first column. This is followed by frequency, relative frequency, and/or percentage columns. The concepts of cumulative frequencies and cumulative relative frequencies are not applicable because the "scores" for qualitative variables are not ordered on any dimension.

As an example of a frequency distribution for a qualitative variable, consider the results of a hypothetical poll of preference for presidential candidates:

Candidate	*f*	*rf*	*%*
Smith	626	.626	62.6
Jones	270	.270	27.0
Undecided	64	.064	6.4
Other	40	.040	4.0

The variable categories are "Smith," "Jones," "Undecided," and "Other." Adding up the individual frequencies, we find that the poll was based on a sample of 1,000 individuals.

2.4

Frequency Graphs

Frequency Graphs for Quantitative Variables

Instead of presenting tables of a frequency analysis, investigators will sometimes report their data graphically. Consider the following portion of the frequency analysis of the aggressiveness scores presented in Section 2.1:

Score	*f*
10	2
9	3
8	4
7	4
4	2

These data can also be presented in the form of a **frequency histogram,** such as appears in Figure 2.1. The horizontal dimension in this graph is called the *X axis* or **abscissa,** and the vertical dimension is called the *Y axis* or **ordinate.** The abscissa lists the score values from low to high, extending from one unit below the lowest score to one unit above the highest score—in this case, from 3 to 11. A label in the form of a name clearly representing the variable under study should appear beneath the score values. Generally speaking, a graph should "stand alone" so that anyone looking at it can readily interpret it. The ordinate represents the frequency with which each score occurred (corresponding to the second column of the frequency table) and is therefore labeled *f*. The bar for a

FIGURE 2.1 Frequency Histogram

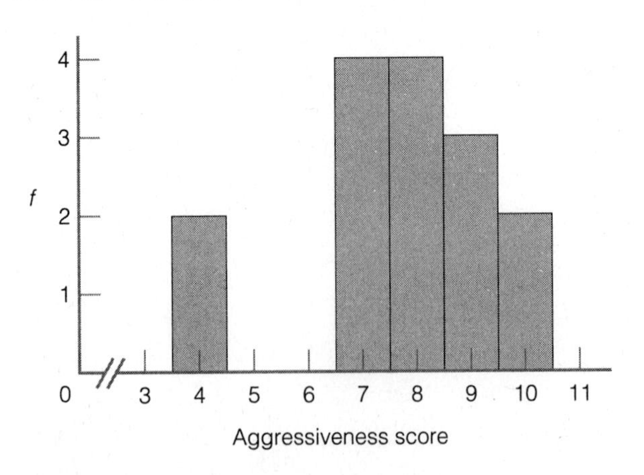

test score of 4 goes up 2 units, indicating that two individuals had scores of 4; the bar for a test score of 7 goes up 4 units, indicating that four individuals had scores of 7; and so forth.

If a variable is continuous, the vertical boundaries of the bar for a given score will represent the real limits of that score. Consider the bar for a score of 4. The leftmost point of the bar represents the lower real limit, 3.5, and the rightmost point of the bar represents the upper real limit, 4.5. The midpoint of the bar corresponds to the midpoint of the real limits 3.5 to 4.5, or 4. Notice also that there is a broken line on the abscissa. This is traditionally done when the abscissa "jumps" from 0 to a larger number and is not drawn to scale. The same principle would hold for the ordinate.

A **frequency polygon** is similar to a frequency histogram and uses the same ordinate and abscissa. A frequency polygon of the same data appears in Figure 2.2. The major difference from the frequency histogram is that bars are not used, but rather, solid dots corresponding to the appropriate frequencies are placed directly above the score values. The dots are then connected by a solid line, forming a polygon when the line is closed with the abscissa, hence the name *frequency polygon*. The similarity of the frequency histogram and frequency polygon can be seen in Figure 2.3, where one has been superimposed on the other.

There are no specific rules governing when a frequency histogram as opposed to a frequency polygon should be used. Frequency polygons are typically used when the variables being reported are continuous in nature, whereas frequency histograms are typically used when the variables being reported are discrete in nature. The major reason is that, from a visual perspective, the frequency polygon tends to highlight the "shape" of the entire distribution more than the

FIGURE 2.2 Frequency Polygon

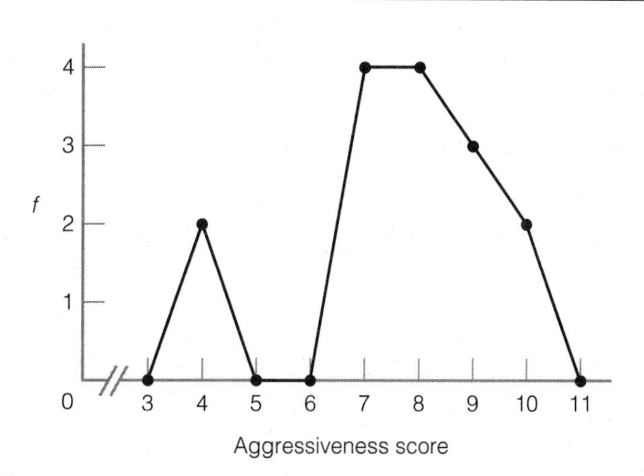

frequency histogram. The frequency histogram, by comparison, tends to highlight the frequency of occurrence of specific scores rather than the entire distribution. The use of a frequency polygon for a continuous variable is thus more consistent with the notion of emphasizing a continuum.

Frequency histograms and frequency polygons can be constructed for grouped as well as ungrouped scores. When the scores are grouped, the abscissa will list the midpoints of the score intervals rather than the individual score values. The procedures for indicating the frequency associated with each interval will then parallel those outlined above.

FIGURE 2.3 Frequency Histogram and Frequency Polygon Superimposed on One Another

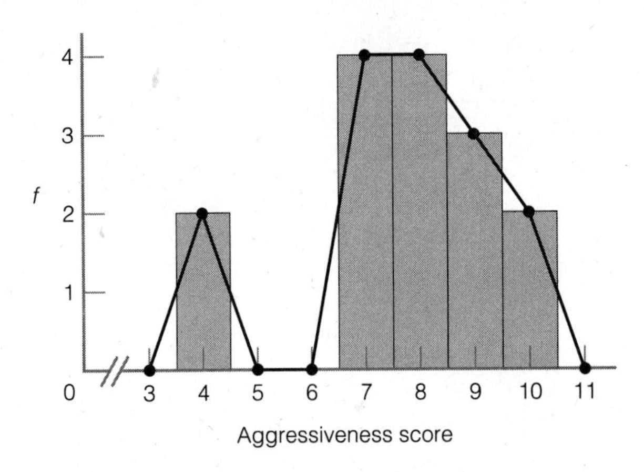

FIGURE 2.4 Bar Graph for Presidential Candidate Preference Poll

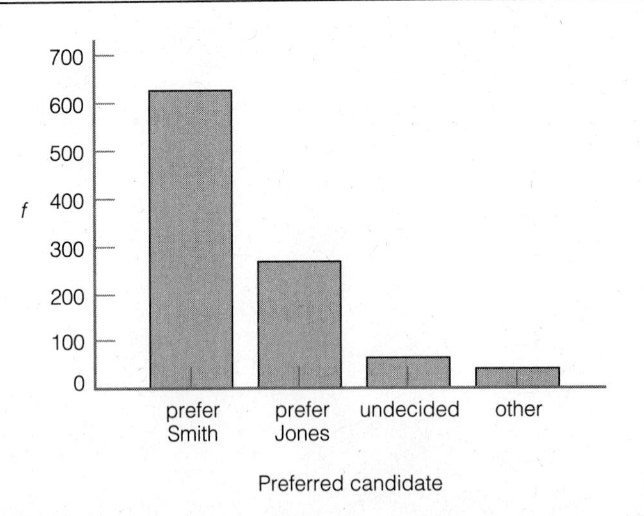

Frequency Graphs for Qualitative Variables Frequency histograms and frequency polygons are used to graph frequency data for quantitative variables. Frequency graphs can also be constructed for qualitative variables. For instance, Figure 2.4 contains a **bar graph** of the data from the poll of preference for presidential candidates referred to in Section 2.3. The values of the variable are listed on the abscissa and the frequencies are listed on the ordinate. The major difference from the frequency histogram is that the bars are drawn such that they do not touch one another. This is because each bar in a bar graph represents a distinct category. Aside from this feature, the basic principles in constructing a bar graph are the same as those for a frequency histogram.

2.5 Misleading Graphs

The presentation of data in graphic form can be highly informative, but it can also be misleading. Consider the case where an investigator has interviewed 100 people in order to determine how many prefer product A over product B or vice versa. Suppose that 45 preferred A over B and 55 preferred B over A. This information could be depicted graphically as in either of the two frequency histograms presented in Figure 2.5.

FIGURE 2.5 Examples of (**a**) a Properly Constructed Graph and (**b**) a Misleading Graph

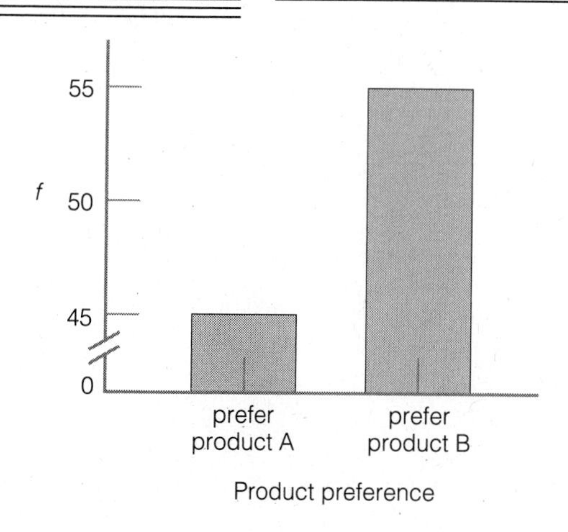

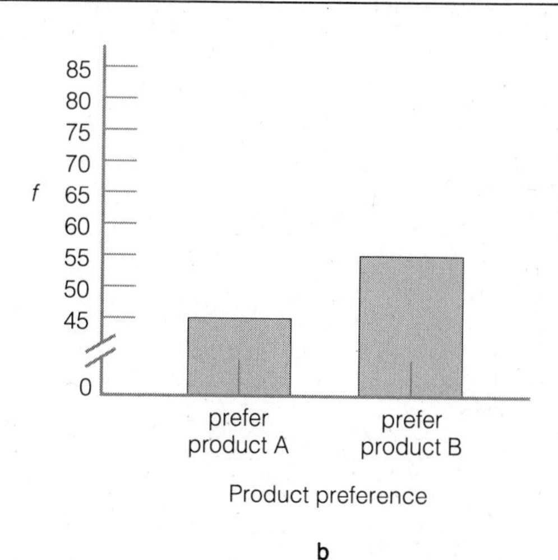

Figure 2.5a makes it clear that product B is preferred over product A. In Figure 2.5b, however, the difference looks much smaller, because the distance between the units on the ordinate is smaller than in Figure 2.5a. As a result, even though the ordinates are the same physical height, a much smaller portion of the ordinate in Figure 2.5b is actually being used to represent the observed frequencies.

Because frequency graphs can be misleading depending on how the abscissa and ordinate are formatted, behavioral scientists have adopted a *"two-thirds high" rule* (American Psychological Association, 1983). This rule states that the ordinate should be presented such that its height at the demarcation for the highest frequency is approximately two-thirds the length of the abscissa. In addition, the ordinate should either start with a frequency of 0 or indicate a "jump" to a larger number by the inclusion of a broken line. These rules ensure uniform, clearly interpretable presentation. One should also be careful to examine the demarcation units on the ordinate. For instance, if you were to look at Figure 2.5a without noting the values of the frequency labels, you might conclude that product B is more preferred over product A than it actually is.

2.6

Graphs of Relative Frequencies, Cumulative Frequencies, and Cumulative Relative Frequencies

It is interesting to note the nature of a graph of relative frequencies as compared to that of frequencies. A polygon of the relative frequencies for the aggressiveness example from Section 2.1 appears in Figure 2.6. Notice that the ordinate is labeled rf and demarcated with relative frequency values. Also notice that the shape of the polygon is identical to the shape of the frequency polygon depicted in Figure 2.2. Similarly, a relative frequency histogram for these data would take the identical shape as the frequency histogram contained in Figure 2.1. This should hardly be surprising since all we have done is divide each frequency by a constant of N.

It is also possible to represent cumulative frequency information graphically. This has been done in Figure 2.7 for the aggressiveness scores example. In this graph, the ordinate is labeled cf and demarcated with cumulative frequency values. Solid dots representing the cumulative frequencies have been placed above the upper real limit of each score value. This contrasts with the approach of placing the dots directly above the score values in frequency and relative frequency graphs. The dots are placed above the upper real limits in Figure 2.7 because a cumulative frequency for a continuous variable encompasses all scores up to the upper real limit of the specified score value. An ad-

FIGURE 2.6 Graph of Relative Frequencies

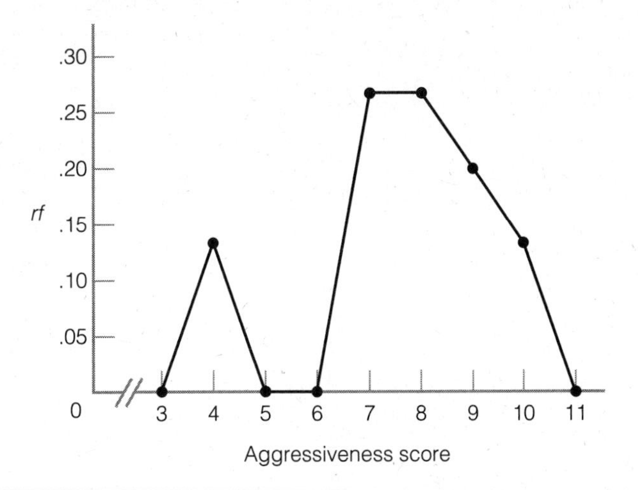

FIGURE 2.7 Graph of Cumulative Frequencies

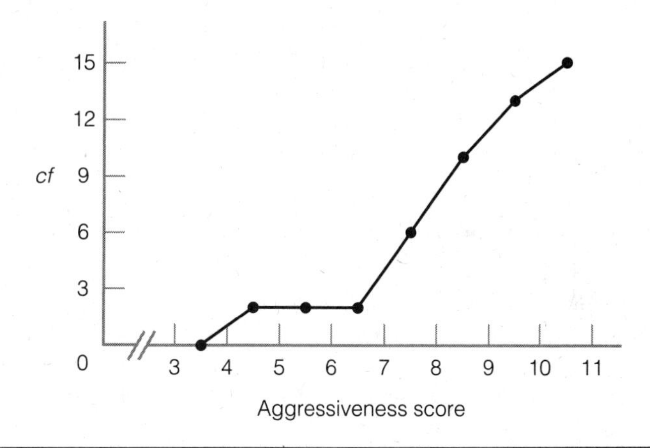

ditional aspect of Figure 2.7 should also be noted: Because the cumulative frequency for a given score value will always be equal to or greater than the cumulative frequency for the preceding score value, the cumulative frequency line will always remain level or increase as it moves from left to right.

Cumulative relative frequencies can also be represented in a graph. Such a graph would take the identical shape as the corresponding cumulative frequency graph, but would have its ordinate labeled *crf* and demarcated with cumulative relative frequency values.

2.7

Probability Distributions

In Section 2.1, we noted that the probability of a certain outcome is equivalent to the relative frequency of that outcome. Thus, given a distribution of scores, the probability of randomly selecting a given score from that distribution equals the relative frequency of that score. However, there is an important conceptual difference between a probability and a relative frequency: *Whereas a relative frequency indicates the proportion of times that some score was previously observed, a probability represents the likelihood of observing that score in the future.* This distinction should be kept in mind as you read the following discussion.

Probability Distributions for Qualitative and Discrete Variables

Consider the case of a qualitative variable, gender, which has two possible values, male and female. Suppose we have a population of 200 individuals, with 150 females and 50 males. The relative frequency for females is 150/200 = .75 and for males it is 50/200 = .25. The probability of randomly selecting a female from this population is .75 and the probability of randomly selecting a male is .25. When the potential values for a qualitative or discrete variable are such that a person can have one and only one score (for example, it is impossible for a person to be both male and female—he or she must be either a male *or* a female), the score values are said to be *mutually exclusive*. When the values considered are also *exhaustive* (that is, they include all possible values that could occur), then the probabilities associated with the individual score values will represent a **probability distribution** with respect to that variable. Figure 2.8 presents a graph of the probability distribution for gender for the population of 200 individuals just described. As with relative frequencies, a graph of probabilities takes the identical shape as the corresponding frequency graph.

Probability Distributions for Continuous Variables

A probability distribution for a continuous variable is conceptualized somewhat differently than that for either a qualitative or discrete variable. Recall from Chapter 1 that a number representing a score on a continuous variable is properly considered from the perspective of the real limits of that number. When we say that a person solved a problem in 30 seconds, we do not mean exactly 30 seconds, but rather somewhere between 29.5 and 30.5 seconds. Because it is always possible, in principle, to have a measuring device that is more accurate than the one we used, we cannot meaningfully talk about the probability of obtaining a score equal to an *exact* value for a continuous variable. Rather, probability is better conceptualized as being associated with a range of values, such as between 29.5 and 30.5.

FIGURE 2.8 Graph of a Probability Distribution for a Qualitative Variable

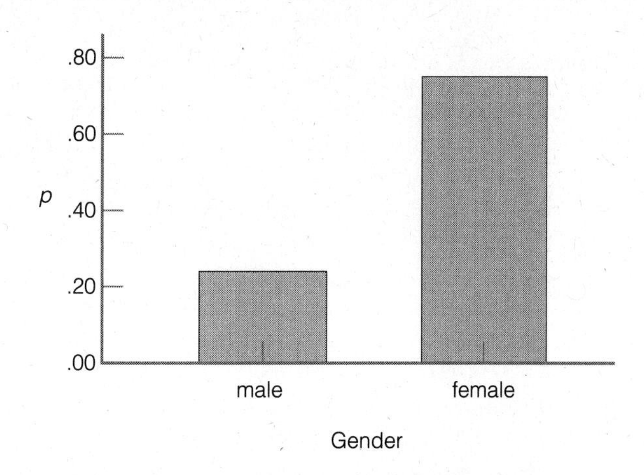

FIGURE 2.9 Graph of a Probability Distribution for a Continuous Variable

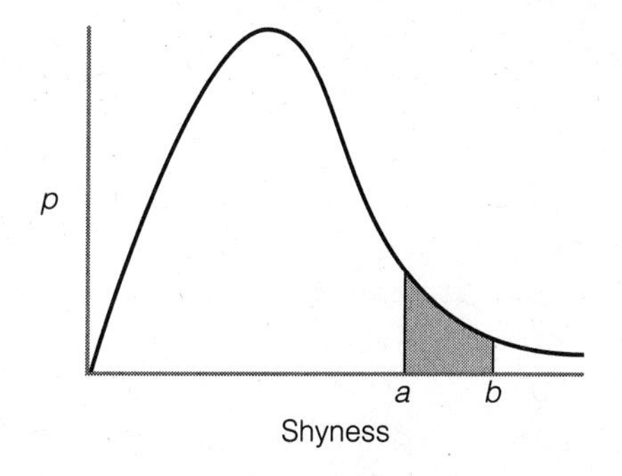

In contrast to qualitative and discrete variables, it is not possible to specify a probability distribution of a continuous variable by listing possible values of the variable and their corresponding probabilities. This is because the number of possible values a continuous variable can have is, in principle, infinite. Statisticians therefore conceptualize a probability distribution of a continuous variable in terms of a **probability density function.**

You can get an intuitive feel for this concept through graphs. Figure 2.9 presents a probability distribution for shyness (a continuous variable) in a population. A continuous probability distribution is always represented as a smooth curve over the abscissa. The abscissa represents the possible values of the continuous variable, with increasingly higher values going from left to right. The ordinate, although not formally demarcated, represents an index of the probability of occurrence of values in the population, with increasingly higher values going from bottom to top.

The key to understanding a probability density function is conceptualizing the probability as an *area* under the curve or, as it is more formally called, the **density curve.** The total area under the curve represents 1.00, or the probability that a given person will have *some* value on the dimension in question. In Figure 2.9, two points, a and b, have been marked on the abscissa. These represent the limits of an interval. The shaded portion between a and b is the area under the curve that corresponds to scores in that interval. The probability of obtaining a value between a and b may thus be represented by this shaded area under the curve. Using appropriate mathematical procedures, it is possible to compute the size of this area based on the case where the total area under the curve equals 1.00. In this way, it is possible to specify the probability of obtaining a set of values falling within some interval for a continuous variable. Specifically, the probability of obtaining a score within a given interval equals the area corresponding to that interval under the density curve. In Figure 2.9, the probability of obtaining a score between points a and b is .07. We will discuss formal procedures for calculating such areas in Chapter 4.

2.8

Empirical and Theoretical Distributions

An important distinction in statistics is that between **empirical distributions** and **theoretical distributions.** Empirical distributions are based on actual measurements collected in the real world. In contrast, theoretical distributions are not constructed by taking actual measurements but, rather, are derived by making assumptions and representing these assumptions mathematically.

One very important type of theoretical distribution that has been studied extensively by statisticians is the **normal distribution.** The normal distribution was so named in the early 19th century by a French mathematician named Quetelet, who believed that it characterized the shape of a large number of phenomena, including height, weight, intelligence, and numerous psychological variables.

There is actually a family of normal distributions, each member of which is precisely defined by a mathematical formula to be given in Appendix 4.1. Figure 2.10 presents some examples of normal distributions. As can be seen, all distributions in this family are symmetrical and are characterized by a "bell shape."

FIGURE 2.10 Examples of Normal Distributions

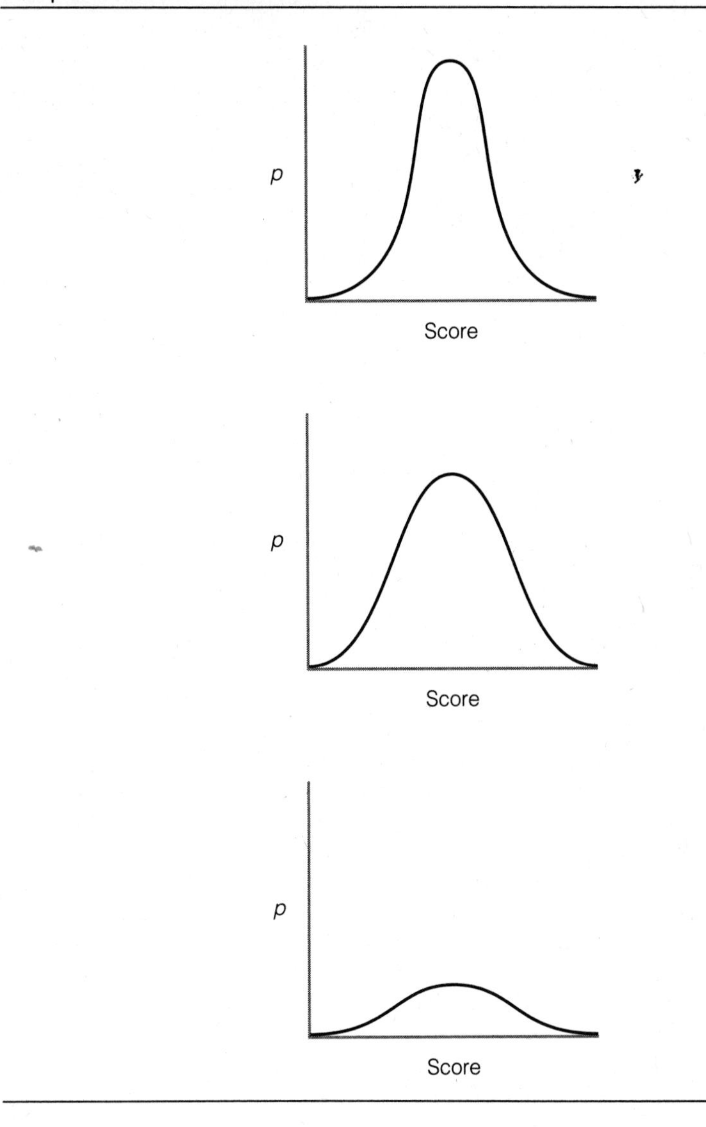

Figure 2.11a presents a relative frequency histogram for an empirical distribution of intelligence test scores for students in a particular high school, and Figure 2.11b presents a theoretical distribution of intelligence scores for this same group. Notice that the latter distribution is normal in shape and closely corresponds to the empirical distribution in Figure 2.11a. Theoretical distributions are frequently used to represent reality when practical limitations make it impossible to construct empirical distributions. For instance, a distri-

FIGURE 2.11 (**a**) Empirical and (**b**) Theoretical Distributions of Intelligence Scores

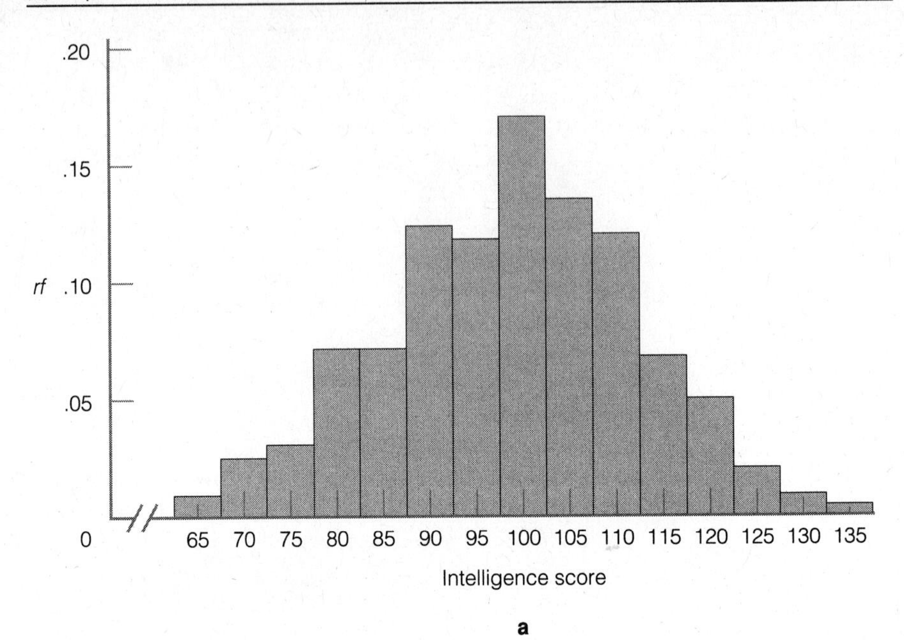

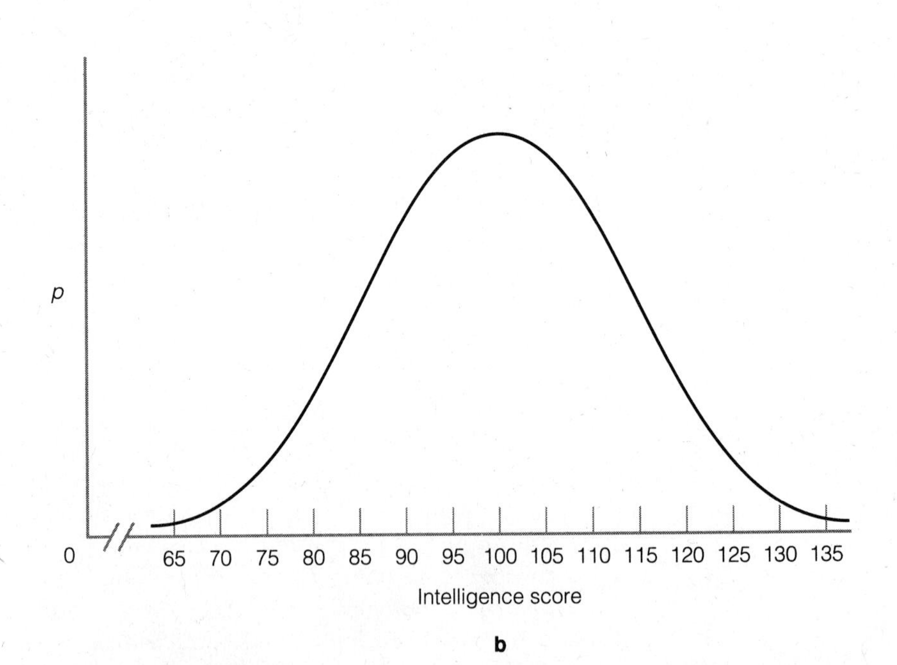

bution of the intelligence scores of *all* adults in the United States could not be constructed empirically. However, we might be able to closely approximate it by reference to the normal distribution. If we are willing to make the assumption that the distribution of intelligence scores is approximately normal, then our knowledge of this theoretical distribution can be used to help us gain insights into the nature of intelligence scores in the real world. As we will see in later chapters, inferential statistics makes extensive use of theoretical distributions.

2.9 Method of Presentation

Many behavioral science journals follow the guidelines laid out in the *Publication Manual of the American Psychological Association* (American Psychological Association, 1983) for presenting the results of statistical analyses. As detailed in that manual, statistical information appears slightly differently in published and unpublished versions of a manuscript. When representing statistical results in "Method of Presentation" sections, the present text will follow the format used by authors when submitting manuscripts to journals for purpose of publication. Where relevant, we will point out differences between this approach and that used in articles as they appear in published form.

It is rare for an investigator to report all of the types of frequency information discussed in this chapter when presenting the results of a frequency analysis. The major reason for this is that journal space is costly; consequently, the data must be presented as efficiently as possible. Therefore, most research reports focus on relative frequencies or percentages. When the total number of scores, N, is also provided, any of the other desired measures can be readily calculated.

An example of a table that might be presented is as follows:

Percentage of Individuals Who Approve or Disapprove of Nuclear Power Plants

Response category	%
Strongly approve	24.6
Moderately approve	19.5
Neither approve nor disapprove	8.4

(table continues)

Response category	%
Moderately disapprove	20.5
Strongly disapprove	27.0

Note. N = 215.

This table reports responses to a question asked of 215 college students concerning their approval or disapproval of nuclear power plants. All of the information is present that would allow the reader to determine the frequencies, relative frequencies, cumulative frequencies, and cumulative relative frequencies, if he or she found it necessary to do so. The relative frequencies are derived by dividing each percentage by 100. Thus, 24.6/100 = .246, 19.5/100 = .195, 8.4/100 = .084, 20.5/100 = .205, and 27.0/100 = .270. The footnote to the table indicates the total number of individuals who responded to the question. In this case, $N = 215$. The number of people in each category can then be obtained by multiplying each relative frequency by N and rounding to the nearest integer. Thus, $.246 \times 215 = 53$, $.195 \times 215 = 42$, $.084 \times 215 = 18$, $.205 \times 215 = 44$, and $.270 \times 215 = 58$. We now have the information to construct a more detailed table:

Response category	*f*	*rf*	*cf*	*crf*
Strongly approve	53	.246	215	1.000
Moderately approve	42	.195	162	.754
Neither approve nor disapprove	18	.084	120	.559
Moderately disapprove	44	.205	102	.475
Strongly disapprove	58	.270	58	.270

This example does not mean that you, as a reader, need to calculate all of this information each time you encounter an incomplete table. Usually what is presented will be sufficient for interpretation. However, ideally if a reader wants to construct a more detailed table (for some reason the author had not anticipated), the information should be present that will allow him or her to do so. The format suggested here accomplishes this.

One difference between published and unpublished versions of a manuscript concerns the way in which statistical symbols are presented. In unpublished reports, these are typed in roman (nonitalicized) letters and underlined, whereas in published form they are italicized but not underlined. Thus, the symbols % and N from the original table would appear as *%* and *N* in a journal article.

Because of cost and space considerations, frequency graphs are typically used only when they are necessary to illustrate major trends in the data that might otherwise be difficult to note. It should be mentioned, however, that given an appropriately presented frequency table, the reader can, if he or she wishes, construct a graph from the table. When frequency graphs are presented, it is important that both the abscissa and the ordinate be clearly labeled and that the other instructions presented earlier in this chapter be followed.

2.10

Example from the Literature

The concept of intelligence is of major interest to behavioral scientists. Many attempts, some of which are controversial, have been made to measure this variable. One of the most widely used tests for adults is the Wechsler Adult Intelligence Scale (WAIS). This test has been extensively studied and applied to several national samples in the United States; one such application is reported in Wechsler (1958). Based on 2,052 individuals, the following frequency analysis characterizes the scores on this test:

Score	*f*	*rf*	*crf*	*Verbal description*
Above 129	29	.014	1.000	Very superior
120–129	150	.073	.986	Superior
110–119	349	.170	.913	Bright normal
100–109	525	.256	.743	Average
90–99	504	.246	.487	Average
80–89	299	.146	.241	Dull normal
70–79	132	.064	.095	Borderline
Below 70	64	.031	.031	Mentally retarded

The verbal descriptions provided on the extreme right are labels often used by clinical psychologists with respect to ranges of scores on the WAIS. Notice that scores labeled "average" (90–109) are the most common. Also, the proportion of individuals with scores above 129 (very superior) is quite small. This is also true of individuals with scores below 70 (mentally retarded). Only 1.4% of the sample had scores above 129 and 3.1% had scores below 70.

In 1926, a behavioral scientist named Cox published an extensive study of eminent men in history. He attempted to estimate the IQ scores that these individuals would have achieved had they lived in a time when such a test could be administered. Some of these estimates are as follows:

Francis Galton—English scientist	200
John Stuart Mill—English philosopher	190
Johann Wolfgang von Goethe—German writer, philosopher	185
Gottfried Wilhelm von Leibniz—German philosopher, mathematician	185
Samuel Taylor Coleridge—English writer, poet	175
John Quincy Adams—American statesman, president	165

David Hume—English philosopher	155
Alfred Tennyson—English poet	155
René Descartes—French philosopher, mathematician	150
Wolfgang Amadeus Mozart—Austrian composer	150
William Wordsworth—English poet, writer	150
Francis Bacon—English philosopher, scientist	145
Charles Dickens—English writer	145
Benjamin Franklin—American inventor, statesman	145
George Frideric Handel—German composer	145
Thomas Jefferson—American statesman, president	145
John Milton—English poet	145
Daniel Webster—American statesman, senator	145

The magnitude of these estimates is impressive, especially in light of the frequency analysis presented in the previous table. These scholars were obviously special individuals.

2.11 Summary

A frequency distribution is a table that conveys information about the frequencies, relative frequencies, cumulative frequencies, and cumulative relative frequencies of a set of scores. When presenting a frequency distribution for a quantitative variable, it is sometimes useful to group the scores. This requires the consideration of three issues: (1) the number of groups to report, (2) the size of the interval, and (3) the beginning of the lowest interval. Frequency information can also be presented graphically using either a frequency histogram or a frequency polygon for quantitative variables, and a bar graph for qualitative variables.

When the potential values on a qualitative or discrete variable are mutually exclusive and exhaustive, the probabilities associated with the individual scores will represent a probability distribution with respect to that variable. Since the probability of an outcome is equivalent to the relative frequency of that outcome, the graph of a probability distribution in these cases will take the same shape as the corresponding relative frequency and frequency graphs. For a continuous variable, a probability distribution is conceptualized in terms of a probability density function. This is represented graphically as a smooth curve over the abscissa.

An important distinction can be made between empirical distributions and theoretical distributions. Empirical distributions are based on actual measurements collected in the real world, whereas theoretical distributions are derived by making assumptions and representing these assumptions mathematically. Normal distributions are a family of symmetrical and bell-shaped theoretical distributions that have been studied extensively by statisticians.

Exercises

Answers to asterisked (*) exercises appear at the back of the book.

Refer to the following information to answer Exercises 1–10.

An employer kept records of how many days her 20 employees reported in sick during the previous year. For these employees, the scores on this variable were as follows:

8	7	6	4	3
6	3	7	6	6
4	6	6	6	7
6	6	8	7	6

***1.** Compute a frequency distribution for the set of scores.

***2.** Compute the relative frequencies, cumulative frequencies, and cumulative relative frequencies.

3. What proportion of employees was sick for 8 days? What proportion of employees was sick for 6 days or less? What proportion of employees was sick for 5 days or less?

***4.** What proportion of employees was sick for 7 days or more? What proportion of employees was sick for more than 7 days? What proportion of employees was sick for 4 days or more? What proportion of employees was sick for more than 4 days?

5. Convert the relative frequencies for the set of scores to percentages.

***6.** Suppose you were to randomly select a score from the 20 scores. What is the probability the selected score would be an 8? What is the probability the selected score would be a 6 or an 8? What is the probability the selected score would be 7 or less?

***7.** Draw a frequency histogram for the set of scores.

8. Draw a frequency polygon for the set of scores.

***9.** Draw a polygon of the relative frequencies. Compare the shape of this graph with that for the frequency polygon from Exercise 8.

10. Draw a graph of the cumulative frequencies.

***11.** How does a grouped frequency distribution differ from an ungrouped frequency distribution?

12. What are the three questions that must be considered when deciding how to group data?

Refer to the following information to answer Exercises 13–22.

A principal in a small school measured the intelligence of students in the fifth grade in his school. The intelligence test scores for these students were as follows:

129	93	104	83	118
99	98	94	100	106
98	109	92	107	122
113	109	119	106	120
103	100	127	101	102
128	111	89	118	117

102	106	95	84	103
110	96	92	119	117
80	108	105	105	103
105	90	108	111	88

*13. Compute a frequency distribution for the set of scores by grouping the scores into five groups.

14. Compute the relative frequencies, cumulative frequencies, and cumulative relative frequencies for the grouped scores.

*15. What proportion of students had scores of 109 or less? What proportion of students had scores of 99 or less?

16. Draw a frequency histogram of the grouped data for the set of scores.

*17. Draw a frequency polygon of the grouped data for the set of scores.

18. Draw a histogram of the relative frequencies for the grouped data. Compare the shape of this graph with that for the frequency histogram from Exercise 16.

*19. Draw a graph of the cumulative frequencies for the grouped data. Compare the general shape of this graph with the cumulative frequency graph for the data on sick days from Exercise 10. What accounts for the similarity in their shapes?

*20. Suppose you wanted to compute a frequency distribution for the data by grouping the intelligence test scores into eight groups. What size interval should you use? With what value should the lowest interval begin?

21. Suppose you wanted to compute a frequency distribution for the data by grouping the scores into 10 groups. What size interval should you use? With what value should the lowest interval begin?

22. Suppose you wanted to compute a frequency distribution for the data by grouping the scores into 15 groups. What size interval should you use? With what value should the lowest interval begin?

23. Redraw the frequency histogram from Exercise 7 such that the ordinate is visually greater than two-thirds the length of the abscissa. Now redraw it such that the ordinate is visually less than two-thirds the length of the abscissa. Compare the three graphs in terms of what they seem to suggest about employee sick days.

Refer to the following information to answer Exercises 24–26.

Suppose you were commissioned to survey a small community to determine the marital status of all adults over 18 years of age. You select a random sample of 50 individuals and ask each individual if they are married (M), divorced (D), widowed (W), or single (S). The data for these individuals are as follows:

M	S	D	W	M	M	S	D	W	M
M	M	S	M	M	M	D	M	M	M
M	W	M	S	S	M	M	M	M	S
M	M	M	M	M	S	S	M	M	M
M	S	D	W	M	M	S	D	W	M

24. Compute a frequency distribution for the set of scores, including relative frequencies and percentages.

25. Draw a bar graph for the set of scores.

26. Draw a bar graph of the relative frequencies. Compare the shape of this graph with that for the bar graph from Exercise 25.

*27. Suppose a realtor is trying to sell someone a house in either neighborhood A or neighborhood B. The realtor knows that the number of people with incomes above $50,000 in neighborhood A is 30, whereas in neighborhood B, it is 15.

a. Draw a bar graph that makes the income difference between these two neighborhoods look large.

b. Draw a bar graph that makes the income difference between these two neighborhoods look small.

*28. What is a probability distribution? Why is the nature of a probability distribution for a qualitative or discrete variable different from that for a continuous variable?

29. What is the difference between an empirical distribution and a theoretical distribution?

*30. A researcher reported the results of a survey of 1,850 people on their attitudes toward capital punishment. One question asked respondents to indicate if they thought the death penalty should be legal. Responses to this question were scored from 1 ("definitely should not be legal") through 5 ("definitely should be legal"), where 3 represents a neutral stance. The results were as shown at right. Compute the frequencies, relative frequencies, cumulative frequencies, and cumulative relative frequencies for the set of responses.

Score	%
5	20
4	30
3	10
2	22
1	18

THREE

MEASURES OF CENTRAL TENDENCY AND VARIABILITY

Describing a set of scores with frequency distributions can be a highly informative way of presenting data about a variable. However, often we are interested in communicating data in a more succinct fashion. Statisticians have developed statistical measures that are useful in characterizing a set of scores for quantitative variables. These will be considered in this chapter.

3.1

Measures of Central Tendency

One piece of information that is useful when trying to describe a set of scores concerns where the scores tend to fall on the numerical scale, or their **central tendency.** A central tendency refers to an *average,* or a score around which other scores tend to cluster. Many indices of central tendency have been proposed and we will consider three of them: the mode, the median, and the mean.

Mode The mode of a distribution of scores is the most easily computed index of central tendency. It is simply the score that occurs most frequently. For the set of scores 6, 8, 8, 8, 10, 10, the mode is 8, since it occurs most frequently (three times). If we were to randomly select one score from a set of scores, the value of that score would most likely be equal to the value of the mode as opposed to any other value, since the mode occurs most frequently. In a graph of a distribution of scores, the modal score will have the highest "peak" in the graph.

Although the mode can be a relatively straightforward index of central tendency, it has several disadvantages. The major problem is that there can be more than one modal score. For example, consider the following scores: 6, 6, 6, 8, 8, 10, 10, 10. Both 6 and 10 are the most frequently occurring scores. In this case, there is ambiguity as to which score is *the* mode because both occur with equal frequency. This set of scores is *bimodal* since it has two modes.

When the mode is not equal to one unique value, it loses some of its effectiveness in characterizing the central tendency of a distribution of scores.

Median Another measure of central tendency is the median. The **median** is the point in the distribution of scores that divides the distribution into two equal parts. In other words, 50% of the scores occur below the median and 50% of the scores occur above the median. In this sense, the median is a measure of central tendency.

There are three different approaches to computing the median. The first approach applies when there is an even number of scores; the second, when there is an odd number of scores; and the third, when there are duplications of the middle score(s), regardless of whether the total number of scores is odd or even. These will be reviewed in turn.

1. *Computation of the median when there is an even number of scores:* Suppose we measured how long it took six students to complete a test. The times for each student are 20, 22, 16, 18, 25, and 27 minutes, respectively. To compute the median, we first order the scores from lowest to highest:

   16 18 20 22 25 27

   The median is the arithmetic average of the two middle scores, or $(20 + 22)/2 = 21$. Figure 3.1a presents this graphically, using a frequency histogram that also shows the real limits of each score. Note that 50% of the scores occur below 21 and 50% of the scores occur above 21.
2. *Computation of the median when there is an odd number of scores:* Suppose we measured the test-taking times of seven students and their scores were 6, 8, 14, 12, 16, 6, and 16 minutes, respectively. To compute the median we order the scores from lowest to highest:

   6 6 8 12 14 16 16

   The median is simply the middle score, 12. Figure 3.1b presents this graphically. It also highlights the importance of considering the median in the context of the real limits of scores. An examination of the values 6, 6, 8, 12, 14, 16, and 16 shows that 50% of the scores are *not* less than 12. Only three of the seven scores are less than 12 and three are greater than 12. With an odd number of scores, the problem is what to do with the middle score. The answer lies in the concept of real limits. The individual who obtained the middle score of 12 minutes is properly conceptualized as scoring somewhere between 11.5 and 12.5 minutes. We do not know exactly where in the interval he or she scored, so the most logical thing to do is to divide the interval in half to define the median. This yields the value of 12. Note in Figure 3.1b that the point 12 divides the distribution

FIGURE 3.1 Illustration of the Median (**a**) When There Is an Even Number of Scores, (**b**) When There Is an Odd Number of Scores, and (**c**) When There Are Duplications of the Middle Score(s)

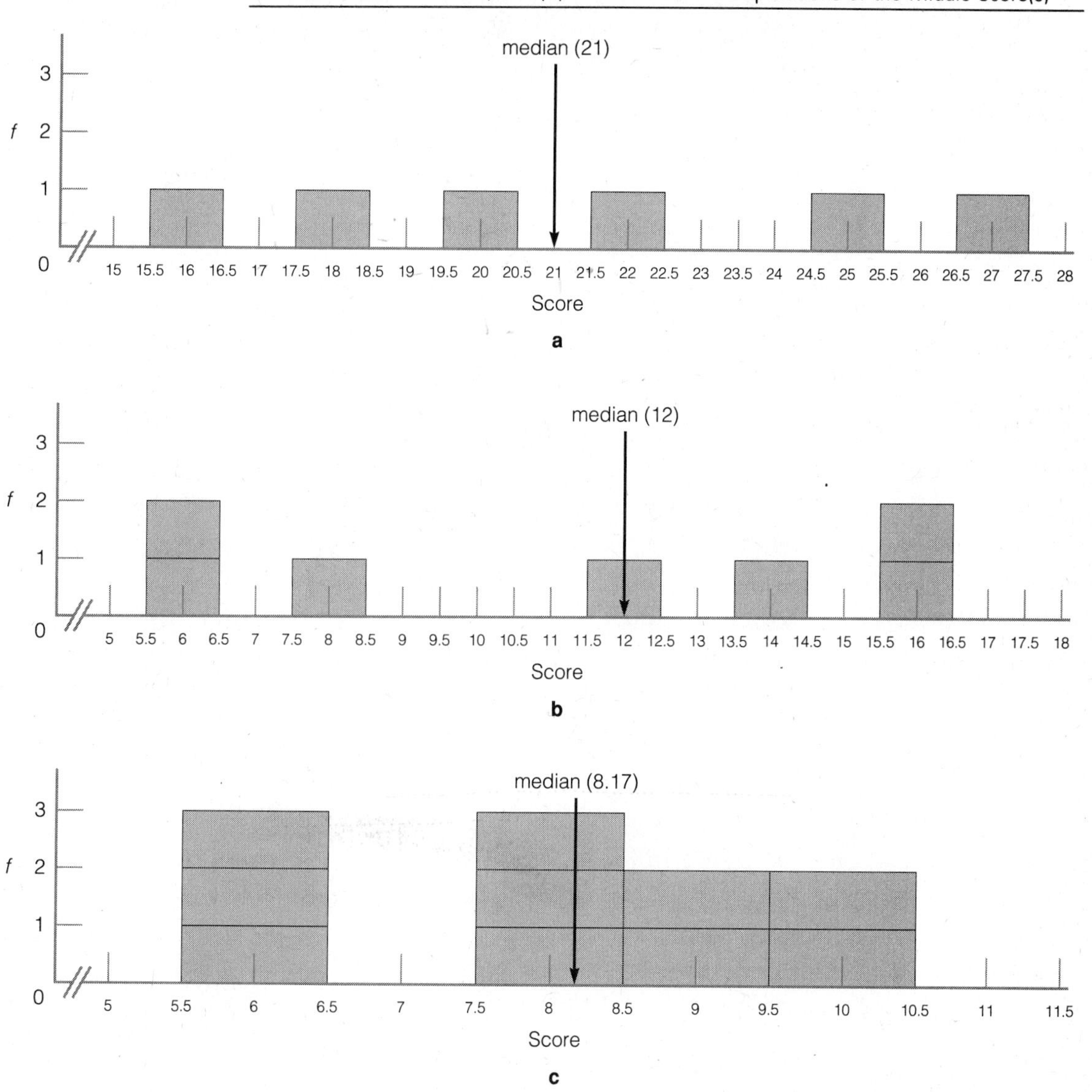

in half. There are 3½ frequency boxes below 12, and 3½ frequency boxes above 12.

3. *Computation of the median when there are duplications of the middle score(s):* This is the most common occurrence in the behavioral sci-

ences. Suppose we measured the test-taking times of 10 students and their scores were 6, 8, 6, 8, 6, 8, 9, 10, 9, and 10 minutes, respectively. We order the scores from lowest to highest:

6 6 6 8 8 8 9 9 10 10

Applying the approach from above for when there is an even number of scores, the arithmetic average of the two middle scores is 8. However, this is not the median—only 3 of the 10 scores are less than 8. Statisticians have developed a formula for computing the median in cases of this type where there are duplications of the middle score(s). This formula is based on the assumption that the median occurs within the real limits of the middle score(s) and is applicable regardless of whether the total number of scores is odd or even. To gain a conceptual understanding of this formula, consider the example in Figure 3.1c, in which there are 10 scores and 8 is the middle score. The median occurs somewhere between 7.5 and 8.5, and represents the point where five scores are less than it and five scores are greater than it. There are already three scores less than 8 since three individuals had a score of 6. The median must therefore be defined so that two more scores are less than it. Individuals with a score of 8, technically, have scores between the real limits 7.5 and 8.5. Since three individuals had a score of 8, and we want to define the median so that two of these three individuals score less than it, we can do so by specifying a score ⅔ (or .67) greater than the lower real limit, or 7.5 + .67 = 8.17. Implicit in this approach is the assumption that the actual scores obtained by the three individuals were equally spaced from one another within the limits 7.5 to 8.5.

The formula for computing the median is

$$Mdn = L + \left[\frac{(N)(.50) - n_L}{n_W}\right] i \qquad [3.1]$$

where *Mdn* represents the median, *L* is the lower real limit of the category containing the median, *N* is the total number of scores in the distribution, n_L is the number of individuals with scores less than *L*, n_W is the number of individuals with scores within the category containing the median, and *i* is the size of the interval of the category containing the median (that is, the difference between the upper real limit and the lower real limit of the category containing the median). The .50 reflects the fact that we are seeking the point that divides the scores such that .50 (50%) of them fall below that point and .50 of them fall above that point. This formula can be applied to either grouped or ungrouped frequency information.

For the last example on test-taking times, the relevant portions of the frequency distribution are as follows:

Score	f	rf	crf
10	2	.20	1.00
9	2	.20	.80
8	3	.30	.60
6	3	.30	.30

To calculate the median using Equation 3.1, we first refer to the cumulative relative frequency column. Starting from the bottom, we move up until we find the first number that is greater than or equal to .50. This number is .60 and occurs in the category for scores of 8. The lower real limit of this category is 7.5, and this represents L. The number of scores *within* the category is 3, and this represents n_W. The number of scores *below* 7.5 is 3; this represents n_L. The interval size, i, is $8.5 - 7.5$, or 1.0. Lastly, the total number of scores, N, is 10. Thus,

$$\begin{aligned} Mdn &= 7.5 + \left[\frac{(10)(.50) - 3}{3}\right](1.0) \\ &= 7.5 + \left(\frac{2.00}{3}\right)(1.0) \\ &= 7.5 + .67 = 8.17 \end{aligned}$$

This is the same value of the median as we stated previously.

Sometimes when there are duplications of the middle score(s), a researcher might want to approximate the median quickly without using Equation 3.1. Under these circumstances, an approximation of the median can be obtained by following the first approach discussed above when N is even and the second approach when N is odd. In our example, $N = 10$, so we will apply the approach for an even number of scores. It will be remembered that this involves taking the arithmetic average of the two middle scores, which in this case is $(8 + 8)/2 = 8$. This value is quite similar to the exact value of the median (8.17) calculated using Equation 3.1.

In summary, when there are no duplications of the middle score(s), the median is equal to the arithmetic average of the two middle scores when N is even and equal to the middle score when N is odd. When there are duplications of the middle score(s), Equation 3.1 will yield the value of the point that equally divides the scores in half, taking into consideration the real limits of the scores.

The median has an important statistical property that underscores its role as a measure of central tendency. Consider the following five scores, which have a median of 3: 1, 2, 3, 7, 8. If we subtract the median from each score and take the absolute values of the resulting differences, we will have a set of unsigned **deviation scores.** Deviation scores are derived by subtracting some constant (in this case, the median) from a set of scores. These deviations are said to be *unsigned* when their absolute values are taken. In our example, the absolute value of how far each score is from the median is

$$|1 - 3| = 2$$
$$|2 - 3| = 1$$
$$|3 - 3| = 0$$
$$|7 - 3| = 4$$
$$|8 - 3| = 5$$

If we add these deviation scores, we obtain a sum of 12. If any value other than the median were to be subtracted from the set of scores, the sum of the unsigned deviation scores would be greater than 12. In fact, across all individuals, scores will always be closer, in an absolute sense, to the median than to any other value. It is this property—the fact that it minimizes the absolute difference between it and the scores in the distribution—that qualifies the median as a measure of central tendency.

STUDY EXERCISE 3.1

A sociologist was interested in determining the median number of years that current employees had been working for a given company. Company records for the 200 employees were used to collect data. The obtained frequency distribution was as follows:

Number of years	*f*	*rf*	*crf*
6	20	.10	1.00
5	20	.10	.90
4	30	.15	.80
3	50	.25	.65
2	50	.25	.40
1	10	.05	.15
0	20	.10	.10

Compute the median number of years in the company.

Answer Using Equation 3.1, we find that

$$Mdn = 2.5 + \left[\frac{(200)(.5) - 80}{50}\right](1.0)$$
$$= 2.5 + \left(\frac{20.00}{50}\right)(1.0)$$
$$= 2.5 + .40 = 2.90$$

Study Exercise 3.1 continued

This means that the average number of years in the company, as defined by the median, was 2.90.

Mean The **mean** of a set of scores is familiar to all of us. It is simply the arithmetic average of the scores. The mean is computed by summing all of the scores and then dividing the sum by the total number of scores. This can be represented by the equation

$$\bar{X} = \frac{\Sigma X}{N} \quad [3.2]$$

where $\bar{X}$ is the symbol used to represent the mean of a set of scores on variable X. As an example, consider the following 10 scores:

2 3 3 4 5 5 5 6 8 9

The sum of the 10 scores is

$$\Sigma X = 2 + 3 + 3 + 4 + 5 + 5 + 5 + 6 + 8 + 9 = 50$$

Hence, the mean is

$$\bar{X} = \frac{\Sigma X}{N}$$

$$= \frac{50}{10} = 5.00$$

The mean has an important statistical property that underscores its role as a measure of central tendency. Consider the following five scores, which have a mean of 3.00: 5, 3, 4, 1, 2. If we subtract the mean from each score and retain the signs of the resulting differences, we will have a set of *signed* deviation scores:

$$5 - 3 = 2$$
$$3 - 3 = 0$$
$$4 - 3 = 1$$
$$1 - 3 = -2$$
$$2 - 3 = -1$$

If we add these deviation scores, we obtain a sum of 0. If any value other than the mean were to be subtracted from the set of scores, the sum of the signed deviation scores would be greater than 0 in absolute value. This is true for any set of scores: The sum of signed deviations about the mean will always equal 0 and will always be less in absolute magnitude than the sum of the signed deviations about any other value. It is this property—the fact that it balances the deviations of scores above it with the deviations of scores below it—that qualifies the mean as an index of central tendency. Note that this is distinct from the median. The mean is the value that minimizes the sum of *signed* deviations, whereas the median is the value that minimizes the sum of *unsigned* deviations.

STUDY EXERCISE 3.2

A researcher developed a test to measure reading ability. The test was intended to be used to help place students in remedial versus advanced reading classes. A major concern of the researcher was how long it would take students to complete the test, since it had to be administered during a short testing period. For 12 students in one class, the following completion times, in minutes, were recorded:

12	13	11	12	11	13
13	11	12	12	12	12

Compute the mean amount of time it took the class to complete the test.

Answer We will compute the mean using Equation 3.2. The first step in applying this equation is to sum the 12 scores:

$$\Sigma X = 12 + 13 + 11 + 12 + 11 + 13 + 13 + 11 + 12 + 12 + 12 + 12 = 144$$

The mean is then obtained as follows:

$$\bar{X} = \frac{\Sigma X}{N} = \frac{144}{12} = 12.00$$

This means that the average amount of time it took the class to complete the test, as defined by the mean, was 12.00 minutes.

Comparison of the Mean, the Median, and the Mode

The mean, the median, and the mode are all indices of central tendency. The mean is the arithmetic average of scores, the median is the point that divides the distribution into halves, and the mode is the most frequently occurring score. Sometimes the mean, the median, and the mode of a set of scores all have the same value. For instance, as illustrated in Figure 3.2, this is always the case with normal distributions. But more often than not, the three measures of central tendency yield different values. Which is the best index of central tendency? Ideally, when trying to characterize a set of scores, it is best to report all three indices. Each represents something slightly different, and the more information we can provide, the better. Most of the inferential statistics used by behavioral scientists make use of the mean, as will be explained in Chapter 7.

FIGURE 3.2 Illustration of the Equivalence of the Mean, Median, and Mode in Normal Distributions

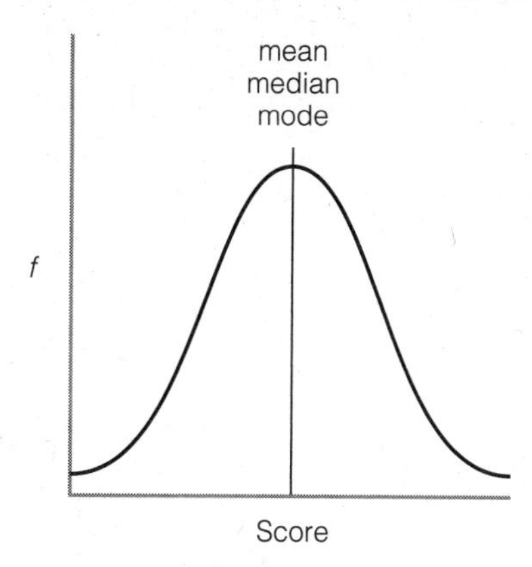

Consequently, the mean is the most frequently encountered measure of central tendency in research reports.

One way of contrasting the mean, the median, and the mode is in terms of a "best guess" interpretation of each. Suppose you know only the mean, median, and mode of a large set of scores. Each individual score is randomly selected in succession and after each selection you are to guess the value of the score. What is your best guess? It depends. First, suppose you want the highest probability of being *exactly* correct. In this case, your best guess is the value of the mode since it is the most frequently occurring score and yields the highest probability of predicting each score exactly. Second, suppose you are not interested in being exactly correct most often, but rather in making the smallest amount of *absolute* error, across all scores. In this case, the *sign* of the error is not important (that is, whether you overpredict or underpredict the value of the score is not critical), but the *size* of the error is important. Here, your best guess is the value of the median. The median is a measure of central tendency in that it minimizes the absolute (unsigned) error across all scores. Finally, if your goal is to minimize *signed* error, then, across all scores, your best guess is the mean. As noted earlier, this is because, across all scores, the amount of signed error will equal 0.

There is no general rule as to which measure of central tendency is best, since it depends on what one is trying to accomplish or communicate. For purely descriptive purposes, the median is a very useful measure because it minimizes unsigned error. For purposes of inferring population parameters from samples, the mean is a superior index, as will be explained in Chapter 7. If one is primarily interested in representing the typical case, the mode is the measure of choice. However, in practice, the mode is used mainly to supplement the median or the mean and is seldom taken as the only measure of central tendency.

It might be instructive to consider an example of when the mean, the median, and the mode can yield very different characterizations of a set of scores.

Suppose you own a business with six employees in addition to yourself. The annual salaries are as follows:

Employee	*Salary ($)*
1	13,000
2	13,000
3	14,000
4	15,000
5	16,000
6	17,000
Yourself	185,000

For the set of seven scores, the mean is 39,000, the median is 15,000, and the mode is 13,000. In characterizing the "average" score, you could paint very different pictures of your company depending on the measure of central tendency used (for example, "Look how generous I am—the average salary is $39,000" as opposed to "Look how small my business is—the average salary is only $13,000").

In general, when there are one or more extreme scores in the distribution (in this case, the salary of $185,000), the information conveyed by the mean can become distorted and the median and the mode will provide better insights into the central tendency of the data. Note that in our example, these latter measures are not affected by the extremity of the seventh score; they would be the same if your salary were $185,000, $17,500, or $250,000. In contrast, the mean is based on the sum of all of the scores in a distribution and is therefore affected by the value of each score. The more extreme a score is relative to the other scores in the distribution, the more it will adjust the mean. Thus, for instance, while the mean salary of the seven employees is $39,000 when your own salary is $185,000, the mean salary would be $15,071 if your salary were $17,500, and $48,286 if your salary were $250,000.

Use of the Mean, the Median, and the Mode

We have presented three measures of central tendency in the context of analyzing quantitative variables. Some qualifications about the appropriateness of the indices for describing central tendencies are now required. When a quantitative variable is measured on approximately an interval level, all three measures of central tendency are meaningful. When a quantitative variable is measured on an ordinal level that departs markedly from interval characteristics, the mean will not be an appropriate index of central tendency and the median and/or mode must be used instead. This is because the mean relies on calculating a sum, and a sum is meaningful only when the intervals between successive categories are approximately equal. Finally, the concepts of mean and median are meaningless for qualitative variables (that is, nominal measures) because these concepts require ordering objects along a dimension. In this case, the mode (that is, the most frequently occurring category) will be the only applicable descriptor of central tendency.

3.2

Measures of Variability

Measures of central tendency indicate where scores tend to cluster in a distribution. A second important characteristic in analyzing quantitative variables is the extent to which scores are alike or different. Consider the two sets of scores in columns 1 and 4 of Table 3.1. In both cases, the mean is 5.00. However, the **variability** of the scores, or the extent to which they are similar or dissimilar, is quite different. While the scores in Set II tend to be alike, the scores in Set I tend to differ from one another. Statisticians have developed a number of indices to measure such variability.

Range One very simple index of variability is the **range,** which is the highest score minus the lowest score. In Set I the range is $8 - 2 = 6$, whereas in Set II it is $6 - 4 = 2$. The range is not a very good index of variability because it can be misleading. Consider the extreme case of 900 scores, of which 899 are equal to 100 and one is equal to 10. The range would be quite large, $100 - 10 = 90$, even though almost every score is identical.

Sum of Squares A second index of variability, and one we will use extensively in later chapters, is called the **sum of squares** and is symbolized as SS. We will develop this measure using the two sets of scores in Table 3.1. The sum of squares, unlike the range, makes use of every score in a distribution and not just the two most extreme scores. We begin by asking how far scores tend to vary from the "typical" score. If we let the typical score be represented by the mean, then we are concerned with how much each score deviates from the mean. Columns 2 and 5 of Table 3.1 present deviation scores in which the mean of each data set (5.00 in both cases) has been subtracted from each original score in that set. If the scores in the distribution are similar, as in Set II, each will be near the mean and

TABLE 3.1 Two Sets of Scores with Different Variability

SET I			SET II		
X	$X - \bar{X}$	$(X - \bar{X})^2$	X	$X - \bar{X}$	$(X - \bar{X})^2$
2	−3	9	4	−1	1
3	−2	4	5	0	0
3	−2	4	5	0	0
5	0	0	5	0	0
7	2	4	5	0	0
7	2	4	5	0	0
8	3	9	6	1	1
$\Sigma X = 35$		SS = 34	$\Sigma X = 35$		SS = 2
$\bar{X} = 5.00$			$\bar{X} = 5.00$		

the deviation scores will therefore be close to 0. In contrast, if, as in Set I, the scores tend to differ from one another, the deviation scores will tend to be relatively large.

We now want to combine the deviation scores in each data set in some way to derive a single numerical index of overall variability. We cannot use the sum of the deviation scores because, as noted in Section 3.1, the sum of signed deviations about the mean will always equal 0. Statisticians have suggested a solution that is very desirable from a statistical perspective. The approach involves first squaring each deviation score and then summing them. Columns 3 and 6 of Table 3.1 do this. Note that the sum of squares for Set I is 34 and the sum of squares for Set II is 2. This reflects the greater variability in the Set I scores than in the Set II scores.

The sum of squares gets its name from the operations performed; it is a shorthand term for the *sum of the squared deviations from the mean.* This quantity can be represented symbolically as

$$SS = \Sigma(X - \bar{X})^2 \qquad [3.3]$$

where SS is an abbreviation for "sum of squares."

Students frequently ask why the mean and not the median is used as the measure of central tendency when defining the sum of squares. There are several reasons. One reason is that when the mean and the median are different, the sum of the *squared* deviations from the mean will always be less than the sum of the *squared* deviations from the median. In fact, it turns out that the sum of squared deviations from the mean will always be less than the sum of squared deviations around any other value; that is, the mean minimizes squared error. A second reason for preferring the mean concerns the important role that means and their sums of squares play when making inferences about populations from sample data. This will be discussed in Chapter 7.

STUDY EXERCISE 3.3

A researcher was interested in the variability of test scores for five children. On a test with 15 points possible, the scores were 10, 8, 6, 4, and 2. Compute the sum of squares for the scores.

Answer Using Equation 3.3, we find that SS = 40:

X	$X - \bar{X}$	$(X - \bar{X})^2$
10	4	16
8	2	4
6	0	0
4	−2	4
2	−4	16
$\Sigma X = 30$		SS = 40
$\bar{X} = 6.00$		

Variance One problem with the sum of squares as an index of variability is that its size depends not only on the amount of variability among scores, but also on the *number* of scores (N). Consider two sets of data, where the scores in Set A are 2, 4, and 6, and the scores in Set B are 4, 4, 4, 4, 4, 6, 6, 6, 6, and 6. We can readily see that the scores in Set A tend to differ from each other more than do the scores in Set B. Nevertheless, the sum of squares for Set A is 8 and the sum of squares for Set B is 10. Thus, although there appears to be greater variability in Set A than in Set B, the sum of squares for Set B is larger than the sum of squares for Set A. This occurs because the sum of squares for Set B is based on 10 scores, whereas the sum of squares for Set A is based on only 3 scores. To avoid inconsistencies of this type, an index that compares the variability of two or more sets of scores should take into account the number of cases within each set. One possibility is to divide the sum of squares by N—that is, to compute the mean squared deviation score. This is called the **variance** and is defined by the formula

$$s^2 = \frac{\text{SS}}{N} \qquad [3.4]$$

The symbol used to represent a variance is a lowercase s with a square operator. Note that the variance for Set A is $8/3 = 2.67$ and the variance for Set B is $10/10 = 1.00$. Consistent with our eyeball interpretation, the mean squared deviation score (the variance) is greater in Set A than in Set B.

For the data in Table 3.1, the variance for Set I is

$$s^2 = \frac{34}{7} = 4.86$$

and the variance for Set II is

$$s^2 = \frac{2}{7} = .29$$

The finding of greater variability in Set I than in Set II is consistent with our earlier results for the sums of squares.

Standard Deviation A fourth index of variability among scores is the **standard deviation.** The standard deviation is the positive square root of the variance and is symbolized by the letter s:

$$s = \sqrt{s^2} \qquad [3.5]$$

In Table 3.1, the standard deviation for Set I is

$$s = \sqrt{4.86} = 2.20$$

and the standard deviation for Set II is

$$s = \sqrt{.29} = .54$$

The standard deviation is probably the most easily interpreted measure of variability among a set of scores. Recall that the variance is the mean squared deviation score. Few of us feel comfortable interpreting squared deviation scores, and by taking the square root of the variance, we are, in essence, eliminating the square and returning to the original unit of measurement. *The standard deviation thus represents an average deviation from the mean.** On the average, the Set I scores reported in Table 3.1 deviate 2.20 units from the mean and the Set II scores deviate .54 unit from the mean. Again, more variability is indicated in Set I than in Set II.

Students learning about the standard deviation frequently ask what value indicates a large standard deviation. The answer is that it depends on what is being measured. For example, suppose we measure the number of children that families have in a given country and find a mean of 4.20 and a standard deviation of 3.00. This represents considerable variability since, on the average, scores deviate three "children" from the mean. In contrast, suppose we measure the annual income for people living in a particular neighborhood and observe a mean of $28,760.40 and a standard deviation of $3.00. In this case, there is very little variability since scores deviate from the mean by an average of only $3.00. When the units are dollars and the concern is annual income, a standard deviation of 3.00 is trivial, but when the units are children and the concern is family size, a standard deviation of 3.00 is substantial.

STUDY EXERCISE 3.4

Compute the variance and the standard deviation for the scores from Study Exercise 3.3.

Answer Since the sum of squares was previously found to equal 40 and there are five scores, the variance is equal to

$$s^2 = \frac{SS}{N}$$

$$= \frac{40}{5} = 8.00$$

The standard deviation is then derived by taking the positive square root of the variance:

$$s = \sqrt{8.00} = 2.83$$

---

* Specifically, the standard deviation represents the positive square root of the arithmetic average of the squared deviations from the mean.

Characteristics and Use of the Sum of Squares, the Variance, and the Standard Deviation

The sum of squares, the variance, and the standard deviation are all useful indices of variability. We will make extensive use of all three in this book. *It will always be the case that SS, s^2, and s are greater than or equal to 0.* These statistics can never be negative, because they are all based on squared deviation scores and any number squared must be nonnegative. When the sum of squares equals 0, the variance and standard deviation will also equal 0. A value of 0 for these three statistics means that there is no variability in the scores; they are all the same. As the values of the three statistics become increasingly greater than 0, more variability among the scores is indicated.

When measurements are taken on approximately an interval level, the sum of squares, the variance, and the standard deviation are all appropriate indices of variability. When measurements are taken on an ordinal level that departs markedly from interval characteristics, these indices are not applicable because they are based on summed deviation scores, and, as noted earlier, a sum is meaningful only when the intervals between successive categories are approximately equal. For such measures, an index of variability would be the range, since it does not rely on a sum. For qualitative variables (that is, nominal measures), there is no index of variability comparable to those discussed above because the categories of scores cannot be ordered (however, see Kirk, 1978, pp. 73–75, for one attempt to measure variability on a qualitative variable).

3.3

Computational Formula for the Sum of Squares

The determination of a sum of squares using Equation 3.3 involves computing a deviation score and a squared deviation score for each individual. Each time one of these scores is calculated, the opportunity exists for rounding error. There is an alternative formula for computing the sum of squares that does not require the computation of deviation scores. This formula is both more efficient and more precise, as it requires fewer steps and presents fewer opportunities for rounding error. The formula is

$$SS = \Sigma X^2 - \frac{(\Sigma X)^2}{N} \qquad [3.6]$$

where ΣX^2 is the *sum of the squared X scores* and $(\Sigma X)^2$ is the *square of the summed X scores.* As demonstrated below, this formula is mathematically equivalent to Equation 3.3.

Consider the seven scores from Set I of Table 3.1. Application of Equation 3.6 requires that we first calculate the sum of the X scores and the sum of the squared X scores:

X	X^2
2	4
3	9
3	9
5	25
7	49
7	49
8	64
$\Sigma X = 35$	$\Sigma X^2 = 209$

Thus,

$$SS = 209 - \frac{(35)^2}{7}$$
$$= 209 - 175 = 34$$

which is identical to the value obtained in Table 3.1 using Equation 3.3.

Given its greater computational ease, we will use Equation 3.6 rather than Equation 3.3 when calculating sums of squares in the remainder of this book.

STUDY EXERCISE 3.5

Compute the sum of squares for the scores from Study Exercise 3.3 using Equation 3.6. Compare your result with the result you obtained in Study Exercise 3.3 using Equation 3.3.

Answer We must first determine the sum of the X scores and the sum of the squared X scores:

X	X^2
10	100
8	64
6	36
4	16
2	4
$\Sigma X = 30$	$\Sigma X^2 = 220$

Thus,

$$SS = \Sigma X^2 - \frac{(\Sigma X)^2}{N}$$
$$= 220 - \frac{(30)^2}{5}$$
$$= 220 - 180 = 40$$

This result agrees with that obtained in Study Exercise 3.3.

3.4

Relationship Between Central Tendency and Variability

Central tendency and variability represent different, independent characteristics of a distribution. For purpose of illustration, we will focus on the mean and the standard deviation.

Suppose we assessed the depression levels of a large sample of "normal" (nonclinical) adults and found that their mean depression score on some measure was 16.00 and the standard deviation was 2.00. A frequency graph for these data might appear as in curve A of Figure 3.3a. Further suppose that we performed the same assessment on a large sample of individuals diagnosed as being clinically depressed and observed the frequency distribution depicted in curve B of Figure 3.3a. This distribution takes the same shape as curve A and has the same standard deviation, but the mean is equal to 32.00 rather than 16.00, indicating a higher overall level of depression. This illustrates an important point: Distributions of scores can have identical variabilities (in this case, standard deviations) but very different central tendencies (in this case, means).

Distributions can also have identical central tendencies but very different variabilities. For instance, Figure 3.3b presents frequency distributions for a large sample of "normal" adults and a large sample of adults diagnosed as being manic-depressive. This is a psychological condition in which cycles of extreme euphoria, sociability, and hyperactivity (mania) alternate with cycles of extreme despair, social withdrawal, and hypoactivity (depression). Although the mean depression rating is 16.00 for each group, the "normal" sample has a relatively small standard deviation of 2.00, indicating that subjects in this group tended to receive depression scores of right around 16 (see curve A), whereas the manic-depressive sample has a standard deviation of 5.00, indicating a relatively higher degree of variability in depression ratings (see curve B). This probably reflects the fact that some members of this group are currently in the manic stage, and are thus not experiencing depression, while others are in the depression stage and still others are in remission. The greater variability of the manic-depressive group is illustrated graphically by the greater spread of curve B relative to curve A.

As illustrated by this second example, measures of variability can help to interpret measures of central tendency. To further demonstrate this point, think about where you would prefer to live if given a choice between a location having a temperature with an annual mean of 70 °F and a standard deviation of 10 °F, and a location having a temperature with an annual mean of 70 °F and a standard deviation of 25 °F.

FIGURE 3.3 Illustration of Distributions Having (**a**) Identical Standard Deviations but Different Means, and (**b**) Identical Means but Different Standard Deviations

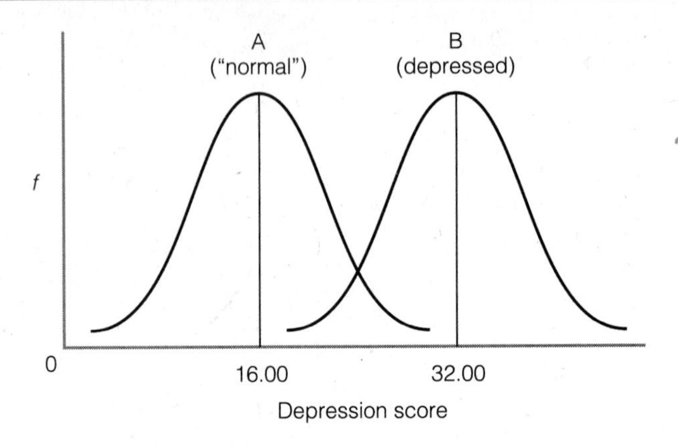

a

A
("normal")
B (manic-depressive)
f
0
16.00
Depression score

b

3.5 Skewness and Kurtosis

Thus far, we have considered two ways in which distributions of scores can differ: (1) central tendency and (2) variability. Two other differences are *skewness* and *kurtosis*. **Skewness** refers to the tendency for scores to cluster on one side of the mean. Figure 3.4 presents two graphs of scores that illustrate this

FIGURE 3.4 Frequency Graphs of (**a**) a Positively Skewed Distribution and (**b**) a Negatively Skewed Distribution

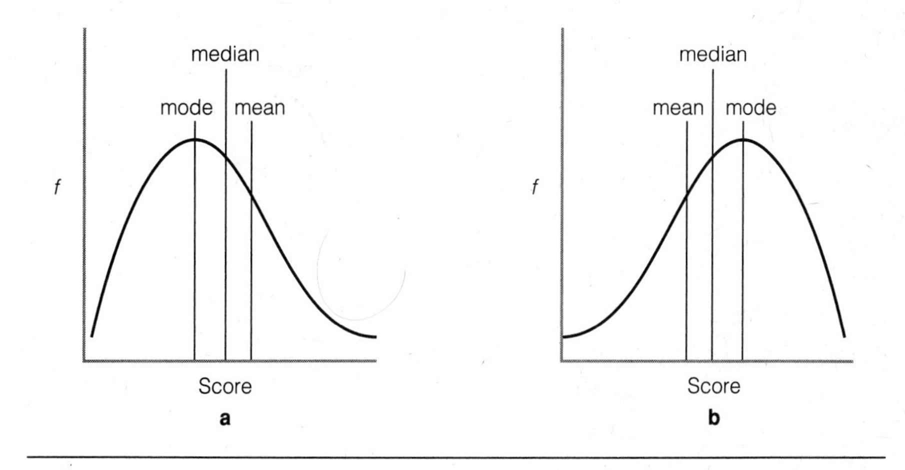

FIGURE 3.5 Frequency Graphs of (**a**) Platykurtic, (**b**) Leptokurtic, and (**c**) Normal Distributions

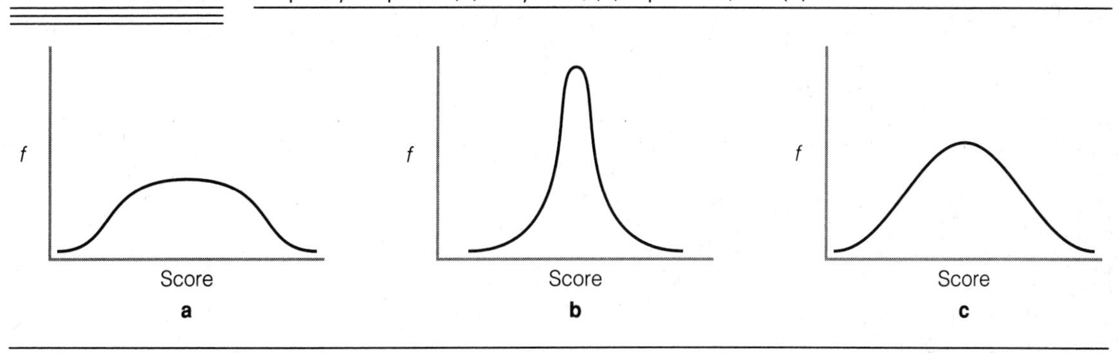

concept. The graph in Figure 3.4a is said to be **positively skewed** because the "tail" is toward the right, or positive, end of the abscissa. In positively skewed distributions, most scores occur below the mean and only a relatively few extreme scores occur above it. The graph in Figure 3.4b represents a distribution that is **negatively skewed,** as the "tail" is toward the left, or negative, end of the abscissa. Most scores in negatively skewed distributions occur above the mean and only a relatively few extreme scores occur below it.

Normal distributions, such as that depicted in Figure 3.2, are not skewed, as equal numbers of scores occur above and below the mean, which is also the median (and the mode). As shown in Figure 3.4, these three measures of central tendency all take on different values in skewed distributions.

Kurtosis refers to the flatness or peakedness of one distribution relative to another. If a distribution is less peaked than another, it is said to be more *platykurtic,* and if it is more peaked than another, it is said to be more *leptokurtic.* It is conventional to label a distribution as being either platykurtic or leptokurtic depending on whether it is more or less peaked than the normal distribution. Figure 3.5 presents graphs of platykurtic, leptokurtic, and normal distributions.

Statisticians have derived numerical indices of skewness and kurtosis similar to the indices of central tendency and variability already discussed. These are rarely used in the behavioral sciences and hence will not be considered here. Interested readers are referred to Ferguson (1976).

3.6 Sample Versus Population Notation

The descriptive statistics we will use most frequently are the mean, the variance, and the standard deviation. Reference will be made to these indices for both samples and populations. It is traditional in statistics to refer to population-derived indices using Greek notation. A sample mean is typically symbolized as $\bar{X}$, whereas a population mean is symbolized as μ (lowercase Greek *m,* called "mu"). Of course, the formula for computing the two indices is identical: Sum the scores and divide by N. The Greek notation, however, makes it explicit that we are describing a population, whereas the $\bar{X}$ notation makes it explicit that we are describing a sample. The Greek notations for a variance and standard deviation are σ^2 and σ (lowercase Greek *s,* called "sigma"), respectively, paralleling the symbol s^2 and s for a sample variance and standard deviation.

3.7 Method of Presentation

When presenting measures of central tendency and variability, most researchers report means and standard deviations. These are undoubtedly the most frequently encountered descriptive statistics. Occasionally, means will be supplemented with reports of the median, especially when the mean can be a misleading index of central tendency (for example, when there are extreme scores). A common format used in behavioral science journals presents means and standard deviations in a table, such as the following:

Means and Standard Deviations for Intelligence Scores of Males and Females

Gender	M	SD	n
Males	101.31	10.62	120
Females	102.48	10.31	115

In this table, M represents the mean and SD represents the standard deviation according to American Psychological Association format. The symbols presented in the text, $\bar{X}$ and s, will also be encountered in some journals. It is also common to report the size of the groups upon which the statistics are based. This is symbolized by n because N is reserved to designate the size of the overall data set when there is more than one group. Thus, in this example, $N = 120 + 115 = 235$.

Note that it is possible to compute the variances and the sums of squares from the information provided. For instance, the variance for males is the square of the standard deviation of 10.62, or $10.62^2 = 112.78$. The sum of squares can then be computed by multiplying the variance by N: $(112.78)(120) = 13{,}533.60$.

3.8 Example from the Literature

Psychologists have studied how we form impressions of others and how we perceive different traits and characteristics. For example, Anderson (1968) asked a large number of individuals to rate different traits in terms of their desirability or favorableness. The ratings were made on 7-point scales ranging from 0 to 6, with higher values indicating that the trait was more desirable. A value of 3 was a neutral evaluation. Ratings less than 3 indicated the trait was undesirable or unfavorable, and ratings greater than 3 indicated the trait was desirable or favorable. The means and standard deviations for six of the traits were as follows:

Trait	M	SD
Sincere	5.73	.55
Honest	5.55	.68
Narrow-minded	.80	.76
Selfish	.82	.80
Cunning	2.62	1.48
Inexperienced	2.62	.79

The first two traits were rated very positively by the individuals in the study. The means were quite high (5.73 and 5.55) and the standard deviations were relatively small, indicating that most individuals rated the traits using the upper points of the scale. The traits "narrow-minded" and "selfish" were similarly rated, but in a negative fashion. The traits "cunning" and "inexperienced" both yielded relatively neutral mean scores (2.62 where 3 is neutral). However, notice the large difference between the standard deviations for the two traits. The small standard deviation for "inexperienced" indicates that people consistently tended to rate this trait near the neutral point, whereas the large standard deviation for "cunning" indicates that there was considerable variability in these ratings, with many individuals perceiving "cunning" as being a positive trait and many individuals perceiving "cunning" as being a negative trait. When averaged, the mean was near the neutral point. By showing that the perceptions of "cunning" were not really all that neutral, but rather exhibited considerable variability across individuals, the standard deviation helps to interpret the mean.

3.9 Summary

Two important characteristics of a distribution of scores are where the scores tend to fall on the numerical scale (central tendency) and the extent to which the scores are alike or different (variability). Central tendency and variability represent different, independent aspects of a distribution. A set of scores can also be described in terms of its skewness and kurtosis.

Three different measures of central tendency are the mean, the median, and the mode. The mean is the arithmetic average of the scores, the median is the point that divides the distribution into halves, and the mode is the most frequently occurring score.

Four different measures of variability are the range, the sum of squares, the variance, and the standard deviation. The range is the difference between the highest score and the lowest score. The sum of squares is the sum of the squared deviations from the mean. By dividing the sum of squares by N, we obtain the variance. The standard deviation is the positive square root of the variance and represents an average deviation from the mean.

Exercises

Answers to asterisked (*) exercises appear at the back of the book. Answers to exercises with two asterisks are also worked out step-by-step in the Study Guide.

1. Identify and define the three measures of central tendency.

2. What are the three approaches to computing the median? When is each appropriate?

****3.** Compute the median for the following scores: 4, 4, 4, 6, 9.

4. Compute the median for the following scores: 2, 3, 5, 5, 7, 12.

5. Given the impact of television on children's attitudes and behavior, an important concern of behavioral scientists is the amount of time children of various ages spend watching. The following data are representative of the weekly viewing times (in hours) of 12-year-olds. Compute the mean, the median, and the mode for these scores.

18	17	22	20	25	20	16
19	18	22	26	23	23	23
24	24	22	21	19	20	20

***6.** For the following scores, compute the mean, the median, and the mode: −3, −3, −2, −2, −2, −1, −1, −1, 0, 0, 0, 0, 1, 1, 1, 2, 2, 2, 3, 3.

7. What are unsigned deviation scores? What are signed deviation scores?

***8.** Compute the mean for the following five scores: 10, 11, 12, 13, 14. Now, generate a new set of five scores by adding a constant of 3 to each original score. Compute the mean for the new scores. Compare the two means. Do the same for another set of five scores generated by subtracting a constant of 10 from each original score. What is the effect on the mean of adding a constant to or subtracting a constant from each score in a set of scores?

9. Suppose you measured the mean amount of time it took 10 children to solve a problem and found it to be 35.31 seconds. You later discovered, however, that your timing device (a watch) was 2 seconds too slow for each child. What would the real mean score be?

***10.** Compute the mean for the following five scores: 10, 20, 30, 40, 50. Now, generate a new set of five scores by multiplying each original score by a constant of 3. Compute the mean for the new scores. Compare the two means. Do the same for another set of five scores generated by dividing each original score by a constant of 10. What is the effect on the mean of multiplying or dividing each score in a set of scores by a constant?

***11.** Repeat the procedures outlined in Exercises 8 and 10, but compute the median rather than the mean. What is the effect on the median of adding a constant to or subtracting a constant from each score in a set of scores? What is the effect on the median of multiplying or dividing each score in a set of scores by a constant? What effects do you think these manipulations have on the mode?

12. What are the advantages and disadvantages of the mean as a measure of central tendency? Of the median? Of the mode?

***13.** Consider two sets of data:

Set I		*Set II*	
20	19	53	55
21	20	54	54
20	21	54	53
10	300	55	54

For which set is the mean a poorer descriptor of central tendency? Why?

14. Identify and define the four measures of variability.

***15.** Under what condition is the range a misleading index of variability?

16. What is the problem with the sum of squares as a measure of variability?

*17. If the variance of a set of scores is 100.00, what must the standard deviation be?

18. If the standard deviation of a set of scores is 7.00, what must the variance be?

*19. Without actually calculating it, what must the standard deviation of the following scores be: 6, 6, 6, 6, 6? What must the variance be? What must the sum of squares be? Why?

*20. Why is the standard deviation more "interpretable" than the variance? That is, what is the advantage of reporting statistics in terms of the standard deviation as opposed to the variance?

21. Compute ΣX^2 and $(\Sigma X)^2$ for the following set of scores: 3, 6, 12, 5, 9, 2, 3, 6, 11.

**22. For the scores in Exercise 21, compute the sum of squares using the defining formula (Equation 3.3). Recalculate the sum of squares using the computational formula (Equation 3.6). Compare the two results. Which approach did you find more efficient?

*23. Compute the range, the sum of squares, the variance, and the standard deviation for the data in Exercise 5.

24. Compute the range, the sum of squares, the variance, and the standard deviation for each of the two sets of data in Exercise 13.

*25. Compute the variance and the standard deviation for the following five scores: 1, 2, 3, 4, 5. Now generate a new set of five scores by adding a constant of 3 to each original score. Compute the variance and standard deviation for the new scores. Compare your results with those obtained for the first set of scores. Do the same for another set of five scores generated by subtracting a constant of 2 from each original score. What is the effect on the variance of adding a constant to or subtracting a constant from each score in a set of scores? What is the effect on the standard deviation?

26. Suppose you measured the weight of 100 people and found that $\bar{X} = 180.29$ and $s = 10.36$. Then you learned that your scale was 1 pound too heavy. What would the correct mean and standard deviation be?

*27. Compute the variance and the standard deviation for the following eight scores: 6, 6, 8, 8, 8, 8, 10, 10. Now generate a new set of eight scores by multiplying each original score by a constant of 3. Compute the variance and the standard deviation for the new scores. Compare your results with those obtained for the first set of scores. Do the same for another set of eight scores generated by dividing each original score by a constant of 2. What is the effect on the variance of multiplying or dividing each score in a set of scores by a constant? What is the effect on the standard deviation?

*28. Generate two sets of scores with equal means but unequal standard deviations.

29. Generate two sets of scores with unequal means but equal standard deviations.

*30. How accurate are eyewitness reports of accidents? Behavioral scientists have studied this question in some detail. In one experiment, subjects viewed a film of an accident in which a car ran a stop sign and hit a parked car. The speed of the car was 30 miles per hour. After viewing the film, subjects were asked to estimate the speed of the car. Fifteen subjects gave the following estimates:

15	18	37	40	25
40	35	35	20	30
30	20	28	32	25

Calculate the mean and the standard deviation for these data. How accurate were the estimates considering the mean score across all subjects? How does the standard deviation help to interpret the mean?

31. An organizational psychologist studied how satisfied employees were in two different companies. All employees were given a job satisfaction test in which scores could range from 1 to 7, with increasingly higher scores indicating increasingly greater satisfaction. The scores were as follows:

Company A	*Company B*
4	4
6	4
5	4
4	4
3	4
2	4

Compute the mean and the standard deviation for each company. Based on the results, compare the nature of employee satisfaction in the two companies. How do the standard deviations help to interpret the means?

***32.** A consultant you hired to assess the public's attitude toward your company told you that the mean evaluation on a 7-point scale (where 1 = "extremely negative" and 7 = "extremely positive") was 5.16 and the standard deviation was −1.43. What would you conclude?

***33.** Given a set of scores for which the mean is 20, the median is 15, and the mode is 12, are these scores skewed? If so, how? If the mean were 20, the median were 25, and the mode were 33, would the scores be skewed? If so, how? If the mean were 20, the median were 20, and the mode were 20, would the scores be skewed? If so, how?

34. What does it mean to say that a distribution of scores is more leptokurtic than another distribution? What does it mean to say that a distribution of scores is more platykurtic than another distribution?

35. Identify each of the following symbols in terms of the index it represents and whether it is a sample or a population value:

a. σ^2
b. $\bar{X}$
c. s
d. σ
e. μ
f. s^2

FOUR

PERCENTILES, PERCENTILE RANKS, STANDARD SCORES, AND THE NORMAL DISTRIBUTION

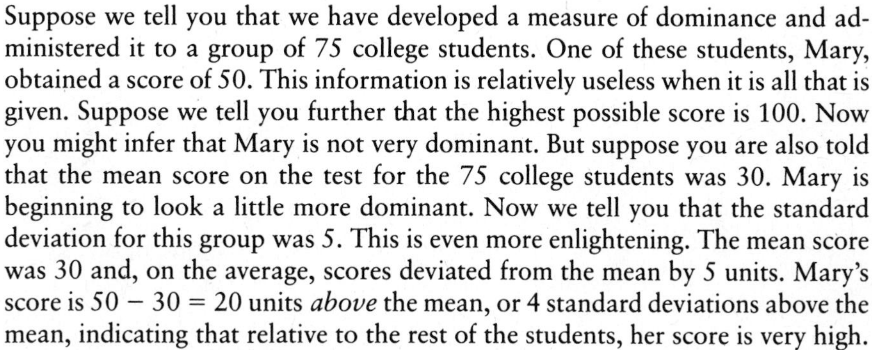

Suppose we tell you that we have developed a measure of dominance and administered it to a group of 75 college students. One of these students, Mary, obtained a score of 50. This information is relatively useless when it is all that is given. Suppose we tell you further that the highest possible score is 100. Now you might infer that Mary is not very dominant. But suppose you are also told that the mean score on the test for the 75 college students was 30. Mary is beginning to look a little more dominant. Now we tell you that the standard deviation for this group was 5. This is even more enlightening. The mean score was 30 and, on the average, scores deviated from the mean by 5 units. Mary's score is $50 - 30 = 20$ units *above* the mean, or 4 standard deviations above the mean, indicating that relative to the rest of the students, her score is very high.

The point to be made is that, in general, observations are meaningful only in relation to other observations. When you are told that a person is 7 feet tall, this is meaningful because you have an intuitive grasp of the distribution of height in humans. You know, for example, that most people are between 5 and 6 feet tall and that someone who is 7 feet tall is unusually tall. In essence, you are intuitively using information about the mean and the standard deviation of height. In the present chapter, we will consider several measures used to identify the location of a specified score within a set of scores.

4.1 Percentiles and Percentile Ranks

One approach that can be used to express the relative standing of a score in a distribution uses percentage as a basis. The percentage of scores in the distribution that occur at or below a given value, X, is the **percentile rank** of that value.* Consider a national survey of 2,000 adults in the United States who

* When percentile ranks are being determined, half of the scores having a value of X are considered to be less than X and half are considered to be greater than X. Thus, a percentile rank is different from a *cumulative percentage,* which, following the logic outlined in Chapter 2, can be obtained by adding the percentage of all scores that are equal to X to the percentage of scores that are less than X.

responded to the question, "Approximately how many hours per week do you watch television?" If 60% of the responses are at or below a score of 30 hours, then the percentile rank of 30 is 60. Thus, one way to convey the relative position of a score is to compute its percentile rank. On the other hand, we might want to know the reverse—namely, what score value is it that a certain percentage of scores are at or below? For instance, we might want to know the number of hours of television viewing that defines the point at which 60% of the reported viewing times are at or below. The score value corresponding to a given percentile rank is referred to as a **percentile.** For instance, the score value corresponding to a percentile rank of 60 is referred to as the 60th percentile. In our example, the 60th percentile is 30. We will now consider how to compute percentiles and percentile ranks.

Computation of Percentiles

Suppose we administered a test designed to measure intelligence to a group of 200 children. A frequency analysis of the scores obtained by the children is presented in Table 4.1. The question of interest is to specify the score that corresponds to a given percentile, *P*. If *P* is 70, then we want to specify the score that defines the 70th percentile. Actually, we have already considered a special case of the formula for answering this question in the form of Equation 3.1 for calculating the median. Recall that the median is that score in the distribution for which 50% of the scores are greater and 50% of the scores are less. The median corresponds to the 50th percentile. Thus, the procedures for calculating the 50th percentile are the same as those for calculating the median. The formula for the median, as presented in Equation 3.1, is

$$Mdn = L + \left[\frac{(N)(.50) - n_L}{n_W}\right] i$$

where *Mdn* represents the median, *L* is the lower real limit of the category containing the median, *N* is the total number of scores in the distribution, n_L is the number of individuals with scores less than *L*, n_W is the number of individuals

TABLE 4.1 Frequency Analysis of Intelligence Scores for 200 Children

Score	*f*	*rf*	*cf*	*crf*
105	9	.045	200	1.000
104	16	.080	191	.955
103	20	.100	175	.875
102	26	.130	155	.775
101	29	.145	129	.645
100	30	.150	100	.500
99	25	.125	70	.350
98	21	.105	45	.225
97	14	.070	24	.120
96	10	.050	10	.050

with scores within the category containing the median, and i is the size of the interval of the category containing the median. The .50 in the numerator refers to the percentile of interest (the 50th), but it is expressed in the form of a proportion. If we were interested in the 70th percentile, the formula for the median would be adapted as follows:

$$X_{70} = L + \left[\frac{(N)(.70) - n_L}{n_W}\right] i$$

where X_{70} represents the score value defining the 70th percentile, L is the lower real limit of the category containing the 70th percentile, N is the total number of scores in the distribution, n_L is the number of individuals with scores less than L, n_W is the number of individuals with scores within the category containing the 70th percentile, and i is the size of the interval of the category containing the 70th percentile. The method of computation is then analogous to that used in computing the median. We start at the bottom of the column of cumulative relative frequencies (see Table 4.1), and move up the column until we find the first number that is greater than or equal to .70 (the percentile of interest, expressed in proportion form). In this case, it is .775 and occurs at the score value 102. The lower real limit (L) of 102 is 101.5. The number of individuals (n_L) whose scores are less than L is 129. The number of individuals (n_W) with scores of 102 (or, more precisely, with scores between 101.5 and 102.5) is 26. Thus,

$$\begin{aligned} X_{70} &= 101.5 + \left[\frac{(200)(.70) - 129}{26}\right](1.0) \\ &= 101.5 + \left(\frac{11.00}{26}\right)(1.0) \\ &= 101.5 + .42 = 101.92 \end{aligned}$$

In this instance, a score value of 101.92 reflects the 70th percentile. Note that this value did not actually occur in the set of scores listed in Table 4.1. This is because the real limits of the numbers were considered, as discussed in Chapter 3. If expressed in terms of the original scale, the 70th percentile would be reported as 102 because 101.92 is rounded to 102.

The general formula for computing a score defining a given percentile is

$$X_P = L + \left[\frac{(N)(P) - n_L}{n_W}\right] i \qquad [4.1]$$

where X_P represents the score value defining the percentile of interest, referred to as the Pth percentile; L is the lower real limit of the category containing the Pth percentile; N is the total number of scores in the distribution; P is the Pth percentile expressed in the form of a proportion; n_L is the number of individuals with scores less than L; n_W is the number of individuals with scores within the category containing the Pth percentile; and i is the size of the interval of the category containing the Pth percentile.

STUDY EXERCISE 4.1

A researcher was interested in determining the size of babies at birth. A group of 1,000 newborn infants was studied and their lengths and weights at birth were measured. The distribution of scores for length (measured to the nearest inch) was as follows:

Length	*f*	*rf*	*cf*	*crf*
24	27	.027	1,000	1.000
23	48	.048	973	.973
22	77	.077	925	.925
21	125	.125	848	.848
20	226	.226	723	.723
19	225	.225	497	.497
18	124	.124	272	.272
17	73	.073	148	.148
16	52	.052	75	.075
15	23	.023	23	.023

Compute the score defining the 90th percentile.

Answer Using Equation 4.1, we find that

$$X_{90} = 21.5 + \left[\frac{(1{,}000)(.90) - 848}{77}\right](1.0)$$

$$= 21.5 + \left(\frac{52.00}{77}\right)(1.0)$$

$$= 21.5 + .68 = 22.18$$

Rounding off, we conclude that 90% of the newborn infants studied were 22 inches long or less. Thus, expressed in terms of the original scale, the 90th percentile in this case is 22.

Computation of Percentile Ranks

Just as Equation 4.1 can be used to compute the score defining a given percentile, it is also possible to specify a formula that will permit the computation of the percentile rank for any given score. This formula is

$$PR_X = \left[\frac{(.5)(n_W) + n_L}{N}\right]100 \qquad [4.2]$$

where PR_X represents the percentile rank of the score X, n_W is the number of individuals with scores equal to X, n_L is the number of individuals with scores less than X, and N is the total number of scores in the distribution. For example, the percentile rank for a score of 101 from Table 4.1 is

$$
\begin{aligned}
PR_{101} &= \left[\frac{(.5)(29) + 100}{200}\right](100) \\
&= \left(\frac{114.50}{200}\right)(100) \\
&= (.572)(100) = 57.20
\end{aligned}
$$

In this instance, a score of 101 reflects a percentile rank of 57.20. That is, 57.20% of the 200 children studied obtained an intelligence score of 101 or less.

STUDY EXERCISE 4.2

For the data presented in Study Exercise 4.1, compute the percentile rank for a length of 20 inches.

Answer Using Equation 4.2, we obtain the following:

$$
\begin{aligned}
PR_{20} &= \left[\frac{(.5)(226) + 497}{1{,}000}\right](100) \\
&= \left(\frac{610.00}{1{,}000}\right)(100) \\
&= (.610)(100) = 61.00
\end{aligned}
$$

The percentile rank for a length of 20 inches is 61.00. That is, 61.00% of the newborn infants in the study were 20 inches long or less at birth.

4.2

Standard Scores

A percentile rank represents one index of the relative position of a score in a set of scores. However, a percentile rank reflects only an *ordinal* measure of relative standing. To say that a score has a percentile rank of 80 is simply to state that 80% of the individuals scored at or below that score. But *how much* lower did these other individuals score? Consider the following scores on two tests:

Test 1	*Test 2*
100	100
99	44
98	43
96	40
96	40
95	39

In both cases, a score of 100 would reflect the same percentile rank. However, the score of 100 on Test 2 is certainly more distinctive than the score of 100 on Test 1. An index of relative standing that would reflect this difference is that of **standard scores.**

The Concept of Standard Scores

Recall the example at the beginning of this chapter concerning the dominance measure. Mary obtained a score of 50 out of 100. When we were told that the average score was 30, a score of 50 took on more meaning: Mary scored 20 units above the mean. With the additional information that the standard deviation was 5, the significance of Mary's score was even clearer. By comparing Mary's score to the mean and standard deviation, we gained considerable insight into its relative position. A standard score does just this: It converts a score from its original, or *raw,* form to a form that takes into consideration its standing relative to the mean and standard deviation of the entire distribution of scores.

Columns 1 and 5 of Table 4.2 present two sets of scores that will be used for illustration. If the data in Table 4.2 represent sample information, the raw scores in each set can be converted to standard scores by the following formula:

TABLE 4.2 Raw Scores and Standard Scores for Two Sets of Scores

SET I				SET II			
X	X^2	$X - \bar{X}$	$\frac{X - \bar{X}}{s}$	X	X^2	$X - \bar{X}$	$\frac{X - \bar{X}}{s}$
1	1	−2	−1.49	1	1	−2	−2.25
3	9	0	0.00	3	9	0	0.00
5	25	2	1.49	3	9	0	0.00
2	4	−1	− .75	3	9	0	0.00
3	9	0	0.00	5	25	2	2.25
4	16	1	.75	3	9	0	0.00
1	1	−2	−1.49	3	9	0	0.00
3	9	0	0.00	3	9	0	0.00
5	25	2	1.49	3	9	0	0.00
3	9	0	0.00	3	9	0	0.00
$\Sigma X = 30$ $\bar{X} = 3.00$	$\Sigma X^2 = 108$		Sum = 0 Mean = 0	$\Sigma X = 30$ $\bar{X} = 3.00$	$\Sigma X^2 = 98$		Sum = 0 Mean = 0

Set I:

$$SS = \Sigma X^2 - \frac{(\Sigma X)^2}{N}$$
$$= 108 - \frac{30^2}{10} = 18$$
$$s^2 = \frac{SS}{N} = \frac{18}{10} = 1.80$$
$$s = \sqrt{s^2} = \sqrt{1.80} = 1.34$$

Set II:

$$SS = \Sigma X^2 - \frac{(\Sigma X)^2}{N}$$
$$= 98 - \frac{30^2}{10} = 8$$
$$s^2 = \frac{SS}{N} = \frac{8}{10} = .80$$
$$s = \sqrt{s^2} = \sqrt{.80} = .89$$

$$\text{Standard score} = \frac{X - \bar{X}}{s} \quad [4.3]$$

If the data in Table 4.2 represent population information, the standard score formula is

$$\text{Standard score} = \frac{X - \mu}{\sigma} \quad [4.4]$$

Note that the only difference between these two formulas is the notation used to represent the mean and standard deviation of the distribution. For purpose of demonstration, we will assume that we are dealing with sample data.

A standard score is the difference between the original score and the mean, divided by the standard deviation. The numerator of the standard score formula reflects the number of units the score is above or below the mean. When the numerator is divided by the standard deviation, the result expresses the number of *standard deviations* the score is above or below the mean. For instance, Mary's score of 50 is $50 - 30 = 20$ units above the mean. The standard deviation was 5 and hence Mary's score is $20/5 = 4$ standard deviations above the mean. In other words, her raw score corresponds to a standard score of 4. Suppose John obtained a dominance score of 25. This score is $25 - 30 = -5$ units below the mean or, in terms of standard deviations, $-5/5 = -1$ standard deviation below the mean. John's standard score is thus -1. *A standard score represents the number of standard deviation units that a score falls above or below the mean.* It summarizes the individual's relative standing, taking into consideration the mean and standard deviation of the distribution.

The standard score equivalents of the raw scores in columns 1 and 5 of Table 4.2 can be found respectively in columns 4 and 8. To obtain these values, we first computed the mean for each data set. Next, we used the computational formula for the sum of squares to separately calculate the sum of squares for Set I and the sum of squares for Set II. The sums of squares were then used to calculate the two standard deviations. Finally, we subtracted the mean for each data set from the constituent scores (see columns 3 and 6) and divided the resulting values by the appropriate standard deviation (see columns 4 and 8). If we compare the standard score for the first raw score in Set I with the standard score for the first raw score in Set II, we find that although the raw scores are the same (both have a value of 1), the standard scores are different, -1.49 as compared to -2.25. The negative signs indicate that in both cases the raw score was below the mean of its respective distribution. For Set I, a score of 1 is only 1.49 standard deviation units below the mean of its distribution, whereas for Set II it is 2.25 standard deviation units below the mean of its distribution. Thus, as reflected by the standard scores, a score of 1 is more distinctive in the context of Set II scores than in the context of Set I scores.

STUDY EXERCISE 4.3

Given a set of scores with $\bar{X} = 20.00$ and $s = 2.00$, what is the standard score corresponding to $X = 17$?

Answer Using Equation 4.3,

$$\text{Standard score} = \frac{X - \bar{X}}{s}$$

$$= \frac{17 - 20.00}{2.00} = -1.50$$

This means that a score of 17 is 1.50 standard deviations below the mean of its distribution.

Properties of Standard Scores

Standard scores have several important properties. A positive standard score indicates that the original score is greater than the mean, and a negative standard score indicates that the original score is less than the mean. A standard score of 0 indicates that the original score is equal to the mean.

If we sum the standard scores for Set I in Table 4.2, the result will equal 0 (see column 4). The same result will occur if we sum the standard scores for Set II (see column 8). In fact, the sum of a set of standard scores will always be 0. This is due to the fact that standard scores reflect signed deviation scores, and as discussed in Section 3.1, the sum of signed deviations about the mean always equals 0. It follows that if the sum of a set of standard scores is always equal to 0, *the mean of a set of standard scores is also always equal to 0.*

It is also the case that if we were to compute the standard deviation for either set of standard scores, the standard deviation would equal 1.00. *The standard deviation (and the variance) of a set of standard scores is always equal to 1.00.*

Use of Standard Scores

As alluded to above, one important use of standard scores is to compare scores on distributions that have different means and standard deviations. For instance, suppose you are contemplating in which of two different fields (management or advertising) to seek employment. Unable to make a decision between the two fields, you decide to enter the field for which you have the greater aptitude. As assessed by the most widely accepted aptitude tests in the respective fields, your management aptitude score turns out to be 73 (out of 100) and your advertising aptitude score turns out to be 82 (out of 100). At first glance, it appears that your advertising aptitude is substantially greater than your management aptitude. However, before rushing off in search of an advertising position, you should examine the two scores in the context of their respective distributions.

Suppose, for instance, that the mean aptitude score for the population of individuals who have previously taken the management test is 62.00 and the standard deviation is 4.00. Further suppose that the mean aptitude score for the population of individuals who have previously taken the advertising test is 78.00 and the standard deviation is 6.00. Now it is not quite so clear that your advertising aptitude is superior to your management aptitude.

Because the mean (0) and standard deviation (1.00) of a set of standard scores are always the same, it is possible to make direct comparisons between different sets of raw scores by converting the scores of interest to standard scores. In our example, the standard score for management aptitude is

$$\text{Standard score} = \frac{73 - 62.00}{4.00} = 2.75$$

and the standard score for advertising aptitude is

$$\text{Standard score} = \frac{82 - 78.00}{6.00} = .67$$

Thus, as compared with other individuals who have taken the two aptitude tests, a management aptitude score of 73 represents relatively greater aptitude than an advertising aptitude score of 82. In the first instance, the obtained aptitude score is 2.75 standard deviation units above the mean of its distribution, whereas in the second instance, the obtained aptitude score is only .67 standard deviation unit above the mean of its distribution.

4.3 Standard Scores and the Normal Distribution

A standard score yields considerable information about the relative position of a score in a distribution. Such scores are even more meaningful when they occur in a normal distribution. In Section 2.8, it was noted that there is a family of normal distributions, each member of which is precisely defined by a mathematical formula.* For our purposes, the important points are that (1) there is a different normal distribution for each unique combination of the distribution mean and the distribution standard deviation, and (2) all normal distributions share a number of characteristics. For instance, as discussed in Chapters 2 and 3, all normal distributions are symmetrical about the mean, all are characterized by a "bell shape," and in all cases, the mean, the median, and the mode are

* The normal distribution formula can be found in Appendix 4.1.

FIGURE 4.1 Proportion of Scores Occurring Between Selected Standard Scores in a Normal Distribution

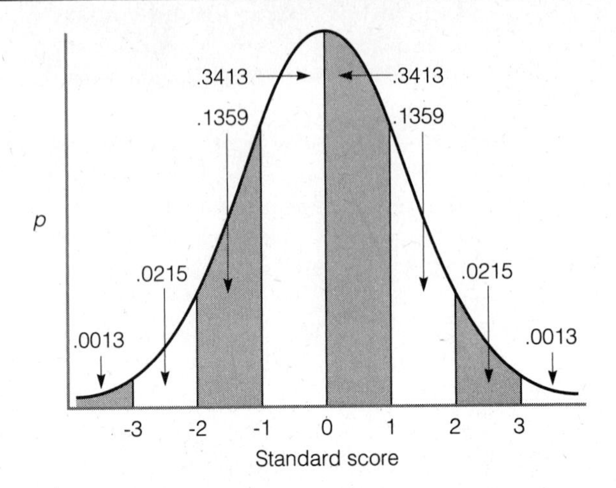

equal. Another important feature of normal distributions is that they are theoretical in nature. If we are able to assume that a set of scores approximates a normal distribution, we can invoke certain statistical properties of normal distributions to aid in the interpretation of the data.

One useful property of normal distributions is that the proportion of scores that occur above or below a given standard score is the same in all such distributions, as is the proportion of scores that occur between two specified standard scores. It is always the case, for example, that .50 of the scores in a normal distribution occur above the mean and .50 of the scores in a normal distribution occur below the mean. Thus, knowing that a set of scores approximates a normal distribution allows us to make probability statements with respect to those scores.

Figure 4.1 presents a normal distribution and the proportion of scores that occur between selected standard scores. The proportion of scores between standard scores of 0 and 1 is .3413, the proportion of scores between standard scores of 1 and 2 is .1359, and the proportion of scores between standard scores of 2 and 3 is .0215. The proportion of scores greater than or equal to a standard score of 3 is .0013. Note that these proportions are symmetrical about the mean, which, as stated previously, will always equal 0 for a set of standard scores. For instance, the proportion of scores occurring between standard scores of 0 and -1 is the same as the proportion of scores occurring between standard scores of 0 and $+1$. Thus, in a normal distribution, .6826, or approximately 68%, of all scores fall between standard scores of -1 and $+1$; .9544, or approximately 95%, of all scores fall between standard scores of -2 and $+2$; and .9974, or over 99%, of all scores fall between standard scores of

−3 and +3. Before proceeding further, take a moment to verify these proportions from the information presented in Figure 4.1.

A standard score in a normal distribution is referred to as a **z score.** Some texts refer to any standard score as a z score. However, in traditional statistics, a z score is used only to represent a standard score *in a normal distribution.* This distinction will be maintained in the present book.

Appendix B presents a table that summarizes the proportion of scores in a normal distribution that are greater than or equal to selected z scores (for example, greater than or equal to 1.00). This table also indicates the proportion of scores in a normal distribution that occur between selected points. Instructions for using the table are presented at the start of Appendix B and should be read at this time.

Let us now explore how the table in Appendix B, coupled with our knowledge of the normal distribution, can give us insights into scores that approximate normality (that is, approximate a normal distribution). For the sake of illustration, try to estimate how many hours per week you spend watching television. This might be easiest to do if you think of each night of the week (Sunday through Saturday) individually and estimate how many hours you typically watch television on those days. Then sum across the 7 days to obtain an estimate of how much you tend to watch television per week. For the authors of this text, the estimate is 14 hours per week. How does this (or your own estimate) compare with that of others? Let us compare our scores with results suggested by national surveys of adults in the United States.

Suppose the mean number of hours of television watched per week by adults in the United States is 25.40 and the standard deviation is 6.10. Suppose also that the scores in this distribution closely approximate a normal distribution. How unique is a score of 14 hours per week? Substituting the more specific z notation for "standard score" in Equation 4.3, the formula for converting a score in a sample to a z score is

$$z = \frac{X - \bar{X}}{s} \tag{4.5}$$

Similarly, the formula for converting a score in a population to a z score is

$$z = \frac{X - \mu}{\sigma} \tag{4.6}$$

In our example,

$$z = \frac{14 - 25.40}{6.10} = -1.87$$

A weekly television viewing time of 14 hours is 1.87 standard deviations *below* the national average. Using column 3 of Appendix B, we find that the proportion of scores less than or equal to a z score of -1.87 is .0307. Thus, 14 hours defines the 3.07th percentile. Stated another way, 3.07% of adults in the United States watch 14 hours or less of television per week. Approximately 1.00 (the total proportion of all scores) $-$ .0307 = .9693, or 96.93%, of adults in the United States must therefore watch 14 hours or more of television per week. Relative to most American adults, the amount of time the authors spend watching television is quite low. Do a similar analysis for your own estimated viewing behavior.

Suppose we want to know what percentage of American adults watch between 30 and 40 hours of television per week. We can use our knowledge of the normal distribution and Appendix B to estimate this. First, we convert the scores of 30 and 40 into z scores:

$$z = \frac{30 - 25.40}{6.10} = .75$$

and

$$z = \frac{40 - 25.40}{6.10} = 2.39$$

Examining Appendix B, we find that .7734 of the scores in a normal distribution are less than or equal to a z score of .75. This is illustrated in Figure 4.2. There are several ways that this value can be obtained from Appendix B. For instance, we could refer to column 3, where we find that .2266 of the scores are greater than or equal to a z score of .75. Thus, 1.00 $-$ .2266 = .7734 of the

FIGURE 4.2 Proportion of Scores Between z Scores of .75 and 2.39

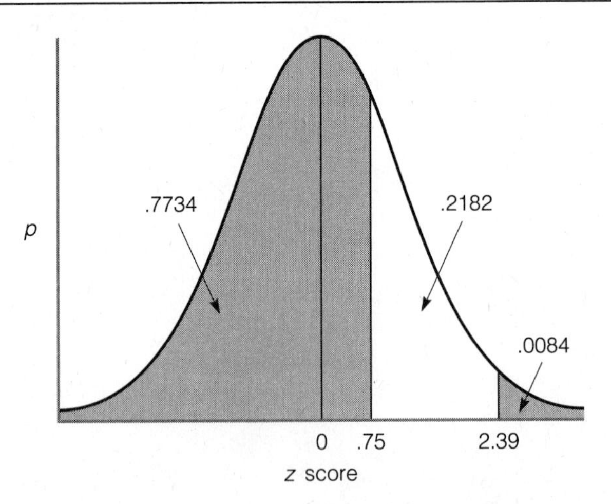

scores must be .75 or less. Alternatively, with the realization that .50 of the scores occur below the mean (z score of 0), we could refer to column 5, where we find that an additional .2734 of the scores fall between a z score of 0 and a z score of .75. The sum of these figures (.50 + .2734) is .7734. Appendix B also indicates that .0084 of the scores are greater than or equal to a z score of 2.39. The most direct way to obtain this value is to refer to column 3. The proportion of scores occurring *between* .75 and 2.39 is 1.00 minus the .0084 of the scores equal to or greater than a z score of 2.39 and minus the .7734 of the scores equal to or less than a z score of .75, or 1.00 − .0084 − .7734 = .2182. Thus, approximately 21.82% of American adults watch between 30 and 40 hours of television per week.

STUDY EXERCISE 4.4

Given a set of scores that are normally distributed with a mean of 100.00 and a standard deviation of 10.00, what proportion of scores are greater than or equal to 120? Less than or equal to 90?

Answer We begin by converting the raw score of interest into a z score. For the question of the proportion of scores greater than or equal to 120,

$$z = \frac{120 - 100.00}{10.00} = 2.00$$

Using column 3 of Appendix B, we find that the proportion of scores greater than or equal to a z score of 2.00 is .0228. Thus, the proportion of scores greater than or equal to 120 is .0228.

For the question of the proportion of scores less than or equal to 90,

$$z = \frac{90 - 100.00}{10.00} = -1.00$$

Again using column 3 of Appendix B, we find that the proportion of scores less than or equal to a z score of −1.00 is .1587. Thus, the proportion of scores less than or equal to 90 is .1587.

In some applications, we might want to reverse the previous procedures and determine the raw scores that define specified proportions of an approximately normal distribution. If the mean and standard deviation of this distribution are known, this can be accomplished by substituting these values and an appropriate value of z into one of the following equations (depending on whether we are dealing with sample data or population data) and solving for X:

$$X = \bar{X} + (z)(s) \quad [4.7]$$

$$X = \mu + (z)(\sigma) \quad [4.8]$$

These equations are merely rearrangements of Equations 4.5 and 4.6, respectively. The value of z of interest is the value that cuts off the specified proportion of the distribution. For instance, in the television-viewing example we might want to identify the number of hours of television viewing at or above which 5% of all weekly viewing times fall. From column 3 of Appendix B we find that .0505 of all scores in a normal distribution are greater than or equal to a z score of 1.64 and .0495 of all scores in a normal distribution are greater than or equal to a z score of 1.65. Thus, the value of z that defines the upper .05, or 5%, of scores in a normal distribution is approximately halfway between these two values, or 1.645. Since the mean in our example is equal to 25.40 and the standard deviation is equal to 6.10, the number of hours of television viewing at or above which 5% of all weekly viewing times fall is

$$X = 25.40 + (1.645)(6.10) = 35.43$$

4.4 Method of Presentation

Percentile information is most commonly encountered in manuals available to educators and helping professionals for interpreting scores on educational and psychological tests. Such manuals generally list selected percentile ranks and corresponding raw score values. For instance, the Miller Analogies Test is designed to measure general aptitude of applicants for graduate and professional schools. The test consists of a series of items in the form of analogies. An example item might appear as follows (where you are to fill in the blank with the appropriate word from those given):

A *book* is to *trees* as a *skirt* is to ________.

(a) shoes (b) sheep (c) dresses (d) women

Scores on the Miller Analogies Test can range from 0 to 100. A list of the scores defining selected percentiles for different groups of graduate students might appear in the test manual as follows:

PERCENTILE RANK	RAW SCORE EQUIVALENT					
	Physical sciences	*Medical science*	*Social sciences*	*English*	*Law*	*Social work*
99	93	92	90	87	84	81
90	88	78	82	80	73	67
80	82	74	76	74	63	61
70	78	67	69	68	58	58
60	74	60	64	65	53	54
50	68	57	61	59	49	50
40	63	53	56	53	45	46
30	58	47	51	46	40	41
20	51	43	46	41	35	37
10	43	34	39	35	30	27
1	28	24	18	7	18	9

Although not all possible score values are included in this table, the relative standing of any test score can nevertheless be determined. Focusing on law students, for instance, we find that a score of 50 has a percentile rank between 50 and 60, a score of 60 has a percentile rank between 70 and 80, and so forth.

Percentile ranks must always be interpreted relative to the group upon which the scores are based. Someone who scores at the 95th percentile on the Miller Analogies Test, for instance, probably has greater general aptitude than someone who scores at the 95th percentile on the Scholastic Aptitude Test (an aptitude test given to applicants for undergraduate schools). This is because only the brighter students who have done well in college tend to take the Miller Analogies Test, whereas the Scholastic Aptitude Test reflects a more general population.

It might be interesting to contrast your own major with those of others in terms of which scores define various percentiles in the above table. For example, the 50th percentile is 68 for graduate students in the physical sciences, 61 for graduate students in the social sciences, 59 for graduate students in English (literature and language), 57 for medical students, 50 for graduate students in social work, and 49 for law students.

To aid in the interpretation of specified scores, most educational and psychological test manuals report standard score equivalents of raw scores in addition to percentile tables. To avoid the confusion associated with decimals and negative values, standard scores are frequently further transformed to *T scores*. *T* scores are directly analogous to standard scores, but instead of having a mean of 0 and a standard deviation of 1.00, they have a mean of 50.00 and a standard deviation of 10.00. The transformation is accomplished through the formula

$$T = 50 + 10(\text{Standard score}) \qquad \textbf{[4.9]}$$

For instance, a standard score of $-.50$ is equivalent to a T score of $50 + (10)(-.50) = 45$.

A portion of a test manual report of raw scores and their standard score and T score equivalents might appear as follows:

Raw score	*Standard score*	*T score*
100	2.90	79
99	2.70	77
98	2.50	75
97	2.30	73
96	2.10	71

In practice, all possible raw score values and corresponding standard and T scores would be included in this listing.

Sometimes T scores are transformed further so that they are also normally distributed. When this is done, the report or manual should state so explicitly. The mathematics of the transformation are complex and can be found in Gulliksen (1960).

4.5 Summary

Percentile ranks, percentiles, and standard scores are measures used to identify the location of a specified score within a set of scores. A percentile rank is the percentage of scores in a distribution that occur at or below a given value. A percentile is a score value corresponding to a given percentile rank. A standard score is the difference between a score in a distribution and the mean of the distribution, divided by the distribution's standard deviation. A standard score thus represents the number of standard deviation units that a score occurs above or below the mean.

A standard score in a normal distribution is referred to as a z score. If we are able to assume that a set of scores approximates a normal distribution, we can use the table presented in Appendix B to determine the proportion of scores that occur above or below a given z score and the proportion of scores that occur between two different z scores. We can also reverse these procedures and determine the raw scores that define specified proportions of a distribution.

Appendix 4.1 The Normal Distribution Formula

The formula for a normal distribution is

$$p(X) = \left(\frac{1}{\sqrt{2\pi\sigma^2}}\right)(e^{-(X-\mu)^2/2\sigma^2})$$

where $p(X)$ is the height of the normal curve associated with a particular value of X, π (pi) is a constant approximately equal to 3.1416, e is a constant approximately equal to 2.7183, μ is the mean of the distribution, and σ^2 is the variance of the distribution. When μ and σ^2 are specified, different values of X can be substituted into the equation and the corresponding values of $p(X)$ obtained. If paired values of X and $p(X)$ are plotted graphically, they will form a normal curve. Thus, as noted in the text, there is a different normal curve for every unique combination of μ and σ^2.

Exercises

Answers to asterisked (*) exercises appear at the back of the book. Answers to exercises with two asterisks are also worked out step-by-step in the Study Guide.

1. What is the relationship between percentile ranks and percentiles?

Refer to the following information to answer Exercises 2–3:

A researcher administered a questionnaire to 500 people that was designed to measure knowledge of positions that two presidential candidates held on major issues. Scores on the test could range from 0 to 10, with higher scores indicating more knowledge. The following frequency analysis resulted:

Score	*f*	*rf*	*cf*	*crf*
8	6	.012	500	1.000
7	44	.088	494	.988
6	51	.102	450	.900
5	49	.098	399	.798
4	200	.400	350	.700
3	98	.196	150	.300
2	52	.104	52	.104

****2.** For the given set of data, compute the scores defining the following percentiles:

a. 20th **b.** 40th **c.** 60th **d.** 80th **e.** 99th **f.** 50th

****3.** Compute the percentile ranks corresponding to the following scores:

a. 7 **b.** 5 **c.** 3 **d.** 2

4. For the following set of data, compute the score defining the 50th percentile: 100, 100, 99, 99, 99, 98, 98, 97, 97, 96, 95, 94.

5. What is the advantage of standard scores as compared to percentile ranks as an index of relative standing?

***6.** Given a distribution with a mean of 20.00 and a standard deviation of 3.00, compute the standard score equivalents of the following scores:

a. 21.87 **b.** 23.00 **c.** 18.91 **d.** 20.08 **e.** 15.63 **f.** 24.30

7. Given a distribution with a mean of −4.00 and a standard deviation of 2.50, compute the standard score equivalents of the following scores:

a. −6.83 **b.** 0 **c.** 2.84 **d.** −6.50 **e.** 1.00 **f.** .87

***8.** For the following distribution, convert a score of 50 to a standard score: 60, 55, 50, 50, 55, 60.

9. For the following distribution, convert a score of 6 to a standard score: 14, 6, 13, 17, 11, 13, 14, 9.

***10.** What does a positive standard score indicate about the original score's position relative to the mean? What does a negative standard score indicate about the original score's position relative to the mean?

11. What are the values of the mean and the standard deviation for any set of standard scores?

***12.** John received 90 out of 100 points on an English exam and 60 out of 100 points on a math exam. The overall class performance was $\bar{X} = 70.00$ and $s = 20.00$ on the English exam, and $\bar{X} = 40.00$ and $s = 3.00$ on the math exam. On which exam was John's performance better (relative to his classmates')?

13. If a person obtained a score of 80 on a test, which one of the following distributions would allow for the most favorable interpretation of that score (assuming that higher values are more favorable)?

- **a.** $\bar{X} = 60.00$, $s = 5.00$
- **b.** $\bar{X} = 60.00$, $s = 10.00$
- **c.** $\bar{X} = 60.00$, $s = 1.00$
- **d.** $\bar{X} = 60.00$, $s = 20.00$

***14.** Describe a situation in which it would be useful to convert scores from two different distributions to standard scores before comparing them.

15. What are the major characteristics of a normal distribution?

16. What is a z score?

17. Given a normal distribution with a mean of 24.87 and a standard deviation of 6.00, compute the z score equivalents of the following scores:

- **a.** 13.78
- **b.** 29.42
- **c.** 26.81
- **d.** 12.87
- **e.** 37.90
- **f.** 33.35

***18.** What proportion of z scores in a normal distribution are:

- **a.** 2.38 or less
- **b.** 1.17 or greater
- **c.** −1.17 or less
- **d.** between 0 and 2.05
- **e.** between −2.05 and 0
- **f.** between .37 and 3.19
- **g.** between −3.19 and −.37
- **h.** between −1.24 and +1.24

***19.** Given a set of normally distributed scores with a mean of 20.00 and a standard deviation of 5.00, what proportion of scores are:

- **a.** 25 or greater
- **b.** 15 or less
- **c.** between 15 and 28
- **d.** between 8 and 32
- **e.** 20 or greater
- **f.** 23 or less

20. Suppose IQ scores in a population are approximately normally distributed with $\mu = 100.00$ and $\sigma = 10.00$. What proportion of individuals have IQ scores of:

- **a.** 100 or greater
- **b.** 100 or less
- **c.** between 105 and 112
- **d.** 103 or less
- **e.** 95 or greater
- **f.** 95 or less

***21.** Lie detectors or polygraphs are used to help determine whether a person has knowledge of a crime. These devices are based on autonomic changes in the nervous system. The assumption is that lying will be reflected in physiological changes that are not under the voluntary control of the individual. Measurements of the individual's physiological response when asked certain questions can be used to make inferences about the veridicality of his or her answers. Lie detection tests typically involve answering a series of neutral questions (for example, "What is your name?" or "Where do you work?") among which the critical questions are embedded. Consider the case where an individual has been asked a series of questions, including a critical item. Physiological measurements in the form of galvanic skin responses are taken for each question. The galvanic skin response scores approximate a normal distribution with a mean of 49.40 and a standard deviation of 3.00. For the critical question, the score was 61.40. Convert this into a standard score and draw a conclusion.

22. A major form of identification in criminal investigations is fingerprints. Fingerprints vary on many different dimensions, one of which is called the ridgecount. Suppose you know that the ridgecounts of people in the United States follow a normal distribution with a mean of 165.00 and a standard deviation of 10.00. Suppose further that a set of fingerprints was found at the scene of a crime and it was determined that the ridgecount was at least 200 (the exact value being in question because of smudging). Finally, suppose that out of 10 suspects, one of them has a ridgecount of 225. What would you conclude and why?

***23.** Given a set of normally distributed scores with $\bar{X} = 100.00$ and $s = 10.00$, what score would correspond to a z score of:

- **a.** 2.86
- **b.** −2.44
- **c.** 0
- **d.** −1.50
- **e.** 1.59
- **f.** .75

*24. Given a normal distribution with a mean of 100.00 and a standard deviation of 5.00, what score would:

a. 33% of the cases be greater than or equal to
b. 5% of the cases be greater than or equal to
c. 50% of the cases be less than or equal to
d. 2.50% of the cases be greater than or equal to
e. 2.50% of the cases be less than or equal to

25. Suppose income in a sample is approximately normally distributed with $\bar{X}$ = \$20,000.00 and s = \$2,000.00. What is the income level defining the:

a. top 2.5% of salaries
b. bottom 2.5% of salaries
c. top 5% of salaries
d. bottom 5% of salaries
e. top 33% of salaries
f. bottom 50% of salaries

*26. Convert each of the following standard scores to a *T* score:

a. .87
b. 2.00
c. 1.56
d. −1.56
e. 0
f. 4.04

FIVE

PEARSON CORRELATION AND REGRESSION: DESCRIPTIVE ASPECTS

5.1

Use of Pearson Correlation

To this point, we have considered ways of summarizing and describing scores on a single variable. Research in the behavioral sciences, however, often involves the measurement of two variables for the same individuals. A common question in this situation concerns the way in which scores on the first variable are related to scores on the second variable. For instance, an investigator interested in the relationship between women's traditionalism and their ideal family size might conduct a study in which each participant is asked to indicate her ideal number of children and to respond to a traditionalism questionnaire. The question of interest is whether there is a relationship between individuals' traditionalism scores and their ideal family size scores. Actually, there are many different ways in which two variables might be related. However, research in the behavioral sciences is most often concerned with *linear* relationships. When both variables under study are quantitative in nature and are measured on a level that approximates interval characteristics, the statistical technique of *Pearson product-moment correlation,* known more simply as **Pearson correlation,** can be used to determine the extent to which they are linearly related. To set the groundwork for an in-depth discussion of this technique, we will now review the basic characteristics of linear relationships.

5.2

The Linear Model

Consider two variables, the number of hours worked (X) and the amount of money paid (Y). Each of four individuals works at a rate of $1 per hour. Their scores on X and Y are as follows:

FIGURE 5.1 Example of a Scatterplot

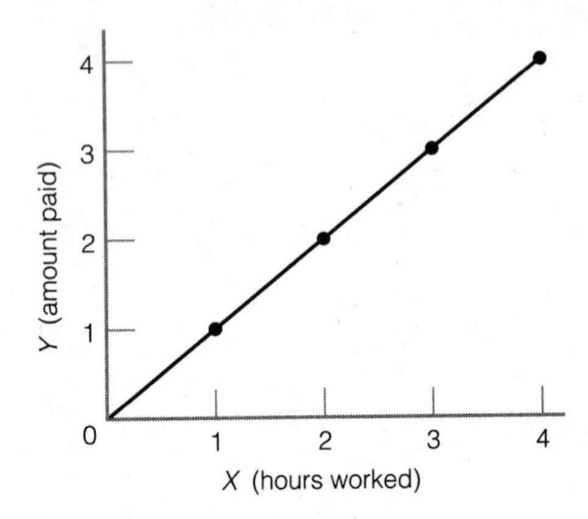

Individual	X (hours worked)	Y (amount paid [$])
1	1	1
2	4	4
3	3	3
4	2	2

The relationship between X and Y is illustrated in Figure 5.1 using a graph called a **scatterplot.** In this scatterplot, the abscissa represents the values of X and the ordinate represents the values of Y. For each individual, we find the value on the X axis corresponding to his or her score on X and the value on the Y axis corresponding to his or her score on Y, and place a dot (or, alternatively, a "×") where the two values intersect. For instance, individual 1 had a score of 1 on X and a score of 1 on Y. We therefore place a dot at the intersection of $X = 1$ and $Y = 1$. The same procedure can be repeated for each individual and the resulting dots connected, as has been done in Figure 5.1. As indicated by the straight line on the scatterplot, there is a linear relationship between X and Y. This relationship can be stated as

$$Y = X$$

In other words, the number of dollars paid equals the number of hours worked.

Suppose the individuals were not paid $1 per hour, but instead were paid $2 per hour. The scores on X and Y would now be as follows:

FIGURE 5.2 Examples of Three Linear Relationships

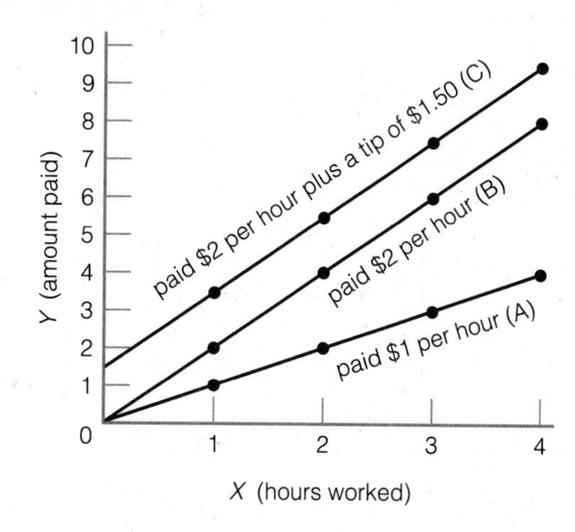

Individual	X *(hours worked)*	Y *(amount paid [$])*
1	1	2
2	4	8
3	3	6
4	2	4

In this case, the relationship between X and Y can be stated as

$$Y = 2.00X$$

In other words, the number of dollars paid equals 2 times the number of hours worked. Figure 5.2 presents a scatterplot of these data (line B) as well as the previous data (line A). Notice that we still have a straight line (and, hence, a linear relationship) but, in the case of $2 per hour, the line rises much faster than in the case of $1 per hour—that is, the **slope** of the line is greater. Technically, the slope of a line indicates the number of units Y changes as X changes by 1 unit. In the case where people are paid $2 per hour, an individual who works 1 hour is paid $2, one who works 2 hours is paid $4, and so forth. When X goes up by 1 unit (for instance, from 1 to 2 hours), Y goes up by 2 units (for instance, from $2 to $4). The slope describing this linear relationship is therefore 2. In contrast, the slope describing the linear relationship $Y = X$ is 1, reflecting the fact that as X changes by 1 unit, so does Y. Thus, linear relationships can differ in terms of the slopes that describe them.

The slope describing a linear relationship can be determined using a simple algebraic formula. This formula involves first selecting the X and Y scores of any two individuals. The slope is computed by dividing the difference between the two Y scores by the difference between the two X scores; in other words, the change in Y scores is divided by the change in X scores. Symbolically,

$$b = \frac{Y_1 - Y_2}{X_1 - X_2} \qquad [5.1]$$

where b represents the slope, X_1 and Y_1 are the X and Y scores for any one individual, and X_2 and Y_2 are the X and Y scores for any other individual. Inserting the scores for individuals 1 ($X = 1$ and $Y = 2$) and 2 ($X = 4$ and $Y = 8$), we find that the slope for line B is

$$b = \frac{2 - 8}{1 - 4} = 2.00$$

This is consistent with what was stated previously.

The value of a slope can be positive, negative, or equal to 0. Consider the following scores:

Individual	*X*	*Y*
1	2	3
2	1	4
3	4	1
4	3	2

Inserting the scores for individuals 2 and 4 into Equation 5.1, we find that the slope is

$$b = \frac{4 - 2}{1 - 3} = -1.00$$

Figure 5.3 presents a scatterplot of this relationship. The relationship is still linear, but now the line moves downward as we move from left to right on the X axis. This downward direction characterizes a negative slope, whereas an upward direction characterizes a positive slope. A slope of 0 would be represented by a horizontal line because such a slope indicates that the value of Y is constant for all values of X.

A positive slope indicates a **positive** or **direct relationship** between X and Y, whereas a negative slope indicates a **negative** or **inverse relationship** between X and Y. In the case of a positive relationship, as scores on X *increase,* scores on Y also *increase.* In the case of a negative relationship, as scores on X *increase,* scores on Y *decrease.* For instance, the slope in the present example is -1.00, meaning that for every unit X increases, Y decreases by 1 unit.

Let us return to the example where individuals are paid $2 per hour worked. Suppose that in addition to this wage, each individual is given a tip of $1.50. Now the relationship between X and Y is

FIGURE 5.3 Example of a Linear Relationship with a Negative Slope

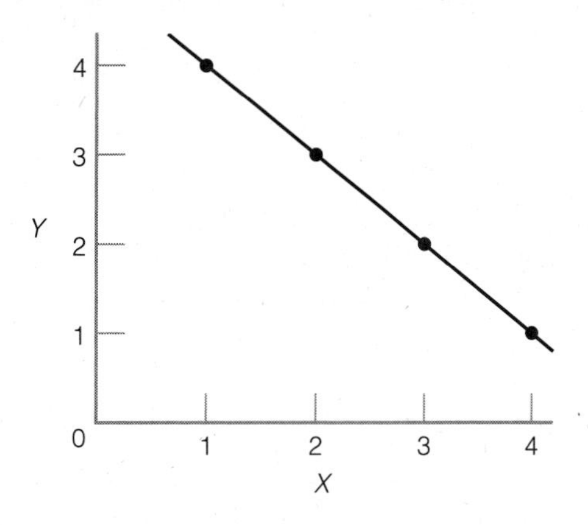

$$Y = 1.50 + 2.00X$$

Line C of Figure 5.2 plots this case for the four individuals. If we compute the slope of this line, we find it to be 2.00, as before. Notice that lines C and B are parallel, but that line C is raised above line B. The amount of separation between these two lines can be measured at the Y axis, where $X = 0$. When $X = 0$, the Y value is 1.50 for line C and 0 for line B. Thus, line C is raised 1.50 units above line B.

The point at which a line intersects the Y axis when $X = 0$ is called the **intercept,** and its value is symbolized by the letter a. Linear relationships can differ in terms of the values of their intercepts as well as in the values of their slopes, as indicated in the previous problem in which the intercept of line C is 1.50 and the intercept of line B is 0. The general form of the **linear model** is thus

$$Y = a + bX \quad [5.2]$$

Any line can be represented in terms of Equation 5.2. There will always be a slope and an intercept that describe the linear relationship between the two variables. Given these values, we can substitute scores on X into the **linear equation** to determine the corresponding scores on Y. For example, the linear equation $Y = 1.50 + 2.00X$ tells us that an individual who worked for 2 hours was paid \$5.50, since the Y score associated with an X score of 2 is

$$\begin{aligned} Y &= 1.50 + 2.00X \\ &= 1.50 + (2.00)(2) = 5.50 \end{aligned}$$

Similarly, an individual who worked for 3 hours was paid $Y = 1.50 + (2.00)(3) = \$7.50$, and an individual who worked for 4 hours was paid $Y = 1.50 + (2.00)(4) = \$9.50$.

STUDY EXERCISE 5.1

Suppose you are told that for a group of 20 students, there is a perfect linear relationship between grade point average (Y) and scores on an intelligence test (X). Suppose you are also told that the equation describing the relationship is

$$Y = 1.00 + .025X$$

If an individual obtained a score of 100 on the intelligence test, what must his or her grade point average be? What must that student's grade point average be if he or she obtained an intelligence test score of 97? Of 108?

Answer The grade point average associated with an intelligence test score of 100 is

$$Y = 1.00 + (.025)(100) = 3.50$$

The grade point average associated with an intelligence test score of 97 is

$$Y = 1.00 + (.025)(97) = 3.42$$

Lastly, the grade point average associated with an intelligence test score of 108 is

$$Y = 1.00 + (.025)(108) = 3.70$$

5.3 The Pearson Correlation Coefficient

It is rare in the behavioral sciences to observe a perfect linear relationship between two variables. Far more common is the situation where, to one degree or another, the relationship between two variables *approximates* a linear one. The extent of linear approximation between two variables is indexed by a statistic known as the **Pearson correlation coefficient.** The *correlation coefficient,* represented by the letter r, can range from -1.00 through 0 to $+1.00$. The *magnitude* of the correlation coefficient, as indexed by its absolute value, indicates the *degree to which a linear relationship is approximated:* The greater r is in either a positive or a negative direction from 0, the better is the approximation. The *sign* of the correlation coefficient indicates the *direction* of the linear approximation. A correlation coefficient of $+1.00$ means the two variables form a perfect linear relationship that is direct in nature (that is, the higher the score an individual obtains on X, the higher the score that individual obtains on Y). A correlation coefficient of -1.00 also means the two variables form a perfect linear relationship, but it is inverse in nature (that is, the higher the score an individual obtains on X, the lower the score that individual obtains on

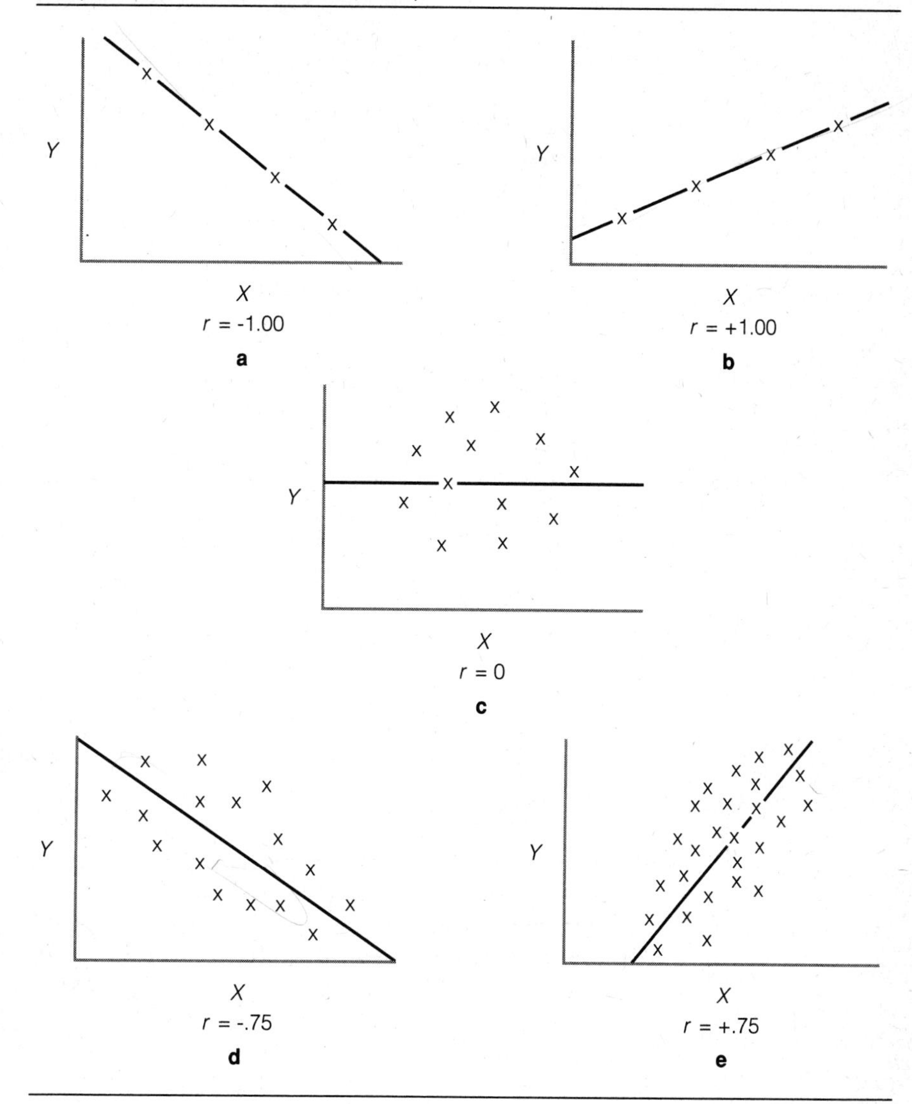

FIGURE 5.4 Scatterplots for (**a**) a Perfect Negative Relationship, (**b**) a Perfect Positive Relationship, (**c**) No Relationship, (**d**) a Strong Negative Relationship, and (**e**) a Strong Positive Relationship (Adapted from Johnson & Liebert, 1977)

Y). A correlation coefficient of 0 means there is no linear relationship between the two variables.

Pearson Correlation and Regression

Figure 5.4 presents some scatterplots for correlation coefficients of different magnitudes. As can be seen, the more strongly the two variables are correlated, the more the data points begin to form a line. In fact, when the relationship between X and Y is perfectly linear (that is, when $r = +1.00$ or -1.00), all the data points fall exactly on a straight line. As discussed in Section 5.2, this line can be represented by an equation of the form $Y = a + bX$. When the correlation is not perfect, the statistical technique of *regression* can be used to identify a line that, although imperfect, will fit the data points better than any other line that we could try to fit to them, as determined by a statistical criterion known as *least squares*. Regression is closely related to Pearson correlation. Together, the two procedures allow us to determine the extent to which two variables approximate a linear relationship and the line that describes this relationship.

We will discuss regression and the least squares criterion in detail in Section 5.5. First, however, we will consider issues related to the calculation and interpretation of the Pearson correlation coefficient.

Calculation of the Pearson Correlation Coefficient

Consider Table 5.1, which presents data illustrating (a) a perfect positive linear relationship, (b) a perfect negative linear relationship, and (c) a complete lack of linear relationship between two variables, X and Y. For each example, we have converted the raw scores on X and Y to standard scores, z_X and z_Y. When a linear relationship is positive, the z scores on X will tend to be similar to the z scores on Y and they will also tend to be alike in sign. In fact, when a positive linear relationship is perfect, the standard scores on X and Y will be identical (see columns 4 and 5 of Table 5.1a). This means, for example, that an individual who has a score 1 standard deviation above the mean on X ($z_X = 1.00$) will also have a score 1 standard deviation above the mean on Y ($z_Y = 1.00$). When the z scores for a given individual are multiplied by one another, the product will be positive (unless $z_X = z_Y = 0$, in which case $z_X z_Y$ will also equal 0). When these products are summed across individuals, a relatively large positive value of $\Sigma z_X z_Y$ will be obtained (see column 6 of Table 5.1a). Positive, though less extreme, values of $\Sigma z_X z_Y$ will similarly be obtained when a nonperfect positive linear relationship exists between two variables.

When a linear relationship is negative, the z scores on X will also tend to be similar to the z scores on Y, but they will generally be opposite in sign. For instance, large positive z scores on X will tend to be associated with large negative z scores on Y. In the case of a perfect negative linear relationship, the standard scores on X and Y will have different signs but will be identical in size (see columns 4 and 5 of Table 5.1b). For example, an individual who has a score 1 standard deviation *above* the mean on X ($z_X = 1.00$) will have a score 1 standard deviation *below* the mean on Y ($z_Y = -1.00$). When the z scores for a given individual are multiplied by one another, the product will be negative (unless $z_X = z_Y = 0$), and when these products are summed across individuals, a relatively large negative value of $\Sigma z_X z_Y$ will be obtained (see column 6 of Table 5.1b). Negative, though less extreme, values of $\Sigma z_X z_Y$ will similarly be obtained when a nonperfect negative linear relationship exists between two variables.

Finally, when there is *no* linear relationship, the z scores on X will bear no consistent relationship to the z scores on Y, either in size or sign. The product of the z scores will be positive for some individuals and negative for others (and for still others, equal to 0), and when summed, the positive and the nega-

TABLE 5.1 Examples of a Perfect Positive Linear Relationship, a Perfect Negative Linear Relationship, and No Linear Relationship

(A) PERFECT POSITIVE LINEAR RELATIONSHIP

Individual	X	Y	z_X	z_Y	$z_X z_Y$
1	8	10	1.41	1.41	2.00
2	7	9	.71	.71	.50
3	6	8	.00	.00	.00
4	5	7	− .71	− .71	.50
5	4	6	−1.41	−1.41	2.00
					$\Sigma z_X z_Y = 5.00$

$$r = \frac{5.00}{5} = +1.00$$

(B) PERFECT NEGATIVE LINEAR RELATIONSHIP

Individual	X	Y	z_X	z_Y	$z_X z_Y$
1	8	6	1.41	−1.41	−2.00
2	7	7	.71	− .71	− .50
3	6	8	.00	.00	.00
4	5	9	− .71	.71	− .50
5	4	10	−1.41	1.41	−2.00
					$\Sigma z_X z_Y = -5.00$

$$r = \frac{-5.00}{5} = -1.00$$

(C) NO LINEAR RELATIONSHIP

Individual	X	Y	z_X	z_Y	$z_X z_Y$
1	3	7	1.12	1.20	1.34
2	1	3	−1.12	− .98	1.10
3	2	4	.00	− .44	.00
4	1	7	−1.12	1.20	−1.34
5	3	3	1.12	− .98	−1.10
					$\Sigma z_X z_Y = 0.00$

$$r = \frac{0.00}{5} = 0.00$$

tive $z_X z_Y$ values will cancel each other out, yielding a $\Sigma z_X z_Y$ value of 0 (see column 6 of Table 5.1c).

To summarize, when a linear relationship is positive, the sum of the products of z scores will also be positive; when a linear relationship is negative, the sum of the products of z scores will also be negative; and when there is a complete lack of a linear relationship, the sum of the products of z scores will be 0. These observations are consistent with the nature of the correlation coefficient as described earlier. We can therefore use the sum of the products of z scores as an index of the correlation between two variables. There is, however, one complication. When the correlation between two variables is nonzero, the value of the sum of z score products is influenced not only by the size of the correlation, but also by the sample size (N). In the case of a positive correlation, for example, the larger the number of observations, the greater will be the sum of the products, everything else being equal. Since we want an index of correlation that is independent of N, we can divide $\Sigma z_X z_Y$ by N. Dividing by N is also advantageous because of a certain property of standard scores. Recall that when the relationship between two variables is positive and perfect, the z scores will be equal to one another. In this case, $\Sigma z_X z_Y = \Sigma z_X^2 = \Sigma z_Y^2$, since $z_X = z_Y$. It turns out that the sum of a set of squared standard scores will always equal N. Thus, when the correlation is positive and perfect, dividing $\Sigma z_X z_Y = N$ by N will always yield a value of $N/N = +1.00$. In the case of a perfect negative relationship, $\Sigma z_X z_Y$ will always equal $-N$ such that a perfect negative correlation will always yield a $\Sigma z_X z_Y / N$ value of $-N/N = -1.00$. Nonperfect linear relationships will yield $\Sigma z_X z_Y / N$ values somewhere between the two extremes of -1.00 and $+1.00$. These properties suggest the following equation for calculating a Pearson correlation coefficient:

$$r = \frac{\Sigma z_X z_Y}{N} \qquad [5.3]$$

Although Equation 5.3 yields an intuitive understanding of the correlation coefficient, it is not computationally efficient. We will now present a formula that is easier from a computational standpoint. This formula is

$$r = \frac{\text{SCP}}{\sqrt{\text{SS}_X \text{SS}_Y}} \qquad [5.4]$$

where SS_X is the sum of squares for variable X, SS_Y is the sum of squares for variable Y, and SCP is the **sum of cross-products.** Recall that the defining formula for the sum of squares for variable X is $\Sigma(X - \bar{X})^2$. This can also be written as

$$\Sigma(X - \bar{X})(X - \bar{X})$$

In other words, we simply multiply each individual's deviation score by itself and then sum these products across individuals. Similarly, the sum of squares for variable Y can be represented as

$$\Sigma(Y - \bar{Y})(Y - \bar{Y})$$

A sum of cross-products is similar to a sum of squares but rather than indicating the extent to which a set of scores vary from their mean, a sum of cross-products indicates the extent to which two sets of scores vary from one another, or *covary.* The sum of cross-products is defined as

$$\text{SCP} = \Sigma(X - \bar{X})(Y - \bar{Y}) \qquad [5.5]$$

In other words, we multiply an individual's deviation score for X by that individual's deviation score for Y and then sum these *cross-products* across individuals.* Unlike a sum of squares, a sum of cross-products can be negative, as scores are not being squared in its calculation.

---

* If we wanted to compute the variance of X or the variance of Y, we would divide the respective sum of squares by N. If we divide the sum of cross-products by N, we obtain a statistic called the *covariance*.

TABLE 5.2 Data and Calculation of SS_X, SS_Y, and SCP for Traditionalism and Ideal Family Size Example

Individual	X	$X - \bar{X}$	$(X - \bar{X})^2$	Y	$Y - \bar{Y}$	$(Y - \bar{Y})^2$	$(X - \bar{X})(Y - \bar{Y})$
1	9	4	16	10	5	25	20
2	7	2	4	6	1	1	2
3	5	0	0	3	−2	4	0
4	3	−2	4	6	1	1	−2
5	1	−4	16	3	−2	4	8
6	1	−4	16	3	−2	4	8
7	3	−2	4	5	0	0	0
8	7	2	4	6	1	1	2
9	5	0	0	1	−4	16	0
10	9	4	16	7	2	4	8
	$\Sigma X = 50$ $\bar{X} = 5.00$		$SS_X = 80$	$\Sigma Y = 50$ $\bar{Y} = 5.00$		$SS_Y = 60$	$SCP = 46$

Phrased in terms of defining formulas, Equation 5.4 is thus equivalent to

$$r = \frac{\Sigma(X - \bar{X})(Y - \bar{Y})}{\sqrt{\Sigma(X - \bar{X})^2 \Sigma(Y - \bar{Y})^2}} \quad [5.6]$$

Consider a group of 10 individuals. Two variables have been measured for each: traditionalism as assessed by responses to a 10-item traditionalism questionnaire (X) and ideal family size (Y). Scores on the traditionalism questionnaire can range from 0 to 10, with higher scores indicating a more traditional orientation. The scores on X and Y are presented in columns 2 and 5 of Table 5.2. The first step in the computation of the correlation coefficient involves the calculation of the sum of squares for the X scores and the sum of squares for the Y scores. This has been done in columns 4 and 7 of Table 5.2 and we find that $SS_X = 80$ and $SS_Y = 60$. The sum of cross-products is computed in column 8 of Table 5.2. For instance, the first individual's deviation score on X is 4 (column 3) and her deviation score on Y is 5 (column 6). Her cross-product score is thus $(4)(5) = 20$. Summing the cross-product scores for the entire data set, we find that SCP = 46. The correlation coefficient can now be calculated using Equation 5.6:

$$r = \frac{46}{\sqrt{(80)(60)}} = .66$$

A correlation coefficient of .66 indicates a strong positive linear relationship between the two variables. The linear trend can be seen in Figure 5.5, which presents the scatterplot of the two variables.

Computational Formula

The expression for the correlation coefficient presented in Equation 5.6 includes the defining formulas for the sum of squares for X, the sum of squares

FIGURE 5.5 Scatterplot for Example on Traditionalism and Ideal Family Size

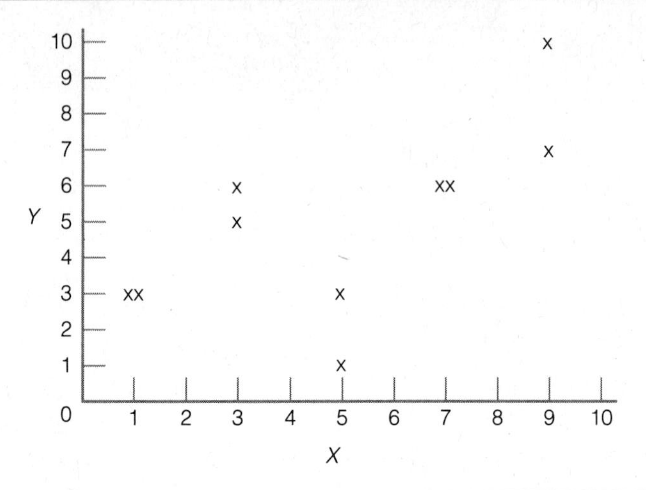

for Y, and the sum of cross-products. However, SS_X can also be derived using the computational formula

$$SS_X = \Sigma X^2 - \frac{(\Sigma X)^2}{N}$$

and SS_Y can be derived using the computational formula

$$SS_Y = \Sigma Y^2 - \frac{(\Sigma Y)^2}{N}$$

Analogous to the computational formulas for SS_X and SS_Y, a computationally efficient formula also exists for SCP:

$$SCP = \Sigma XY - \frac{(\Sigma X)(\Sigma Y)}{N} \qquad [5.7]$$

where ΣX is the sum of individuals' X scores, ΣY is the sum of individuals' Y scores, and ΣXY is the sum of the products of individuals' X scores multiplied by their Y scores. Substituting these symbols into the general expression for a correlation coefficient, $r = SCP/\sqrt{SS_X SS_Y}$, yields the following computational formula for the correlation coefficient:

$$r = \frac{\Sigma XY - \frac{(\Sigma X)(\Sigma Y)}{N}}{\sqrt{\left(\Sigma X^2 - \frac{(\Sigma X)^2}{N}\right)\left(\Sigma Y^2 - \frac{(\Sigma Y)^2}{N}\right)}} \qquad [5.8]$$

TABLE 5.3 Computational Formula Approach for Correlation Coefficient for Traditionalism and Ideal Family Size Example

Individual	X	Y	X^2	Y^2	XY
1	9	10	81	100	90
2	7	6	49	36	42
3	5	3	25	9	15
4	3	6	9	36	18
5	1	3	1	9	3
6	1	3	1	9	3
7	3	5	9	25	15
8	7	6	49	36	42
9	5	1	25	1	5
10	9	7	81	49	63
	$\Sigma X = 50$	$\Sigma Y = 50$	$\Sigma X^2 = 330$	$\Sigma Y^2 = 310$	$\Sigma XY = 296$

The advantage of this formula over Equation 5.6 is that it is both more efficient and more precise as it requires fewer steps and presents fewer opportunities for rounding error.

The calculation of intermediate statistics for Equation 5.8 is demonstrated in Table 5.3 for the traditionalism and ideal family size data from Table 5.2. As can be seen, $\Sigma X^2 = 330$, $\Sigma Y^2 = 310$, and $\Sigma XY = 296$. Thus,

$$r = \frac{296 - \dfrac{(50)(50)}{10}}{\sqrt{\left(330 - \dfrac{50^2}{10}\right)\left(310 - \dfrac{50^2}{10}\right)}}$$

$$= \frac{296 - 250}{\sqrt{(80)(60)}} = .66$$

This is the same value of r as was calculated previously using Equation 5.6.

STUDY EXERCISE 5.2

A director of a political campaign was interested in the relationship between voters' perceptions that his candidate supported labor unions (X) and their willingness to vote for her (Y). Individuals were asked to indicate on a 10-point scale the extent to which they thought the candidate supported labor unions, with higher numbers indicating greater perceived support. They also indicated on a similar 10-point scale their willingness to vote for the candidate. The scores on these variables are presented below for nine individuals. Compute the Pearson correlation coefficient for these data.

Individual	X	Y
1	6	5
2	7	4
3	8	4
4	8	5
5	8	3
6	9	6
7	9	7
8	7	5
9	10	6

Answer The correlation coefficient is most readily calculated using Equation 5.8. The intermediate statistics necessary for the application of this equation are computed as follows:

Individual	X	Y	X^2	Y^2	XY
1	6	5	36	25	30
2	7	4	49	16	28
3	8	4	64	16	32
4	8	5	64	25	40
5	8	3	64	9	24
6	9	6	81	36	54
7	9	7	81	49	63
8	7	5	49	25	35
9	10	6	100	36	60
	$\Sigma X = 72$	$\Sigma Y = 45$	$\Sigma X^2 = 588$	$\Sigma Y^2 = 237$	$\Sigma XY = 366$

Thus,

$$r = \frac{\Sigma XY - \frac{(\Sigma X)(\Sigma Y)}{N}}{\sqrt{\left(\Sigma X^2 - \frac{(\Sigma X)^2}{N}\right)\left(\Sigma Y^2 - \frac{(\Sigma Y)^2}{N}\right)}}$$

$$= \frac{366 - \frac{(72)(45)}{9}}{\sqrt{\left(588 - \frac{72^2}{9}\right)\left(237 - \frac{45^2}{9}\right)}}$$

$$= \frac{6}{\sqrt{(12)(12)}} = .50$$

A correlation coefficient of .50 indicates a relatively strong positive linear relationship between the two variables.

5.4

Correlation and Causation

The fact that two variables are correlated does not necessarily imply that one variable *causes* the other to vary as it does. It is entirely possible for two variables to be related to one another, but for no causal relationship to exist between them. In fact, there are many reasons why two variables, X and Y, might be correlated. Three possibilities of interest to us are that (1) X might cause Y, (2) Y might cause X, or (3) some additional variable(s) might cause both X and Y. For instance, while we might find a positive correlation between the number of hours college students spend working for pay and the number of campus organizations they belong to, it is unlikely that working *causes* students to join organizations or that membership in organizations *causes* students to work. Rather, the correlation between hours of work and group membership is probably attributable to the desire to achieve and related personality characteristics—as the desire to achieve increases, individuals might work more as they pursue their financial and occupational goals and join more organizations as a means of achieving in the social realm. As this example illustrates, one must be cautious when drawing causal inferences from correlational analyses.

5.5

Regression

We noted earlier that when two variables are perfectly correlated, all the data points will fall exactly on a straight line defined by an equation of the form $Y = a + bX$. However, we will almost certainly never encounter such a situation in the behavioral sciences. In fact, correlations for the types of variables typically studied will seldom exceed −.40 or +.40 and will often be considerably smaller. When the two variables are not perfectly correlated, the statistical technique of **regression** can be used to identify a line that, although imperfect, will fit the data points better than any other line that we could try to fit to them, as determined by a criterion known as *least squares,* to be discussed shortly. This line serves to describe the nature of the linear relationship between the two variables.

We will begin our discussion of regression by returning to the example on traditionalism and ideal family size introduced in Section 5.3. It will be remembered that the correlation between these two variables was found to be .66, indicating a strong direct linear relationship. This linear trend can be seen in Figure 5.5. The linear relationship between X and Y can be formally represented by a **regression line** taking the general form

$$\hat{Y} = a + bX \qquad [5.9]$$

This equation, known as the **regression equation,** is similar to the linear model, $Y = a + bX$, but since the data points in our example do not form a straight line, different Y scores might be associated with the same X score. For instance, one of the research participants had a traditionalism (variable X) score of 3 and indicated a preference for five children (variable Y). A second individual also had a traditionalism score of 3, but perceived six children, rather than five children, as the ideal family size. Because of this and related issues, the symbol $\hat{Y}$ (read "*predicted* Y") is used in the framework of regression to indicate the value of Y that is *predicted* to be paired with a specified value of X. This value of Y is the point on the line $a + bX$ corresponding to a person's score on X.

Calculation of the Slope and Intercept

The slope and intercept of a regression line are readily calculated using formulas available for this purpose. We will present the formulas and then discuss the logic underlying them. For the slope, the formula is *

$$b = \frac{\text{SCP}}{\text{SS}_X} \qquad [5.10]$$

Turning again to the example on traditionalism and ideal family size, we find that

$$b = \frac{46}{80} = .58$$

since SCP was previously calculated to be 46 and SS_X was previously calculated to be 80.

The formula for computing the intercept is

$$a = \bar{Y} - b\bar{X} \qquad [5.11]$$

Inserting the value of b from above and the values of $\bar{Y}$ and $\bar{X}$ from Table 5.2 for the traditionalism and ideal family size example, we find that

$$a = 5.00 - (.58)(5.00) = 2.10$$

The regression equation describing the relationship between traditionalism (X) and ideal family size (Y) is thus

$$\hat{Y} = 2.10 + .58X$$

The regression line described by this equation has been drawn in Figure 5.6. Note that this line intersects the Y axis at the value of the intercept (2.10). Further, the slope of this line is such that when X increases by 1 unit, Y increases by .58 unit. Lines with arrowheads have been drawn emanating from the data points and extending to the regression line. These illustrate visually the rationale in defining the values of the slope and intercept: The slope and intercept are defined so as to *minimize the squared vertical distances that the data points, considered collectively, are from the regression line.* This is what we mean when we say that the regression line fits the data points better than any

* An alternative formula for the slope is $b = r(s_Y/s_X)$.

FIGURE 5.6 Scatterplot and Regression Line for Example on Traditionalism and Ideal Family Size

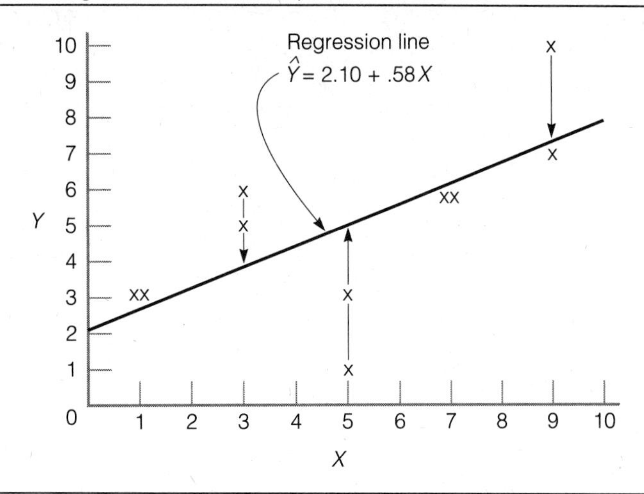

TABLE 5.4 Scores on X and Y, Predicted Scores, and Discrepancy Scores

Individual	X	Y	$\hat{Y}$	$Y - \hat{Y}$
1	9	10	7.32	2.68
2	7	6	6.16	− .16
3	5	3	5.00	−2.00
4	3	6	3.84	2.16
5	1	3	2.68	.32
6	1	3	2.68	.32
7	3	5	3.84	1.16
8	7	6	6.16	− .16
9	5	1	5.00	−4.00
10	9	7	7.32	− .32

other line that we could try to fit to them. This is accomplished in our example by the equation $\hat{Y} = 2.10 + .58X$. Any other linear equation would result in a larger sum of squared distances from the generated line.

The criterion for deriving the values of the slope and intercept is formally known as the **least squares criterion.** It can be illustrated algebraically as well as visually. Earlier we noted that a regression equation can be used to identify the value of Y that is predicted to be paired with an individual's score on X. This is accomplished by substituting a person's score on X into the regression equation. In the traditionalism and ideal family size example, this equation is

$$\hat{Y} = 2.10 + .58X$$

Let us substitute the X scores of each of the 10 research participants into this equation. These scores have been reproduced in column 2 of Table 5.4. The score on X for the first individual is 9. The predicted Y score ($\hat{Y}$) for this individual is $2.10 + (.58)(9) = 7.32$. The score on X for the second individual is 7.

The predicted Y score for this individual is $2.10 + (.58)(7) = 6.16$. The remaining predicted scores are obtained similarly.

Columns 3 and 4 of Table 5.4 present individuals' actual Y scores and their predicted Y scores based on the regression equation. Inspection of these scores indicates that there are discrepancies between Y and $\hat{Y}$. These discrepancies are formally defined in the last column of Table 5.4. Note that the discrepancies are generally rather small, reflecting the strong linear relationship ($r = .66$) that exists between X and Y. If the correlation had been smaller, there would be larger discrepancies between actual and predicted Y scores. The least squares criterion concerns itself with the *squares* of the discrepancy scores and formally defines the values of the slope and intercept so as to minimize the sum of these squares; that is, the least squares criterion defines the regression line such that $\Sigma(Y - \hat{Y})^2$ is minimized.

The Standard Error of Estimate

Unless two variables are perfectly correlated, some degree of error will result when scores on Y are predicted from scores on X using a regression equation. The amount of error for a given individual can be represented by the discrepancy between that person's actual and predicted Y scores. However, the usefulness of discrepancy scores as a summary measure of predictive error is limited by the fact that the sum of discrepancies between actual Y scores and Y scores predicted from the regression equation will always equal 0. This is demonstrated in column 5 of Table 5.4 for the traditionalism and ideal family size example. A more useful index of predictive error is provided by the **standard error of estimate.** This is defined as

$$s_{YX} = \sqrt{\frac{\Sigma(Y - \hat{Y})^2}{N}} \qquad [5.12]$$

where s_{YX} represents the standard error of estimate and all other symbols are as previously presented.

Equation 5.12 is conceptually similar to the defining formula for a standard deviation estimate, which, phrased in terms of Y, is

$$s_Y = \sqrt{\frac{\Sigma(Y - \bar{Y})^2}{N}}$$

The only difference between the two equations is in the numerator. For the standard deviation, the numerator reflects the deviation of the Y scores from their mean. For the standard error of estimate (Equation 5.12), the numerator reflects the deviation of the Y scores from the predicted Y scores. The standard error of estimate thus represents an average error across individuals in predicting scores on Y from the regression equation.

While Equation 5.12 has the advantage of being conceptually clear, there is a more efficient computational formula available for calculating the standard error of estimate. This formula is

$$s_{YX} = s_Y\sqrt{1 - r^2} \qquad [5.13]$$

Phrased in terms of the sum of squares of Y rather than the standard deviation, this is equivalent to

$$s_{YX} = \sqrt{\frac{SS_Y(1 - r^2)}{N}} \qquad [5.14]$$

Since SS_Y was previously calculated to be 60, the value of the standard error of estimate for the traditionalism and ideal family size example is, according to Equation 5.14,

$$s_{YX} = \sqrt{\frac{60(1 - .66^2)}{10}} = 1.84$$

indicating that, on the average, predicted Y (ideal family size) scores deviate from actual Y scores by 1.84 units.

There are two perspectives in interpreting the standard error of estimate. First, its absolute magnitude is meaningful. In the present example, the average error in predicting scores on Y from scores on X was calculated to be 1.84 units. Given the range of possible ideal family sizes, this degree of error is not unreasonable. Second, the standard error of estimate can be compared with the standard deviation of Y. The standard deviation of Y indicates what the average error in prediction would be if one were to predict for each individual a Y score equal to the mean of Y. If X helps to predict Y, then the standard error of estimate will be smaller than the standard deviation of Y, and the better the predictor X is, the smaller the standard error of estimate will be. In the present example, $SS_Y = 60$ and $N = 10$, so $s_Y = \sqrt{SS_Y/N} = \sqrt{60/10} = 2.45$. The reduction in error from 2.45 (the standard deviation of Y) to 1.84 (the standard error of estimate) in predicting scores on Y when scores on X are considered reflects the strong linear relationship ($r = .66$) that exists between traditionalism and ideal family size.

STUDY EXERCISE **5.3**

Compute the regression equation and the standard error of estimate for the data in Study Exercise 5.2.

Answer The sum of cross-products and the sum of squares for X were calculated in Study Exercise 5.2 to equal 6 and 12, respectively. Applying Equation 5.10, the slope of the regression line is

$$b = \frac{SCP}{SS_X} = \frac{6}{12} = .50$$

Since $\Sigma X = 72$, $\Sigma Y = 45$, and $N = 9$, the mean of X is $72/9 = 8.00$ and the mean of Y is $45/9 = 5.00$. Applying Equation 5.11, the intercept of the regression line is

$$a = \bar{Y} - b\bar{X} = 5.00 - (.50)(8.00) = 1.00$$

This yields a regression equation of $\hat{Y} = 1.00 + .50X$.

The standard error of estimate can be calculated using Equation 5.14. For $SS_Y = 12$, $r = .50$, and $N = 9$, this is found to equal

Study Exercise 5.3 continued

$$s_{YX} = \sqrt{\frac{SS_Y(1 - r^2)}{N}}$$

$$= \sqrt{\frac{12(1 - .50^2)}{9}} = 1.00$$

This indicates that, on the average, predicted Y (willingness to vote for the candidate of interest) scores deviate from actual Y scores by 1.00 unit.

5.6

Summary

Pearson correlation is based on the linear model and indexes the extent to which two quantitative variables that are measured on a level that approximates interval characteristics are linearly related. The correlation coefficient can range from -1.00 through 0 to $+1.00$. The magnitude of the correlation coefficient indicates the degree to which the variables approximate a linear relationship, and the sign of the correlation coefficient indicates whether this relationship is direct or inverse. The fact that two variables are correlated does not necessarily imply that one variable causes the other to vary as it does, as both variables might be caused by some additional variable(s).

The least squares criterion can be used to identify the slope and intercept of the line that fits the data points better than any other line that we could try to fit to them. This line, known as the regression line, serves to describe the nature of the linear relationship between the two variables and can be used to identify the value of Y that is predicted to be paired with an individual's score on X. This is accomplished by substituting a person's score on X into the regression equation. A measure of the average error across individuals in predicting scores on Y in this manner is provided by the standard error of estimate.

Exercises

Answers to asterisked (*) exercises appear at the back of the book. Answers to exercises with two asterisks are also worked out step-by-step in the Study Guide.

1. Draw a scatterplot for the following data:

Individual	*X*	*Y*
1	10	9
2	8	7
3	6	5
4	9	8
5	10	8
6	8	8
7	5	6

***2.** What information is conveyed by the slope of a line?

3. What information is conveyed by the intercept of a line?

4. What is the general form of the linear model?

***5.** Given a perfect linear relationship between two variables, X and Y, and a slope of 3.00, by how many units will Y change if X changes by 1 unit? If X changes by 2 units? If X changes by 7 units?

***6.** What information is conveyed by the magnitude of a correlation coefficient?

7. What information is conveyed by the sign of a correlation coefficient?

***8.** For each pair of correlation coefficients, indicate which coefficient represents a better approximation to a linear relationship:

a. +.37 or +.18
b. −.37 or −.18
c. +.52 or −.76
d. 0 or +.26
e. 0 or −.44
f. +.61 or +1.07

***9.** Draw a scatterplot for two variables that are negatively correlated.

10. Draw a scatterplot for two variables that are positively correlated.

***11.** Give an example of two variables that you think are positively correlated.

12. Give an example of two variables that you think are negatively correlated.

****13.** Compute the Pearson correlation coefficient for the following data:

Individual	*X*	*Y*
1	3	7
2	8	9
3	3	3
4	2	8
5	6	8
6	6	9
7	8	6
8	5	4
9	7	2
10	2	4

14. Compute the Pearson correlation coefficient for the following data:

Individual	*X*	*Y*
1	6	10
2	7	9
3	7	7
4	8	8
5	8	8
6	8	8
7	8	8
8	9	9
9	9	7
10	10	6

15. Why must one be cautious when drawing causal inferences from correlational analyses?

***16.** Give an example of two variables that are probably correlated with one another but that are not causally related.

***17.** What is the general form of the regression equation? How does this differ from the linear model?

***18.** Explain the least squares criterion.

****19.** Compute the regression equation that describes the relationship between X and Y for the data in Exercise 13.

***20.** What is the predicted Y score for individual 1 in Exercise 13? What is the predicted Y score for individual 5? What is the predicted Y score for individual 10?

21. Compute the regression equation that describes the relationship between X and Y for the data in Exercise 14.

22. What is the predicted Y score for individual 2 in Exercise 14? What is the predicted Y score for individual 4? What is the predicted Y score for individual 6?

23. What information is conveyed by the standard error of estimate?

***24.** Compute the standard error of estimate for the data in Exercise 13.

25. Compute the standard error of estimate for the data in Exercise 14.

SIX

PROBABILITY

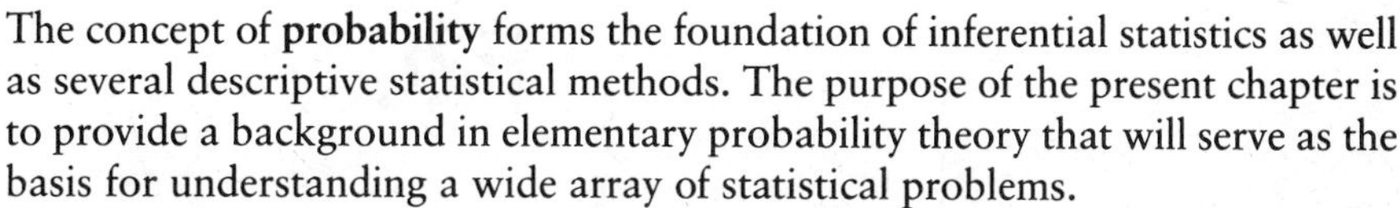

The concept of **probability** forms the foundation of inferential statistics as well as several descriptive statistical methods. The purpose of the present chapter is to provide a background in elementary probability theory that will serve as the basis for understanding a wide array of statistical problems.

When we flip a coin, either of two possible outcomes can result: We can obtain a head or we can obtain a tail. In probability theory, the act of flipping the coin is called a **trial** and each unique outcome is called an **event.** There are two different conceptualizations of probability. The first is the classical viewpoint and is based on logical analysis. The probability of an event, A, is formally defined as the number of observations favoring event A divided by the total number of possible observations:

$$p(A) = \frac{\text{number of observations favoring event } A}{\text{total number of possible observations}} \qquad \textbf{[6.1]}$$

In the coin flipping example, there are two possible observations. On any given trial there is one observation favoring a head and one observation favoring a tail. The probability of obtaining a head is 1/2 or .50. Suppose you randomly select an individual from a population consisting of 60 males and 40 females. What is the probability you will select a male? There are 60 observations favoring the event "male," and a total of 100 observations. According to Equation 6.1, the probability of selecting a male is 60/100 = .60.

A second conceptualization of probability focuses on the long run. According to this interpretation, if we flip a coin a large number of times, then over the long run, the proportion of heads relative to the total number of observations should approach .50. Over an infinite number of trials, the proportion of heads will be .50.

We will develop additional concepts in probability theory using the following example. Suppose an investigator classifies all married men in a community in terms of two variables: (1) satisfaction with one's marriage, and (2) satisfaction with one's job. The first variable has two values, men who are satisfied with their marriages versus men who are dissatisfied with their marriages. The second variable also has two values, men who are satisfied with their jobs versus men who are dissatisfied with their jobs. These two variables can be combined to classify each man into one of four groups: (1) those who

TABLE 6.1 Contingency Table for Marital and Job Satisfaction Example

JOB SATISFACTION	MARITAL SATISFACTION: *Satisfied with marriage*	MARITAL SATISFACTION: *Dissatisfied with marriage*	TOTALS
Satisfied with job	156	54	210
Dissatisfied with job	52	148	200
TOTALS	208	202	410

are satisfied with both their marriages and their jobs, (2) those who are satisfied with their jobs but not their marriages, (3) those who are satisfied with their marriages but not their jobs, and (4) those who are dissatisfied with both their marriages and their jobs. The results of this study can be displayed in a **contingency table** (also called a *frequency* or *crosstabulation table*), as in Table 6.1.

Each unique combination of variables in a contingency table is referred to as a **cell.** The entries within the cells represent the number of individuals who are characterized by the corresponding values of the variables. In the present example, 156 men are satisfied with both their marriages and their jobs, 54 men are satisfied with their jobs but not their marriages, 52 men are satisfied with their marriages but not their jobs, and 148 men are dissatisfied with both their marriages and their jobs. The numbers in the last column and bottom row of Table 6.1 indicate how many individuals have each separate characteristic. For example, 210 men are satisfied and 200 men are dissatisfied with their jobs. Also, 208 men are satisfied and 202 men are dissatisfied with their marriages. These frequencies are referred to as **marginal frequencies** and are the sum of the frequencies in the corresponding row (for example, $156 + 54 = 210$) or column (for example, $156 + 52 = 208$). Finally, the number in the lower right-hand corner indicates the total number of observations there were in the study (in this case, 410).

6.1 Probabilities of Simple Events

In the language of probability theory, the variable of job satisfaction has two possible outcomes: (1) being satisfied with one's job or (2) being dissatisfied with one's job. These outcomes are said to be **mutually exclusive** because it is impossible for both outcomes to occur for a given individual: If a person is classified as being satisfied with his job, he cannot also be classified as being dissatisfied with his job. The variable of marital satisfaction also has two possible outcomes: (1) being satisfied with one's marriage or (2) being dissatisfied with one's marriage. These outcomes are also mutually exclusive.

If we were to select a married man at random from the community in the preceding example, what would the probability be that he is satisfied with his job? According to Equation 6.1, it would be the number of men who are satisfied with their jobs (210) divided by the total number of men (410):

$$p(\text{satisfied with job}) = \frac{210}{410} = .512$$

The probability that a man is dissatisfied with his job would be the number of men who are dissatisfied with their jobs (200) divided by the total number of men (410):

$$p(\text{dissatisfied with job}) = \frac{200}{410} = .488$$

Note that the sum of these two probabilities is equal to 1.00 (.512 + .488 = 1.00). Given a set of outcomes that are mutually exclusive and **exhaustive** (that is, which includes all potential outcomes that could occur), the sum of the probabilities of the outcomes will always equal 1.00. As noted in Chapter 2, such a set of outcomes and their associated probabilities is known as a **probability distribution.**

STUDY EXERCISE 6.1

For the data in Table 6.1, what is the probability of randomly selecting a man who is satisfied with his marriage? A man who is dissatisfied with his marriage?

Answer Applying Equation 6.1, the probability that a man is satisfied with his marriage is

$$p(\text{satisfied with marriage}) = \frac{208}{410} = .507$$

and the probability that a man is dissatisfied with his marriage is

$$p(\text{dissatisfied with marriage}) = \frac{202}{410} = .493$$

Thus, for these data, the likelihood that a man is satisfied with his marriage is roughly equal to the likelihood that he is dissatisfied with his marriage.

6.2

Conditional Probabilities

A useful concept for analyzing the data in Table 6.1 is that of a **conditional probability.** A conditional probability indicates the likelihood that an event will occur given that some other event occurs. In the context of Table 6.1, this might be expressed as follows: Given that a person is satisfied with his job, what is the probability that he is satisfied with his marriage? A total of 210 men are satisfied with their jobs. Of these 210 men, 156 are satisfied with their marriages. The probability that a man is satisfied with his marriage given that he is satisfied with his job is thus

$$p(\text{satisfied with marriage}|\text{satisfied with job}) = \frac{156}{210} = .743$$

The general symbolic form for a conditional probability is

$$p(A|B) = \frac{\text{number of observations favoring } \textit{both} \text{ event } A \textit{ and } \text{event } B}{\text{number of observations favoring event } B} \qquad [6.2]$$

where $p(A|B)$ is read "the probability of event A, given event B." The conditional probability in the above example was computed by considering only those individuals who are satisfied with their jobs (event B) and then deriving the proportion of these individuals who are also satisfied with their marriages (event A).

Conditional probabilities are useful for understanding relationships between events. This can be illustrated by considering the concept of **independence.** An event A is said to be independent of an event B if $p(A) = p(A|B)$, that is, if the occurrence of event A is unrelated to the occurrence of event B.* Consider the two events "being satisfied with one's marriage" and "being satisfied with one's job." The probability of the first event was computed in Study Exercise 6.1 as .507 while the conditional probability of being satisfied with one's marriage given that one is satisfied with his job was computed above as .743. It is clear that being satisfied with one's marriage is not independent of being satisfied with one's job for the married men in this town. The occurrence of event A in this case *is* related to the occurrence of event B in that its probability is substantially raised (from .507 to .743) given event B.

It is important to realize that even though two events are related (nonindependent), this does not necessarily mean that one causes the other. We will return to this issue in Chapter 9.

---

* If two events are independent, not only does $p(A) = p(A|B)$, but the reverse is also true; that is, $p(B) = p(B|A)$.

STUDY EXERCISE 6.2

For the data in Table 6.1, compute the conditional probability of being dissatisfied with one's marriage given being dissatisfied with one's job.

Answer A total of 200 men are dissatisfied with their jobs, of whom 148 are also dissatisfied with their marriages. Thus, applying Equation 6.2, the conditional probability of being dissatisfied with one's marriage given being dissatisfied with one's job is

$$p(\text{dissatisfied with marriage}|\text{dissatisfied with job}) = \frac{148}{200} = .740$$

6.3 Joint Probabilities

Another type of probability an investigator might be interested in is a **joint probability.** This refers to the likelihood of observing each of two events. A joint probability can be represented as

$$p(A, B) = \frac{\text{number of observations favoring } \textit{both} \text{ event } A \textit{ and } \text{event } B}{\text{total number of observations}} \qquad [6.3]$$

where $p(A, B)$ stands for the probability of *both* event A *and* event B occurring. For the data in Table 6.1, we might want to determine the probability that an individual is satisfied with his marriage (event A) *and* satisfied with his job (event B). Since there are 156 such individuals out of the total of 410 men in the study, the joint probability of being satisfied with both one's marriage and one's job is

$$p(\text{satisfied with marriage, satisfied with job}) = \frac{156}{410} = .380$$

STUDY EXERCISE 6.3

For the data in Table 6.1, compute the joint probability of being dissatisfied with both one's marriage and one's job.

Answer There are 148 individuals who are dissatisfied with both their marriages and their jobs out of a total of 410 men. This yields a joint probability of being dissatisfied with both one's marriage and one's job of

Study Exercise 6.3 continued

$$p(\text{dissatisfied with marriage, dissatisfied with job}) = \frac{148}{410} = .361$$

according to Equation 6.3.

6.4 Adding Probabilities

A fourth probability of interest focuses on the likelihood of observing *at least one* of two events. For the data in Table 6.1, we might be interested in specifying the probability that an individual is satisfied with his marriage or satisfied with his job, or both. One way to determine this would be as follows. First determine the probability that an individual is satisfied with his marriage (previously calculated to be .507) and the probability that an individual is satisfied with his job (previously calculated to be .512). Next, sum these two values (.507 + .512 = 1.019). This does *not* represent the probability that a person is satisfied with at least his marriage or satisfied with at least his job, however, because we have counted men who are satisfied with both their marriages and their jobs twice. So we subtract this value (previously calculated to be .380) out. This yields a probability of 1.019 − .380 = .639. In general notation, the probability of observing *at least one* of event A and event B is

$$p(A \text{ or } B) = p(A) + p(B) - p(A, B) \tag{6.4}$$

STUDY EXERCISE 6.4

For the data in Table 6.1, compute the probability of being dissatisfied with one's marriage or being dissatisfied with one's job, or both.

Answer The proportion of men who are dissatisfied with their marriages is .493, the proportion of men who are dissatisfied with their jobs is .488, and the proportion of men who are dissatisfied with both their marriages and their jobs is .361. Applying Equation 6.4, the probability of being dissatisfied with one's marriage or being dissatisfied with one's job, or both, is

$$\begin{aligned} p(\text{dissatisfied with marriage or dissatisfied with job}) &= .493 + .488 - .361 \\ &= .620 \end{aligned}$$

6.5

Relationships Among Probabilities

Among the probabilities we have discussed are the probability of a simple event, conditional probability, and joint probability. Each of these can provide information that is useful to an investigator. The probability of a simple event refers to the likelihood that a given event will occur. A conditional probability tells us the likelihood that a given event will occur given some other event. Finally, a joint probability indicates the likelihood of two events occurring together. Let us explore the insights such concepts can give.

Suppose you are a counselor and are discussing with a couple their possible decision to have a child. You are exploring the advantages and disadvantages of this decision. You know that the incidence of a certain type of birth defect is generally relatively low. In fact, the probability of the defect occurring is estimated to be .001. However, it is also the case that the offspring of couples who are in racial group X are particularly susceptible to the defect. The conditional probability that a baby will have the birth defect given that its parents are in racial group X is estimated to be .020. In this instance, the occurrence of the birth defect is not independent of the race of the parents and your advice to the couple might differ depending on their race.

Let us consider another example. As you might have suspected, there exist certain mathematical relationships among probabilities of simple events, conditional probabilities, and joint probabilities. One important relationship is characterized by the following equation:

$$p(A, B) = p(B)p(A|B) \qquad [6.5]$$

In our original problem, we were concerned with men's satisfaction with their marriages and jobs. Let event A be "satisfied with one's marriage" and event B be "satisfied with one's job." Suppose we were particularly interested in men who are satisfied with both their marriages and their jobs. Equation 6.5 states that the probability that someone is satisfied with *both* his marriage *and* his job [$p(A, B)$] is equal to the probability that he is satisfied with his job [$p(B)$] weighted by the conditional probability that being satisfied with his job implies that he is also satisfied with his marriage [$p(A|B)$]. This can be demonstrated by substituting the relevant probabilities from our earlier calculations into Equation 6.5. In Section 6.3 we found that $p(A, B) = .380$, in Section 6.1 we found that $p(B) = .512$, and in Section 6.2 we found that $p(A|B) = .743$. Indeed, consistent with Equation 6.5,

$$.380 = (.512)(.743)$$

The fact that $p(A, B)$ is low implies that men in our study are generally not satisfied with both their marriages and their jobs. The probability $p(B)$ is moderate, indicating that only about half of the men are satisfied with their jobs.

The conditional probability, $p(A|B)$, is high, which suggests that if the incidence of job satisfaction were to increase, the incidence of marital satisfaction might also increase. Equation 6.5 can thus provide insights into the nature of a joint probability.

It is possible to derive a number of other formulas that relate different probabilities to one another. An extended discussion of these is beyond the scope of this book. However, for the sake of completeness, we will list the formulas that are most frequently used in the behavioral sciences:

1. In Equation 6.5, it was noted that $p(A, B) = p(B)p(A|B)$. If A and B are independent, then $p(A|B) = p(A)$. It follows by simple substitution and rearrangement of terms that *when A and B are independent,*

$$p(A, B) = p(A)p(B) \quad [6.6]$$

2. In Equation 6.4, it was noted that $p(A \text{ or } B) = p(A) + p(B) - p(A, B)$. If A and B are mutually exclusive, then $p(A, B)$ must equal 0. It follows that *when A and B are mutually exclusive,*

$$p(A \text{ or } B) = p(A) + p(B) \quad [6.7]$$

3. Since $p(A, B) = p(B)p(A|B)$ (Equation 6.5), it is possible to express a conditional probability in terms of the following formula, often referred to as *Bayes' theorem:*

$$p(A|B) = \frac{p(A, B)}{p(B)} \quad [6.8]$$

Equations 6.1–6.8 have wide applicability to a number of problems. As an example, suppose a student applied to two medical schools, A and B, and suppose further that she assessed the probability that she would get into school A as .200 and the probability that she would get into school B as .300. What is the probability that the student will be admitted to *at least one* of the two schools? Let A be "being admitted to school A" and B be "being admitted to school B." According to Equation 6.4, the probability of observing at least one of event A and event B is

$$p(A \text{ or } B) = p(A) + p(B) - p(A, B)$$

Assuming A and B are independent, we can use Equation 6.6 to compute $p(A, B)$:

$$\begin{aligned} p(A, B) &= p(A)p(B) \\ &= (.200)(.300) = .060 \end{aligned}$$

Then, using Equation 6.4, we find the probability that the student will be admitted to at least one of the two schools to be

$$p(A \text{ or } B) = .200 + .300 - .060 = .440$$

For an additional application of probability principles, see the Box on p. 131.

BOX BELIEFS AND PROBABILITY THEORY

A major concern of social psychologists has been the relationships among beliefs that an individual holds. How are a person's beliefs structured? If we change a belief, will this produce changes in other beliefs? If so, can we predict these changes?

One approach to the study of relationships among an individual's beliefs has been the application of probability theory. In this approach, beliefs are characterized as subjective probabilities (that is, probability judgments made by an individual). The belief that "nuclear energy is dangerous" is characterized by the individual's subjective probability that this statement is true. Some individuals might believe that nuclear energy is *definitely* dangerous, whereas others might believe that nuclear energy is *probably* dangerous. Still others might believe that it is not dangerous at all. In accord with probability theory, a subjective probability can range from .00 to 1.00. A subjective probability of .00 means that the individual is completely certain that a given statement is *not* true, whereas a subjective probability of 1.00 means that the individual is completely certain that a given statement *is* true. Increasing numbers between these values represent increasing degrees of certainty that the belief statement is true.

Some psychologists have suggested that, if beliefs are characterized as subjective probabilities, it may be possible to predict and understand these beliefs using probability theory. For example, consider the two statements

B: Nuclear power plants are dangerous.

$A|B$: Nuclear power plants should be shut down if they are dangerous.

It is possible to measure an individual's subjective probability concerning each of these statements. If subjective probabilities are organized in accord with probability theory, then knowledge of these two subjective probabilities should allow us to predict the individual's belief in the statement

A, B: Nuclear power plants are dangerous and they should be shut down.

Recall from Equation 6.5 that

$$p(A, B) = p(B)p(A|B)$$

Thus, the belief in statement A, B should be predictable from the product of the beliefs in statements B and $A|B$. If the individual believes that the probability nuclear power plants are dangerous is .90, and if he also believes with a probability of .95 that nuclear power plants should be shut down given that they are dangerous, then his belief that nuclear power plants are dangerous and should be shut down should be $(.95)(.90) = .855$.

Research on the relationship between subjective and actual probabilities has provided some interesting insights into belief organization and change. Although there is not always a strong correspondence between them, the degree of agreement has been striking in many respects. Interested readers are referred to Wyer and Goldberg (1970) for a discussion of this research.

6.6

Sampling With Versus Without Replacement

The preceding discussion of probability theory has important implications for researchers who draw random samples from populations. When cases are sampled from a population, two different situations are possible. First, a given case can be randomly selected, the measurement of interest taken, and the case then returned to the population. Then, with the population fully intact, the random selection procedure can be performed again. This process is called **sampling with replacement** because each case is "replaced" back into the population before another case is randomly selected. In contrast, **sampling without replacement** involves selecting a case at random and then, without replacing this case, selecting another case at random, and so on. The method of sampling—with versus without replacement—can affect the probability of observing some event.

Consider the data in Table 6.1 and suppose we are interested in determining the probability of randomly selecting first someone who is satisfied with his job and then someone who is dissatisfied with his job. In this case, event B is "selecting someone who is satisfied with his job as the first case" and event A is "selecting someone who is dissatisfied with his job as the second case."

Let us first consider the case of sampling with replacement. Before any case is selected, the probability of selecting someone who is satisfied with his job as the first case $[p(B)]$ is $210/410 = .512$. After this has been done, the case is returned to the population. This means that the outcome of the first random selection in no way affects the outcome of the second random selection—that is, the two events are independent. The probability of selecting someone who is dissatisfied with his job as the second case $[p(A)]$ is thus $200/410 = .488$. According to Equation 6.6, the joint probability of two independent events is

$$p(A, B) = p(A)p(B)$$

In our example,

$$p(A, B) = (.488)(.512) = .2499$$

Now consider the case of sampling without replacement. It is no longer the case that the outcome of the first selection is independent of the second selection. The probability that we will select someone who is satisfied with his job as the first case and someone who is dissatisfied with his job as the second case is now equal to

$$p(A, B) = p(B)p(A|B)$$

as per Equation 6.5. The probability of selecting someone who is satisfied with his job as the first case $[p(B)]$ is, as before, $210/410 = .512$. Since 200 of the 409 remaining men are dissatisfied with their jobs, the probability of selecting someone who is dissatisfied with his job as the second case given that we selected someone who is satisfied with his job as the first case $[p(A|B)]$ and did not replace him is $200/409 = .489$. Hence,

$$p(A, B) = (.512)(.489) = .2504$$

a value that is slightly greater than the probability computed on the basis of sampling with replacement.

If the size of the sample is small relative to the size of the population (for example, a ratio of 1 case in the sample for every 20 cases in the population), sampling with versus without replacement will not affect probabilities appreciably. Such was the case in the previous example, as the sample size was 1 and the population size was 410. However, when the size of the sample relative to the size of the population is large, the different sampling procedures can produce very different results. This should be kept in mind when interpreting the results of probability studies using sampling procedures.

6.7 Counting Rules

We have defined a probability as the number of observations favoring some event divided by the total number of observations. For some complex events, it would be very tedious to enumerate all possible outcomes in order to compute a probability. For example, if we wanted to state the number of ways we could seat five different people in each of five positions at a table in a study of social interaction, the task of enumerating each sequence would be quite laborious. Fortunately, there are some *counting rules* that permit such computations to be easily performed.

Suppose we have a set of three objects, A, B, and C, and we want to specify for this set all possible subsets of two of the objects. There are two different conditions under which we can attempt this task. One condition is where the ordering of the two objects matters, such that the subset "AB" is considered different from the subset "BA," since the two objects occur in a different order. The other condition is where ordering does not matter, and "AB" and "BA" are considered instances of the same event. The lists of all possible subsets of two objects under the two conditions are as follows:

Permutations (*ordering matters*)		*Combinations* (*ordering does not matter*)
AB	*BA*	*AB*
AC	*CA*	*AC*
BC	*CB*	*BC*

As can be seen, a **permutation** of a set of objects or events is an *ordered* sequence, whereas a **combination** of a set of objects or events is a sequence in which the internal ordering of elements is irrelevant. The general notation for the number of permutations (P) of n things taken r at a time is

$$_nP_r$$

In our problem, $n = 3$ since there are three objects, A, B, and C; and $r = 2$ since we are concerned with how we can order these three objects "taking two at a time." The general notation for the number of *combinations* (C) of n things taken r at a time is

$${}_{n}C_{r}$$

Before specifying the formulas for permutations and combinations, we must introduce one other concept, that of a **factorial** of a number. The factorial of a number, n, is formally defined as

$$n! = (n)(n-1)(n-2)(n-3)\ldots(1) \qquad [6.9]$$

where $n!$ stands for "n factorial." Consider the following examples:

$$6! = (6)(5)(4)(3)(2)(1) = 720$$
$$5! = (5)(4)(3)(2)(1) = 120$$
$$3! = (3)(2)(1) = 6$$

In factorial notation, the expression $0!$ equals 1.

The formula for calculating the number of *permutations* of n things taken r at a time is

$${}_{n}P_{r} = \frac{n!}{(n-r)!} \qquad [6.10]$$

This constitutes our first counting rule. In our problem, we have three objects (A, B, and C) and we want to specify the number of ordered sequences of two objects that can be derived from these. Using Equation 6.10, we find

$${}_{3}P_{2} = \frac{3!}{(3-2)!} = \frac{(3)(2)(1)}{(1)} = 6$$

This is consistent with the list presented in the previous table—there are six possible permutations of three objects taken two at a time (AB, BA, AC, CA, BC, CB).

The formula for calculating the number of *combinations* of n things taken r at a time is

$${}_{n}C_{r} = \frac{n!}{(n-r)!r!} \qquad [6.11]$$

This constitutes our second counting rule. For the problem with three objects taken two at a time, we find

$${}_{3}C_{2} = \frac{3!}{(3-2)!2!} = \frac{(3)(2)(1)}{(1)(2)(1)} = 3$$

Thus, consistent with the list presented in the previous table, there are three possible combinations of three objects taken two at a time (AB, AC, BC).

Appendix C presents factorials for the numbers 0 through 20. The use of this appendix can greatly simplify the calculation of the number of possible permutations or combinations of a set of objects or events.

A third useful counting rule can be specified as follows: If any one of r_i mutually exclusive and exhaustive events can occur on trial i, then the number of different sequences of events that can occur across n trials is $(r_1)(r_2) \ldots (r_n)$. Suppose you toss a coin on the first trial (two possible outcomes) and roll a die on the second trial (six possible outcomes). Then the total number of different sequences that can result is $(2)(6) = 12$ (for example, a head and a 1, a head and a 2, and so on).

Although these three counting rules might seem abstract, they can be used to good effect in a number of instances. Consider the following example on crime and statistics by Zeisel and Kalven (1978), which makes use of the third counting rule:

> This simple example comes from a Swedish trial on a charge of overtime parking. A policeman had noted the position of the valves of the front and rear tires on one side of the parked car, in the manner pilots note directions: one valve pointed, say, to one o'clock, the other to six o'clock, in both cases to the closest "hour." After the allowed time had run out, the car was still there, with the two valves still pointing toward one and six o'clock. In court, however, the accused denied any violation. He had left the parking place in time, he claimed, but had returned to it later, and the valves just happened to come to rest in the same positions as before. The court had an expert compute the probability of such a coincidence by chance. The answer was that the probability is 1 in 144 (12 × 12), because there are twelve positions for each of two wheels. In acquitting the defendant, the judge remarked that if all four wheels had been checked and found to point in the same directions as before, then the coincidence claim would have been rejected as too improbable and the defendant convicted: four wheels with twelve positions each can combine in 20,736 (= 12 × 12 × 12 × 12) different ways so the probability of a chance repetition of the original position would be only 1 in 20,736. Actually, these formulas probably understate the probability of a chance coincidence because they are based on the assumption that all four wheels rotate independently of each other, which, of course, they do not. On an idealized straight road all rotate together, in principle. It is only in the curves that the outside wheels turn more rapidly than the inside wheels, but even then the front and rear wheels on each side will presumably rotate about the same amount.

For those readers who attend horse races, the first counting rule could prove to be illuminating. A common bet at horse races is the "trifecta," or predicting which horses will come in first, second, and third in a given race. In a 10-horse race, how many ways are there that the positions of finish of 3 horses could be ordered, considering the entire field? Using our first counting rule,

$$
\begin{aligned}
{}_{10}P_3 &= \frac{10!}{(10-3)!} = \frac{(10)(9)(8)(7)(6)(5)(4)(3)(2)(1)}{(7)(6)(5)(4)(3)(2)(1)} \\
&= (10)(9)(8) = 720
\end{aligned}
$$

Only 1 of these 720 orders will actually occur. Thus, in the absence of any information about the horses, the probability of accurately predicting a trifecta in a 10-horse race is 1/720 or .0014. Even for individuals with racetrack savvy, betting a trifecta is likely to be a losing proposition.

6.8

The Binomial Expression

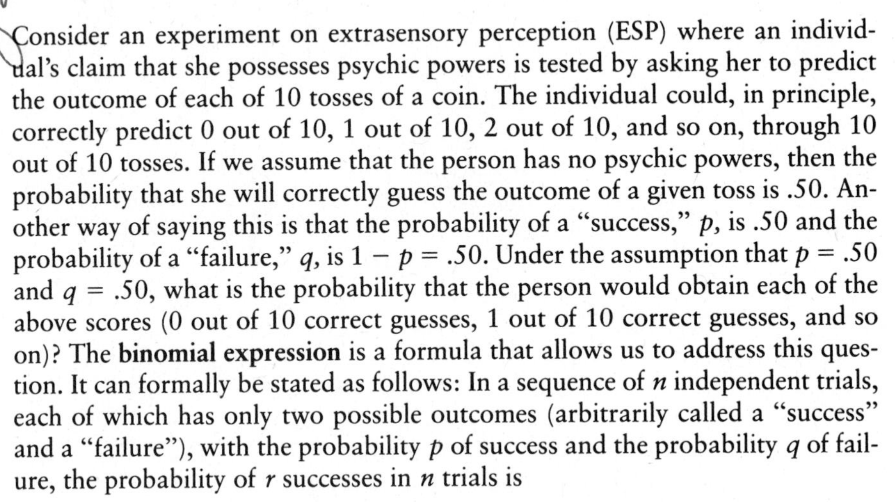

Consider an experiment on extrasensory perception (ESP) where an individual's claim that she possesses psychic powers is tested by asking her to predict the outcome of each of 10 tosses of a coin. The individual could, in principle, correctly predict 0 out of 10, 1 out of 10, 2 out of 10, and so on, through 10 out of 10 tosses. If we assume that the person has no psychic powers, then the probability that she will correctly guess the outcome of a given toss is .50. Another way of saying this is that the probability of a "success," p, is .50 and the probability of a "failure," q, is $1 - p = .50$. Under the assumption that $p = .50$ and $q = .50$, what is the probability that the person would obtain each of the above scores (0 out of 10 correct guesses, 1 out of 10 correct guesses, and so on)? The **binomial expression** is a formula that allows us to address this question. It can formally be stated as follows: In a sequence of n independent trials, each of which has only two possible outcomes (arbitrarily called a "success" and a "failure"), with the probability p of success and the probability q of failure, the probability of r successes in n trials is

$$p(r \text{ successes}) = \left[\frac{n!}{r!(n-r)!}\right] p^r q^{n-r} \qquad [6.12]$$

As an example, let us compute the probability of correctly predicting 10 out of 10 coin tosses. In this instance, the number of coin tosses, or trials (n), is 10 and the number of successes (r) is also 10. For $p = .50$ and $q = 1 - p = .50$, we find that

$$\begin{aligned} p(10 \text{ correct}) &= \left[\frac{10!}{10!(10-10)!}\right](.50^{10})(.50^{10-10}) \\ &= (1)(.001)(1) = .001 \end{aligned}$$

The probability of correctly predicting 9 out of 10 coin tosses is

$$\begin{aligned} p(9 \text{ correct}) &= \left[\frac{10!}{9!(10-9)!}\right](.50^{9})(.50^{10-9}) \\ &= (10)(.002)(.50) = .010 \end{aligned}$$

Column 2 of Table 6.2 presents the probability for each possible score in the ESP experiment under the assumption that $p = q = .50$.

TABLE 6.2 Probabilities of Various Scores in ESP Experiment

Number of correct predictions	*Probability*	*Probability of making the indicated number or more correct predictions*
10	.001	.001
9	.010	.011
8	.044	.055
7	.117	.172
6	.205	.377
5	.246	.623
4	.205	.828
3	.117	.945
2	.044	.989
1	.010	.999
0	.001	1.000

Use of the Binomial Expression in Hypothesis Testing

Before undertaking the ESP experiment, we will want to specify what scores are consistent with the conclusion that the individual possesses psychic powers. We will begin by assuming that she does *not* possess such ability. If this is true, we would expect her to accurately predict the outcome of the coin toss .50 of the time, or 5 out of 10 trials. We realize, of course, that due to chance, she might be slightly more accurate than this figure even if she does not actually possess the claimed powers. Thus, we would be unlikely to conclude that she is paranormal even if she were to accurately predict 6, or even 7, of the 10 coin tosses.

However, as the number of correct predictions increases, the probability that these outcomes are due to chance alone decreases. We note from column 3 of Table 6.2 that the probability of correctly guessing 8 or more tosses is $.044 + .010 + .001 = .055$, that the probability of correctly guessing 9 or more tosses is $.010 + .001 = .011$, and that the probability of correctly guessing all 10 tosses is .001. At some point, the probability that a set of outcomes is due to chance alone will be so small that, lacking evidence of chicanery, we will accept the claim of psychic powers.

In the context of **hypothesis testing,** the proposal, or *hypothesis,* that the individual does not possess psychic ability is called a **null hypothesis.** Assuming the null hypothesis is true, we can specify an expected result of an investigation. In our example, this takes the form that the individual will accurately predict 5 of the 10 coin tosses. If our observations are so discrepant from the expected result that the difference cannot be attributed to chance, we will reject the null hypothesis in lieu of a competing proposal, referred to as an **alternative hypothesis.** In our example, the alternative hypothesis is that the individual possesses the claimed psychic ability. On the other hand, if the observed result is similar enough to the outcome stated in the null hypothesis such that it can reasonably be attributed to chance, we will *fail to reject* the null hypothesis.

The problem thus becomes one of determining what constitutes chance versus nonchance results under the assumption that the null hypothesis is true.

This distinction is made with reference to a probability value known as an **alpha level.** For example, when the alpha level is .05, a result is defined as *nonchance* (that is, reflective of factors other than chance) if the probability of obtaining that result, assuming the null hypothesis is true, is less than .05.

Applying the .05 criterion to our example, we decide that correctly predicting 9 or more coin tosses will cause us to conclude that the results are consistent with an individual having psychic powers.* Fewer correct predictions will be attributed to chance guessing. Try flipping a coin 10 times and see if you can meet this criterion.

The Binomial and Normal Distributions

In experiments such as the ESP experiment, it is usually the case that more than 10 trials are used. Suppose 500 trials had been used and 280 correct predictions were made. Assuming $p = .50$, is this sufficient evidence for us to conclude that an individual possesses psychic power or can this outcome be attributed to chance? The application of the binomial expression would be an arithmetical nightmare in this instance. Fortunately, when certain conditions are met the normal distribution can be used to obtain a very close approximation to relevant binomial probabilities. For instance, Figure 6.1 presents a histogram of the probability distribution in Table 6.2 for the 10-trial ESP experiment. A normal distribution has been superimposed over the exact binomial probabilities of the score values. Note the similarity in the shapes of the distributions. It turns out that the binomial and normal distributions are closely related with the correspondence between them depending on the values of n and p. The correspondence improves as n increases and as p becomes closer to .50. Table 6.3 illustrates the relationship between the two distributions for our ESP example.

We will demonstrate how the similarity between normal and binomial probabilities can be applied to hypothesis testing by returning to the question raised earlier: Is the fact that an individual correctly predicts 280 of 500 coin tosses sufficient evidence for us to conclude that she possesses psychic powers, or can this outcome be attributed to chance? To answer this question, we must first determine the mean and standard deviation of the relevant binomial distribution. The mean of a binomial distribution is equal to

$$\mu = np \qquad [6.13]$$

where n is the number of trials and p is the probability of success. In the present instance,

$$\mu = (500)(.50) = 250.00$$

The standard deviation of a binomial distribution is equal to

$$\sigma = \sqrt{npq} \qquad [6.14]$$

* There may, however, be alternative explanations for such an outcome. All we really know is that something other than chance is probably operating and this "something" might be psychic ability. However, the observed outcome could also be the result of something else. This issue is considered in Chapter 9.

FIGURE 6.1 Relationship Between the Binomial and Normal Distributions for $n = 10$ and $p = .50$

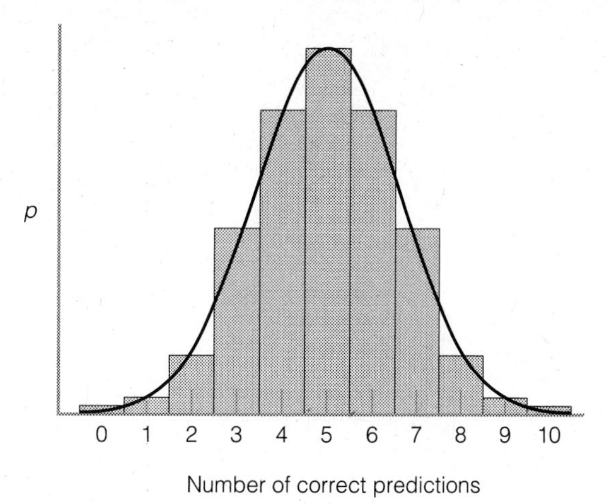

TABLE 6.3 Comparison of Binomial Probabilities and Corresponding Normal Approximations for ESP Example

Number of correct predictions	*Binomial probability*	*Normal approximation*
10	.001	.002
9	.010	.011
8	.044	.044
7	.117	.114
6	.205	.205
5	.246	.248
4	.205	.205
3	.117	.114
2	.044	.044
1	.010	.011
0	.001	.002

where $q = 1 - p$. In the present example,

$$\sigma = \sqrt{(500)(.50)(.50)} = 11.18$$

Since the distribution of the number of correct predictions for $n = 500$ and $p = .50$ will approximate a normal distribution, we can convert a score of 280 into a z score using Equation 4.6:

$$z = \frac{X - \mu}{\sigma}$$

$$= \frac{280 - 250.00}{11.18} = 2.68$$

Referring to Appendix B, we find that the probability of obtaining a z score of 2.68 or greater is .0037. This is less than the criterion value (alpha level) of .05. Thus, 280 correct predictions is consistent with the hypothesis that the person possesses psychic powers.

STUDY EXERCISE 6.5

A person who claims to have psychic powers tries to predict the outcome of a roll of a die on each of 100 trials. He correctly predicts 21 rolls. Using an alpha level of .05 as a criterion, what should we conclude about the person's claim?

Answer Since there are six numbers on a die, the probability of success on any one trial is $p = 1/6 = .167$, and the probability of failure is $q = 1 - p = 1 - .167 = .833$. The mean of the relevant binomial distribution is thus

$$\mu = np = (100)(.167) = 16.70$$

and the standard deviation is

$$\sigma = \sqrt{npq} = \sqrt{(100)(.167)(.833)} = 3.73$$

The binomial probability associated with a score of 21 can be approximated by converting this score into a z score and deriving the corresponding normal probability from Appendix B. A score of 21 translates into a z score of

$$z = \frac{X - \mu}{\sigma} = \frac{21 - 16.70}{3.73} = 1.15$$

From Appendix B, the probability of obtaining a z score of 1.15 or greater is .1251. Since .1251 is not less than the criterion value of .05, we conclude that the person's performance does not allow us to say that he has the claimed powers. In other words, there is insufficient evidence to conclude that he performed at an above-chance level.

6.9 Summary

Among the probabilities of interest to behavioral scientists are the probability of a simple event, conditional probability, and joint probability. The probability of a simple event refers to the likelihood that a given event will occur. A conditional probability tells us the likelihood that a given event will occur given that some other event occurs. A joint probability indicates the likelihood of observing each of two events. A fourth probability of interest focuses on the likelihood of observing at least one of two events. These concepts can be used to gain considerable insights into relationships between events.

When cases are sampled from a population, two different situations are possible. In sampling with replacement, each case is "replaced" back into the population before another case is randomly selected. In contrast, sampling without replacement involves selecting a case at random and then, without replacing this case, selecting another case at random, and so on. Whether sampling is done with versus without replacement can affect the probability of observing some event.

The number of subsets of a set of objects or events can be specified using either of two counting rules. First, we can determine the number of permutations, or ordered sequences. Second, we can determine the number of combinations, or sequences in which the internal ordering of elements is irrelevant. A third counting rule specifies the number of different sequences of mutually exclusive and exhaustive events that can occur across trials.

The binomial expression can be used to determine the probability of various numbers of "successes" and "failures" across trials in situations where trials are independent and each has only two outcomes. One important use of the binomial expression is in hypothesis testing. This involves the specification of an expected result of an investigation under the assumption that some proposal, known as a null hypothesis, is true. If the observations are so discrepant from the expected result that the difference cannot be attributed to chance, the null hypothesis will be rejected in lieu of a competing proposal, known as an alternative hypothesis. Otherwise, we will fail to reject the null hypothesis. The decision as to whether to reject the null hypothesis is made with reference to a probability value known as an alpha level.

When the application of the binomial expression is not practical, under certain conditions the normal distribution can be used to obtain a very close approximation to relevant binomial probabilities. Given the mean and standard deviation of the binomial distribution of interest, our knowledge of the normal distribution can thus be used to test whether an observed outcome is due to chance or nonchance factors.

Exercises

Answers to asterisked (*) exercises appear at the back of the book. Answers to exercises with two asterisks are also worked out step-by-step in the Study Guide.

1. What are the two conceptualizations of probability?

Refer to the following information to answer Exercises 2–10.

The Equal Rights Amendment (ERA) has been the subject of considerable controversy in the United States. The proposed amendment to the constitution concerns equal rights for both genders and consists of three simple statements: "Equality of rights under the law shall not be denied or abridged by the United States or by any State on account of sex. The Congress shall have power to enforce, by appropriate legislation, the provisions of this article. This amendment shall take effect two years after the date of ratification." A behavioral scientist was interested in the extent to which people in a particular city favored or opposed the ERA and, accordingly, interviewed a random sample of 350 people from this city. The number of males and females supporting or opposing the ERA is summarized in the following contingency table:

	GENDER	
POSITION	*Male*	*Female*
Favors ERA	60	120
Opposes ERA	115	55

***2.** What is the marginal frequency for people who favor the ERA?

3. What is the marginal frequency for females?

***4.** How many observations are represented?

***5.** For the given data, what is the probability that an individual favors the ERA? What is the probability that an individual opposes the ERA? What is the probability that an individual is a male? A female?

***6.** What is the probability that an individual favors the ERA given that the individual is a male? What is the probability that an individual favors the ERA given that the individual is a female? What is the probability that an individual opposes the ERA given that the individual is a male? What is the probability that an individual opposes the ERA given that the individual is a female?

***7.** Are being a male and favoring the ERA independent? Why or why not? Be specific.

***8.** What is the probability that an individual is a male who favors the ERA? What is the probability that an individual is a female who opposes the ERA?

***9.** How many joint probabilities are represented? Name them.

***10.** What is the probability that an individual is a male or favors the ERA, or both? What is the probability that an individual is a female or opposes the ERA, or both?

Refer to the following information to answer Exercises 11–16.

A behavioral scientist interviewed a random sample of 500 people in order to study the relationship between individuals' political party identification and their attitudes toward legal abortions. The following contingency table was observed:

	ATTITUDE	
PARTY IDENTIFICATION	*Favors legal abortions*	*Opposes legal abortions*
Democrat	160	40
Republican	40	160
Independent	75	25

11. For the given data, what is the probability that an individual favors legal abortions? What is the probability that an individual opposes legal abortions? What is the probability that an individual is a Democrat? A Republican? An Independent?

12. What is the probability that an individual favors legal abortions given that the individual is a Democrat? What is the probability that an individual identifies with the Democratic party given that the individual favors legal abortions? Why are these probabilities different?

13. Are being a Democrat and opposing legal abortions independent? Why or why not? Be specific.

14. What is the probability that an individual is a Republican who favors legal abortions? What is the probability that an individual is a Republican who opposes legal abortions?

15. How many joint probabilities are represented? Name them.

16. What is the probability that an individual is a Democrat or favors legal abortions, or both? What is the probability that an individual is a Democrat or opposes legal abortions, or both? What is the probability that an individual is a Republican or favors legal abortions, or both? What is the probability that an individual is an Independent or opposes legal abortions, or both?

*17. If the probability of some event, B, is .542 and the probability of some second event, A, given event B, is .896, what is the probability of observing *both* event A *and* event B?

18. If two events, A and B, are independent and the probability of A is .40 and the probability of B is .30, what is the probability of observing *both* event A *and* event B?

19. If two events, A and B, are mutually exclusive and the probability of A is .278 and the probability of B is .581, what is the probability of observing *at least one* of event A and event B?

*20. If the probability of some event, B, is .349 and the probability of observing B along with some second event, A, is .180, what is the probability of event A, given event B?

21. Explain the difference between sampling with versus without replacement.

22. Explain the difference between a permutation and a combination.

*23. Compute the following permutations:

a. ${}_5P_3$ b. ${}_6P_5$ c. ${}_4P_2$ d. ${}_5P_5$ e. ${}_4P_4$

*24. Compute 5!. Compare your answer with the answer to part **d** of Exercise 23. Compute 4!. Compare your answer with the answer to part **e** of Exercise 23. What generalization can you make from this?

25. Suppose you are doing an experiment that involves showing five different slides to a group of subjects. You are concerned that the order in which you present the slides could affect the outcome of the experiment. In how many ordered sequences can the slides be presented? How might you deal with the problem of order effects in this instance?

*26. Compute the following combinations:

a. ${}_5C_3$ b. ${}_6C_5$ c. ${}_4C_2$ d. ${}_5C_5$ e. ${}_4C_4$

27. A researcher was interested in the effects of four different independent variables on a dependent variable. For practical reasons, she could study only two of the variables. How many different combinations of two variables are there that she could potentially study? Suppose she could study three variables. How many different combinations of three variables are there that she could potentially study?

*28. How many different sequences of events can occur on three rolls of a die?

29. How many different sequences of events can occur if one card is drawn from each of three standard bridge decks?

*30. If a student simply made random responses on a 10-item true-false test, what is the probability he or she would:

a. get 9 correct
b. get 8 correct
c. get 7 or more correct
d. get 1 correct
e. get 2 correct

31. Sixty-five percent of people eligible to vote in the 1976 presidential election did not do so. If we obtain a random sample of 10 eligible voters and X is the number of people who did not vote, compute the probability distribution for possible values of X.

32. Define each of the following:

a. null hypothesis
b. alternative hypothesis
c. alpha level

*33. What factors influence the correspondence between the binomial and normal distributions?

*34. Calculate the mean and standard deviation of a binomial distribution for $n = 150$ and $p = .60$.

35. Calculate the mean and standard deviation of a binomial distribution for $n = 156$ and $p = .80$.

**36. Twenty percent of individuals who seek psychotherapy will exhibit a return to normal personality irrespective of whether they receive treatment (a phenomenon called spontaneous recovery). A researcher finds that a particular type of psychotherapy is successful with 30 out of 100 clients. Using an alpha level of .05 as a criterion, what should she conclude about the effectiveness of this psychotherapeutic approach?

37. A student takes a 10-item multiple-choice quiz having four response options for each question and gets five items correct. Using an alpha level of .05 as a criterion, can we conclude that the student performed at an above-chance level?

SEVEN

ESTIMATION AND SAMPLING DISTRIBUTIONS

Suppose you wanted to describe the annual income of all married women in the United States. It would be impossible to contact each of these women in order to determine this information. Instead, you might decide to select a random sample and then try to make generalizations about the population based on the sample. The problem of using sample data to make inferences about populations is fundamental to statistical techniques used in the behavioral sciences. This chapter is concerned with issues in estimating population parameters from sample statistics.

7.1

Finite Versus Infinite Populations

Estimating population parameters can be accomplished with reference either to small, finite populations, or to populations that are infinite in size (or that are so large that for all practical purposes they can be considered infinite). Most applications of statistics in the behavioral sciences are concerned with very large or infinite-sized populations. Behavioral science research is typically conducted with the goal of explaining the behavior of large numbers of individuals, often including people of the past and the future as well as the present. As such, the statistics in most behavioral science disciplines are applicable to extremely large, if not infinite, populations.

Given that behavioral science research is concerned with very large or infinite-sized populations, it is impossible for investigators to select truly random samples from such populations. Recall from Chapter 1 that a random sample requires listing all members of the population and then using a random number table to select a sample. Because this is not practical for very large populations, behavioral scientists typically select a set of individuals to study, and then *assume* that these individuals are randomly drawn from *some* population. Exactly what population the individuals are assumed to represent depends, in part, on the characteristics of the subjects in the study. If a learning experiment was conducted on 200 college students at a midwestern university, the investigator might want to generalize his or her results to people in general. In this case, the population is conceptualized as consisting of "all people" and the college students are said to represent a random sample from this population. Obviously this is a very questionable conceptualization. Perhaps the population should be conceptualized in terms of "all college students." Or maybe it should be conceptualized in terms of "all college students at midwestern universities." Or maybe it should be conceptualized in even more specific terms. Behavioral scientists sometimes disagree about what the appropriate population is for purposes of generalization based on an investigation of a small set of individuals. When interpreting research, you should always keep in mind who the results should generalize to (in other words, exactly what the relevant population is).

For the sake of illustration, the examples in this chapter will use small, finite populations. Sampling from a small, finite population *with replacement* (that is, where each sample member is returned to the population before the next sample member is selected) is analogous to sampling from a very large or infinite population, as behavioral science research is conceptualized as doing. Accordingly, when developing principles for estimating population parameters using finite populations, sampling will be done with replacement to mimic the case of infinite populations.*

7.2 Estimation of the Population Mean

Consider the case of 100 families in a small town. An investigator wants to describe the inhabitants in terms of the average number of children in each family. Because of practical limitations, the investigator is unable to interview all 100 families, so she instead resorts to a sample. For the sake of convenience, let us say that the sample size will be 10. In this case, the *population* is the 100 families in the town and the *sample* is the 10 families who are selected to be interviewed. Table 7.1 lists the number of children in each family of the population. Although the 100 family sizes are unknown to the investigator, the fact remains that there *are* 100 families in the population and that each one has the indicated number of children. In other words, if the investigator *were* able to interview all 100 families she would obtain the scores (family sizes) listed in Table 7.1. The true population mean, μ, is 3.50, and the variance, σ^2, is 2.09. Again, the investigator is unaware of the values of μ and σ^2.

Let us randomly select 10 families from this set of 100 and let these represent the 10 families interviewed by the investigator. Suppose the scores for these families are found to be 3, 4, 4, 5, 2, 4, 1, 1, 4, 3, with a mean equal to

* For a discussion of statistical applications to finite populations, see Hays (1981).

TABLE 7.1 Number of Children in Families for a Hypothetical Population

3	6	5	4	4	1	3	4	4	5
4	4	6	4	3	3	5	3	4	4
4	3	4	3	5	3	4	3	5	2
2	3	2	5	4	2	3	5	2	3
0	2	4	3	2	4	2	2	1	3
4	3	3	2	4	5	4	4	2	4
3	5	4	5	2	3	5	3	4	3
2	4	3	7	5	1	4	4	5	4
5	2	2	3	7	4	7	3	3	1
2	0	7	4	3	6	5	2	3	1

$\mu = 3.50$
$\sigma^2 = 2.09$

3.10. Note that this value is *not* equal to the true population mean. The fact that a sample statistic may not equal the value of its corresponding population parameter is said to be the result of **sampling error.** The term "sampling error" reflects the fact that sample values are likely to differ from population values because they *are* based on only a portion of the overall population, and does not imply that mistakes have been made in the collection and analysis of the data. The amount of sampling error can be represented as the difference between the value of a sample statistic (in the present context, $\bar{X}$) and the value of the corresponding population parameter (in the present context, μ). In the above example, the amount of sampling error is $3.10 - 3.50 = -.40$. In practice, an investigator does not know the value of the population parameter, in which case it is impossible to compute the exact amount of sampling error that occurs.

In the absence of any other information, the sample mean that one observes is one's "best guess" about the value of the population mean. One justification is based on an important statistical property of the mean. This property can be illustrated with reference to the family size example above. Suppose in addition to the first sample, we select another random sample of size 10 from the population of 100 families. The scores for this sample are 1, 3, 3, 4, 3, 7, 5, 1, 3, 4, with a mean of 3.40. Suppose this process is repeated over and over until we have computed the mean score for *every possible random sample of size 10* that could be selected from this population. Table 7.2 illustrates 15 of the many sample means that would be observed and the amount of sampling error that occurs in each instance (that is, $\bar{X} - \mu$). If we were to compute the average (mean) amount of sampling error that resulted across all possible samples of size 10, we would find that the average would equal 0. In other words, some of the sample means overestimate the true population mean while others underestimate it. Across all samples of size 10, however, the overestimations cancel the underestimations and the *average* amount of error is 0. This property of the sample mean makes it a useful estimate of the population mean: On the average, as indexed by the differences between the sample means and the population mean, the amount of sampling error across all possible samples

TABLE 7.2 Illustrative Mean Scores for Samples of Size 10 Randomly Selected from the Population Represented in Table 7.1

Sample scores	$\bar{X}$	$\bar{X} - \mu$
3, 4, 4, 5, 2, 4, 1, 1, 4, 3	3.10	− .40
1, 3, 3, 4, 3, 7, 5, 1, 3, 4	3.40	− .10
3, 5, 4, 3, 2, 4, 5, 4, 3, 2	3.50	.00
0, 2, 7, 2, 5, 2, 5, 5, 4, 3	3.50	.00
6, 3, 4, 4, 4, 2, 3, 3, 4, 3	3.60	.10
3, 3, 4, 4, 5, 3, 2, 2, 3, 3	3.20	− .30
7, 4, 7, 3, 3, 3, 5, 4, 5, 2	4.30	.80
5, 4, 5, 2, 2, 3, 4, 3, 5, 3	3.60	.10
6, 7, 4, 4, 4, 5, 4, 4, 2, 2	4.20	.70
2, 2, 3, 3, 1, 2, 3, 2, 3, 4	2.50	−1.00
3, 4, 3, 4, 3, 4, 3, 4, 3, 4	3.50	.00
1, 3, 3, 2, 4, 3, 4, 3, 2, 0	2.50	−1.00
4, 6, 4, 3, 3, 4, 4, 5, 4, 1	3.80	.30
2, 2, 2, 2, 5, 5, 5, 3, 5, 4	3.50	.00
6, 7, 7, 4, 3, 0, 5, 5, 4, 2	4.30	.80

of a given size will always equal 0. In statistical terms, the sample mean is said to be an *unbiased estimator* of the population mean. An **unbiased estimator** of a population parameter is one whose average (mean) over all possible random samples of a given size equals the value of the parameter.

7.3

Estimation of the Population Variance and Standard Deviation

In the previous section, we selected a sample of size 10 from the population of 100 families and computed a mean of 3.10. The variance of the 10 scores in this sample is 1.69. This value is different from the true population variance ($\sigma^2 = 2.09$) and, again, reflects sampling error. Unlike the sample mean, however, the sample variance is *not* our best guess of the true population variance. Statisticians have determined that the sample variance is a *biased estimator* of the population variance in that it underestimates (is smaller than) the population variance across all possible samples of a given size.

Equation 3.4 defines the variance of a set of scores as the sum of squares divided by N:

$$s^2 = \frac{SS}{N}$$

As noted above, this is not our best estimate of the population variance since it is biased. However, an unbiased estimator can be obtained from sample data by making a modification in the denominator of Equation 3.4:

$$\hat{s}^2 = \frac{SS}{N - 1} \qquad [7.1]$$

where $\hat{s}^2$ (read "s-hat squared") is the symbol used to represent a sample estimate of the population variance, or as it is more commonly called, a **variance estimate.** Equation 7.1 is identical to Equation 3.4 except that $N - 1$ is used in the denominator instead of N. This "correction" involving the subtraction of 1 from N serves to make the variance estimate larger than the sample variance and, hence, corrects for the tendency of the sample variance to underestimate the population variance.

Consider the 10 scores from the sample described at the beginning of Section 7.2: 3, 4, 4, 5, 2, 4, 1, 1, 4, 3. It was noted above that the variance of these scores is 1.69. This reflects the fact that the sum of squares for the data set is 16.90. Applying Equation 7.1, the variance estimate is

$$\hat{s}^2 = \frac{SS}{N - 1}$$

$$= \frac{16.90}{10 - 1} = 1.88$$

Our best guess as to the value of the population variance, based on these data, would thus be 1.88. Again, we recognize that there is sampling error involved and that the variance estimate will usually not exactly equal the true population variance. Note, however, that the variance estimate ($\hat{s}^2 = 1.88$) in our example is closer to the population variance ($\sigma^2 = 2.09$) than is the sample variance ($s^2 = 1.69$).

The sample standard deviation is defined as the positive square root of the variance, or $\sqrt{s^2}$. By the same token, the estimate of the population standard deviation is the positive square root of $\hat{s}^2$, or

$$\hat{s} = \sqrt{\hat{s}^2} \qquad [7.2]$$

In the above example, $s = \sqrt{1.69} = 1.30$ and $\hat{s} = \sqrt{1.88} = 1.37$. Our best guess about the standard deviation of the population is thus 1.37. The symbol $\hat{s}$ is referred to as the **standard deviation estimate.***

---

* Technically, $\hat{s}$ is not an unbiased estimator of σ, and a correction factor is necessary to make $\hat{s}$ an unbiased estimator (Hays, 1981, p. 189). However, if $N > 10$, the amount of bias in $\hat{s}$ tends to be small. Given this and the fact that the statistics to be used in later chapters circumvent potential problems introduced by this bias, we will use $\hat{s}$ as the sample estimate of the population standard deviation.

STUDY EXERCISE 7.1

A sample of 10 individuals took an intelligence test on which scores could range from 0 to 150. This sample yielded a mean score of 100.00 and a sum of squares of 50. Compute the variance and the standard deviation for the sample, and estimate the variance and the standard deviation in the population.

Answer The sample variance is the sum of squares divided by N, whereas the variance estimate is the sum of squares divided by $N - 1$. Thus,

$$s^2 = \frac{50}{10} = 5.00 \qquad \hat{s}^2 = \frac{50}{9} = 5.56$$

The respective standard deviations are the square roots of the variances:

$$s = \sqrt{5.00} = 2.24 \qquad \hat{s} = \sqrt{5.56} = 2.36$$

The preceding relationships are summarized in the following table:

Statistical term	*Sample value*	*Population value*	*Sample estimate of population value*
Mean	$\bar{X} = \frac{\Sigma X}{N}$	$\mu = \frac{\Sigma X}{N}$	$\bar{X} = \frac{\Sigma X}{N}$
Variance	$s^2 = \frac{SS}{N}$	$\sigma^2 = \frac{SS}{N}$	$\hat{s}^2 = \frac{SS}{N - 1}$
Standard deviation	$s = \sqrt{s^2}$	$\sigma = \sqrt{\sigma^2}$	$\hat{s} = \sqrt{\hat{s}^2}$

If we had scores for *all* members of a population, we would use the formulas for population values. This situation is extremely rare in the behavioral sciences. If we had scores for a subset of a population and were interested only in describing that subset *without making inferences to the population,* we would use the formulas for sample values. This situation is also extremely rare in the behavioral sciences. By far the most common occurrence in the behavioral sciences involves the estimation of population parameters from sample data. To accomplish this, we use the formulas for sample estimates of population values.

Some texts do not distinguish between the sample variance and sample standard deviation, on the one hand, and the variance estimate and standard deviation estimate, on the other. Rather, the sample estimates, $\hat{s}^2$ and $\hat{s}$, are introduced as *the* measures of variability for a sample. For some applications, this might lead to confusion, and hence, the more traditional distinctions are maintained here.

7.4

Degrees of Freedom

An important concept in statistical estimation is that of *degrees of freedom*. Suppose you are told that a friend received a score of 95 on a statistics exam. Suppose you are also told that there were 100 points possible and that your friend missed 5 points. You have been given three pieces of information:

1. Your friend received a score of 95 on a test.
2. The test had 100 points possible.
3. Your friend missed 5 points.

If you had been told any two of the above pieces of information, you could have deduced the third (a score of 95 out of 100 means 5 points were missed). Thus, you have actually been given two independent pieces of information and one piece of information that is dependent on (follows from) the other two. Which two of the three are called "independent" and which of the three is called "dependent" are arbitrary. The fact of the matter is that only two pieces of information are independent. In statistics, the phrase **degrees of freedom** is used to indicate the number of pieces of information that are "free of each other" in the sense that they cannot be deduced from one another. In the above example, there are two degrees of freedom, since there are two pieces of independent information.

Statistical indices such as the sum of squares and the variance have a certain number of degrees of freedom since they are based on a certain number of pieces of information (scores on the variable of interest) that are independent of one another. Let's consider the sum of squares. When computing a sum of squares it is necessary to derive deviation scores about the mean. Consider the following four scores: 8, 10, 10, 12. The mean of these scores is 10.00, and if we were to compute the deviation scores of the first three scores, we would obtain -2, 0, and 0. Recall from Chapter 3 that the sum of deviation scores about the mean will always equal 0. If you are told that there are four scores in a distribution and that the deviation scores for three of these are -2, 0, and 0, you know that the last deviation score must be 2. The last deviation score to be computed is not *free to vary* but is determined by the previous deviation scores. Thus, there are four scores and $4 - 1 = 3$ degrees of freedom in this example. This is illustrative of the fact that a sum of squares around a sample mean will always have $N - 1$ degrees of freedom associated with it.

The above leads us to a basic principle in estimation techniques: *As the degrees of freedom associated with an estimate increase, the accuracy of the estimate also tends to increase.* Of concern here is the distinction between the number of degrees of freedom involved as opposed to the total number of pieces of information. When an estimate is based on a large number of independent pieces of information, then it will tend to be relatively accurate. However, if the pieces of information tend to be dependent on one another, there is greater room for error. Technically, the accuracy of a variance estimate is not a function of the sample size (N), but rather is a function of the degrees of freedom ($N - 1$)—that is, the number of independent pieces of information—used in calculating such an estimate.

In later chapters, we will encounter types of sums of squares in which deviations are taken around entities other than the sample mean. As will be discussed, the degrees of freedom for these sums of squares will typically be equal to quantities other than $N - 1$.

Irrespective of its specific computation, any sum of squares divided by its associated degrees of freedom is referred to as a mean sum of squares, or **mean square.** Symbolically,

$$\text{MS} = \frac{\text{SS}}{\text{df}} \qquad [7.3]$$

where MS represents a mean square, SS represents the sum of squared deviations around some entity, and df represents the degrees of freedom. As we have seen, the mean square derived by dividing the sum of squares around a sample mean by its corresponding degrees of freedom, $N - 1$, is formally known as a variance estimate.

To summarize thus far, a sample statistic may differ from the value of its corresponding population parameter because of sampling error. The mean and the variance estimate are unbiased estimators of the population mean and population variance, respectively. An unbiased estimator is one whose average (mean) across all possible random samples of a given size is equal to the value of the population parameter. In general, as the degrees of freedom associated with a sample statistic increase, the more accurate that statistic will be in estimating the corresponding population parameter.

7.5

Sampling Distribution of the Mean and the Central Limit Theorem

As previously noted, a sample mean will often differ from the corresponding population mean because of sampling error. For a given population of scores, a useful index of the degree to which sampling error affects the accuracy of the sample mean as an estimator of the population mean can be obtained by selecting all possible samples of a specified size, calculating the mean score for each sample, and then computing the standard deviation of the sample means across all samples.

For the sake of illustration, suppose we are trying to estimate the mean number of children for the population of 100 families alluded to earlier from a sample of 10 cases. As indicated in Table 7.1, the overall population mean is 3.50. If we randomly select a sample of 10 families, we might observe a mean score of, say, 3.10. Although we would not actually do so in practice, suppose we were to select another random sample of size 10 from the 100 families and find the mean to be 3.70. Further suppose that we repeated this process over and over until we computed the mean score for all possible samples of size 10. We would then have a distribution of "scores" composed of the mean scores for all possible samples of size 10. Such a distribution of scores, consisting of the means for all possible samples of a given size, is called a **sampling distribution of the mean.** If we wanted, we could compute a standard deviation for these scores just as we could for any other set of scores. This would constitute our measure of sampling error. We could also compute a mean for these scores (that is, the mean of the sample means), if we so desired.

In practice, we would never actually calculate a sampling distribution of the mean. It is not necessary to do so because statisticians have determined that all sampling distributions of this type possess certain common characteristics that allow us to readily identify the extent of sampling error, as well as other important information. Thus, a sampling distribution of the mean can be formally defined as a *theoretical distribution consisting of the mean scores of all possible samples of a given size that could be drawn from a population.*

The mean and standard deviation of a sampling distribution of the mean, as well as its shape, are defined by an important formulation called the **central limit theorem.** This theorem can be stated as follows:

> *The sampling distribution of the mean for a population having a mean of μ and a standard deviation of σ has a mean of μ, a standard deviation of $\sigma/\sqrt{N}$ (and, thus, a variance of σ^2/N), and approaches a normal distribution as the sample size on which it is based, N, becomes larger.*

The Mean of the Sampling Distribution of the Mean

As stated in the central limit theorem, *the mean of a sampling distribution of the mean is always equal to the population mean* (of the raw scores). The reason for this can be stated intuitively. When we select our samples of a given size, some of the sample means will overestimate the true population mean and others will underestimate it. However, when we average all of these, the underestimations will cancel the overestimations, with the result being the true population mean. If we were, for instance, to compute the mean of all possible sample means in the family size example, it would equal the population mean of 3.50. This characteristic of a sampling distribution of means is very important for the statistical concepts to be developed later.

The Standard Deviation of the Sampling Distribution of the Mean

The standard deviation of a sampling distribution of the mean is called the **standard error of the mean.** The standard error of the mean reflects the accuracy with which sample means estimate a population mean. Recall from Chapter 3 that a standard deviation represents an average deviation from the mean of a distribution. If the mean in this case is the true population mean, μ, then the standard error of the mean represents the average deviation of the sample means from the population mean. If the standard error of the mean is small, then all of the sample means based on a given N tend to yield the same result and all are close to the population mean. If the standard error of the mean is large, then the sample means based on N tend to yield very different results, some of which will and some of which will not be close to the population mean.

If the standard error of the mean for samples of size 10 is 1.30 in the family size example, this would mean that on the average, sample means differ 1.30 units (in this case, "children") from the true population mean. This indicates a sizeable amount of error, and one would have to be concerned with how similar the particular observed sample mean is to the actual population mean. If, on the other hand, the standard error of the mean for samples of size 10 is only .10, this would mean that, on the average, sample means deviate only .10 unit (or "child") from the true population mean. You would now have reason to believe that the sample mean is a reasonably accurate estimator of the population mean. The utility of the concept of the standard error of the mean is that statisticians have developed a method for estimating it based on sample data. This helps us to interpret the sample mean relative to the population mean.

As defined by the central limit theorem, the standard deviation of a sampling distribution of the mean is equal to

$$\sigma_{\bar{X}} = \frac{\sigma}{\sqrt{N}} \qquad [7.4]$$

where $\sigma_{\bar{X}}$ is the symbol used to represent the standard error of the mean, σ is the standard deviation of scores in the population, and N is the sample size. Equation 7.4 indicates that two factors influence the size of the standard error of the mean. The first is the sample size. As the sample size increases, the standard error becomes smaller, other things being equal. If in the example on family sizes we had used a sample size of 99 instead of 10, it is clear that the sampling distribution would consist of sample means that are very similar to one another, since each would be based on 99 of the same 100 scores. This would result in a relatively small value of $\sigma_{\bar{X}}$. However, with smaller sample sizes, there is likely to be greater variability among the sample means. The second factor that influences the size of the standard error is the variability of scores in the population. As σ decreases, the standard error becomes smaller, other things being equal. To take an extreme example, if all of the scores in the population were identical (that is, if $\sigma = 0$), then every sample would yield a mean exactly equal to the population mean. For instance, if all scores were equal to 4, then the mean of any sample would also be equal to 4, as would the population mean. This would result in a value of $\sigma_{\bar{X}}$ of 0. However, when there is considerable variability in the population, the mean of a given sample might be influenced by the inclusion of one or more extreme scores. This will result in greater variability among the sample means and, consequently, a larger standard error of the mean.

In practice, we are rarely in a position where the population standard deviation, σ, is known. Rather, we typically have data based on one sample. We must therefore estimate the standard error of the mean by substituting the sample estimate of the population standard deviation, $\hat{s}$, into Equation 7.4 for σ:

$$\hat{s}_{\bar{X}} = \frac{\hat{s}}{\sqrt{N}} \quad [7.5]$$

where $\hat{s}_{\bar{X}}$ is the estimate of the standard error of the mean and all other terms are as previously defined. Let us consider a numerical example.

Suppose we have two populations, A and B. A sample of 10 individuals is interviewed from population A and a sample of 10 individuals is interviewed from population B. Both samples are randomly chosen. For each sample, a measure of each person's age is obtained. These data are reported in Table 7.3.

TABLE 7.3 Numerical Example for Estimating the Standard Error of the Mean

SAMPLE A		SAMPLE B	
X	X^2	X	X^2
19	361	26	676
21	441	30	900
20	400	34	1,156
19	361	35	1,225
20	400	25	625
21	441	30	900
21	441	24	576
20	400	30	900
20	400	30	900
19	361	36	1,296
$\Sigma X = 200$	$\Sigma X^2 = 4{,}006$	$\Sigma X = 300$	$\Sigma X^2 = 9{,}154$
$\bar{X} = 20.00$		$\bar{X} = 30.00$	

Sample A:

$$SS = \Sigma X^2 - \frac{(\Sigma X)^2}{N}$$

$$= 4{,}006 - \frac{200^2}{10} = 6$$

$$\hat{s}^2 = \frac{SS}{N-1} = \frac{6}{9} = .67$$

$$\hat{s} = \sqrt{\hat{s}^2} = \sqrt{.67} = .82$$

$$\hat{s}_{\bar{X}} = \frac{\hat{s}}{\sqrt{N}} = \frac{.82}{\sqrt{10}} = .26$$

Sample B:

$$SS = \Sigma X^2 - \frac{(\Sigma X)^2}{N}$$

$$= 9{,}154 - \frac{300^2}{10} = 154$$

$$\hat{s}^2 = \frac{SS}{N-1} = \frac{154}{9} = 17.11$$

$$\hat{s} = \sqrt{\hat{s}^2} = \sqrt{17.11} = 4.14$$

$$\hat{s}_{\bar{X}} = \frac{\hat{s}}{\sqrt{N}} = \frac{4.14}{\sqrt{10}} = 1.31$$

BOX POLLS AND RANDOM SAMPLES

Most of us have encountered the concept of sampling error in the context of political polls published in newspapers. A consideration of sampling procedures used by major polling agencies and how sampling error affects their results may be useful. Recall from Chapter 1 that a random sample requires a list of every member of the population. Such a list is often not feasible, or even possible, to construct. For example, no master list of names of all people in the United States currently exists, and even if it did, such a list would become dated very quickly. Because of this, many polling agencies use a variant of random sampling, called *area sampling,* when doing national polls. In this method, the United States is first divided into a large number of homogeneous geographic regions. A subset of these regions is then randomly selected using a random number table. Each of these regions is then subdivided into a large number of geographic areas (based on census tracts or districts). A subset of these areas is then randomly selected. Within each area a list of the city or town streets is compiled. A subset of streets is then randomly selected. Finally, on each of the chosen streets, the house addresses are listed, and a subset of these is, in turn, randomly selected. The interviewer then approaches each house and randomly selects an individual within it. This completes the process. The overall region is selected at random. The tract is selected at random. The streets are selected at random. The houses are selected at random. And the individuals are selected at random. The end result is that every individual ideally has an equal chance of being included in the sample.

Although this is true in theory, it does not really hold in practice. The chosen dwelling might be vacant. Or the individual within the dwelling may not agree to participate in the poll. No nationwide survey has ever reached every person designated in a random sample and, accordingly, special "correction" factors must be adopted. A detailed consideration of these is presented in Gallup (1976). The process of area sampling is an extremely expensive one and is usually carried out only by major polling firms. Many smaller companies, newspapers, and magazines obtain samples in ways that call their representativeness into question. The results of public opinion polls must always be interpreted in light of their sampling procedures.

One of the most famous polling errors in presidential elections was the prediction made by the

We begin by calculating the mean score for each sample. For sample A it is 20.00 and for sample B it is 30.00. We next calculate the standard deviation estimate for each sample, which turns out to be .82 for sample A and 4.14 for sample B. The respective estimates of the standard error of the mean are then calculated by dividing these standard deviation estimates by the square root of N. This yields a value of .26 for sample A and a value of 1.31 for sample B.

These analyses indicate that the mean of sample A estimates the mean of population A better than the mean of sample B estimates the mean of population B. Since the samples are identical in size, $N = 10$, the difference in the estimated standard errors of the mean is probably due to differential variability of scores in the populations. This is reflected in the different sizes of the standard deviation estimates. For sample A, $\hat{s} = .82$, while for sample B, $\hat{s} = 4.14$.

Literary Digest in 1936. The polling procedure used by this magazine had accurately predicted previous elections. It consisted of mailing out millions of postcard ballots to individuals who were listed in the phone book or who were on lists of automobile owners. This system was effective as long as the more affluent (people with phones and cars) were as equally likely to vote Democratic or Republican as the less affluent (people without phones and cars). With the advent of the New Deal, however, a shift in the American electorate occurred with the more affluent tending to vote Republican and the less affluent tending to vote Democratic. The result was that the *Literary Digest* sample was overrepresented with Republican voters, leading to a prediction that the Republican candidate (Landon) would defeat the Democratic candidate (Roosevelt) by a margin of 57% to 43%. Of course, Landon did not win and was, in fact, defeated by Roosevelt by a margin of 62.5% to 37.5%.

Even when random procedures are used, a sample may not be representative of its population, as unrepresentative samples can still occur by chance. What factors influence whether a random sample will truly reflect the population? One factor noted in this chapter is the size of the sample. In general, the larger the sample size, the closer a sample estimate will be to the corresponding population parameter. The Gallup agency reports margins of error (ranges that have a high probability of containing the actual population values) associated with its polls and these are directly related to the size of the sample. For example, for a typical analysis of political opinions of people in the United States (for instance, the percentage of people favoring a candidate), the following margins of error, in percentage points, occur:

Sample size	*Margin of error*
1,500	2.5
750	4.0
400	6.0
200	8.0
100	11.0

For samples based on $N = 1{,}500$, the observed percentage will be accurate within approximately 2.5 percentage points. For samples based on $N = 100$, the margin of error is much larger (approximately 11.0 percentage points).

STUDY EXERCISE 7.2

Compute the estimated standard error of the mean for the scores in Study Exercise 7.1.

Answer The estimated standard error of the mean is the standard deviation estimate divided by the square root of the sample size:

$$\hat{s}_{\bar{X}} = \frac{\hat{s}}{\sqrt{N}}$$

$$= \frac{2.36}{\sqrt{10}} = .75$$

An Example of an Empirical Sampling Distribution of the Mean

We can demonstrate some of the above points by considering a contrived example with a small population, using sampling with replacement. Consider a population consisting of four scores: 2, 4, 6, and 8. The mean of this population, μ, is $(2 + 4 + 6 + 8)/4 = 5.00$, and the standard deviation, σ, is 2.24.

We can construct an empirical sampling distribution of the mean for samples of size 2 by listing all possible samples of this size and their corresponding means. This has been done in columns 1 and 2 of Table 7.4. The mean of the 16 sample means is 5.00. Thus, the mean of the sampling distribution equals μ. The standard deviation of the 16 sample means can be derived using the usual procedures for a standard deviation. As shown in Table 7.4, the standard deviation of the sample means in this instance is 1.58. This is the standard error of the mean, $\sigma_{\bar{X}}$, and indicates how far, on the average, the sample

TABLE 7.4 All Possible Random Samples and Sample Means for Samples of Size 2 (Based on Minium, 1970)

Population: 2, 4, 6, 8

Sample	*Sample mean*	$\left(\text{Sample mean}\right)^2$
2, 2	2.00	4.00
2, 4	3.00	9.00
2, 6	4.00	16.00
2, 8	5.00	25.00
4, 2	3.00	9.00
4, 4	4.00	16.00
4, 6	5.00	25.00
4, 8	6.00	36.00
6, 2	4.00	16.00
6, 4	5.00	25.00
6, 6	6.00	36.00
6, 8	7.00	49.00
8, 2	5.00	25.00
8, 4	6.00	36.00
8, 6	7.00	49.00
8, 8	8.00	64.00
	sum = 80.00	sum = 440.00

$$\text{mean} = \frac{80.00}{16} = 5.00$$

$$\text{sum of squares of sample means} = 440.00 - \frac{80.00^2}{16} = 40.00$$

$$\text{variance of sample means} = \frac{40.00}{16} = 2.50$$

$$\text{standard deviation of sample means} = \sqrt{2.50} = 1.58 = \sigma_{\bar{X}}$$

means deviate from μ. This same value would result if we computed $\sigma_{\bar{X}}$ using Equation 7.4 since, in this example, $\sigma_{\bar{X}} = \sigma/\sqrt{N} = 2.24/\sqrt{2} = 1.58$.

Suppose that instead of a sample size of 2, we construct a sampling distribution for a sample size of 3. Table 7.5 lists all possible samples of size 3 and their corresponding means. The mean of these sample means is, as before, equal to 5.00, the value of μ. Due to space limitations, we have not calculated the standard deviation of the 64 sample means. If we were to do so, we would find that it equals 1.29, the same value we would obtain if we used the formula $\sigma_{\bar{X}} = \sigma/\sqrt{N} = 2.24/\sqrt{3} = 1.29$. Note that the standard error of the mean for samples of size 3 ($\sigma_{\bar{X}} = 1.29$) is smaller than the standard error of the mean for samples of size 2 ($\sigma_{\bar{X}} = 1.58$). This is because larger sample sizes will generally lead to more accurate mean estimates, everything else being equal.

TABLE 7.5 All Possible Random Samples and Sample Means for Samples of Size 3 (Based on Minium, 1970)

Population: 2, 4, 6, 8

Sample	*Sample mean*	*Sample*	*Sample mean*	*Sample*	*Sample mean*
2, 2, 2	2.00	4, 4, 6	4.67	6, 8, 2	5.33
2, 2, 4	2.67	4, 4, 8	5.33	6, 8, 4	6.00
2, 2, 6	3.33	4, 6, 2	4.00	6, 8, 6	6.67
2, 2, 8	4.00	4, 6, 4	4.67	6, 8, 8	7.33
2, 4, 2	2.67	4, 6, 6	5.33	8, 2, 2	4.00
2, 4, 4	3.33	4, 6, 8	6.00	8, 2, 4	4.67
2, 4, 6	4.00	4, 8, 2	4.67	8, 2, 6	5.33
2, 4, 8	4.67	4, 8, 4	5.33	8, 2, 8	6.00
2, 6, 2	3.33	4, 8, 6	6.00	8, 4, 2	4.67
2, 6, 4	4.00	4, 8, 8	6.67	8, 4, 4	5.33
2, 6, 6	4.67	6, 2, 2	3.33	8, 4, 6	6.00
2, 6, 8	5.33	6, 2, 4	4.00	8, 4, 8	6.67
2, 8, 2	4.00	6, 2, 6	4.67	8, 6, 2	5.33
2, 8, 4	4.67	6, 2, 8	5.33	8, 6, 4	6.00
2, 8, 6	5.33	6, 4, 2	4.00	8, 6, 6	6.67
2, 8, 8	6.00	6, 4, 4	4.67	8, 6, 8	7.33
4, 2, 2	2.67	6, 4, 6	5.33	8, 8, 2	6.00
4, 2, 4	3.33	6, 4, 8	6.00	8, 8, 4	6.67
4, 2, 6	4.00	6, 6, 2	4.67	8, 8, 6	7.33
4, 2, 8	4.67	6, 6, 4	5.33	8, 8, 8	8.00
4, 4, 2	3.33	6, 6, 6	6.00	sum =	320.00
4, 4, 4	4.00	6, 6, 8	6.67		

$$\text{mean} = \frac{320.00}{64} = 5.00$$

The Shape of the Sampling Distribution of the Mean

As stated in the central limit theorem, the sampling distribution of the mean approaches a normal distribution as the sample size on which it is based becomes larger. Of crucial importance is the fact that this holds *regardless of the shape of the underlying population*. When the sample size is greater than around 30, the approximation is usually quite close. For sample sizes of 30 or less, the approximation is less exact, though in many instances, particularly if the population is not highly skewed, the fit will be reasonable. The relationship of the sampling distribution of the mean to the normal distribution is very important in statistical theory and will be referred to extensively in later chapters.

The above points are illustrated in Figure 7.1, which presents graphs of frequency distributions for three different populations and corresponding frequency graphs of empirical sampling distributions of the mean for each of two sample sizes, $N = 10$ and $N = 30$. Note that in all cases, the sampling distribution is bell-shaped and symmetrical. The most frequently occurring sample mean is the true population mean, with very deviant sample means (from the true population mean) being infrequent. This is true even for population C even though all scores in the population occur with equal frequency. The frequency of highly deviant sample means is much less when the sample size is larger (for each population, compare the graphs for $N = 10$ and $N = 30$) and when the population variability is lower (for each sample size, compare the graphs for population A, which has a relatively small standard deviation, and population B, which has a relatively large standard deviation). This is consistent with principles developed earlier in the chapter.

The Sampling Distribution of the Mean Summarized

The major points about the sampling distribution of the mean can be summarized as follows:

1. There is a different sampling distribution for every sample size. For instance, a sampling distribution of the mean for random samples of size 10 is different from a sampling distribution of the mean for random samples of size 90.
2. The mean of a sampling distribution of the mean equals the population mean of the raw scores.
3. The standard deviation of a sampling distribution of the mean is called the standard error of the mean.
4. The standard error of the mean gets smaller as the sample size increases and as the variability of scores in the population decreases.
5. The sample mean is a more accurate estimator of the population mean when the standard error of the mean is small (tends toward 0) than when the standard error of the mean is large.
6. The standard error of the mean can be estimated by dividing the standard deviation estimate by the square root of the sample size.
7. The sampling distribution of the mean approaches a normal distribution as the sample size becomes larger. This is true regardless of the shape of the underlying population.

FIGURE 7.1 Frequency Distributions for Three Populations and Corresponding Sampling Distributions for Two Sample Sizes (Adapted from Johnson & Liebert, 1977)

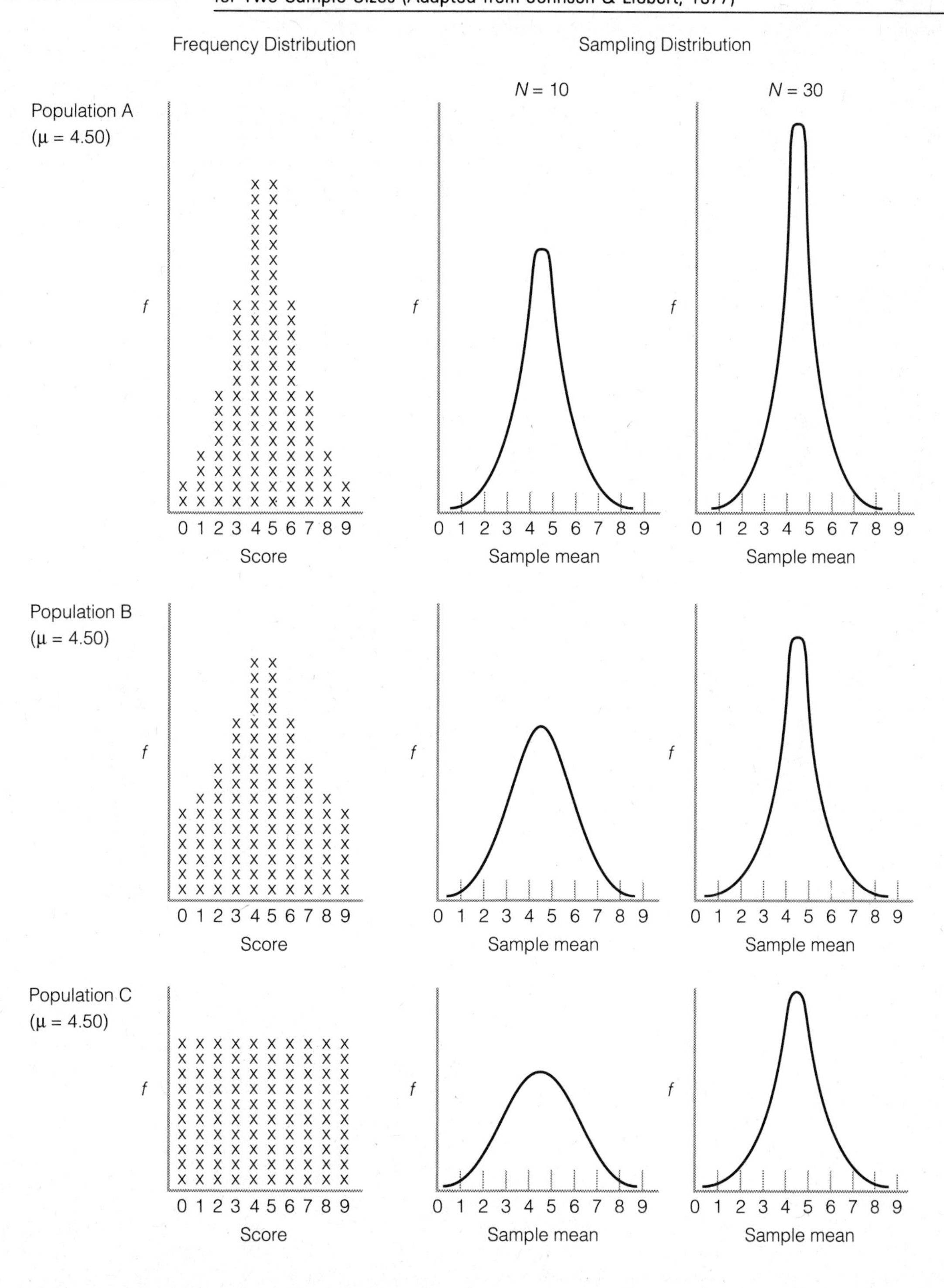

7.6

Types of Sampling Distributions

Although we have focused on the sampling distribution of the mean, it is possible to conceptualize sampling distributions of other statistics, such as the median, the mode, or the variance. In the case of the median, for example, the sampling distribution reflects the medians of all possible random samples of size N selected from a population. Given the same population, the sampling distribution of the mean will show less variability (that is, it will have a smaller standard error) than either the sampling distribution of the median or the sampling distribution of the mode. It is for this reason that the mean is usually preferred by statisticians as a measure of central tendency.

7.7

Summary

Whenever behavioral scientists study samples, they must deal with the problem of sampling error, or the fact that a sample statistic might differ from the value of its corresponding population parameter. An unbiased estimator is one whose mean over all possible random samples equals the value of the population parameter being estimated. A key concept for understanding the nature of sampling error is that of sampling distributions. A sampling distribution of the mean refers to a distribution of mean scores based on all possible random samples of a given size. As stated in the central limit theorem, the mean of a sampling distribution of the mean will always be equal to the true population mean.

The central limit theorem also defines the standard deviation and the shape of a sampling distribution of the mean. The standard deviation, called the standard error of the mean, indicates the average amount that the means of random samples of size N deviate from the true population mean. The size of the standard error of the mean is influenced by two factors, the sample size (N) and the variability of scores in the population (σ). The shape of a sampling distribution of the mean will approach normality as N increases, regardless of the shape of the underlying population. The fit is particularly good when N is greater than around 30. These properties are crucial for inferential statistics and will be referred to extensively in later chapters.

Exercises

Answers to asterisked (*) exercises appear at the back of the book. Answers to exercises with two asterisks are also worked out step-by-step in the Study Guide.

1. Under what circumstance is sampling from a finite population analogous to sampling from an infinite population?

*2. What is sampling error?

3. Why, in the absence of any other information, is the observed sample mean one's "best guess" about the value of the population mean?

4. What is a biased estimator? What is an unbiased estimator?

5. Why is it necessary to divide the sum of squares by $N - 1$ rather than N when computing the variance estimate?

*6. What does the term "degrees of freedom" refer to? Why are there $N - 1$ degrees of freedom associated with a sum of squares?

7. What is a mean square? What is the relationship between a mean square and a variance estimate?

*8. What is a sampling distribution of the mean? How is a sampling distribution of the mean different from a frequency distribution as discussed in Chapter 2?

*9. What is the central limit theorem?

*10. In general terms, what value will the mean of a sampling distribution of the mean always be equal to? Why?

11. Define the standard error of the mean. What information does it convey?

*12. Distinguish a standard error of the mean from a standard deviation of a set of raw scores.

*13. If a sampling distribution of the mean has an associated standard error of the mean equal to 0, what does this imply about sample means drawn from the relevant population?

*14. If a sampling distribution of the mean has an associated standard error of the mean equal to 0, what does this imply about the variability of scores in the population (σ)?

15. What two factors influence the size of the standard error of the mean?

**16. For the following sample data, compute the mean and the estimated standard error of the mean: 2, 3, 3, 4, 4, 4, 5, 5, 5, 5, 6, 6, 6, 7, 7, 8.

17. For the following sample data, compute the mean and the estimated standard error of the mean: 4, 4, 4, 4, 4, 4, 5, 5, 5, 5, 6, 6, 6, 6, 6, 6.

18. Compare the estimated standard error of the mean in Exercise 16 with that in Exercise 17. Which sample mean is probably a better estimate of its population mean? Why?

*19. A random sample of size 30 is drawn from each of two populations. For population A, $\mu = 10.00$ and $\sigma = 5.00$. For population B, $\mu = 8.00$ and $\sigma = 7.00$. Which sample mean is probably a better estimate of its population mean? Why?

20. For each of two populations, $\mu = 10.00$ and $\sigma = 6.00$. A random sample of size 20 is taken from population A and a random sample of size 40 is taken from population B. Which sample mean is probably a better estimate of its population mean? Why?

**21. Given a population with three scores, 2, 4, and 6, specify all possible samples of size 2 one could obtain from this population using sampling with replacement. (*Hint:* There are nine of them.) Compute the mean for each sample. Next compute the mean across the nine sample means. How does this result compare with the population mean, μ? What principle does this illustrate?

22. Match each symbol in the first column with the appropriate term in the second column:

Symbol	*Term*
1. $\hat{s}_{\bar{X}}$	a. standard error of the mean
2. s	b. standard deviation in a sample
3. $\sigma_{\bar{X}}$	c. variance in a sample
4. σ^2	d. variance in a population
5. s^2	e. standard deviation in a population
6. σ	f. variance estimate
7. $\hat{s}^2$	g. estimated standard error of the mean
8. $\hat{s}$	h. standard deviation estimate

23. Discuss the relationship between the shape of the sampling distribution of the mean, sample size, and the shape of the underlying population.

*24. Why is the mean usually preferred to the median and the mode as a measure of central tendency by statisticians?

EIGHT

HYPOTHESIS TESTING: INFERENCES ABOUT A SINGLE MEAN

In the present chapter we consider basic principles of hypothesis testing. This material is central to all of inferential statistics and should be studied carefully. We will rely on several concepts developed in previous chapters and you should make sure that you are familiar with these concepts before covering the present material:

standard score
z score
normal distribution
probability of an event
sampling distribution of the mean
central limit theorem

You may wish to return to previous chapters and review these concepts.

8.1

A Simple Analogy for Principles of Hypothesis Testing

Some of the basic steps of hypothesis testing in behavioral science research can be characterized by a simple analogy. Suppose you were given a coin and asked to determine whether it was a fair coin, or biased toward heads or tails. You might respond to this task by conducting a simple experiment. If you assume that the coin is fair, then you would expect that flipping the coin a large number of times would result in heads about one-half the time, since the probability of a head is .50. So, you flip the coin 100 times and count the number of heads that occur. Suppose the tosses resulted in heads 52 times. This does not correspond *exactly* to what you would expect based on a probability of .50, but it certainly is close (52 out of 100 as compared with 50 out of 100). Because the result is not very discrepant from the expected result, and because you know that your observed result could have deviated somewhat from the expected result just by chance, you might conclude that the coin is not biased. But suppose the result of 100 flips had yielded 65 heads? Or 95 heads? At some point, the discrepancy from the expected result becomes too great to attribute it to chance and, at this point, you would reject your original assumption that the coin is fair.

In behavioral science research, the process of **hypothesis testing** is very similar to that outlined above. The investigator begins by stating a proposal, or *hypothesis,* that is assumed to be true (the coin is fair). Based on this assumption, an expected result is specified (we should obtain 50 heads out of 100 flips). The data are collected and the observed result is compared with the expected result (52 heads versus an expectation of 50 heads). If the observed result is so discrepant from the expected result that the difference cannot be attributed to chance, then the original hypothesis is rejected. Otherwise, it is not rejected.

8.2 Statistical Inference and the Normal Distribution

Suppose an investigator is interested in the intelligence of the type of student who attends Victor University. Specifically, she wants to know if the mean intelligence level of such students is different from the typical intelligence level of college students in general. Suppose that past research concerning a particular intelligence test has shown that a score of 100 represents the performance of the typical college student. The question of interest, then, is whether the mean intelligence test score for the population of students who attend Victor University is different from 100.

Competing Hypotheses

We can state this issue more precisely in terms of two competing hypotheses regarding the population mean. First, the mean intelligence test score for students who attend Victor University may, in fact, equal the value of interest, 100. This possibility can be stated in terms of a **null hypothesis:**

$$H_0: \quad \mu = 100$$

where H_0 is the symbol for a null hypothesis and μ represents the true population mean for the students who attend Victor University. Second, the true population mean for these students may not equal 100 but rather may be higher or lower than 100. We can state this in terms of an **alternative hypothesis:**

$$H_1: \quad \mu \neq 100$$

where H_1 is the symbol for an alternative hypothesis. The task at hand is to choose between these two competing hypotheses.

The investigator decides to do this by collecting some data. A random sample of 50 students currently attending Victor University is administered the intelligence test. The mean score for the sample is found to be 105.00. These data would seem to support the second hypothesis stated above, since the sample mean is not equal to 100, but rather is equal to 105.00. However, we know from Chapter 7 that a sample mean may not be an accurate descriptor of the population mean because of sampling error. Maybe the true population mean is 100 and we observed a sample mean of 105.00 because of sampling error. We need to test the viability of this possibility. Our knowledge of sampling distributions and the normal distribution will help us in this regard.

Analysis of Sampling Distributions

In the coin flipping example, we tested whether the coin was fair by assuming that such was the case and, based on this assumption, specifying an expected result which we then compared with the observed result. We will do the same for the problem with intelligence scores. Assume that the true population mean for Victor University students is equal to 100. If we select a random sample of 50 students from the population, we would expect the mean score of the sample to be near 100. We would not expect it to be exactly 100 because of sampling error (just as we would not expect flipping a fair coin 100 times to yield exactly 50 heads and 50 tails). How much sampling error can we reasonably expect to have and still be confident that μ equals 100? If the sample mean were equal to 101.00, could this be due to sampling error? What about 110.00?

We can be quite specific on this matter by making reference to a sampling distribution of means based on all possible random samples of size 50. Recall that the mean of a sampling distribution of the mean equals the true population mean. We have assumed, for purposes of our statistical test, that this equals 100. Also recall that the standard deviation of the sampling distribution of the mean is called the standard error of the mean ($\sigma_{\bar{X}}$) and, per Equation 7.4, is equal to the population standard deviation divided by the square root of the sample size. In the present context, $\sigma_{\bar{X}}$ represents how much, on the average, sample means based on $N = 50$ deviate from the true population mean. Suppose we knew the value of the population standard deviation to be 17.68 and, thus, the value of the standard error of the mean to be*

$$\sigma_{\bar{X}} = \frac{\sigma}{\sqrt{N}}$$

$$= \frac{17.68}{\sqrt{50}} = 2.50$$

This indicates that, on the average, sample means based on $N = 50$ deviate 2.50 units from the true population mean.

* In practice, the standard error of the mean will usually not be known and will have to be estimated using the procedures discussed in Chapter 7. We will consider principles of hypothesis testing when estimates of the standard error are used in Section 8.9.

If the true population mean is 100 and the standard error of the mean is 2.50, how much sampling error can we reasonably expect in our data? Recall from Chapter 7 that a sampling distribution of means based on a relatively large N is approximately normally distributed. Also recall from Chapter 4 that 68.26% of all scores in a normal distribution occur between 1 standard deviation below the mean and 1 standard deviation above the mean. About 68% of all means in a sampling distribution of means will therefore occur between 1 standard error (standard deviation) below μ and 1 standard error above μ. Between what two scores will 95% of the sample means in a sampling distribution occur? If we consult column 2 of the z table in Appendix B, we find that 95% of the scores in a normal distribution fall between 1.96 standard deviations below the mean and 1.96 standard deviations above the mean. If 95% of all sample means fall between ± 1.96 standard errors of the mean, the probability that we would observe a sample mean *outside* this range is less than .05 (since less than 5% of the sample means are outside this range). This is a highly unlikely event. The mean of our sample *is* outside this range. It is $105.00 - 100 = 5.00$ units or 2.00 standard errors (since the standard error is 2.50 and $5.00/2.50 = 2.00$) above the hypothesized population mean. Given this, we might reasonably conclude that the observed sample mean cannot be attributed to sampling error and that the assumption that $\mu = 100$ is untenable. We would therefore reject the null hypothesis and, given that the sample mean is 105.00, conclude that the mean intelligence test score for the population of students who attend Victor University is greater than the typical score of 100 for college students in general.

We can formally summarize the above steps as follows:

1. Translate the research question into two competing hypotheses, a null hypothesis and an alternative hypothesis.

   In the present example, the null hypothesis states H_0: $\mu = 100$ and the alternative hypothesis states H_1: $\mu \neq 100$.

2. Assuming the null hypothesis is true, state an expected result in the form of a range of values within which the sample mean would be expected to fall. This is expressed in terms of the standard (z) scores, called **critical values,** that determine the endpoints of this range. The set of all standard scores more extreme than the critical values (that is, less than the negative critical value or greater than the positive critical value) is called a **rejection region** and constitutes an unexpected result.

   In the present example, the critical values are -1.96 and $+1.96$; thus, an expected result is defined by standard scores in the range -1.96 to $+1.96$, and an unexpected result (the rejection region) is defined by all standard scores less than -1.96 or greater than $+1.96$.

3. Characterize the mean and standard deviation of the sampling distribution of the mean, assuming the null hypothesis is true.

If we assume that the null hypothesis is true in the present example, then μ equals 100. It follows that the mean of the sampling distribution will also equal 100, since the mean of a sampling distribution of the mean is always equal to the population mean.

The standard deviation of the sampling distribution is the standard error of the mean. This was previously calculated to equal 2.50 in our example.

4. Convert the observed sample mean into a z value to determine how many standard errors it is from μ, *assuming the null hypothesis is true*. This is accomplished by the formula

$$z = \frac{\bar{X} - \mu}{\sigma_{\bar{X}}} \qquad [8.1]$$

where $\bar{X}$ is the observed sample mean, μ is the population mean assuming the null hypothesis is true, and $\sigma_{\bar{X}}$ is the standard error of the mean.

In the present example,

$$z = \frac{105.00 - 100}{2.50} = 2.00$$

5. **a.** Compare the observed result with the expected result. If the observed result falls within the rejection region, reject the null hypothesis. Otherwise, do not reject the null hypothesis.
b. If the null hypothesis is rejected, compare the observed sample mean ($\bar{X}$) with the value of μ stated in the null hypothesis. If the observed sample mean is *greater* than the stated μ, conclude that the actual population mean is greater than the stated population mean. If the observed sample mean is *less* than the stated μ, conclude that the actual population mean is less than the stated population mean.*

In the present example, the observed z value of 2.00 exceeds the critical value of +1.96. This suggests that the observed difference between the sample mean and the hypothesized population mean is too large to be attributed to sampling error, and the null hypothesis is therefore rejected. Since the observed sample mean of 105.00 is greater than the stated μ of 100, the appropriate conclusion is that the mean intelligence test score of the population of students who attend Victor University is greater than 100.

* Since the observed value of z will be positive if the sample mean is greater than the stated value of μ and negative if the sample mean is less than the stated value of μ, an alternative approach is to examine the sign of the observed z score.

8.3

Defining Expected and Unexpected Results

A major step in the preceding hypothesis testing procedure was the specification of what constitutes expected and unexpected results under the assumption that the null hypothesis is true. As discussed above, this is stated in terms of positive and negative critical values and a corresponding rejection region. Rejection regions are determined with reference to a probability value known as an **alpha level.** For example, when the alpha level is .05, a result is defined as being "unexpected" if the probability of obtaining that result, assuming the null hypothesis is true, is less than .05.

As noted previously, 95% of all sample means in the example pertaining to intelligence test scores at Victor University will fall between −1.96 and +1.96 standard errors of the true population mean, which, for purposes of hypothesis testing, is assumed to equal 100. Thus, there is less than a .05 probability of observing a sample mean outside this range. For an alpha level of .05, then, an unexpected result includes any sample mean occurring more than 1.96 standard errors below or more than 1.96 standard errors above the value of μ represented in the null hypothesis.

Although alpha can be set at levels other than .05, it is traditional practice to adopt an alpha level of .05 in behavioral science research. We will discuss the issue of setting the alpha level in more detail shortly.

STUDY EXERCISE 8.1

Suppose an investigator is interested in whether the mean score on an aptitude test for students who attend rural elementary schools is different from the typical score of elementary school students in general. The typical score for elementary school students in general is known to be 50. A random sample of 25 rural elementary school students is obtained, the aptitude test administered, and the mean score for the sample found to be 56.00. The population standard deviation, σ, is known to equal 10.00. Test the viability of the hypothesis that $\mu = 50$.

Answer We begin by explicitly stating the null and alternative hypotheses:

$$H_0: \quad \mu = 50$$
$$H_1: \quad \mu \neq 50$$

Next, we state an expected result under the assumption that the null hypothesis is true. For an alpha level of .05, this includes any z score between −1.96 and +1.96. An unexpected result thus consists of all z scores less than −1.96 or greater than +1.96. Since $\sigma = 10.00$,

$$\sigma_{\bar{X}} = \frac{\sigma}{\sqrt{N}} = \frac{10.00}{\sqrt{25}} = 2.00$$

According to Equation 8.1, the observed z score is thus

$$z = \frac{\bar{X} - \mu}{\sigma_{\bar{X}}} = \frac{56.00 - 50}{2.00} = 3.00$$

Since 3.00 is greater than +1.96, we reject the null hypothesis and, based on a comparison of the observed sample mean (56.00) with the value of μ stated in the null hypothesis (50), we conclude that the actual population mean for rural elementary school students is greater than the typical score of 50 for elementary school students in general.

8.4

Failing to Reject Versus Accepting the Null Hypothesis

In the Victor University example, the observed sample mean was inconsistent with a sampling error interpretation of the data and the null hypothesis was rejected. However, if the observed sample mean had been close to 100, say 100.50, would we have accepted the null hypothesis? The answer is "no."

When a researcher obtains a result that is consistent with the null hypothesis (that is, when it falls within the range defined by the critical values rather than in the rejection region), he or she does not accept the null hypothesis as being true. Rather, the researcher *fails to reject* the null hypothesis. There is a subtle distinction here, which is very important. In principle, *we can never accept the null hypothesis as being true via our statistical methods; we can only reject it as being untenable.* Consider the null and alternative hypotheses from the Victor University example:

$$H_0: \quad \mu = 100$$
$$H_1: \quad \mu \neq 100$$

Note that the null hypothesis is stated such that the true population mean must equal one and only one value (100), whereas the alternative hypothesis is stated such that the true population mean could potentially equal any of an infinite number of values (for example, 100.50, 101.00, 97.60, or 112.30—anything but 100). When we observe a highly discrepant sample mean and reject the null hypothesis, we are saying that it is unlikely that the true population mean equals 100 and, in this sense, we "accept" the alternative hypothesis. In contrast, because of sampling error, we can never unambiguously conclude that the true population mean is equal to any one specific value based on sample data. If the observed sample mean is not extremely discrepant from the hypothesized population value of 100, we can say only that the sample mean is too close for us to confidently conclude that the true population mean does *not* equal 100. We cannot say that μ is equal to 100, but we also cannot confidently say that it is not. Even an observed sample mean of exactly 100 does not prove that the *population* mean is equal to 100, as sampling error can produce a sample mean of 100 even when the population mean is equal to a much smaller or much larger value. Thus, when the observed value of z falls within the range defined by the critical values, we *fail to reject* the null hypothesis.

8.5 Statistical and Real-World Significance

If the null hypothesis is rejected, the results of a statistical test are said to be *significant*. If the null hypothesis is not rejected, the phrase *nonsignificant* is used instead. It is important to realize that these terms are meant to apply only to the *statistical* outcome. We have seen several instances in the popular press where studies are quoted as reporting "significant" findings when the researchers simply intended to signify rejection of a null hypothesis. The real-world implications and significance of a particular rejection of a null hypothesis in a study are very different matters than the statistical rejection of a null hypothesis. A finding that a null hypothesis about the fairness of a coin was rejected, although "statistically significant," certainly is not significant in terms of its real-world implications for human behavior. Possible misunderstanding can be avoided by using the terms **statistically significant** and **statistically nonsignificant** rather than the more general versions when referring to statistical outcomes.

8.6 Type I and Type II Errors

Once an investigator has drawn a conclusion with respect to the null hypothesis, that conclusion can be either correct or in error. Two types of errors are possible: (1) rejection of the null hypothesis when it is true (called a **Type I error**), and (2) failure to reject the null hypothesis when it is false (called a

TABLE 8.1 Two Types of Errors in Hypothesis Testing

		TRUE STATE OF AFFAIRS	
		H_0 is true	*H_0 is false*
DECISION ON THE BASIS OF THE STATISTICAL ANALYSIS	*Reject H_0*	Type I error (Probability = α)	Correct decision (Probability = $1 - \beta$)
	Fail to reject H_0	Correct decision (Probability = $1 - \alpha$)	Type II error (Probability = β)

FIGURE 8.1 Illustration of Type I and Type II Errors (Adapted from McCall, 1980)

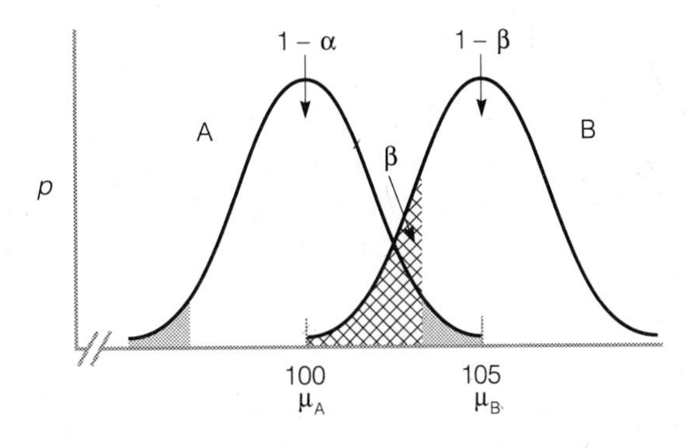

Type II error). These are illustrated in Table 8.1. The probability of making a Type I error is equal to the alpha level—in most cases, .05. As can be seen, alpha is symbolized by α, the lowercase Greek *a*. The probability of making a Type II error is traditionally called **beta** and is represented by β, the lowercase Greek *b*.

The nature of decision errors can be explained with reference to Figure 8.1. Consider a population of scores in which the null hypothesis that $\mu = 100$ is true. Distribution A represents a sampling distribution based on $N = 50$. The point marked "100" represents the population mean. A Type I error would occur if we selected a sample from this population and concluded, based on the sample mean, that μ did not equal 100. If the alpha level was .05, this would occur, on the average, only 5 times out of 100, since only 5% of the sample means would occur outside the range -1.96 to $+1.96$ standard errors. Thus, the probability of a Type I error equals the alpha level, and is indicated by the shaded areas of Figure 8.1. The probability, given a true null hypothesis, of *not* making a Type I error—that is, of failing to reject the null hypothesis—is defined by the area labeled $1 - \alpha$.

Now consider the probability of a Type II error. Suppose the mean of a population of scores is 105 and we are testing the null hypothesis that $\mu = 100$. Distribution B in Figure 8.1 presents the sampling distribution of the mean based on $N = 50$. The point marked "105" represents the true population mean. Note that if the observed sample mean occurred anywhere within the cross-hatched area, we would fail to reject the null hypothesis when it is, in fact, false. This cross-hatched area represents the probability of a Type II error, or β. The area labeled $1 - \beta$ defines the probability that an investigator will correctly reject the null hypothesis when it is false, and this probability is called the **power** of the statistical test.

The above concepts can be illustrated intuitively with an electronics example. Suppose you are listening to a set of earphones and trying to decide if you hear a particular signal. There is static on the earphones, which makes this difficult for you. You have been told that you should hear the signal within 30 seconds. One type of error you could make is to say you heard the signal, when, in fact, it did not occur. This is analogous to a Type I error. Suppose that making such an error would lead to negative consequences. You would want to be very sure of yourself. Only if you are very certain you heard the signal would you say you heard it. This is similar to setting a low alpha level (for example, .05) in an investigation.

On the other hand, there is another type of error you could make—saying you did not hear the signal, when, in fact, it was there. This corresponds to a Type II error. The ability not to miss the signal corresponds to the power of a statistical test. If you have a very sensitive ear, you will be likely to detect the signal when it occurs (high power). However, if you do not have a sensitive ear, you will be more likely to miss the signal (low power).

Notice that the value of the alpha level directly affects the power of the statistical test. If you are very conservative about saying you heard the signal (setting a very low alpha level), then this decreases the likelihood that you will say the signal is there when, in fact, it is (that is, decreases the power).

8.7

Effects of Alpha and Sample Size on the Power of Statistical Tests

As previously noted, the alpha level in an investigation reflects the probability of making a Type I error. The tradition of adopting a low, or *conservative,* alpha level in behavioral science research evolved from experimental settings where a certain kind of error was very important and had to be avoided. An example of such an experimental setting is that of testing a new drug for medical purposes, with the aim of ensuring that the drug is safe for the general adult population. In this case, deciding that a drug is safe when, in fact, it tends to produce adverse reactions in a large proportion of adults is an error that is certainly to be avoided. Under these circumstances, the hypothesis that "the medicine is unsafe," or its statistical equivalent, would be cast as the null hypothesis and a low alpha level selected so as to avoid making the costly error. With a conservative alpha level, the medical researcher takes little risk of concluding that the drug is safe (H_1) when actually it is not (H_0). Thus, the practice of setting conservative alpha levels evolved from situations where one kind of error was extremely important and had to be avoided if possible. By casting such errors as Type I errors in the hypothesis testing framework and then setting a low alpha level, the risk of committing the error could be minimized.

Several researchers (Cohen, 1977; Greenwald, 1975) have argued that behavioral scientists have been preoccupied with Type I errors at the expense of Type II errors. The alpha level directly affects the power of the statistical test (and hence the probability of making a Type II error), with more conservative alpha levels yielding less powerful tests. The argument is that for some behavioral science research, it is hard to justify that a Type I error should have the drastic character implied by a low alpha level. It is not necessarily worse, the argument goes, to conclude falsely that there is a difference between a mean and a hypothesized value (that is, to make a Type I error) than it is to conclude falsely that there is not a difference (that is, to make a Type II error). The issue concerns the balance between the risk of placing a false finding in the body of scientific knowledge versus the risk of letting an existing difference go undetected and, thus, unreported. By setting alpha at a less conservative level, we reduce the risk of the latter type of error, albeit at the expense of the former. The issue of setting one's alpha level is one that enjoys much controversy. Interested readers are referred to Kirk (1972) for a detailed discussion of this issue.

In terms of the power of a statistical test, not only is the alpha level important, but so too is the sample size: The larger the sample size, the more powerful the statistical test will be, everything else being equal. This can be seen with reference to the formula for the inferential z test (Equation 8.1):

$$z = \frac{\bar{X} - \mu}{\sigma_{\bar{X}}}$$

which can also be written

$$z = \frac{\bar{X} - \mu}{\dfrac{\sigma}{\sqrt{N}}}$$

In this equation, as N becomes larger, $\sigma_{\bar{X}}$ becomes smaller so z also becomes larger, other things being equal. As z becomes larger, it is more likely that we will reject the null hypothesis when it is false, thereby increasing the power of the statistical test.

In summary, a researcher typically has control over the alpha level he or she selects as well as the sample size to be used in the investigation. The power of one's test can be increased by selecting larger sample sizes and larger alpha levels. Selecting a larger sample size must be evaluated relative to the practical concerns of the increased costs of obtaining more individuals in the investigation. In addition, increasing the alpha level must be evaluated relative to the importance of making a Type I versus a Type II error. There comes a point when increasing the power of a test is of diminishing value, since at that point the test is sufficiently powerful to draw a conclusion with an appropriate degree of confidence. As a rough guide, investigators generally attempt to achieve statistical power (the probability of correctly rejecting the null hypothesis when it is false) in the range of .80 to .95, depending on the nature of the proposition being investigated.

Estimation of the power of a statistical test can be accomplished through special tables developed for this purpose. In future chapters, we will encounter such tables, which are based on principles of power analysis discussed in Cohen (1977). These tables can also be used to estimate the sample sizes necessary to achieve desired levels of power, given the value of alpha. In most instances, researchers choose to adopt the traditional alpha level of .05. We will follow this convention in the remainder of this book.

8.8

Directional Versus Nondirectional Tests

In each of the preceding examples, the alternative hypothesis has been *nondirectional.* It has stated that the true population mean is *either* higher *or* lower than the value specified in the null hypothesis. A *directional* alternative hypothesis, in contrast, is one that specifies that a population mean is different from a given value and also indicates the direction of that difference. For instance, the alternative hypothesis

$$H_1: \quad \mu > 100$$

states that the true population mean is *greater* than 100.

A **directional test** is one designed to detect differences from a hypothesized population mean score (for example, $\mu = 100$) in one direction only. Suppose, for example, that a counselor is concerned about the reading skills of freshman students at her college due to poor preparation. She decides to administer a well-known reading test to a sample of incoming freshmen; if they score significantly lower than the national test average of 112, she will consider instituting a remedial program. In this case, the investigator would state a directional alternative hypothesis concerning the population mean of incoming freshmen:

$$H_1: \quad \mu < 112$$

FIGURE 8.2 Rejection Regions for (**a**) Directional and (**b**) Nondirectional Tests (From Witte, 1980)

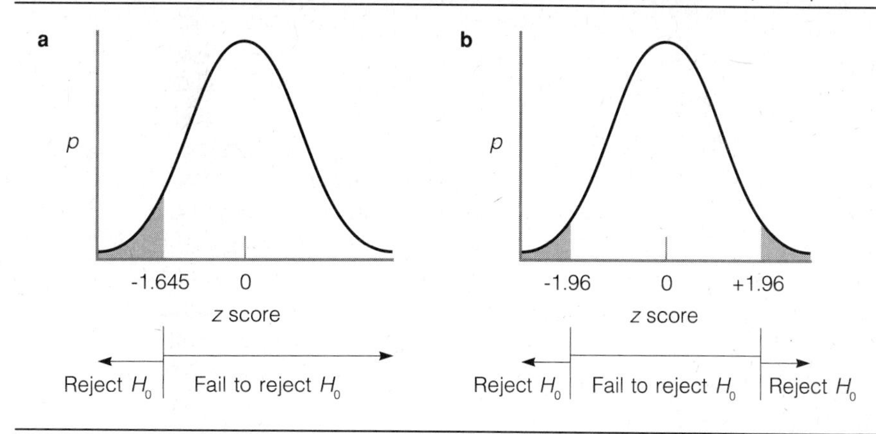

That is, the investigator wants the null hypothesis, H_0: $\mu = 112$, to be rejected only if the mean reading test score for the *population* of incoming freshmen is less than the national average of 112, since this will indicate a deficiency in reading skills. The investigator's concern is only with detecting a mean *lower* than the national average.

Figure 8.2a presents a rejection region associated with only the lower end of a sampling distribution. Any score occurring in the rejection region would lead to the rejection of the null hypothesis. Assuming an alpha level of .05, the rejection region is defined by z scores of less than -1.645, as indicated in column 3 of Appendix B.* Since they focus on only one tail of the distribution, directional tests are often referred to as **one-tailed tests.**

Figure 8.2b presents a rejection region for a **nondirectional test**—one that is designed to detect differences either above or below the hypothesized population mean. In this case, the .05 alpha level is "split" such that .025 of the scores occur at the upper end of the distribution and .025 of the scores occur at the lower end of the distribution. Thus, nondirectional tests are also called **two-tailed tests.** The z scores that define the rejection region are now -1.96 and $+1.96$. This can be verified using column 4 of Appendix B. Note in Figure 8.2 that the directional test is more sensitive than the nondirectional test to freshman scores being lower than the national average. If the null hypothesis is false due to scores being lower than the national average, the observed sample mean is more likely to fall in the rejection region of the directional test (z scores less than -1.645) than in the corresponding rejection region of the nondirectional test (z scores less than -1.96), since it is broader. Thus, we will be more likely to correctly reject the null hypothesis with the directional test than with

---

* Specifically, Appendix B indicates that .0505 of scores in a normal distribution are less than or equal to a z score of -1.64 and that .0495 of scores in a normal distribution are less than or equal to a z score of -1.65. Therefore, .05, or 5%, of scores in a normal distribution must be less than or equal to a z score of $[-1.64 + (-1.65)]/2 = -1.645$.

the nondirectional test. In other words, in this example, the directional test is *more powerful* than the nondirectional test.

In general, a directional test will always be more powerful than a corresponding nondirectional test if the actual population mean and the hypothesized population mean differ in the specified direction. However, if the actual population mean differs from the hypothesized population mean in the opposite direction from that stated in the alternative hypothesis, a nondirectional test will be more powerful than its directional counterpart. This is because the null hypothesis cannot be rejected if the observed z score falls in the tail that does not contain the critical value. For instance, if the mean reading test score for the population of freshmen in the present example were greater than 112, perhaps yielding a sample mean corresponding to a z score of 2.36, the null hypothesis would be rejected if a nondirectional test were used (since 2.36 exceeds the critical value of $+1.96$) but not if a directional test were used (since 2.36 does not fall in the rejection region bounded by -1.645).

When there is *exclusive* concern that the population mean differs from a value in a specified direction, a directional test should be used, as it will be more powerful than a nondirectional test. This concern must be stated before the data analysis is performed. Never compute a statistical test and then, based on the results, decide that a directional hypothesis should be used. This defeats the logic of the hypothesis testing procedures. If concern is not with a specific direction of difference, a nondirectional test should be used, as such tests ensure an equal likelihood of rejecting null hypotheses that are false because the actual population mean is less than the hypothesized value of μ and null hypotheses that are false because the actual population mean is greater than the hypothesized value of μ.

8.9 Statistical Inference Using Estimated Standard Errors: The One-Sample *t* Test

An important property of the statistical test we developed in Section 8.2 was that the standard error of the mean, $\sigma_{\bar{X}}$, was known. It is more common that the standard error of the mean is not known, and that it must be estimated from sample data.

Suppose an investigator administered to a random sample of 100 college students an attitude scale designed to measure attitudes toward living in dormitories. Scores on the scale can range from 1 to 7, with increasing numbers representing increasing levels of favorability toward living in dormitories. A score of 1 indicates a very unfavorable attitude; a score of 4, a neutral attitude; and a score of 7, a very favorable attitude. The investigator is interested in whether the mean attitude score for the population is different from 4, the score representing a neutral feeling. The null and alternative hypotheses are

$$H_0: \quad \mu = 4$$
$$H_1: \quad \mu \neq 4$$

Note that a nondirectional alternative hypothesis is stated because the investigator is interested in whether the mean attitude is unfavorable (less than 4) or favorable (greater than 4).

The data collected from the sample of 100 students yielded a mean of 4.51 and a standard deviation estimate of 1.94. Although the sample mean is consistent with the alternative hypothesis, a test for the possibility that the observed value is due to sampling error is necessary. But, unlike previous problems, we do not know the value of the standard error of the mean, $\sigma_{\overline{X}}$. However, we can estimate $\sigma_{\overline{X}}$ from the sample data using Equation 7.5:

$$\hat{s}_{\overline{X}} = \frac{\hat{s}}{\sqrt{N}}$$
$$= \frac{1.94}{\sqrt{100}} = .19$$

Given this estimate, you might reason that we can modify Equation 8.1 from $z = (\overline{X} - \mu)/\sigma_{\overline{X}}$ to $z = (\overline{X} - \mu)/\hat{s}_{\overline{X}}$ by simply substituting the estimated standard error, $\hat{s}_{\overline{X}}$, for the known standard error, $\sigma_{\overline{X}}$. Unfortunately, some complications would result from this. Since $\hat{s}_{\overline{X}}$ is calculated from sample data, it is subject to sampling error, whereas $\sigma_{\overline{X}}$ is not. Stated differently, the estimate of the standard error will tend to vary from sample to sample; one could select several different samples from the same population and obtain several different values of $\hat{s}_{\overline{X}}$. This is true even for samples having the same mean. Since the value yielded by the ratio $(\overline{X} - \mu)/\hat{s}_{\overline{X}}$ is influenced by the size of $\hat{s}_{\overline{X}}$, it would thus be possible to observe different z scores for the same value of $\overline{X}$.

Statisticians have developed procedures for dealing with this problem. Although the mathematics are complex, the idea can be stated in general terms: When $\sigma_{\overline{X}}$ is known and we calculate a z score for the sample mean, we are calculating the *exact* number of standard errors that the sample mean is from the mean of the sampling distribution, μ. If we substitute $\hat{s}_{\overline{X}}$ for $\sigma_{\overline{X}}$ and calculate a z score, we are calculating the *estimated* number of standard errors that the sample mean is from μ. The probability that a sample mean will be a certain number of *actual* standard errors from μ is not the same as the probability that a sample mean will be a certain number of *estimated* standard errors from μ. A distribution other than the normal distribution is required to determine this probability, and this distribution is called the ***t* distribution.**

The t distribution is very similar to the normal distribution in standardized form, as it is bell-shaped and symmetrical and has a mean of 0. However, unlike the normal distribution, the t distribution is influenced by the number of scores in the sample. Thus, there is a different t distribution for every sample size. Figure 8.3 presents comparisons of t and normal distributions for sample

FIGURE 8.3 Comparisons of t and Normal Distributions for Sample Sizes of 5, 15, and 30

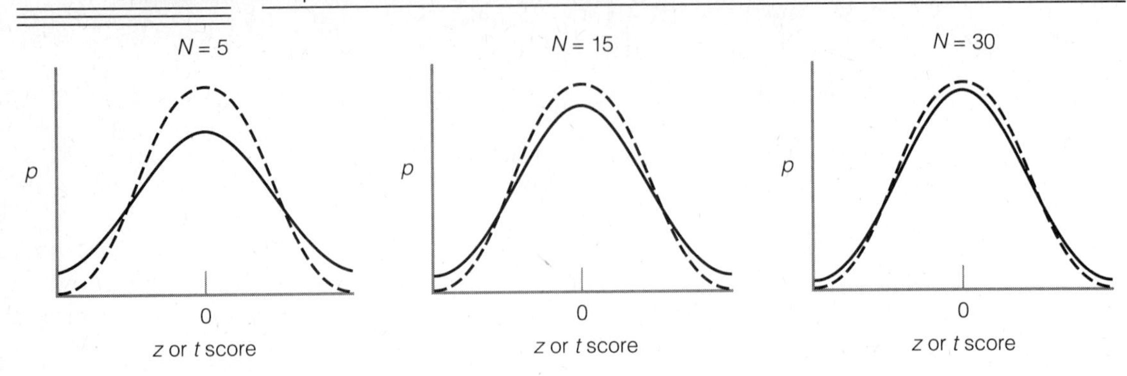

Solid lines represent the t distribution; broken lines represent the normal distribution.

sizes of 5, 15, and 30. When $N > 30$, the normal and t distributions are similar. However, when the sample size is 30 or less, the shapes of the two distributions differ. For example, given an interval occurring at the end of a distribution, the proportion of scores occurring in that interval will be greater in the t distribution than in the normal distribution. Stated in terms of probability, this means that the probability associated with an interval of scores at the end of the distribution will be greater in the t distribution than in the normal distribution. The smaller the sample size, the greater this discrepancy will be. As illustrated in Figure 8.3, this is because the tails of the t distribution become increasingly "fatter" than the tails of the normal distribution as N decreases.

A ***t* value,** then, is analogous to a z score except that it represents the number of *estimated* standard errors a sample mean is from μ. The formula for calculating a t value is

$$t = \frac{\bar{X} - \mu}{\hat{s}_{\bar{X}}} \qquad [8.2]$$

which is the same as the z formula but with $\hat{s}_{\bar{X}}$ substituted for $\sigma_{\bar{X}}$ in the denominator. This is formally known as the **one-sample *t* test.** Applying this test to the example concerning attitudes toward dormitories,

$$t = \frac{4.51 - 4}{.19} = 2.68$$

Just as statisticians have studied the normal distribution extensively, a considerable amount of information is also known about the t distribution. For example, it is possible to specify the probability of obtaining a t value greater than or equal to any specified t score. This is done in a similar manner as is done with z scores in the normal distribution. As noted earlier, there is a different t distribution for each sample size. Technically, there is a different t distribution depending on the degrees of freedom associated with the t statistic, which, in the one-sample case that we are considering, equal $N - 1$.

Because of the complexity of the t distribution, Appendix D presents a table of critical t values (that is, values of t that define rejection regions) corresponding only to selected alpha levels. Instructions for using this table are presented in Appendix D. In the current example, the critical values of t that define the rejection region for an alpha level of .05, nondirectional test, and $N - 1 = 100 - 1 = 99$ degrees of freedom are approximately -1.987 and $+1.987$.* Thus, observed t values less than -1.987 or greater than $+1.987$ would lead us to reject the null hypothesis. The observed value of t in our problem is 2.68, and we therefore reject the null hypothesis. Given that the observed sample mean of 4.51 is greater than the hypothesized population mean of 4, we conclude that attitudes toward living in dormitories, on the average, are favorable.

Numerical Example

Several years ago the legal speed limit for motor vehicles on U.S. highways was changed from 65 miles per hour to 55 miles per hour. One of the reasons for this change was the fuel savings that would result if people drove at the lower speed. To determine whether people adhere to the 55-mile-per-hour limit, suppose a county performed an investigation in which the speed of 25 drivers was monitored in selected highway locations. The observed speeds are contained in Table 8.2. As can be seen, $\bar{X} = 58.00$ and $\hat{s} = 3.34$ for this sample.

The population from which this sample was drawn was conceptualized as all people who drive in this particular county. Can we conclude from these data that the average speed of this population is, in fact, greater than the 55-mile-per-hour limit? We begin by specifying a null hypothesis and an alternative hypothesis:

$$H_0: \quad \mu = 55$$
$$H_1: \quad \mu > 55$$

The alternative hypothesis dictates a directional test since we are interested only in whether people drive faster, on the average, than 55 miles per hour. For an alpha level of .05, directional test, and $N - 1 = 25 - 1 = 24$ degrees of freedom, the critical t value, taken from Appendix D, is 1.711. If the observed value of t is greater than 1.711, then the null hypothesis will be rejected.

Since $\hat{s} = 3.34$ and $N = 25$,

$$\hat{s}_{\bar{X}} = \frac{\hat{s}}{\sqrt{N}}$$
$$= \frac{3.34}{\sqrt{25}} = .67$$

* Since there are no entries in Appendix D for df = 99, we must **interpolate**—that is, estimate the relevant critical values from the critical values associated with the next lowest and the next highest degrees of freedom that are listed in the t table. In the present example, the closest listed degrees of freedom below and above 99 are 60 and 120. For an alpha level of .05, nondirectional test, the respective critical values are ± 2.000 and ± 1.980, a difference of .020 unit. Since 99 is $39/60 = .65$ of the way between 60 and 120, the positive critical value for df = 99 can be interpolated as $2.000 - (.65)(.020) = +1.987$, thus yielding an estimated negative critical value of -1.987. The same general strategy can be applied whenever the degrees of freedom of interest are not included in a statistical table.

TABLE 8.2 Data and Calculations for Driving Speed Investigation

X	X^2
55	3,025
60	3,600
60	3,600
55	3,025
57	3,249
60	3,600
55	3,025
58	3,364
63	3,969
54	2,916
65	4,225
56	3,136
61	3,721
58	3,364
55	3,025
57	3,249
59	3,481
53	2,809
59	3,481
65	4,225
56	3,136
61	3,721
55	3,025
54	2,916
59	3,481
$\Sigma X = 1{,}450$	$\Sigma X^2 = 84{,}368$
$\bar{X} = 58.00$	

$$SS = \Sigma X^2 - \frac{(\Sigma X)^2}{N}$$

$$= 84{,}368 - \frac{1{,}450^2}{25} = 268$$

$$\hat{s}^2 = \frac{SS}{N-1} = \frac{268}{25-1} = 11.17$$

$$\hat{s} = \sqrt{\hat{s}^2} = \sqrt{11.17} = 3.34$$

Thus,

$$t = \frac{\bar{X} - \mu}{\hat{s}_{\bar{X}}}$$

$$= \frac{58.00 - 55}{.67} = 4.48$$

Since the observed t of 4.48 exceeds 1.711, we reject the null hypothesis and conclude that people who drive in this county, on the average, drive faster than the 55-mile-per-hour speed limit.

Although the statistical analyses for this study are consistent with the conclusion that drivers in this particular county drive faster than 55 miles per hour, the conclusion is not definitive. Beyond the statistical analyses, one must consider the research design in order to draw an appropriate conclusion. For example, if the measurements of speed were taken during one day only, perhaps there was something unique about that particular day relative to other days that caused drivers to speed. Driving speeds are affected by weather conditions such as temperature and wind current. Or maybe the day on which the measurements were taken was a holiday, resulting in an unusual number of out-of-county drivers passing through the county on their way home from a long trip. In the final analysis, interpretation of one's data is a function of the results of statistical analyses *and* the research design used to collect the data. Factors related to this issue will be considered in detail in the next chapter.

Assumptions of the *t* Test

The one-sample *t* test is appropriate when the variable being studied is quantitative in nature and measured on a level that approximates interval characteristics. In addition, the test is based on the following assumptions:

1. The observations are independently and randomly sampled from the population of interest. In most applications, independence of observations means that the scores on the variable are provided by different individuals.
2. The scores on the variable are normally distributed in the population. This is known as the **normality assumption.**

These assumptions are important because they assure that the sampling distribution of $(\bar{X} - \mu)/\hat{s}_{\bar{X}}$ corresponds to the theoretical *t* distribution. This, in turn, assures that the incidence of Type I errors will be equal to alpha and that the incidence of Type II errors will be equal to beta, as previously discussed. This has obvious implications for the accuracy of the inferences to be drawn from the test.

While it is essential that the assumption of independent and random observations be met for the test to be valid, under some conditions the one-sample *t* test is **robust** to violations of the normality assumption. When we say that a test is *robust* to violations of a distributional assumption, we mean that the frequency of Type I and Type II errors, and, thus, the accuracy of our conclusions, is relatively unaffected as compared to when the assumption is met. As you might expect, the robustness of a test is influenced by several factors, including sample size (in general, robustness increases as sample size increases), the *degree* of violation (in general, robustness decreases as violations become more severe), and the *form* of violation. In the present context, for instance, a population might be positively skewed, negatively skewed, bimodal, leptokurtic, platykurtic, and so forth.

When a test is robust to violations of an assumption, it can be appropriately applied even when that assumption is violated. The difficulty lies in establishing that the test is indeed robust for the specific circumstances under study. We will discuss procedures for assessing violations of distributional assumptions and their inherent problems in Chapter 9, at which time we will return to the issue of the robustness of the one-sample t test.

8.10 Confidence Intervals

Suppose we are trying to estimate the population mean on an intelligence test for students at a large university. We do so by selecting a random sample of 100 students and administering the test to them. Suppose the mean score for the sample is 107.00. If we wanted to estimate the true population mean with one value, our best guess would be 107.00. But we would also not be very confident that the true population mean is exactly 107.00, because of sampling error.

Another approach to estimating the population mean would be to specify a range of values that we are relatively confident the population mean is within (for example, between 100 and 110). The larger the range of values we specify, the more likely it is that the true population mean will be contained within it. Statisticians have developed a procedure for specifying such a range of values based on sampling distributions and probability theory.

The interval to be constructed is called a **confidence interval.** The values that define the boundaries of the interval are called the **confidence limits.** The degree of confidence we have that the population mean is contained within the confidence interval is stated in terms of a probability or a percentage. The confidence interval most commonly used by researchers in the behavioral sciences is the *95% confidence interval.*

The construction of confidence intervals differs somewhat depending on whether the standard error of the mean is known or has to be estimated from sample data. As discussed below, this difference involves the use of $\sigma_{\bar{X}}$ and z versus $\hat{s}_{\bar{X}}$ and t in the formula for computing confidence intervals, according to whether $\sigma_{\bar{X}}$ is known or unknown.

Confidence Intervals When $\sigma_{\bar{X}}$ Is Known

Confidence intervals are conceptualized with respect to sampling distributions. In the present example, the relevant sampling distribution is the sampling distribution of the mean for samples of size 100. The central limit theorem tells us that this distribution will be approximately normally distributed when $\sigma_{\bar{X}}$ is known. When this is the case, we can thus invoke our knowledge of the area under the normal curve to determine the desired confidence interval. To illustrate the procedure, we will continue with the intelligence test example.

Suppose $\sigma_{\bar{X}}$ is known to equal 2.00. Given that the sampling distribution is approximately normal in shape, we know that approximately 68% of all sample means based on $N = 100$ will fall between 1 standard error below μ and 1 standard error above μ, or, in terms of raw scores, between $(1.00)(\sigma_{\bar{X}}) = (1.00)(2.00) = 2.00$ units below μ and 2.00 units above μ. Similarly, we know that 95% of all scores in a normal distribution occur between -1.96 and $+1.96$ standard deviations (in this case, standard errors) from the mean. Thus, 95% of all sample means based on $N = 100$ will fall between $(-1.96)(\sigma_{\bar{X}}) = (-1.96)(2.00) = -3.92$ and $(1.96)(\sigma_{\bar{X}}) = (1.96)(2.00) = +3.92$ raw score units of μ.

In practice, we will not know the value of μ, and, in fact, μ is what we are trying to estimate based on data from our sample. Consequently, confidence intervals are calculated around the observed sample mean, $\bar{X}$, rather than μ. Establishing confidence intervals around $\bar{X}$ as opposed to μ makes sense when it is remembered that in the absence of additional information, $\bar{X}$ is one's "best guess" about the value of μ. The observed sample mean in our example is 107.00. Thus, 95% of all sample means based on $N = 100$ in this problem are estimated to fall between

$$107.00 - (1.96)(2.00) = 107.00 - 3.92 = 103.08$$

and

$$107.00 + (1.96)(2.00) = 107.00 + 3.92 = 110.92$$

The values 103.08 and 110.92 constitute the confidence limits and the range of 103.08 to 110.92 is the confidence interval.

The above calculations reflect the general formula

$$\text{CI} = \bar{X} - (z)(\sigma_{\bar{X}}) \text{ to } \bar{X} + (z)(\sigma_{\bar{X}}) \qquad [8.3]$$

where CI is the abbreviation for "confidence interval," $\bar{X} - (z)(\sigma_{\bar{X}})$ is the *lower confidence limit,* and $\bar{X} + (z)(\sigma_{\bar{X}})$ is the *upper confidence limit.* In this equation, z is the z score corresponding to 1.00 minus the confidence level. When using Appendix B to find the value of the desired z, always refer to the nondirectional column (column 4), as the goal is to establish equivalent ranges below and above $\bar{X}$. In the example just given, we were concerned with the 95% confidence interval, so 1.00 minus the confidence level was $1.00 - .95 = .05$ and the appropriate value of z was thus 1.96.

The use of a single sample mean in the construction of a confidence interval raises an important interpretational issue. Suppose we draw all possible random samples of size N from a population and for each sample mean, compute the 95% confidence interval using Equation 8.3. What we would find is that 95% of these confidence intervals would contain the value of μ, and 5% of the confidence intervals would not. *A confidence claim reflects a long-term performance of an extended number of confidence intervals across all possible random samples of a given size.* In practice, only one confidence interval is constructed and that one interval either contains the population mean or does not contain the population mean. We never know for sure whether a particular confidence interval contains μ. However, most applications that use confidence intervals are concerned with the 95% confidence level. With such a confidence level, we can be reasonably certain that the observed confidence interval does contain μ, since μ will be included in 95% of the 95% confidence intervals constructed from samples of size N.

Sometimes investigators will construct 99% confidence intervals (in which case $z = 2.575$) rather than 95% confidence intervals. The confidence level selected should depend on how important it is not to have a "false" interval (that is, one that does not contain μ). The lower the confidence level (for example, 80% as opposed to 95%), the more likely it is that false intervals will be observed. On the other hand, everything else being equal, the higher the degree of confidence (for example, 99% as opposed to 95%), the wider the confidence interval. The problem with wide confidence intervals is that as the width of the confidence interval increases, the larger is the range of values that might contain μ. At some point, the range of the confidence interval becomes so large that the utility of the confidence interval is diminished. In our intelligence example, the 95% interval is 103.08 to 110.92, whereas the 99% interval is

$$\begin{aligned} \text{CI} &= \bar{X} - (z)(\sigma_{\bar{X}}) \text{ to } \bar{X} + (z)(\sigma_{\bar{X}}) \\ &= 107.00 - (2.575)(2.00) \text{ to } 107.00 + (2.575)(2.00) \\ &= 101.85 \text{ to } 112.15 \end{aligned}$$

Since the width of a confidence interval is influenced by $\sigma_{\bar{X}}$ in addition to the confidence level, it follows that the sample size, N, will also affect how wide the interval is. As N becomes larger, $\sigma_{\bar{X}}$ becomes smaller, meaning that as N becomes larger, the width of the confidence interval decreases, other things being equal.

Confidence Intervals When $\sigma_{\bar{X}}$ Is Unknown

The structure of the formula for confidence intervals when $\sigma_{\bar{X}}$ is unknown is the same as when $\sigma_{\bar{X}}$ is known, except that the t distribution is used in place of the z distribution and $\hat{s}_{\bar{X}}$ is used in place of $\sigma_{\bar{X}}$. This formula can be represented as

$$\text{CI} = \bar{X} - (t)(\hat{s}_{\bar{X}}) \text{ to } \bar{X} + (t)(\hat{s}_{\bar{X}}) \qquad [8.4]$$

where t is the nondirectional t value corresponding to 1.00 minus the confidence level and all other terms are as previously defined. When determining the value of t to be used in Equation 8.4, the appropriate degrees of freedom are, as usual, $N - 1$.

For the driving speed example presented in Section 8.9, where $N = 25$, $\bar{X} = 58.00$, and $\hat{s}_{\bar{X}} = .67$, the 95% confidence interval is

$$\begin{aligned} \text{CI} &= \bar{X} - (t)(\hat{s}_{\bar{X}}) \text{ to } \bar{X} + (t)(\hat{s}_{\bar{X}}) \\ &= 58.00 - (2.064)(.67) \text{ to } 58.00 + (2.064)(.67) \\ &= 56.62 \text{ to } 59.38 \end{aligned}$$

Thus, there is a high probability that the true value of the population mean is within the range of 56.62 to 59.38.

The calculation of confidence intervals bears a relationship to aspects of hypothesis testing. Consider the case of testing the following null hypothesis:

$$H_0: \quad \mu = 100$$

Suppose this hypothesis is tested for $N = 25$, $\bar{X} = 110.00$, and $\hat{s}_{\bar{X}} = 2.00$ at an alpha level of .05, nondirectional test, and that the null hypothesis is rejected. The 95% confidence interval about the mean is

$$\begin{aligned} \text{CI} &= 110.00 - (2.064)(2.00) \text{ to } 110.00 + (2.064)(2.00) \\ &= 105.87 \text{ to } 114.13 \end{aligned}$$

Note that the value of μ stated in the null hypothesis (100) is *not* contained within this interval. As it turns out, any null hypothesis that specifies a population mean value outside the above interval would be rejected based on the sample data. Further, any null hypothesis that specifies a population mean within the interval would not be rejected. Because of this, confidence intervals provide the researcher with more information than the standard hypothesis testing procedures discussed earlier in this chapter. Specifically, the interval indicates to the investigator all null hypotheses that would or would not be rejected with a nondirectional test. Consequently, some behavioral scientists have advocated the reporting of confidence intervals rather than the results of formal hypothesis tests about specific values of μ when providing statistical results. Nevertheless, the large majority of research reports that you will encounter will use the formal hypothesis testing procedures presented earlier rather than the present strategy, formally known as **interval estimation.**

Before concluding our discussion of confidence intervals, we should note that confidence intervals can be constructed for many parameters other than the mean, although these are rarely encountered in practice.

STUDY EXERCISE 8.2

An investigator administered a reading test to a sample of 30 students and found $\bar{X} = 83.00$ and $\hat{s} = 17.35$. Calculate the 95% and 99% confidence intervals.

Answer Since the population standard deviation is unknown, the standard error of the mean must be estimated using Equation 7.6:

$$\hat{s}_{\bar{X}} = \frac{\hat{s}}{\sqrt{N}} = \frac{17.35}{\sqrt{30}} = 3.17$$

Since $\text{df} = N - 1 = 30 - 1 = 29$, the t value used in determining the 95% confidence interval in this case is 2.045. Applying Equation 8.4, we find that the 95% confidence interval is

$$\begin{aligned} \text{CI} &= \bar{X} - (t)(\hat{s}_{\bar{X}}) \text{ to } \bar{X} + (t)(\hat{s}_{\bar{X}}) \\ &= 83.00 - (2.045)(3.17) \text{ to } 83.00 + (2.045)(3.17) \\ &= 76.52 \text{ to } 89.48 \end{aligned}$$

The t value used in determining the 99% confidence interval is 2.756, so the 99% confidence interval is

Study Exercise 8.2 continued

$$CI = 83.00 - (2.756)(3.17) \text{ to } 83.00 + (2.756)(3.17)$$
$$= 74.26 \text{ to } 91.74$$

8.11

Method of Presentation

The *Publication Manual of the American Psychological Association* (American Psychological Association, 1983) states that when presenting the results of a statistical test, "give the symbol, degrees of freedom, value, and probability level. . . . In addition, give the mean, standard deviation, or other descriptive statistic to clarify the nature of the effect" (p. 80). We will illustrate this procedure in the case of the one-sample *t* test by focusing on the driving speed example discussed in Section 8.9. The results for this problem might be reported as follows:

Results

A one-sample t test was performed comparing the sample mean against a hypothesized population mean of 55. The sample mean of 58.00 was found to be significantly different from this value, t(24) = 4.48, p < .0005, one-tailed, suggesting that the mean driving speed in the monitored county is greater than 55 miles per hour.

The heading "Results" identifies this as a results section. Other sections of a research report are similarly headed "Method" and "Discussion." A fourth main section, the introduction, is not formally labeled, as this section always appears at the very beginning of a research article.

The first sentence states the type of test that was conducted and the value of μ specified in the null hypothesis. It should be noted that the null and alternative hypotheses are not formally written out when reporting the results of a statistical test. This is because the hypothesis testing steps discussed earlier in this chapter are implied whenever a statistical test is reported, so writing out each step would be redundant and, consequently, a poor use of journal space.

The second sentence begins by specifying the value of the observed sample mean and then presents selected aspects of the statistical analysis, per American Psychological Association requirements. The symbol t indicates that a

t test was performed. This is followed, in parentheses, by the degrees of freedom associated with the relevant t distribution. Next comes the observed value of t computed using Equation 8.2. The statement "$p < .0005$" indicates that the probability of obtaining a t value as extreme as the one observed in the study, assuming the null hypothesis is true, is less than .0005. The value associated with p is referred to as a **probability level** or a **significance level.** This is distinct from the alpha level—an alpha level reflects the researcher's decision as to how extreme the results of a statistical test should be before the null hypothesis is rejected, whereas a significance level represents the probability of obtaining a result as extreme as the one that was actually observed. The terminology *one-tailed* is used to indicate that a directional test was used. If this is not explicitly stated, a nondirectional test is assumed.

In the present example, the observed t value of 4.48 not only exceeds the critical value of 1.711 defining the (directional) .05 rejection region, but it also exceeds the critical values of 2.064, 2.492, 2.797, and 3.745, respectively defining the .025, .01, .005, and .0005 rejection regions. The significance level is thus .0005. The advantage of reporting significance levels rather than alpha levels is that different readers might have different ideas concerning where alpha should be set, and if the probability associated with the observed result is specified, readers can immediately determine whether the null hypothesis would have been rejected had a more conservative (lower) alpha level been adopted.

As discussed earlier in this chapter, alpha is typically set equal to .05 by convention. If a different value of alpha is employed, justification for this should be provided in the report.

A results section takes the same general form when the null hypothesis is not rejected as when it is. As before, the observed sample mean, the degrees of freedom, and the observed t value are reported. However, the statement of the significance level is replaced by the notation "*ns*." This is an abbreviation for *not significant* and indicates that the test was statistically nonsignificant—that is, that the probability of obtaining the observed result was *not* less than the adopted alpha level.

8.12

Examples from the Literature

Accuracy of Subjective Life Expectancies

A subjective life expectancy is an individual's estimate of what age he or she expects to live to. One interesting question is how subjective life expectancies relate to actual life expectancies as represented by actuarial predictions. In a study of this issue, Robbins (1988) asked 49 female and 27 male college students to indicate their expected age in response to the question, "Approximately how long do you expect to live?" These estimates were then compared with actuarial predictions from the National Center for Health Statistics. Results showed that the mean subjective life expectancy for females (77.2 years) did not significantly differ from the actuarial prediction of 79.2 years, $t(48) = -.85$, *ns*. The mean subjective life expectancy of 77.6 years for males, on the other hand, was significantly greater than the actuarial prediction of 72.4 years, $t(26) = 2.49$, $p < .02$, thus indicating that males tend to overestimate their life spans.

Validation of a Priming Procedure

According to a cognitive principle known as the *availability heuristic,* the more readily instances of a class of objects come to mind, the larger that class of objects is judged to be. For instance, it has been found that people estimate that more words begin with the letter *K* than have *K* in the third position when, in actuality, the reverse is true. This bias presumably results from the fact that it is easier to think of words that start with *K* than words that have *K* as the third letter.

One factor that has been hypothesized to influence the ease with which instances of a class of objects come to mind and subsequent frequency judgments is the recency with which that class has been cognitively activated, or primed, through prior exposure; the more recent the exposure, the more available that information should be in memory and the greater the estimate of the class size should be. According to this perspective, this should be true even if the prior exposure occurs on an unconscious, or subliminal, level. In a test of this proposition, Gabrielcik and Fazio (1984) asked 15 college students to judge the frequency with which the letter *T* appears in the English language. Subjects had previously been assigned to conditions where they were either subliminally exposed to a series of 40 words containing the letter *T* (primed condition) or subliminally exposed to strings of asterisks (control condition). If priming has the predicted effect on frequency judgments, subjects in the primed condition should estimate the letter *T* to be more common than should subjects in the control condition.

Crucial to this experiment was the establishment that the presentation of the 40 *T* words was indeed subliminal. This was accomplished in a preliminary study by asking eight undergraduate subjects to view a series of 40 words flashed before them for 1/500 of a second each. Unbeknownst to the subjects, each of these words contained one or more *T*s. After each presentation, subjects were given a slip of paper containing four words and instructed to circle the word they had just viewed, and to make a guess if uncertain. If the presentation of the words was indeed subliminal, subjects' responses to the recognition test should have been no better than chance. For a 40-item test having four response options, a chance result is $(40)(.25) = 10.00$ correct responses. The subjects' actual mean of 9.13 correct responses was compared with this value using a one-sample *t* test. This test was found to be statistically nonsignificant, $t(7) < 1$ (the specific value of *t* was not reported), *ns*, thus suggesting that word exposure of 1/500 of a second might be sufficiently short to be subliminal. With this established, Gabrielcik and Fazio were able to proceed with the main experiment. Consistent with their hypothesis, it was found that subjects primed with the subliminal *T* words estimated the letter *T* to appear significantly more frequently than did the control subjects.

8.13 Summary

In this chapter, we have considered the basic logic underlying hypothesis testing. In doing so, we have introduced two statistical tests for testing a hypothesized mean value. These tests are used whenever an investigator wants to test the viability of an assertion that a given population has a specific mean score on some variable. The z test is used when the standard error of the mean is known, whereas the t test is used when the standard error of the mean must be estimated.

The logic of hypothesis testing begins with the specification of a null hypothesis (a hypothesis that is tentatively accepted as being true for purposes of the statistical test) and an alternative hypothesis. The alternative hypothesis can be either directional or nondirectional, depending on the nature of the question being asked. Next, an alpha level, usually .05, is specified and, based on this (and in the case of the t test, the degrees of freedom), a rejection region is defined. The data are collected and analyzed by converting the observed sample mean into a z or t value, as appropriate. This observed z or t value is compared with the critical z or t values that define the rejection region, and a conclusion is drawn. The null hypothesis is never accepted. Rather, we either reject it or fail to reject it.

When drawing a conclusion from a statistical test, we recognize that there is a possibility of error. The two types of errors we can make are (1) rejecting a true null hypothesis (Type I error), and (2) failing to reject a false null hypothesis (Type II error). The probability of making a Type I error is indicated by alpha (α) and the probability of making a Type II error is indicated by beta (β). The probability that an investigator will correctly reject the null hypothesis is called the power of the test and is indicated by $1 - \beta$.

One approach to estimating the value of a population parameter is to specify a range of values that has a high probability of containing the true population score. Such a range of values is called a confidence interval. When the standard error of the mean is known, confidence intervals about the mean can be determined from our knowledge of the area under the normal curve. When the standard error of the mean is estimated from sample data, the t distribution is used instead.

Exercises

Answers to asterisked (*) exercises appear at the back of the book. Answers to exercises with two asterisks are also worked out step-by-step in the Study Guide.

Exercises to Review Concepts

1. Define each of the following:
   a. null hypothesis
   b. alternative hypothesis
   c. critical values
   d. rejection region
   e. alpha level

*2. Why is it necessary to assume the null hypothesis is true in the context of hypothesis testing?

3. Summarize the five steps involved in hypothesis testing for the one-sample z or t test.

*4. Why can we never "accept" the null hypothesis in traditional statistical tests?

*5. An economist is interested in whether high school students in a certain geographical area save more or less than $100 per month toward their college education. Translate this question into a null hypothesis and an alternative hypothesis.

*6. Given H_0: $\mu = 4$, H_1: $\mu \neq 4$, $\bar{X} = 7.31$, $\sigma = 14.78$, and $N = 64$, calculate the observed value of z.

*7. Test the viability of the null hypothesis for the problem in Exercise 6 and draw a conclusion about the actual value of μ.

8. Given H_0: $\mu = 10$, H_1: $\mu \neq 10$, $\bar{X} = 3.80$, $\sigma = 2.77$, and $N = 81$, calculate the observed value of z.

9. Test the viability of the null hypothesis for the problem in Exercise 8 and draw a conclusion about the actual value of μ.

10. Why is it important to use the phrases "statistically significant" and "statistically nonsignificant" rather than the phrases "significant" and "nonsignificant" when discussing the results of statistical tests?

11. Define each of the following:
   a. Type I error
   b. Type II error
   c. alpha
   d. beta
   e. power

*12. What is the relationship between alpha and the probability of a Type I error? What is the reason for this relationship?

*13. What is the relationship between power and the probability of a Type II error? What is the reason for this relationship?

*14. What is the relationship between alpha and power? What is the reason for this relationship?

15. What effect does sample size have on the power of a statistical test?

16. Under what circumstances should a directional rather than a nondirectional test be used? Why? Under what circumstances should a nondirectional rather than a directional test be used? Why?

17. When is the t distribution used instead of the normal distribution to test hypotheses about population means?

18. Under what condition does the t distribution closely approximate the normal distribution?

*19. State the critical value(s) of t that would be used to reject the null hypothesis for a one-sample t test at an alpha level of .05 under each of the following conditions:
   a. H_0: $\mu = 3$, H_1: $\mu > 3$, $N = 20$
   b. H_0: $\mu = 3$, H_1: $\mu < 3$, $N = 20$
   c. H_0: $\mu = 3$, H_1: $\mu \neq 3$, $N = 20$
   d. H_0: $\mu = 3$, H_1: $\mu > 3$, $N = 6$
   e. H_0: $\mu = 3$, H_1: $\mu < 3$, $N = 10$
   f. H_0: $\mu = 3$, H_1: $\mu \neq 3$, $N = 10$

*20. Suppose you wanted to test whether a sample was drawn from a population with $\mu = 100$ on an intelligence test. Calculate the t value and evaluate the viability of the null hypothesis for each of the following cases:

a. $\bar{X} = 101.00$, $\hat{s} = 10.00$, $N = 10{,}000$
b. $\bar{X} = 101.00$, $\hat{s} = 10.00$, $N = 100$
c. $\bar{X} = 101.00$, $\hat{s} = 2.00$, $N = 100$

In each of the above cases, the difference between the sample mean ($\bar{X} = 101.00$) and the hypothesized population mean ($\mu = 100$) is the same. Why is the result in part **a** different from the result in part **b**? Why is the result in part **b** different from the result in part **c**?

21. A researcher administered a measure of life satisfaction to a sample of 30 individuals and found a mean of 121.00 and a standard deviation estimate of 10.77. Test the viability of the hypothesis that the true population mean equals 110 using a nondirectional test.

22. What are the assumptions underlying the one-sample *t* test? Why are these assumptions important?

23. What is a confidence interval?

***24.** A researcher administered a test of mathematical ability to a sample of 169 students. The mean score for the sample was 74.40. The standard deviation for the population, σ, was known to be 13.00. Compute the 95% and 99% confidence intervals.

***25.** In Exercise 24 suppose the sample size were 200 instead of 169. Compute the 95% and 99% confidence intervals in this case. What is the effect of increasing *N* on the width of the intervals?

26. In Exercise 24 suppose σ were equal to 7.00 instead of 13.00. Compute the 95% and 99% confidence intervals in this case. What is the effect of decreasing the size of σ on the width of the intervals?

****27.** Compute the 95% and 99% confidence intervals for Exercise 21.

***28.** What is a probability level or significance level? How does this differ from the alpha level?

Exercises to Apply Concepts

****29.** Population trends in the United States have shown a decrease in the fertility rate during recent years. Presently, the fertility rate is below the zero population growth level. The fertility rate corresponding to zero population growth is 2.11 (that is, if couples average 2.11 children, then the size of the population will remain stable). One factor that has been shown to be related to family size is religion. A researcher was interested in whether Catholics in the United States are having children at a rate consistent with zero population growth. The following data on the number of children were obtained for a representative sample of Catholics. Test the viability of the hypothesis that Catholics are reproducing at a zero population growth rate using a nondirectional test, draw a conclusion, and report your results using the principles developed in the Method of Presentation section.

Number of children				
4	5	2	3	4
6	3	3	2	2
2	4	0	3	3
1	3	4	5	0
2	1	8	2	2

30. A large number of studies have investigated the effects of marijuana on human physiology and behavior. It is commonly believed by laypersons that marijuana does affect physiological processes in the human. One of the more common beliefs is that smoking marijuana makes one hungry (gives one the "munchies"). Weil, Zinberg, and Nelson (1968) examined this influence empirically. Ten subjects participated in the study. Each received a high dose of marijuana by smoking a potent marijuana cigarette. The subjects were males, 21 to 26 years of age, all of whom smoked tobacco cigarettes regularly but had never tried marijuana. The precise physiological mechanism that causes hunger is not well understood by psychologists. However, one factor that is often associated with hunger is blood sugar. Several theorists have suggested that this mechanism may be the cause of marijuana-induced hunger. The present experiment measured the level of blood sugar for each subject before he smoked marijuana and again 15 minutes after he smoked marijuana. A "change score" was then computed for each subject by subtracting the amount of sugar in the blood after smoking marijuana from the amount of sugar in the blood before smoking marijuana. Hypothetical data regarding blood sugar levels (measured in mg/100 ml) representative of the results of the study are presented below. If marijuana has no effect on blood sugar level, we would expect the mean change score in the population to equal 0. Test the viability of this hypothesis using a nondirectional test, draw a conclusion, and report your results using the principles developed in the Method of Presentation section.

Change score (before − after)	
14	−2
−2	−6
6	−2
−2	−18
−2	−6

2

THE ANALYSIS OF BIVARIATE RELATIONSHIPS

NINE

RESEARCH DESIGN AND STATISTICAL PRELIMINARIES FOR ANALYZING BIVARIATE RELATIONSHIPS

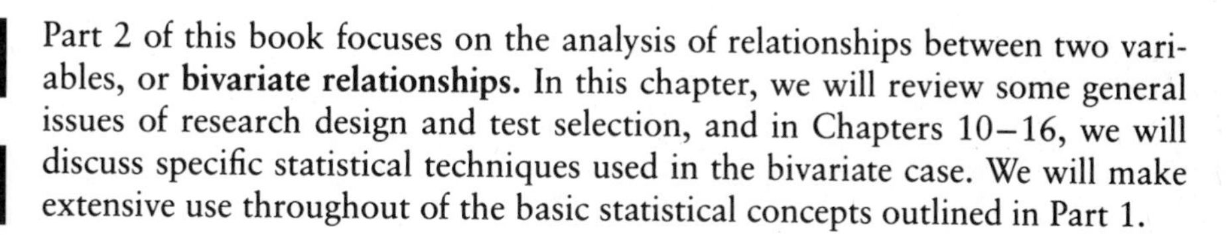

Part 2 of this book focuses on the analysis of relationships between two variables, or **bivariate relationships.** In this chapter, we will review some general issues of research design and test selection, and in Chapters 10–16, we will discuss specific statistical techniques used in the bivariate case. We will make extensive use throughout of the basic statistical concepts outlined in Part 1.

9.1

Principles of Research Design: Statistical Implications

In order to facilitate an understanding of the use of statistics in interpreting research, it will be instructive to review general principles that guide research design. As noted in Chapter 1, statistics and research design are highly interwoven with one another, and a consideration of design principles will lead to a better understanding of the statistics considered in later chapters.

Two Strategies of Research

When studying the relationship between two variables, the investigator is essentially interested in determining how the different values of one variable are associated with the values of another variable. For instance, if one were interested in studying the relationship between gender (male versus female) and mathematical ability, the concern would be with whether the different values on the variable of gender are associated with different values on the variable of mathematical ability (for example, whether males have higher mathematical aptitude than females).

Behavioral scientists use two general strategies for assessing the relationship between variables. First, they may use what is referred to as an **experimental strategy,** where a set of procedures or manipulations is performed in order to *create* different values of the independent variable for the research participants. The relationship of these manipulations to the dependent variable is then examined. For example, if the independent variable is test anxiety, the researcher might create three different values on this variable by telling one-third of the research participants that their performance on a test is very important and will reveal many aspects of their personal competencies (thus creating high test anxiety), telling another third that the task is unimportant and will not reflect on them personally (thus creating low test anxiety), and not addressing the issue of the test's importance with the final third. In this instance, the variable of test anxiety has three values, or *levels,* and each research participant can be distinguished with respect to which value describes that participant.

Notice that in this example, the third group of participants was not actually exposed to the independent variable. A group of this type is formally known as a **control group.** The advantage of including a control group when utilizing an experimental strategy is that it provides a *baseline* for evaluating the effects of the experimental manipulation. Suppose, for instance, that in the above study we found that high test-anxious subjects scored higher than low test-anxious subjects on the relevant dependent variable. If a control group were not incorporated into the design, we would be unable to determine if this was due primarily to high test anxiety *increasing* scores on the dependent variable, low test anxiety *decreasing* scores on the dependent variable, or some combination of the two. However, by including a control group, we can compare the dependent variable scores of each experimental group with those that naturally occur in the absence of the manipulation and thus determine the extent to which each value of the independent variable influences performance on the behavior of interest.

In contrast to an experimental strategy, an **observational strategy** does not involve the process of actively creating values on an independent variable, but rather involves measuring differences in values that naturally exist in the research participants. For instance, a person's gender might be measured in a questionnaire based on response to the question, "What is your sex, male or female?" In this instance the researcher is not using a set of manipulations to create values on a variable, but rather is measuring the values that naturally exist.

Experimental and observational strategies are often used in concert. For example, a study might investigate the effects of a person's gender *and* test anxiety on test performance. Gender would be indexed by an observational strategy and test anxiety could be indexed by the manipulations noted above. The effects of both of these variables on test performance might then be examined.

A dependent variable is always measured in the "observational" sense noted above. This is because we are trying to determine how the dependent measure responds to the manipulations of the independent variable or varies with the naturally existing values of the independent variable. In this context, it is important to note that many of the statistical techniques that we will be discussing in subsequent chapters are equally applicable to the experimental and the observational situations.

Random Assignment to Experimental Groups

A major goal of research design is to control for alternative explanations. Consider an experiment in which an investigator is interested in the effects of alcohol on reaction time. Two mixed-gender groups of subjects from a small college are used. One-half respond to a reaction time task while under the influence of alcohol and the other half respond to the same reaction task while not under the influence of alcohol. The task involves pressing a button when a certain slide appears in a series of slides shown sequentially. The investigator computes the mean reaction time in each group. Suppose the mean reaction time in the alcohol condition was 2.45 seconds, whereas in the no-alcohol group the mean reaction time was .98 second. It appears that alcohol has had an effect. However, there are alternative explanations. First, it may be the case that the alcohol did *not* affect reaction time, and that the difference between means is simply the result of subjects in the alcohol condition having slower reaction times than the other subjects, independent of alcohol. If the study had been conducted without giving alcohol to any of the subjects, perhaps the results would still have yielded means of 2.45 and .98, respectively.

In order to control for this possibility, investigators typically assign subjects to groups using random assignment procedures similar to those used for selecting random samples (for instance, random number tables). If the condition in which a given subject participates is determined on a completely random basis, then it is no more likely that subjects assigned to the alcohol condition will have slower reaction times than those assigned to the no-alcohol condition, independent of alcohol. Thus, **random assignment** helps to control for alternative explanations such as the above. Of course, random assignment is feasible only when an investigator is using a manipulative experimental strategy to "create" values on an independent variable. If gender is the independent variable, we cannot randomly assign subjects to the conditions "male" and "female." By definition, males are males and females are females. Thus, random assignment is not possible when an observational strategy is employed.

It is important to note that random assignment *does not guarantee* that the research groups will not differ beforehand on the dependent variable. Rather, it is *unlikely* that they will. There is always the chance that, even with random assignment, the various groups will differ on the dependent variable. In general, the larger the number of subjects randomly assigned to each condition, the less likely it is that this will happen. Consequently, one should be cautious when interpreting studies with small numbers of subjects.

Reducing Sampling Error

A second approach to controlling for alternative explanations focuses on sampling error. Consider the experiment on alcohol and reaction time. In this study, the investigator conceptualizes the two groups of subjects as random samples from two populations: (1) a population of individuals who are under the influence of alcohol and (2) a population of similar individuals who are not under the influence of alcohol. The two populations are assumed to be similar to each other in all respects except one—the presence or absence of alcohol. This is a reasonable assumption given the random assignment of subjects to experimental conditions. If the alcohol has no effect on reaction time, then the two populations are, for all intents and purposes, similar in *all* respects and one would expect the mean reaction time scores for the two populations to be the same. Stated in formal terms, we would expect

$$\mu_A = \mu_N$$

where μ_A represents the population reaction time mean for individuals under the influence of alcohol, and μ_N represents the population reaction time mean for individuals not under the influence of alcohol. If the alcohol *does* have an effect on reaction time, then we would expect the population means to be different, or

$$\mu_A \neq \mu_N$$

Suppose the mean reaction time in the alcohol condition was found to be 2.45 and the mean reaction time in the no-alcohol condition was found to be .98. This would appear to be consistent with the notion that $\mu_A \neq \mu_N$. However, we know from Chapter 7 that a sample mean does not usually equal the population mean, due to sampling error. Perhaps the two population means are equal and the results of the experiment are nothing more than the result of sampling error. Ideally, we would like to minimize sampling error so as to rule out this interpretation as an alternative explanation. We will now consider procedures that can be used to accomplish this.

Recall from Chapter 7 that two factors that influence the accuracy of a sample mean as an estimate of μ are the size of the sample (N) and the variance of scores in the population (σ^2). One way an investigator can reduce sampling error is to increase the sample sizes for the various groups. In general, the larger the N in each group, the less the sampling error. Obviously, there are practical limitations to the number of subjects that can participate in a research study. Research is expensive, and the time and effort involved in collecting data can be quite consuming. As such, researchers will sometimes have to settle for relatively small sample sizes.

A second procedure for reducing sampling error is to define the populations so that the variances of scores in the groups (σ^2) will be relatively small. Consider the alcohol and reaction time example. There is some evidence to indicate that males, in general, have slightly faster reaction times than females on tasks similar to that used in the experiment. The population of individuals in the alcohol condition consists of both males and females, as does the population of individuals in the no-alcohol condition. The presence of both males and females within the populations yields more variability in reaction time scores than if either gender were considered separately. This can be illustrated with the following hypothetical sets of reaction time scores:

Males	*Females*	*Males and females combined*	
1.10	1.30	1.10	1.30
1.20	1.40	1.20	1.40
1.30	1.50	1.30	1.50
$\mu_M = 1.20$	$\mu_F = 1.40$	$\mu_C = 1.30$	
$\sigma^2_M = .007$	$\sigma^2_F = .007$	$\sigma^2_C = .017$	

Note that the reaction time for males is slightly faster, on the average, than the reaction time for females. Note also that the variance of the combined groups ($\sigma^2_C = .017$) is greater than the variance of either group considered separately ($\sigma^2_M = .007$ and $\sigma^2_F = .007$).

One procedure for reducing sampling error would be to restrict the experiment to female subjects. This would have the effect of yielding a smaller population variance within the experimental and control groups relative to a study using both males and females. Unfortunately, to reduce σ^2, we have had to restrict the generalizability of the results of the study to females only. Behavioral scientists frequently find themselves in such trade-off situations. The researcher must weigh the theoretical and applied benefits of reducing sampling error against the cost of reducing the generalizability of one's results.

Control of Confounding and Disturbance Variables

Much of behavioral science research is designed to study relationships between independent and dependent variables. In order to draw unambiguous inferences about such relationships, it is necessary to control other variables in the research setting. In the previous sections we have implicitly considered two basic types of variables a researcher must control. In this section, we will make the nature of these variables explicit.

One type of variable that a researcher seeks to control is a *confounding variable*. Suppose a researcher wanted to test for the existence of sex discrimination and did so by examining the relationship between the gender of a job applicant and a person's decision to hire that individual. An experiment might be designed in which the résumés of 50 applicants (25 males and 25 females) are gathered from personnel files. A group of personnel directors is then asked to read each résumé and rate on a 20-point scale the likelihood that they would hire each applicant. Suppose the appropriate statistical analysis indicated there was, in fact, a relationship between the applicant's gender and the likelihood-of-hiring ratings, with males being more likely to be hired than females. Would this be evidence for sex discrimination? Not necessarily. The strongest evidence for sex discrimination would occur when males are chosen over females who are equally or more qualified. However, in the above experiment, it may have been the case that the females were, in fact, less qualified than the males. Rather than the judgments being a function of the applicants' gender, they may have simply reflected the applicants' qualifications, which just happened to be related to their gender. If this were the case, the test for sex discrimination would be ambiguous since the result could be attributed either to sex discrimination *or* to differences in the quality of applicants. In this experiment, the qualifications of the applicants represent a **confounding variable.** *A confounding variable is one that is related to the independent variable (the presumed influence) and that affects the dependent variable (the presumed effect), rendering a relational inference between the independent variable and the dependent variable ambiguous.* In a study on the effect of alcohol on reaction time, the failure to randomly assign subjects to conditions could result in differences in reaction times from one group to another due to individual differences. Random assignment is one method for controlling confounding variables defined by individual differences.

A second type of variable that must be controlled in research aimed at drawing relational inferences is a **disturbance variable.** *A disturbance variable is one that is unrelated to the independent variable (and hence, not confounded with it) but that influences the dependent variable.* As a result, a disturbance variable increases sampling error by increasing variability within groups. In the alcohol and reaction time study, the gender of the subject was an example of a disturbance variable. Disturbance variables serve to obscure or mask a relationship that exists between the independent and dependent variables. To use an electronics analogy, they create "noise" in a system where we are trying to detect a "signal" and the more "noise" there is, the harder it is to detect the "signal."

There are several procedures that behavioral scientists use to control for confounding and disturbance variables. Three of the most common strategies, (1) holding a variable constant, (2) matching, and (3) random assignment to experimental groups, will be discussed below. The first of these is applicable to both confounding and disturbance variables, while the last two can be applied only to confounding variables.

Consider the following example. In the sociological literature, a relationship has been established between a woman's religion and how many children she wants to have in her completed family (that is, her ideal family size). Generally speaking, Catholics want more children than Protestants, who in turn want more children than Jews. Sociologists have interpreted this in terms of the effects of the religious doctrine to which these women are exposed. Another interpretation is possible, however. Catholics tend to come from larger families than Protestants, who tend to come from larger families than Jews. It may not be religion that influences family size desires but rather the fact that those people who are raised in large families prefer large families and those people who are raised in small families prefer small families. Thus, religion and the size of the family raised in might be confounded with one another. How might family size background be controlled?

We saw earlier with the alcohol and reaction time example how **holding a variable constant** could be used to control for a disturbance variable (subjects' gender). This procedure can also be used to control for a confounding variable, such as family size background. For example, a study might be undertaken with Catholics, Protestants, and Jews who all come from families with two children. In this case, family size background and religion are *not* related since family size background has been held constant. Differences in ideal family size among the three religious groups cannot be attributed to differences in family size background since everyone in the study comes from the same-sized family. The variables are no longer confounded.

As noted earlier, the major disadvantage of holding a variable constant is that it may restrict the generalizability of the results. If the above study is conducted only on individuals who come from families with two children, would

BOX CONFOUNDING AND DISTURBANCE VARIABLES

The identification of confounding and disturbance variables is critical for evaluating any research design. Huck and Sandler (1979) have presented an interesting and very readable collection of 100 studies that have received attention in the popular press or professional forums. For each one, they provide a description and elaborate some confounding and disturbance variables that could affect the interpretation of the results. An example from their book illustrates their approach and underscores the importance of controlling such variables.

Problem The following story appeared about an advertisement in a weekly news magazine as well as in the local newspapers—you may have seen it yourself. It seems that the Pepsi-Cola Company decided that Coke's three-to-one lead in Dallas was no longer acceptable, so they commissioned a taste-preference study. The participants were chosen from Coke drinkers in the Dallas area and asked to express a preference for a glass of Coke or a glass of Pepsi. The glasses were not labeled "Coke" and "Pepsi" because of the obvious bias that might be associated with a cola's brand name. Rather, in an attempt to administer the two treatments (the two beverages) in a blind fashion, the Coke glass was simply marked with a "Q" and the Pepsi glass with an "M." Results indicated that more than half chose Pepsi over Coke. Besides a possible difference in taste, can you think of any other possible explanation for the observed preference of Pepsi over Coke?

Solution After seeing the results of the Pepsi experiment, the Coca-Cola Company conducted the same study, except that Coke was put in both glasses. Participants preferred the letter "M" over the letter "Q," thus creating the plausible rival hypothesis that letter preference rather than taste preference could have accounted for the original results. [In other words, the type of beverage might have been confounded with the type of letter used to label the glasses.] Since no statistical tests were given, another plausible rival hypothesis is that of instability; that is, we don't know whether "more than half" means 51% or 99% or how much confidence we should place in the finding. Flipping a coin 100 times is almost sure to result in either heads or tails occurring more than half the time.

Strangest of all was the fact that the same design error of using one letter exclusively for each brand was repeated in a second study conducted by Pepsi. In a feeble attempt to demonstrate that their initial results were not biased by the use of an "M" or a "Q," Pepsi duplicated their first study, this time using an "L" for Pepsi and an "S" for Coke! Clearly, these three studies indicate that there is sometimes more in advertisements than meets the eye (or the taste buds).

the results generalize to individuals who come from families with five children? Perhaps religion influences family size desires when one comes from a relatively small family (two children), but not when one comes from a relatively large family (five children). When we hold a variable constant, we have no way of knowing the extent to which the results will generalize across the different levels of the variable that is held constant. One way to circumvent this would be to design a study in which one studied the effect of religion at each of several levels of family size background (for example, one child, two children, three children, and so on), separately. In fact, this procedure is part of a research strategy called *factorial designs,* which is discussed in Chapter 18.

A second strategy that is used to control for confounding variables is a technique called **matching.** In this approach, an individual in one group is "matched" with an individual in each of the other groups such that all of them have the same value on the confounding variable. This strategy is different from holding a variable constant, since within a group the confounding variable can vary considerably. However, for each individual in one group, there is a comparable individual in each of the other groups who has the same value on the confounding variable. As an example, 15 subjects (5 in each group) having the following family size backgrounds might be selected for inclusion in the religion and ideal family size study:

Catholics	*Jews*	*Protestants*
3	3	3
4	4	4
2	2	2
3	3	3
1	1	1
$\bar{X} = 2.60$	$\bar{X} = 2.60$	$\bar{X} = 2.60$

Note that the average family size background is identical in all three groups. Thus, any differences in the mean *ideal* family size among the three groups cannot be attributed to differences in family size background. Again, religion and family size background are now unconfounded. Further, the problem of restricted generalizability of results that is encountered when holding a variable constant is not applicable because a range of family size backgrounds is used. Unfortunately, however, in practice it is often difficult to identify appropriate variables to serve as a basis for matching, and, once identified, to readily complete the matching process.

A final strategy for dealing with confounding variables is random assignment to experimental groups. As discussed above, when used in this context random assignment helps to control for confounding variables due to individuals' backgrounds. However, since random assignment cannot be applied to observational groups, observational independent variables will always be confounded with all other variables that are naturally related to them (except, of course, for those variables that can be controlled through being held constant or matching).

An Electronics Analogy

Many of the concepts discussed thus far can be illustrated intuitively using the electronics analogy from Chapter 8. Suppose you are listening through a set of earphones and trying to decide if you hear a particular signal. There is a good deal of static on the earphones. In research design, the static corresponds to disturbance variables and your goal is to eliminate it to the extent that you can. Suppose you do some repairs and eliminate a large portion of the static. This is analogous to controlling for disturbance variables. There is still another problem, however—there are two other signals very much like the one you must detect. If you hear them, you will think your signal has occurred when, in fact, it has not. These other signals represent confounding variables. A mechanical device you hook up completely eliminates one of these signals, and turns the other one into static (that is, a disturbance variable). The additional static is relatively minor, so you decide that you have done everything possible to assure accurate identification of the target signal.

However, you are confronted with yet another problem, as was discussed in Chapter 8. There are two types of errors you can make: (1) claiming you heard the signal when, in fact, it did not occur or (2) claiming you did not hear the signal when, in fact, it did occur. These correspond to Type I and Type II errors. Suppose you determine that falsely saying the signal was present would be very detrimental, and that falsely saying it did not occur would be of little import. In this case, you would want to ensure that you minimize the first type of error. This would correspond to setting a low alpha level in your study. With this in mind, you proceed to listen.

Problems with Inferring Causation

The statistics that will be developed in this text are designed to indicate to a researcher whether there is a relationship between two variables in the context of the research conducted. It must be emphasized that these statistics say nothing about whether two variables are *causally* related. It is entirely possible for two variables to be related to one another, but for no causal relationship to exist between them. A good example of this concerns shoe size and verbal ability. A random sample of people living in the United States would probably reveal a moderate relationship between shoe size and verbal ability: The larger one's foot, the greater one's verbal ability is likely to be. Does a causal relationship exist between these variables? Certainly not. It turns out that this relationship is due to a confounding variable, age. A sizeable proportion of a random sample of people in the United States would be children. When children are very young, they have small feet and also have little verbal ability. As they grow older, their feet become larger and their verbal skills increase. If we were to remove the influence of age, there would be no relationship between shoe size and verbal ability.

Most statistics and research design books emphasize this point only when considering correlational analysis. However, the issue holds for *all* statistics that assess the relationship between variables. *The ability to make a causal inference between two variables is a function of one's research design, not the statistical technique used to analyze the data that are yielded by that research design.* Since observational independent variables will always be confounded with all other variables that are naturally related to them, causal inferences are typically not possible when an observational research strategy is used. When an experimental research strategy is used, inferences of causation can be made only when confounding variables are controlled. The fact that two variables being related does not necessarily imply causality is important to keep in mind when interpreting statistics in the context of behavioral science research.

Between- Versus Within-Subjects Designs

When deciding on the type of statistical analysis that should be used to analyze the relationship between two variables, one must be aware of a distinction concerning how a given variable is measured. This distinction also has implications for how a researcher might design an investigation and, hence, is considered here. Consider an experiment where the investigator is interested in the relationship between two variables, type of drug and learning. Specifically, the investigator wants to know if two types of drugs, A and B, differentially affect performance on a learning task. Fifty subjects are randomly assigned to one of two conditions. In the first condition, subjects are administered drug A and then read a list of 15 words. They are subsequently asked to recall as many of the words as possible. A learning score is derived by counting the number of words correctly recalled (hence, scores can range from 0 to 15). In the second condition, subjects read the same list of 15 words and respond to the same recall task after being administered drug B. The relative effects of the drugs on learning are determined by comparing the responses of the two groups.

In this experiment, the investigator is studying the relationship between two variables: (1) type of drug and (2) learning as measured on a recall task. Type of drug is the independent variable and the learning measure is the dependent variable. The independent variable is such that subjects who received drug A did *not* receive drug B and those who received drug B did *not* receive drug A—that is, the two groups included different individuals. A variable of this type is known as a *between-subjects* variable because the values of the variable are "split up" between subjects instead of occurring completely within the same individuals. Research designs involving between-subjects independent variables are referred to as **between-subjects designs** or **independent groups designs.**

Now consider a similar experiment that is conducted in a slightly different fashion. A group of 25 subjects are administered drug A and then given the learning task. One month later, the same 25 subjects return to the experiment and are given the learning task after being administered drug B. The performance of these subjects under the influence of drug B is then compared with their earlier performance while under the influence of drug A. Note that in this experiment the 25 subjects who received drug A also received drug B—that is, the same individuals participated in both conditions. A variable of this type is known as a *within-subjects* variable. Research designs involving within-subjects independent variables are referred to as **within-subjects designs, correlated groups designs,** or **repeated measures designs.**

As illustrated by the above example, between-subjects designs and within-subjects designs are both viable strategies when the independent variable is experimental in nature. However, as we will see in later chapters, the two approaches require different statistical procedures. For many observational independent variables, only between-subjects designs are applicable. Consider the variable of gender, which has two levels, male and female. This variable is, by definition, a between-subjects variable. Individuals who are in the group called "male" by definition cannot also be in the group called "female."

The relative advantages and disadvantages of designing an investigation using a between-subjects versus within-subjects design can be illustrated in the context of the example relating type of drug (A or B) to learning. One advantage of the within-subjects approach is that it is more economical in terms of subjects. For instance, in our example, half the number of subjects were required to achieve the same per-condition sample size (25) when a within-subjects design was used as when a between-subjects design was used (25 versus 50). Subject economy is particularly important when a large amount of time, effort, or expense is necessary to recruit and/or train research participants.

A second advantage of within-subjects designs concerns the control of confounding variables, such as we discussed earlier. In the ideal experiment, the conditions would be such that the individuals in the two groups (drug A versus drug B) are identical in all respects except one—the type of drug they are given. If differences between the two groups occur in learning, then there would be one and only one logical explanation: The difference in drugs caused the differences in learning. In the between-subjects design, individuals are randomly assigned to one of the two conditions. The random assignment constitutes an attempt to "equalize" the two groups on all variables except that of the drug. If individuals are randomly assigned, it is unlikely that the individuals in one condition will, for example, be more intelligent (on the average) than the individuals in the other condition. But the key word here is "unlikely." Although it is *unlikely* that between-group differences in intelligence will occur, it *could* happen, due to chance factors. When this does occur, the differential intelligence in the two conditions could make one drug appear superior when it is not. Or, it might offset the effects of the superior drug, making it appear as if there is no difference between the drugs when there is.

In contrast, this cannot happen with within-subjects designs. The same individuals who receive drug A also receive drug B. Since the same individuals are used, intelligence must be the same in both conditions (unless it changed in the 1 month separating the administration of drugs). This is true not only of intelligence but also of all individual differences that might otherwise render interpretation ambiguous. Thus, the within-subjects design can offer considerably more experimental control than the between-subjects design.

This last statement must be qualified by a number of additional considerations. One problem with within-subjects designs is the fact that the treatment in the first condition may have **carry-over effects** that influence performance in the second condition. For example, the effects of drug A may not have worn off completely when drug B is administered. This could make interpretation of the experiment ambiguous. Carry-over effects are not necessarily restricted to the independent variable of interest. For example, performance on the dependent measure during the second condition in our hypothetical experiment might be better than in the first condition. Instead of reflecting any difference in drugs, this may simply reflect the fact that the subjects are taking the recall test for the second time and are more familiar with it. This increased familiarity, not the drugs, could produce learning differences.

When the investigator is confident that no carry-over effects will occur, a within-subjects research design is usually superior to a between-subjects design. When carry-over effects are possible, a between-subjects design may be more appropriate. We will return to the issue of within-subjects versus between-subjects designs in later chapters.

There is another type of research design, called a **matched-subjects design,** in which different individuals are used in the different conditions, but are treated as if a within-subjects design is in force. In this case, a given subject in one condition is "matched" with a subject in each of the other conditions who has characteristics similar to the first subject. The data are then treated as if these different subjects represent the same individual. As noted earlier, matching is difficult to accomplish on a practical level. For this and other reasons (see Thorndike, 1942), there are some serious problems with this strategy and its use is recommended only under restricted circumstances.

STUDY EXERCISE 9.1

For each of the following studies, indicate whether the independent variable is between-subjects or within-subjects in nature, and whether an experimental or an observational research strategy is involved.

Study I

A researcher was interested in the effects of television viewing on aggressive behavior in children. Fifty children were identified as watching television less than 5 hours per week (low viewers), 50 children were identified as watching television between 5 and 10 hours per week (moderate viewers), and 50 children were identified as watching television more than 10 hours per week (high viewers). For each child, a measure of aggressiveness was determined by interviewing the child's teacher and classmates. Children were subsequently rated as being low in aggressiveness, moderate in aggressiveness, or high in aggressiveness.

Answer This study is concerned with the relationship between a quantitative independent variable (amount of television viewing) and a quantitative dependent variable (amount of aggressiveness). Amount of television viewing is a between-subjects variable because the 50 children who are low viewers are not the same children as the 50 children who are moderate viewers, who, in turn, are not the same children as the 50 children who are high viewers. Since the children characteristically watch low, moderate, or high amounts of television in their daily lives, this study involves an observational research strategy.

Study II

An investigator was interested in the effects of music on problem-solving performance. Two hypotheses are possible: (1) music helps to relax an individual and should therefore facilitate problem-solving performance, or (2) music serves to distract the individual and, hence, it should interfere with problem-solving performance. One hundred individuals each tried to solve 10 problems with soft background music playing. Three weeks later, the same individuals returned and tried to solve 10 similar problems; this time, however, there was no background music. The number of problems correctly solved by each individual under each of the two conditions was computed separately and compared.

Answer In this study, the independent variable is a qualitative variable with two values (background music versus no background music). The dependent variable is the number of problems solved and constitutes a quantitative variable. The independent variable represents a within-subjects variable, because the same 100 individuals participated in both the background music and no-music conditions. Since the presence or absence of background music was manipulated by the investigator, this study involves an experimental research strategy.

9.2 Selecting the Appropriate Statistical Test to Analyze a Relationship: A Preview

Parametric Versus Nonparametric Statistics

Many of the statistical techniques that we will discuss involve the analysis of means, variances, and standard deviations. Such techniques are called **parametric statistics.** Parametric statistics require quantitative dependent variables and are usually applied when these variables are measured on approximately an interval level. Parametric statistics also require assumptions about the distribution of scores within the populations of interest. These assumptions will be made explicit as the techniques are introduced. In contrast, there is also a class of statistics called **nonparametric statistics** that do not require assumptions about distributional properties of scores and that are appropriate for all four levels of measurement.

There is currently some controversy among behavioral scientists concerning the use of parametric and nonparametric statistics. Those arguing for nonparametric analyses hold that measures used in the behavioral sciences often depart radically from interval level characteristics. Further, the application of parametric analyses, they argue, is inappropriate when distributional assumptions are not met. Those arguing for parametric analyses note that parametric-based analyses are more refined and powerful than current nonparametric methods. They argue that many of these techniques are *robust* to violations of the distributional assumptions required of them. The term **robust** means that even though the assumptions of the technique are violated, the frequencies of Type I and Type II errors and, thus, the accuracy of one's conclusions, are relatively unaffected as compared with conditions under which the assumptions are met. Because of this, some statisticians, such as Bohrnstedt and Carter (1971), have concluded that "when one has a variable which is measured at the ordinal level, parametric statistics not only can be, but should be, applied."

A complete discussion of this issue is beyond the scope of this book and interested readers are referred to Bohrnstedt and Carter (1971), Boneau (1960), Lord (1953), Stevens (1951), and an excellent collection of readings in Kirk (1972).

Robustness of Statistical Tests

The concept of robustness is important in inferential statistics. As noted above, robustness refers to the extent that conclusions drawn on the basis of a statistical test (for example, rejection of the null hypothesis) are unaffected by violations of the assumptions underlying the test. Mathematicians who develop inferential tests use assumptions for several reasons. Sometimes an assumption will be used to simplify mathematical derivations so as to make the statistics more *manageable*. Other times, an assumption will be used because it characterizes what is likely to be the case in the real world; that is, the assumption is *credible*. As an example, a statistical test that assumes each intelligence test score occurs with equal frequency would have little applicability to most real-world problems. The assumption of a normal distribution, with a large proportion of central scores and few extreme scores, is much more credible.

The most common distributional assumptions that we will encounter in future chapters are *normality* and *homogeneity of variance*. As discussed in Chapter 8, the normality assumption requires that scores on the variable of interest be normally distributed in the population from which they are drawn. The homogeneity of variance assumption is applicable when two or more groups of scores are being considered in the research design and requires that the variances of scores be equal, or *homogeneous,* in the populations underlying each of the samples. Note that these assumptions relate to the populations from which the samples were drawn rather than to the samples themselves. Thus, the determination of violations of distributional assumptions requires sophisticated statistical procedures.

In fact, a number of methods have been proposed for assessing violations of the normality and homogeneity of variance assumptions. Unfortunately, each of these has problems associated with it. For instance, the most commonly cited tests for homogeneity of variance (Bartlett's test, Cochran's test, and Hartley's F_{max} test) have been found to be unsatisfactory when the population data are not normally distributed. More generally, existing tests of normality and homogeneity of variance suffer from the fact that they are insensitive to the *severity* of distributional violations.

This is problematic because distributional assumptions will almost always be violated to one degree or another. Technically speaking, for instance, the assumption of homogeneous variances is violated if the population variances in a three-group design are a relatively similar 1.23, 1.37, and 1.60, because the assumption requires that the population variances be *exactly equal.* Assuming that a given test of homogeneity of variance were sensitive enough to detect this discrepancy, it would be unable to differentiate it from a more extreme situation, where, for instance, the three population variances are 1.68, 4.27, and 12.39.

While an inferential test (we are talking here about the one-sample t test and related procedures, *not* tests of distributional assumptions) might be robust to violation of the homogeneity of variance assumption in the first of the above situations, it would be less likely to be robust to violation of this assumption in the second situation. Nevertheless, since existing tests of homogeneity of variance merely tell us that the assumption is violated without specifying the severity of the violation, we wll be unable to differentiate the two occurrences.

When an inferential test is robust to violations of an assumption, it can be appropriately applied even when that assumption is violated. This is because, by definition, the frequencies of Type I and Type II errors will be similar to what they would be under conditions where no violations occurred. Thus, the accuracy of the test's conclusions will be relatively unaffected.

The general strategy used by statisticians to test robustness can be illustrated with reference to the one-sample t test discussed in Chapter 8. When determining the effects of distributional violations on Type I errors (that is, when the null hypothesis is true), the following steps would be followed:

1. Given a true null hypothesis and a specified alpha level, for instance, .05, the critical values of t are identified such that only 5% of the scores in the t distribution exceed these values *when the assumptions of the test have been satisfied.* This is accomplished using Appendix D. The alpha level in this case is called the **nominal significance level.**
2. By empirical or mathematical means, the *actual proportion* of t scores that exceed the critical values when the null hypothesis is true and *one of the assumptions is not satisfied* is determined. This proportion represents the **"actual" significance level** for when an assumption is violated in the manner being studied.
3. The nominal significance level and the "actual" significance level are compared. If the two are very similar, then the test is robust to the violation of the assumption in question. If there is a large discrepancy between the two, then the test is not robust.

A similar procedure would be followed when the null hypothesis is false, except, of course, that the focus would be on Type II errors.

As you might expect, the results of studies investigating the robustness of inferential tests are quite complex. This is because, as noted in Chapter 8, robustness is influenced by several different factors, including sample size, the degree of violation, and the form of violation. For instance, there are many ways in which the homogeneity of variance assumption can be violated in the three-group situation. Focusing on the *form* of violation, it might be the case either that two of the population variances are the same but different from the third or that all three population variances differ from one another. In terms of the *degree* of violation, differences in population variances can range in magnitude from slight to large.

While a given statistical test might be quite robust under one set of circumstances, its robustness might substantially decrease under somewhat different conditions. For example, statisticians have found that in many instances, even marked violations of the normality assumption will not seriously affect the validity of the one-sample t test as long as the sample size is larger than around 10 (Pearson & Please, 1975). While the test continues to be robust against various forms of minor to moderate violations when sample size decreases, marked violations under these circumstances can seriously affect the frequency of both Type I and Type II errors.

Given the numerous ways that factors can combine to influence robustness, it will be impossible to discuss robustness with any degree of precision when considering the robustness of specific tests in future chapters. Further complicating matters is the insensitivity of existing distributional tests to the severity of distributional violations. Thus, not only is it difficult to determine whether an inferential test is robust for the specific circumstances under study, but it is also difficult even to accurately identify these underlying circumstances. For instance, as discussed earlier, while existing tests of homogeneity of variance can identify sets of population variances as being nonhomogeneous, they are incapable of identifying the magnitude of the differences between these variances. Our strategy in future chapters will therefore be to provide general characterizations of the robustness of the tests that are presented. More detailed discussion can be found in Jaccard and Becker (1988). If a researcher has reason to believe that a distributional assumption has been violated to the extent that a parametric test is no longer robust, one possible solution is to use a nonparametric alternative. These will be discussed in Chapter 16.

Selection of a Statistical Test

It is impossible to state any precise rules for determining the statistical test to apply in a given situation because it would always be possible to find exceptions whereby a different form of analysis might be more appropriate. Nevertheless, we can specify the most common research designs and the types of analyses typically used in the context of these designs. In Chapters 10–16, we will consider the most common statistical tests for analyzing bivariate relationships and, in each chapter, specify the factors to be considered in deciding to apply the test. As we will see, the essence of the proposed framework rests on distinguishing between qualitative and quantitative variables and between within-subjects and between-subjects designs. The required steps are: (1) identify the independent and dependent variables; (2) classify each as being either a qualitative or a quantitative variable; (3) classify the independent variable as being either a between-subjects or a within-subjects variable; and (4) note the number of levels that occur for each variable. Chapter 18 will formally discuss procedures for choosing a statistical test to analyze one's data.

Our discussion of statistics relevant to the analysis of bivariate relationships will consistently focus on three questions:

1. Given sample data, can we infer that a relationship exists between two variables in the population?
2. If so, what is the *strength* of the relationship?
3. If so, what is the *nature* of the relationship?

9.3 Summary

The results of a statistical analysis must be interpreted in the context of the research design used to generate the data. Behavioral scientists use two general types of research design. An experimental strategy involves performing a set of procedures in order to create different values of the independent variable for the research participants. In contrast, an observational strategy does not involve the process of actively creating values on the independent variable, but rather involves using a set of procedures that will allow one to measure differences in values that naturally exist in the research participants.

A major goal of research is to control for alternative explanations. This involves the control of confounding variables and disturbance variables. A confounding variable is one that is related to the independent variable and that influences the dependent variable, rendering a relational inference between the independent variable and the dependent variable ambiguous. A disturbance variable is one that is unrelated to the independent variable but that influences the dependent variable. Techniques for controlling confounding and disturbance variables include random assignment to groups, holding a variable constant, matching, and the use of between-subjects versus within-subjects designs.

In Chapters 10–16, statistics for analyzing bivariate relationships will be examined. An important feature of these statistics is the assumptions they make about the distribution of scores within the populations of interest. To the extent that conclusions drawn from a statistical test are relatively unaffected by the violation of these assumptions, the test is said to be robust. The issue of robustness will be examined for each test considered in future chapters.

Exercises

Answers to asterisked (*) exercises appear at the back of the book.

1. What is the difference between an experimental research strategy and an observational research strategy?

***2.** What is a control group? What is the advantage of including a control group in the research design?

For each of the studies described in Exercises 3–6, indicate whether an experimental or an observational research strategy is involved. Also, indicate whether a control group was used and, if so, identify the nature of this group. Justify your answers.

***3.** Morrow and Davidson (1976) studied the effects of race on family-size decisions. These investigators interviewed a total of 300 people, 100 of whom were black, 100 of whom were Mexican-American, and 100 of whom were white. Each person was asked the number of children they wanted to have in their completed family. The average number of desired children was compared for the three groups.

***4.** Sears (1969) reviewed studies on the relationship between an individual's gender and his or her political party preference. In one investigation, 75 males and 75 females were interviewed. Respondents were asked whether they considered themselves Democrats, Republicans, or Independents. The frequencies with which males identified with the three classifications were compared with the corresponding frequencies for females.

5. Steiner (1972) reviewed studies on the effects of the presence of others on problem-solving performance. In one study, a group of 100 female volunteers served as participants. At the first session, each subject was seated alone in a room and given a math problem to solve. The amount of time it took to solve the problem was measured. Two weeks later, the subject returned and solved another problem, but this time there was an observer present watching her. The amount of time it took to solve the problem was again measured. For each of the two conditions, observer present versus observer absent, the mean problem-solving time was computed across the 100 subjects. These means were then compared.

6. Harvath (1943) was interested in the effects of noise on problem-solving performance. A group of 150 individuals participated in the study. Subjects were randomly assigned to two conditions. Seventy-five subjects tried to solve 30 problems while a steady "buzz" was present in the background. The other 75 subjects tried to solve the same 30 problems but with no background noise. The number of correct solutions was computed for each subject and the average number of correct solutions compared for individuals in the noise versus the no-noise condition.

7. What is a random assignment? Why is random assignment important?

***8.** What are the limitations on the use of random assignment?

***9.** What procedures are available for reducing sampling error?

10. What are confounding variables? How can confounding variables be controlled?

11. What are disturbance variables? How can disturbance variables be controlled?

***12.** There is a moderately strong relationship between individuals' height and the length of their hair. In general, the taller one is, the shorter one's hair tends to be. Is there a causal relationship between these two variables? If so, what is its nature? If not, what additional variable might account for the observed relationship?

***13.** What are the advantages of a within-subjects research design as compared to a between-subjects research design? What are the disadvantages?

***14.** Indicate whether the independent variables in the studies described in Exercises 3–6 are between-subjects or within-subjects in nature.

15. How do parametric and nonparametric statistics differ?

16. What is meant by the term *robust?* Why is robustness important?

17. What is the normality assumption? What is the homogeneity of variance assumption?

***18.** What is the major problem with existing tests of distributional assumptions?

***19.** What three general factors influence the robustness of a statistical test?

TEN

INDEPENDENT GROUPS t TEST

10.1

Use of the Independent Groups *t* Test

The statistical technique developed in this chapter is called the **independent groups *t* test.** It is typically used to analyze the relationship between two variables when

1. the dependent variable is quantitative in nature and is measured on a level that approximates interval characteristics;
2. the independent variable is *between-subjects* in nature (it can be either qualitative or quantitative)*; and
3. the independent variable has two and only two levels.

Let us consider an example of an experiment that meets these conditions. When a friend describes a stranger to you, you may form an impression as to whether you would like that individual. One question of interest to behavioral scientists has been whether the order in which information is provided about a stranger influences the kind of impression formed about him. Consider the following procedures. Twenty subjects read a verbal description of a stranger and then rate on a 7-point scale the extent to which they like or dislike that person. The scale is such that the higher the number, the more the stranger is liked. The stranger is described by six adjectives: intelligent, sincere, honest, conceited, rude, and nervous. The first three characteristics are positive in nature while the last three characteristics are negative. Ten of the subjects are randomly selected and given the description in the order listed above—that is, first the positive traits and then the negative traits. The other 10 subjects are given the same description, but in reverse order—that is, first the negative traits and

* Matched-subjects designs are analyzed as if the independent variable is within-subjects in nature, using the procedures in Chapter 11.

TABLE 10.1 Data and Calculations for Order-of-Information and Likeability Experiment

PRO–CON		CON–PRO	
X	X^2	X	X^2
7	49	1	1
4	16	5	25
3	9	2	4
5	25	5	25
5	25	3	9
6	36	3	9
5	25	3	9
5	25	2	4
4	16	4	16
6	36	2	4
$\Sigma X_1 = 50$	$\Sigma X_1^2 = 262$	$\Sigma X_2 = 30$	$\Sigma X_2^2 = 106$
$\bar{X}_1 = 5.00$		$\bar{X}_2 = 3.00$	

$$SS_1 = \Sigma X_1^2 - \frac{(\Sigma X_1)^2}{n_1} = 262 - \frac{50^2}{10} = 12$$

$$\hat{s}_1^2 = \frac{SS_1}{n_1 - 1} = \frac{12}{9} = 1.33$$

$$SS_2 = \Sigma X_2^2 - \frac{(\Sigma X_2)^2}{n_2} = 106 - \frac{30^2}{10} = 16$$

$$\hat{s}_2^2 = \frac{SS_2}{n_2 - 1} = \frac{16}{9} = 1.78$$

then the positive traits. Thus, we have two groups reflecting different orders of presentation:

Group 1 (pro–con): intelligent, sincere, honest, conceited, rude, nervous

Group 2 (con–pro): nervous, rude, conceited, honest, sincere, intelligent

The likeability ratings for the two conditions are presented in columns 1 and 3 of Table 10.1. If the order of information has no effect on the type of impression formed, we would expect that, on the average, the two sets of ratings would not differ. However, if the order of information does matter, then the average likeability ratings should differ between the conditions.

In this experiment, the order in which trait information is presented is the independent variable, and the likeability of the stranger is the dependent variable. The independent variable has two levels (pro–con versus con–pro) and is between-subjects in nature. The dependent variable is quantitative in nature. Given these conditions, the independent groups t test is the statistical technique that would typically be used to analyze the relationship between the variables.

10.2

Inference of a Relationship Using the Independent Groups *t* Test

Null and Alternative Hypotheses

The first question to be addressed is whether a relationship exists between the independent variable and the dependent variable. We begin by stating this question in terms of null and alternative hypotheses. This is accomplished with reference to population means. Because we are interested in generalizing the results of our study beyond just those subjects who participated in it, we conceptualize the subjects as representing random samples from very large populations of similar individuals. In the experiment, there are two populations of interest: (1) individuals who read a verbal description of a stranger in which traits are presented in an order from pro to con, and (2) individuals who read a verbal description of a stranger in which traits are presented in an order from con to pro. Given that individuals were randomly assigned to groups, and if we assume that the order of information does *not* matter, then we would expect the population means for the two groups to be equal. The null hypothesis will thus take the form

$$H_0: \quad \mu_1 = \mu_2$$

where μ_1 is the population mean for the first group and μ_2 is the population mean for the second group. The null hypothesis posits no relationship between the independent variable (order of information) and the dependent variable (likeability). It states that it does not matter what the value of the independent variable is, because the mean score on the dependent variable is the same at both levels. The alternative hypothesis states that there *is* a relationship between the two variables since the value of the independent variable *does* influence the average score on the dependent variable:

$$H_1: \quad \mu_1 \neq \mu_2$$

We have now restated the question in the context of two competing hypotheses, a null hypothesis (there is no relationship between the variables) and an alternative hypothesis (there is a relationship between the variables). The next step is to choose between these two hypotheses. Examine the sample means in Table 10.1. The mean likeability score for the pro–con group is 5.00, while the mean likeability score for the con–pro group is 3.00. The two means are not equal and would therefore appear to be consistent with the alternative hypothesis. However, we know from Chapter 7 that a sample mean may not reflect the true value of its population mean due to sampling error. Thus, the observed difference between the two sample means may not reflect the influence of the order of information on likeability, but rather may reflect sampling error. Our task is to determine whether this is a reasonable interpretation of the observed difference between sample means.

Sampling Distribution of the Difference Between Two Independent Means

In order to test the sampling error interpretation, we will use logic directly analogous to that developed in Chapter 8 on hypothesis testing. As an initial step, we must develop the concept of a **sampling distribution of the difference between two independent means.** Consider two large populations whose mean scores on a variable are equal ($\mu_1 = \mu_2$). Now suppose we select a random sample of size 10 from each population and compute the mean score in each sample as well as the difference between the means. We might find a result such as that illustrated in the first row of Table 10.2. In this table, $\bar{X}_1$ represents the sample mean for population 1, $\bar{X}_2$ represents the sample mean for population 2, and $\bar{X}_1 - \bar{X}_2$ represents the difference between them. Suppose we repeat this process again, yielding a second pair of means and a second mean difference, as shown in Table 10.2. In principle, we could do this for all possible random samples of size 10, yielding a distribution of mean differences.* Table 10.2 presents examples of some differences that might be observed. This distribution of the differences between two means represents a sampling distribution of the difference between two independent means. It is directly analogous to a sampling distribution of the mean. However, now the concern is with a distribution of scores representing the difference between two means. As we did in Chapter 7, we can compute the mean and standard deviation of this sampling distribution. If we were to do so, we would find that many of the properties of a sampling distribution of the mean also hold for a sampling distribution of the difference between two independent means. For instance, just as the mean of a sampling distribution of the mean is always equal to the population mean, *the mean of a sampling distribution of the difference between two independent means is always equal to the difference between the population means.* Consider the case where two population means are equal and hence their difference is 0 (that is, $\mu_1 - \mu_2 = 0$), as in the present example. If we were to generate a sampling distribution of the difference between means for these populations, the mean of the differences would equal 0. The underlying principle is much the same as that developed in Chapter 7. When we repeatedly select samples and compute the difference between means, some of the differences will overestimate the true mean difference while others will underestimate it. When we average all of these, the underestimations will cancel the overestimations, with the result being the true population difference between means.

* In practice, we would never actually do this. It is unnecessary to do so because, as discussed below, the important information about a distribution of mean differences can be derived *mathematically*.

TABLE 10.2 Illustrative Sample Means and Mean Differences from a Sampling Distribution of the Difference Between Two Independent Means

$\bar{X}_1$	$\bar{X}_2$	$\bar{X}_1 - \bar{X}_2$
5.00	3.00	2.00
4.30	3.60	.70
5.60	6.40	−.80
4.80	5.40	−.60
4.70	4.70	00
5.20	5.90	−.70
6.00	5.70	.30
4.30	3.90	.40
5.20	6.50	−1.30
6.10	7.50	−1.40
5.00	3.90	1.10
5.20	5.90	−.70
4.70	4.10	.60
4.90	4.40	.50
5.00	5.90	−.90
6.20	6.40	−.20
4.30	2.50	1.80
5.10	6.40	−1.30
5.70	6.60	−.90
5.10	4.10	1.00
4.90	5.40	−.50
4.90	5.40	−.50

The standard deviation of the sampling distribution of the difference between two independent means is called the *standard error of the difference between two independent means* or, more simply, the **standard error of the difference.** Like the standard error of the mean, the standard error of the difference indicates how much sampling error will occur, on the average. Statisticians have developed a formula that allows us to compute the standard error of the difference from the population standard deviations. The formula is

$$\sigma_{\bar{X}_1-\bar{X}_2} = \sqrt{\frac{\sigma_1^2}{n_1} + \frac{\sigma_2^2}{n_2}} \qquad [10.1]$$

where $\sigma_{\bar{X}_1-\bar{X}_2}$ is the standard error of the difference, σ_1^2 is the population variance for group 1, σ_2^2 is the population variance for group 2, n_1 is the sample size for group 1, and n_2 is the sample size for group 2.* Note the similarity of this formula to the one for the standard error of the mean (Equation 7.4):

* As we noted in Chapter 3, when more than one group is involved, n refers to the sample size of a particular group, whereas N refers to the total number of subjects in the study. In the case of two groups, $N = n_1 + n_2$. The independent groups t test does not require that sample sizes be equal.

$$\sigma_{\bar{X}} = \frac{\sigma}{\sqrt{N}} = \sqrt{\frac{\sigma^2}{N}}$$

Analogous to the standard error of the mean, the size of the standard error of the difference is influenced by two factors: (1) the sample sizes (n_1 and n_2) and (2) the variability of scores in the populations (σ_1^2 and σ_2^2). Following the logic outlined in Section 7.5, the standard error of the difference becomes smaller as the sample sizes increase and the variability of scores in the populations decreases.

Pooled Variance Estimate

If we know the values of σ_1^2, σ_2^2, n_1, and n_2, we can compute the value of the standard error in Equation 10.1. In practice, we typically know the values of n_1 and n_2, but we do not know the values of σ_1^2 and σ_2^2. It is therefore necessary to estimate them from sample data. This is possible because of the fact that the test we will ultimately apply assumes that the two population variances are equal, or *homogeneous*. This assumption, known as the assumption of **homogeneity of variance,** redefines the estimation problem. Instead of estimating σ_1^2 and σ_2^2 separately, our goal is to estimate σ^2, the variance of both populations. In other words, it is assumed that $\sigma_1^2 = \sigma_2^2 = \sigma^2$. The quantity σ^2 can best be estimated by combining, or *pooling,* the variance estimates from the two samples to obtain a **pooled variance estimate.** By pooling the variance estimates from two independent samples, we increase the degrees of freedom on which the estimate of σ^2 is based and thereby obtain a better estimate. The simplest method of pooling the two variance estimates is to compute their (unweighted) mean. This is, in fact, what is done when n_1 and n_2 are equal. However, if one of the groups has a larger sample size than the other, it makes sense to give the variance estimate from the group with the larger n more "weight" in determining the pooled variance estimate, since the larger the sample size, the greater the degrees of freedom and, thus, the better the estimate, as discussed in Chapter 7. This is accomplished by the following equation:

$$\hat{s}^2_{\text{pooled}} = \frac{(n_1 - 1)\hat{s}_1^2 + (n_2 - 1)\hat{s}_2^2}{n_1 + n_2 - 2} \quad [10.2]$$

where $\hat{s}^2_{\text{pooled}}$ represents the pooled estimate of σ^2 (that is, the pooled variance estimate), $\hat{s}_1^2$ is the variance estimate for group 1, $\hat{s}_2^2$ is the variance estimate for group 2, and n_1 and n_2 are as previously defined.* Examine the right-hand side of Equation 10.2. We have multiplied the variance estimate for each group by its degrees of freedom, summed the products for the two groups, and then divided the resulting quantity by the total number of degrees of freedom, $(n_1 - 1) + (n_2 - 1) = n_1 + n_2 - 2$.† This has been done so that the contributions of $\hat{s}_1^2$ and $\hat{s}_2^2$ to $\hat{s}^2_{\text{pooled}}$ will be proportional to their degrees of freedom. For example, if $\hat{s}_1^2$ has twice as many degrees of freedom as $\hat{s}_2^2$, it will contribute

---

* When $n_1 = n_2$, Equation 10.2 reduces to $\hat{s}^2_{\text{pooled}} = (\hat{s}_1^2 + \hat{s}_2^2)/2$.

† The same general procedure can be used with more complex designs to pool the variance estimates of more than two groups.

twice as much to $\hat{s}^2_{\text{pooled}}$. The meaning of Equation 10.2 becomes clearer if we rephrase it as follows:

$$\hat{s}^2_{\text{pooled}} = \frac{(\text{df}_1)(\hat{s}^2_1) + (\text{df}_2)(\hat{s}^2_2)}{\text{df}_{\text{TOTAL}}} \qquad [10.3]$$

where df_1 represents the degrees of freedom for the variance estimate for sample 1, df_2 represents the degrees of freedom for the variance estimate for sample 2, and df_{TOTAL} represents the total degrees of freedom, or $\text{df}_1 + \text{df}_2$. In short, Equations 10.2 and 10.3 characterize $\hat{s}^2_{\text{pooled}}$ as the mean of $\hat{s}^2_1$ and $\hat{s}^2_2$ after these have been adjusted by their respective degrees of freedom.

We will demonstrate the application of Equation 10.2 by returning to the example on the effect of order of information on likeability ratings discussed at the beginning of this chapter. Looking at Table 10.1, we find that $n_1 = 10$, $n_2 = 10$, $\hat{s}^2_1 = 1.33$, and $\hat{s}^2_2 = 1.78$. Thus,

$$\hat{s}^2_{\text{pooled}} = \frac{(10 - 1)(1.33) + (10 - 1)(1.78)}{10 + 10 - 2} = 1.56$$

Estimated Standard Error of the Difference

Under the assumption of homogeneity of variance, $\hat{s}^2_{\text{pooled}}$ can be substituted for σ^2_1 and σ^2_2 in Equation 10.1 to yield the following formula for the estimated standard error of the difference:

$$\hat{s}_{\bar{X}_1-\bar{X}_2} = \sqrt{\frac{\hat{s}^2_{\text{pooled}}}{n_1} + \frac{\hat{s}^2_{\text{pooled}}}{n_2}} \qquad [10.4]$$

where $\hat{s}_{\bar{X}_1-\bar{X}_2}$ represents the estimated standard error of the difference between two independent means and all other terms are as previously defined. For the example on the effect of order of information on likeability ratings,

$$\hat{s}_{\bar{X}_1-\bar{X}_2} = \sqrt{\frac{1.56}{10} + \frac{1.56}{10}} = .56$$

Thus, the mean differences constituting the sampling distribution of the difference between two independent means are estimated to deviate an average of .56 unit from the true difference between the two population means.

If we wished, we could combine the above two steps into one by integrating Equations 10.2 and 10.4 as follows:

$$\hat{s}_{\bar{X}_1-\bar{X}_2} = \sqrt{\frac{\hat{s}^2_{\text{pooled}}}{n_1} + \frac{\hat{s}^2_{\text{pooled}}}{n_2}}$$

$$= \sqrt{\hat{s}^2_{\text{pooled}}\left(\frac{1}{n_1} + \frac{1}{n_2}\right)}$$

$$= \sqrt{\left[\frac{(n_1 - 1)\hat{s}^2_1 + (n_2 - 1)\hat{s}^2_2}{n_1 + n_2 - 2}\right]\left(\frac{1}{n_1} + \frac{1}{n_2}\right)} \qquad [10.5]$$

Since a variance estimate is equivalent to the corresponding sum of squares divided by the sample size minus 1, this can also be expressed as

$$\hat{s}_{\bar{X}_1-\bar{X}_2} = \sqrt{\left[\frac{(n_1 - 1)\left(\frac{SS_1}{n_1 - 1}\right) + (n_2 - 1)\left(\frac{SS_2}{n_2 - 1}\right)}{n_1 + n_2 - 2}\right]\left(\frac{1}{n_1} + \frac{1}{n_2}\right)}$$

$$= \sqrt{\left(\frac{SS_1 + SS_2}{n_1 + n_2 - 2}\right)\left(\frac{1}{n_1} + \frac{1}{n_2}\right)} \qquad [10.6]$$

where SS_1 is the sum of squares for the first group and SS_2 is the sum of squares for the second group. The decision to calculate $\hat{s}_{\bar{X}_1-\bar{X}_2}$ using Equation 10.5 (or alternatively, Equations 10.2 and 10.4) or Equation 10.6 merely depends on whether one is dealing with variance estimates or sums of squares.

The *t* Test We now have the background to formally test whether the observed difference between sample means in likeability ratings (5.00 versus 3.00) can be attributed to sampling error or whether it reflects a true relationship between order of information and likeability. To do so, we will adapt the hypothesis testing steps outlined in Chapter 8 as follows:

1. Translate the research question into a null hypothesis and an alternative hypothesis.

   In the present example, the null hypothesis states that there is no relationship between order of information and likeability. The alternative hypothesis states that there is a relationship between the two variables. Expressed in terms of population means, the null and alternative hypotheses are respectively

$$H_0: \mu_1 = \mu_2$$
$$H_1: \mu_1 \neq \mu_2$$

   While the alternative hypothesis is nondirectional in this instance, directional alternative hypotheses are also possible. These would be phrased in terms of the mean of one population being larger than the mean of the other. As with the one-sample *t* test, directional tests should be used only when there is exclusive concern with a specific direction of difference. This is because a directional test is less powerful than the corresponding nondirectional test if the two populations differ in the direction opposite to that stated in the directional alternative hypothesis. The rationale for this follows that outlined in Section 8.8 for the one-sample *t* test.

   Researchers often use subscripts based on the first letter of a condition's name rather than the more general "1" and "2" notation to identify specific conditions. For instance, we could state the null and alternative hypotheses for the order-of-information and likeability experiment as

$$H_0: \mu_P = \mu_C$$
$$H_1: \mu_P \neq \mu_C$$

where the subscript "P" denotes the pro–con ordering and the subscript "C" denotes the con–pro ordering. The advantage of this approach is that the particular condition being represented is immediately identifiable. We will adopt this more precise notational system where appropriate in the remainder of this book.*

2. Assuming the null hypothesis is true, state an expected result in the form of a range of values within which the difference between sample means would be expected to fall. This is expressed in terms of the critical values of t that define the endpoints of this range. The set of all t values more extreme than the critical values constitutes an unexpected result or, more formally, the rejection region.

   The critical values of t are determined by reference to the appropriate t distribution in Appendix D. The degrees of freedom for the independent groups t test are equal to $n_1 + n_2 - 2$, reflecting the fact that the degrees of freedom associated with the first sample are $n_1 - 1$, the degrees of freedom associated with the second sample are $n_2 - 1$, and $(n_1 - 1) + (n_2 - 1) = n_1 + n_2 - 2$.† Following the logic outlined in Chapter 8, by convention alpha is typically set equal to .05.

   The t distribution in the present example has $n_1 + n_2 - 2 = 10 + 10 - 2 = 18$ degrees of freedom. For an alpha level of .05, nondirectional test, and 18 degrees of freedom, Appendix D defines the critical values of t as ± 2.101. An expected result is therefore defined by t scores in the range -2.101 to $+2.101$, and an unexpected result is defined by all t scores less than -2.101 or greater than $+2.101$.

3. Characterize the mean and standard deviation of the sampling distribution of the difference between two independent means, assuming the null hypothesis is true.

   If we assume that the null hypothesis is true in the present example, then μ_1 and μ_2 will be equal, so $\mu_1 - \mu_2$ will equal 0. It follows that the mean of the sampling distribution will also equal 0, since the mean of a sampling distribution of the difference between two independent means is always equal to the difference between the population means.

   The standard deviation of the sampling distribution is the standard error of the difference. This was estimated previously to equal .56 in our example.

---

* So that students can become familiar with the various statistical formulas, these will consistently be presented using the more general numerical subscripts.

† It will be remembered that $n_1 + n_2 - 2$ is also the degrees of freedom associated with the pooled variance estimate. This reflects the fact that (as we will see shortly) the value of the t statistic in the independent groups case is dependent on the value of $\hat{s}_{\bar{X}_1 - \bar{X}_2}$, which, in turn, is dependent on the value of the pooled variance estimate.

4. Convert the observed difference between sample means into a t value to determine how many estimated standard errors it is from $\mu_1 - \mu_2$, *assuming the null hypothesis is true.* This is accomplished by the formula

$$t = \frac{(\bar{X}_1 - \bar{X}_2) - (\mu_1 - \mu_2)}{\hat{s}_{\bar{X}_1 - \bar{X}_2}} \qquad [10.7]$$

where $\bar{X}_1 - \bar{X}_2$ is the observed difference between sample means, $\mu_1 - \mu_2$ is the hypothesized difference between population means, and $\hat{s}_{\bar{X}_1 - \bar{X}_2}$ is the estimated standard error of the difference.

In most instances, the null hypothesis will state that the population means are identical, so $\mu_1 - \mu_2$ will equal 0. However, other null hypotheses are also possible. For instance, a given null hypothesis might state that the mean of the first population is 10 units higher than the mean of the second population. This could be tested by setting $\mu_1 - \mu_2$ equal to 10 in Equation 10.7. Any other hypothesized difference between two population means can be similarly assessed.

When $\mu_1 - \mu_2$ is hypothesized to equal 0, Equation 10.7 reduces to

$$t = \frac{\bar{X}_1 - \bar{X}_2}{\hat{s}_{\bar{X}_1 - \bar{X}_2}} \qquad [10.8]$$

This is the most commonly used version of the independent groups t test. However, for purpose of illustration, we will employ the more general formula represented by Equation 10.7 whenever an independent groups t test is called for in the remainder of this chapter.

In the present example, $\bar{X}_1 - \bar{X}_2 = 5.00 - 3.00 = 2.00$, $\hat{s}_{\bar{X}_1 - \bar{X}_2} = .56$, and $\mu_1 - \mu_2$ is hypothesized to equal 0. Hence,

$$t = \frac{(\bar{X}_1 - \bar{X}_2) - (\mu_1 - \mu_2)}{\hat{s}_{\bar{X}_1 - \bar{X}_2}}$$

$$= \frac{2.00 - 0}{.56} = 3.57$$

5. Compare the observed result with the expected result. If the observed result falls within the rejection region, reject the null hypothesis.* Otherwise, do not reject the null hypothesis.

In the present example, the observed t value of 3.57 exceeds the critical value of +2.101. This suggests that the observed mean difference is too large to be attributed to sampling error, and the null hypothesis is therefore rejected. Order of information is related to likeability, and we have answered the first of our three questions apropos of the analysis of bivariate relationships.

---

* If the null hypothesis is rejected, the nature of the relationship between the independent and dependent variables can be determined using the procedure discussed in Section 10.4.

STUDY EXERCISE 10.1

In order to test for sex discrimination by women against other women, a researcher had 13 women read an essay that was presented as having been written by "John McKay" and then rate the essay for the quality of the writing style. The ratings were made on a scale from 1 to 10, with higher numbers indicating greater perceived quality. An additional 13 women read and rated the identical essay, but were led to believe that the author was "Joan McKay." The following summary data were observed. Use an independent groups *t* test to test for the existence of a relationship between the supposed gender of the author and the perceived quality of the essay.

Male author	*Female author*
$n_M = 13$	$n_F = 13$
$\bar{X}_M = 7.20$	$\bar{X}_F = 6.10$
$SS_M = 15$	$SS_F = 18$

Answer The null and alternative hypotheses are

$$H_0: \mu_M = \mu_F$$

$$H_1: \mu_M \neq \mu_F$$

where the subscripts "M" and "F" represent the male author and the female author conditions, respectively. For an alpha level of .05, nondirectional test, and $n_1 + n_2 - 2 = 13 + 13 - 2 = 24$ degrees of freedom, the critical values of *t* from Appendix D are ± 2.064. Thus, if the observed *t* is less than -2.064 or greater than $+2.064$ we will reject the null hypothesis.

Since we are dealing with sums of squares, the standard error of the difference is estimated using Equation 10.6:

$$\hat{s}_{\bar{X}_1-\bar{X}_2} = \sqrt{\left(\frac{SS_1 + SS_2}{n_1 + n_2 - 2}\right)\left(\frac{1}{n_1} + \frac{1}{n_2}\right)}$$

$$= \sqrt{\left(\frac{15 + 18}{13 + 13 - 2}\right)\left(\frac{1}{13} + \frac{1}{13}\right)} = .46$$

The observed difference between sample means is $\bar{X}_1 - \bar{X}_2 = 7.20 - 6.10 = 1.10$ and the hypothesized difference between population means is $\mu_1 - \mu_2 = 0$. Using Equation 10.7, we compute the observed value of *t:*

$$t = \frac{1.10 - 0}{.46} = 2.39$$

Since 2.39 is greater than +2.064, we reject the null hypothesis and conclude that there is a relationship between the supposed gender of the author and the perceived quality of the essay.

Assumptions of the *t* Test

As noted at the beginning of this chapter, the independent groups t test is appropriate when the dependent variable is quantitative in nature and measured on approximately an interval level. The test also requires that Equation 10.7 yield scores distributed as t. This is true when the following assumptions are met:

1. The samples are independently and randomly selected from their respective populations.*
2. The scores in each population are normally distributed.
3. The scores in the two populations have equal variances; that is, $\sigma_1^2 = \sigma_2^2$. As discussed previously, this is referred to as the assumption of homogeneity of variance.

For the test to be valid, it is essential that the assumption of independent and random selection be met. However, under certain conditions, the independent groups t test is robust to violations of the normality and homogeneity of variance assumptions. A detailed discussion of this issue can be found in Jaccard and Becker (1988). In general, the test is more robust when sample sizes are equal and relatively large ("relatively large" being a function of how severely the assumptions are violated in the specific populations under study) than when sample sizes are discrepant and/or small. Thus, a desirable practice, when possible, is to employ an equal number of subjects in each group, with values of n being as large as circumstances reasonably permit.

10.3

Strength of the Relationship

If we reject the null hypothesis and conclude that a relationship exists between the independent and dependent variables, it becomes meaningful to ask how strong the relationship is. There are many different statistics one can use to address this question, and statisticians are in disagreement as to which one is best. We will develop the general logic of these approaches using an index called *eta-squared* and then discuss the advantages and disadvantages of different measures. We will initially derive eta-squared using procedures that are computationally inefficient but that best illustrate the concept. A computational formula will then be presented. We will use the experiment on the relationship between order of information and likeability as the example to develop the logic of the approach.

---

* In the context of statistical assumptions, a population is conceptualized as consisting of *all* individuals who meet the criteria for inclusion in a particular research condition. A population of scores is similarly conceptualized as the scores that would theoretically be obtained if *all* individuals meeting the criteria for participation actually provided a score on the dependent variable.

TABLE 10.3 Computation of SS_{TOTAL} for Order-of-Information and Likeability Experiment

Individual	*Condition*	*Likeability rating (X)*	X^2
1	Pro–Con	7	49
2	Pro–Con	4	16
3	Pro–Con	3	9
4	Pro–Con	5	25
5	Pro–Con	5	25
6	Pro–Con	6	36
7	Pro–Con	5	25
8	Pro–Con	5	25
9	Pro–Con	4	16
10	Pro–Con	6	36
11	Con–Pro	1	1
12	Con–Pro	5	25
13	Con–Pro	2	4
14	Con–Pro	5	25
15	Con–Pro	3	9
16	Con–Pro	3	9
17	Con–Pro	3	9
18	Con–Pro	2	4
19	Con–Pro	4	16
20	Con–Pro	2	4
		$\Sigma X = 80$	$\Sigma X^2 = 368$

$$SS_{TOTAL} = \Sigma X^2 - \frac{(\Sigma X)^2}{N}$$

$$= 368 - \frac{80^2}{20} = 48$$

The second and third columns of Table 10.3 present the condition in which each of the 20 individuals participated and their corresponding scores on the dependent variable. Looking at column 3, we see that there is variability in the likeability ratings—some individuals said they like the hypothetical stranger more than did other individuals. Our goal is to analyze this variability and determine what proportion of it is associated with the independent variable. Because of certain desirable statistical properties to be discussed shortly, we will accomplish this using the sum of squares as our measure of variability.

Sum of Squares Total

As a first step, we need to derive a numerical index of the amount of variability in the likeability ratings. This involves computing the sum of squares for the dependent variable across all individuals in the experiment, which can be accomplished using the standard formula for a sum of squares. This has been done in columns 3 and 4 of Table 10.3. This sum of squares is called the **sum of squares total** (symbolized SS_{TOTAL}) since it represents the total amount of variability that exists in the data. In this instance, $SS_{TOTAL} = 48$.

TABLE 10.4 Computation of SS_{ERROR} for Order-of-Information and Likeability Experiment

Condition	X	T	$X_n = X - T$	X_n^2
Pro–Con	7	1	6	36
Pro–Con	4	1	3	9
Pro–Con	3	1	2	4
Pro–Con	5	1	4	16
Pro–Con	5	1	4	16
Pro–Con	6	1	5	25
Pro–Con	5	1	4	16
Pro–Con	5	1	4	16
Pro–Con	4	1	3	9
Pro–Con	6	1	5	25
Con–Pro	1	−1	2	4
Con–Pro	5	−1	6	36
Con–Pro	2	−1	3	9
Con–Pro	5	−1	6	36
Con–Pro	3	−1	4	16
Con–Pro	3	−1	4	16
Con–Pro	3	−1	4	16
Con–Pro	2	−1	3	9
Con–Pro	4	−1	5	25
Con–Pro	2	−1	3	9
	$\Sigma X = 80$ $\bar{X} = G = 4.00$		$\Sigma X_n = 80$	$\Sigma X_n^2 = 348$

$$SS_{ERROR} = \Sigma X_n^2 - \frac{(\Sigma X_n)^2}{N}$$

$$= 348 - \frac{80^2}{20} = 28$$

Treatment Effects and Variance Extraction

The next step in the analysis is to determine what effect the independent variable had on the dependent variable. This is accomplished by comparing the mean score in each condition with the mean score for both conditions combined. Since it is based on all individuals in the study, the overall mean is known as the *grand mean* (symbolized G). As indicated in column 2 of Table 10.4, the grand mean in this instance is 4.00. The mean likeability rating for the pro–con condition was previously found to be 5.00. Thus, the effect of having the traits presented in the pro to con order was to raise the likeability ratings, on the average, 1 unit above the grand mean. For the con–pro condition, the mean likeability score was previously found to be 3.00. The effect of having the traits presented in the con to pro order was thus to lower the likeability ratings, on the average, 1 unit below the grand mean. These effects are formally called *treatment effects* (symbolized T), and are defined as the difference between a given group mean and the grand mean:

$$T_1 = \bar{X}_1 - G = 5.00 - 4.00 = 1.00$$

$$T_2 = \bar{X}_2 - G = 3.00 - 4.00 = -1.00$$

The treatment effects for the two groups have been indicated in column 3 of Table 10.4.

We can use the treatment effects to derive the strength of the relationship between the independent and dependent variables. This will require first removing, or *nullifying,* the influence of the independent variable on the dependent variable. For individuals in the pro–con condition, the effect of the order of information was to raise scores, on the average, 1 unit *above* the grand mean. If we want to nullify this effect, we can *subtract* 1 unit from each of these individuals' original likeability scores. Consider the first subject in Table 10.4. The likeability rating for this subject is 7. To nullify the effect of being in the pro–con condition, we subtract 1 from his score, yielding a *nullified score* (symbolized X_n) of 6. This process has been repeated in column 4 of Table 10.4 for each subject in the pro–con condition.

On the other hand, the effect of the order of information for individuals in the con–pro condition was to lower scores, on the average, 1 unit *below* the grand mean. If we want to nullify this effect, we simply *add* 1 unit (that is, *subtract* 1 *negative* unit) to each of these individuals' original likeability scores. The first subject listed in Table 10.4 in the con–pro condition has a likeability rating of 1. To nullify the effect of being in the con–pro condition, we add 1 to his score, yielding a nullified score of 2. This process has been repeated for each subject in the con–pro condition in column 4 of Table 10.4.

When you examine the nullified dependent variable scores in column 4, note that there is less variability than in the original dependent variable scores contained in column 2. This is because we have removed a certain amount of variability that was associated with the independent variable. Note, however, that there is still variability in the X_n scores, due to factors other than the order of information. In statistical terminology, the remaining variability is called **unexplained variance** or **error variance** and reflects the influence of disturbance variables, as discussed in Chapter 9.

We can derive a numerical index of how much error variance remains after we have removed the influence of the independent variable by computing a sum of squares for the nullified scores. This can be accomplished by applying the standard formula for a sum of squares to the X_n scores, as has been done in columns 4 and 5 of Table 10.4. This sum of squares is formally referred to as the **sum of squares error** (symbolized SS_{ERROR}). In this instance, $SS_{ERROR} = 28$.

Whereas the sum of squares total indexes the *total* amount of variability in the dependent variable, the sum of squares error indexes the amount of *unexplained* variability in the dependent variable—that is, variability that remains after the effects of the independent variable have been removed. If we subtract the sum of squares error from the sum of squares total, we will obtain an index of **explained variance**—that is, variability associated with (explained by) the independent variable. This index is referred to as the **sum of squares explained** (symbolized $SS_{EXPLAINED}$). In our example, $SS_{EXPLAINED} = 48 - 28 = 20$.

Since $SS_{TOTAL} - SS_{ERROR} = SS_{EXPLAINED}$, it is also the case that

$$SS_{TOTAL} = SS_{EXPLAINED} + SS_{ERROR} \qquad [10.9]$$

In other words, the total variability in the dependent variable, as represented by SS_{TOTAL}, can be split up, or *partitioned,* into two components, one ($SS_{EXPLAINED}$) reflecting the influence of the independent variable and one (SS_{ERROR}) reflecting the influence of disturbance variables. It is this property of sums of squares—the fact that they can be added to and subtracted from one another and otherwise algebraically manipulated—that makes the sum of squares the preferred measure of variability when analyzing the strength of the relationship between variables.*

Eta-Squared The stronger the influence of the independent variable on the dependent variable, the larger the sum of squares explained should be relative to the sum of squares total. By dividing the former by the latter, we can calculate the proportion of variability in the dependent variable that is explained by the independent variable. This statistic, the proportion of explained variance, is called **eta-squared** and is defined by the formula

$$\text{eta}^2 = \frac{SS_{EXPLAINED}}{SS_{TOTAL}} \qquad [10.10]$$

where eta^2 stands for "eta-squared" and the other terms are as previously defined.† Eta-squared indexes the strength of the relationship between the independent and dependent variables since it represents the proportion of variability in the dependent variable that is associated with the independent variable. Eta-squared can range from 0 to 1.00. As eta-squared approaches 1.00, the relationship between the variables is stronger, and as eta-squared approaches 0, the relationship between the variables is weaker. In our example,

$$\text{eta}^2 = \frac{20}{48} = .42$$

Thus, 42% of the variability in the dependent variable (likeability) is explained by the independent variable (order of information).

Since eta-squared represents the proportion of variability in the dependent variable that is associated with the independent variable, 1.00 minus eta-squared must represent the proportion of variability in the dependent variable that is *not* associated with the independent variable—that is, the proportion of variability that is due to disturbance variables. In our example, $1.00 - .42 = .58$, or 58%, of the variability in likeability ratings is attributable to disturbance variables.

* In contrast, neither standard deviation estimates nor variance estimates can be manipulated in this manner. This is because they are averages (involving deviation or squared deviation scores) and averages cannot be meaningfully added to or subtracted from one another or otherwise algebraically manipulated.

† It should be noted that eta-squared corresponds to the square of a statistic called the *point-biserial correlation coefficient.*

Standards differ considerably among researchers on the substantive interpretation of eta-squared. Much depends on the specific area of application. In the author's (JJ) own research area on the relationship between attitudes and behavior, an eta-squared smaller than .20 represents a "weak" relationship, an eta-squared between .20 and .50 represents a "moderate" relationship, and an eta-squared larger than .50 represents a "strong" relationship between variables. In other contexts, different interpretations might apply. The behavior of organisms is complex, and rarely are there only a small number of variables that determine that behavior. To the extent that a behavior is determined by a multitude of variables, the explanation of as little as 5–10% of the variance in the dependent measure is, in some respects, a considerable amount. Typically, research in the behavioral sciences produces relatively small values of eta-squared. In the remainder of this book, we will conceptualize an eta-squared of less than .10 as constituting a weak effect, an eta-squared between .10 and .25 as constituting a moderate effect, and an eta-squared of greater than .25 as constituting a strong effect. The values of eta-squared that we will report will generally be inflated relative to those observed in actual behavioral science research.

Computational Formula for Eta-Squared

While it is possible to derive eta-squared with Equation 10.10 using the approach discussed above, there is a much simpler computational formula for eta-squared that is used in practice. This formula is

$$\text{eta}^2 = \frac{t^2}{t^2 + \text{df}} \qquad [10.11]$$

where t is the observed t value and df is the corresponding degrees of freedom.

For the experiment on the order of information and likeability,

$$\text{eta}^2 = \frac{3.57^2}{3.57^2 + 18} = .41$$

This value agrees, within rounding error, with our earlier result and represents a strong effect.

Alternative Measures of the Strength of the Relationship

We have chosen to use eta-squared as the measure of the strength of the relationship between two variables because of its relationship to certain statistics discussed in later chapters and because it possesses desirable statistical properties (Kennedy, 1970; Kesselman, 1975). However, it must be emphasized that eta-squared describes the strength of the relationship between two variables in a set of *sample* data. Estimating the strength of the relationship in the *population* requires a somewhat different approach. A number of different estimation procedures have been proposed, the application of which are controversial among statisticians. *Eta-squared is a biased estimator in that it tends to slightly overestimate the strength of the relationship in the population across random samples.* Hays (1981) has argued for a relatively conservative estimate (that is, one that is likely to underestimate rather than overestimate the true strength of the relationship) called *omega-squared.* The method of estimation one should use is not agreed upon by statisticians. For a discussion of the statistical properties of various indices of strength of association, see Carrol and Nordholm (1975), Fisher (1950), Glass and Hakstian (1969), Haggard (1958), and Kesselman (1975).

Any index of the strength of the relationship derived from sample data will be subject to sampling error and must be interpreted accordingly. An observed sample eta-squared of .50 could result from a population with a value of eta-squared that is quite different. The confidence one has in the accuracy of a sample eta-squared as an estimate of the corresponding population parameter is a function, in part, of the overall sample size. The larger the N, the better the estimate. In fact, for Ns less than 30, the amount of sampling error can be rather sizeable (Carrol & Nordholm, 1975). For this reason, indices of the strength of association must be interpreted with caution. In light of the above, we advocate the use of eta-squared only as a heuristic index that will aid the researcher in appreciating relationships contained within the data. The measure should *not* be used as an estimate of the strength of the relationship in the population because, in our opinion, point estimation procedures (the identification of one specific value) are not very useful in this respect. Instead, an interval estimation approach (the identification of a range of values) seems more reasonable.

Analogous to the construction of confidence intervals for the mean discussed in Chapter 8, it is possible to construct confidence intervals for eta-squared. These provide the investigator with ranges of values that, with a specified degree of confidence, contain the population value of eta-squared. The mathematics for calculating such intervals are complex and are discussed by Fleishman (1980).

Eta-Squared Following a Nonsignificant Test

The foregoing discussion clearly demonstrates the utility of examining eta-squared following a significant statistical test. Given that it indexes the strength of the relationship between two variables in a set of sample data, the calculation of eta-squared can also be useful even when the statistical test is nonsignificant. This relates to the fact that, as discussed in Chapter 8, power increases as does the size of the data set. Thus, when sample sizes are small, power will tend to be low and we will be relatively unlikely to reject a false null hypothesis. By examining eta-squared, we can gain additional insight into the situation. If the null hypothesis is not rejected and eta-squared is small, the statistical decision is reinforced. However, if the null hypothesis is not rejected and eta-squared is relatively large, we might want to collect additional data since this is an indication that the two variables may be related but that the sample sizes were not large enough to yield a significant result with the desired degree of consistency. With this in mind, we recommend that eta-squared be calculated following nonsignificant as well as significant statistical tests, and we will follow this strategy throughout the book.

10.4

Nature of the Relationship

We have now considered the first two questions apropros of the analysis of bivariate relationships: (1) Is there a relationship between the independent and dependent variables? and (2) If so, what is the strength of the relationship? The final question concerns the nature of the relationship. Although this will be determined by examining sample means, the conclusion is intended to apply to the population means. In the experiment on order of information and likeability of a stranger, the mean likeability score is 5.00 in the pro–con condition and 3.00 in the con–pro condition. Since the mean of the pro–con condition is greater than the mean of the con–pro condition, we conclude that the nature of the relationship between order of information and likeability is such that when people are presented information going from positive to negative, they will form more favorable impressions of a stranger (at least in terms of likeability) than when the order of information goes from negative to positive.

10.5

Methodological Considerations

Several methodological features should be noted about the experiment on the order of information and likeability. These will help place the results of the statistical test in proper context. First, the investigator attempted to control for a number of confounding variables by randomly assigning subjects to the two experimental conditions. Without this random assignment, differences between the two groups could be attributed to individual difference factors rather than to the experimental treatment. For example, psychologists have shown that some people have a tendency to view everyone in a positive light, whereas others tend to view everyone in a negative light. Without random assignment, it is possible that the pro–con condition could have contained more of the former individuals and the con–pro condition more of the latter. With random assignment, this is unlikely.

Another issue concerns the extent to which disturbance variables were controlled in the experiment. These have the effect of creating sampling error, the amount of which is reflected in the standard error of the difference. In this experiment, this was estimated to be .56, suggesting that on the average, the difference between any two sample means based on random samples of size 10 will deviate .56 unit from the true population difference between means. Given a potential range of 6 (since likeability ratings were made on a scale ranging from 1 to 7), this suggests a relatively small amount of sampling error.

A second perspective on sampling error is gained by examining eta-squared. In this experiment, 42% of the variability in the dependent variable is associated with the independent variable and 58% is due to disturbance variables. The picture that emerges is one of the disturbance variables (and the consequent sampling error) not overwhelming any "signal" produced by the independent variable.

The experiment has several limitations from the standpoint of generalizability. First, it used college students from one university. Would the results generalize to other types of individuals as well? Second, the experiment used only one pro–con list and one con–pro list. Maybe it was something about the particular adjectives used that produced the experimental outcome. Would similar results occur if different adjectives were used? These and other questions of generalizability can be addressed through additional experimentation. The important point is that one should always consider such issues when trying to draw conclusions from research results. Statistics and research design go hand in hand when interpreting the results of an investigation.

10.6 Numerical Example

In Western society, individuals have long sought methods that will help them to relax. One such method is that of meditation. Maharishi Mahesh Yogi and his followers have introduced a simple relaxation technique called transcendental meditation (TM). An individual who practices TM meditates twice a day for 15 to 20 minutes. During meditation the individual repeats, subvocally, a mantra, which is a meaningless, two-syllable sound. The mantra is different for each individual and is given to the meditator by a trained teacher of transcendental meditation. Proponents of TM emphasize the importance of the mantra in the meditation process. However, it has been suggested that the mantra is not essential to achieve optimal levels of relaxation during meditation and that the same results could be achieved by using the word *One* instead. Let us consider a hypothetical experiment designed to test this assertion.

Twenty individuals served as experimental subjects. Ten were randomly assigned to the mantra condition, and 10 were randomly assigned to the "One" condition. Each subject practiced meditation using the appropriate technique for 2 weeks prior to the test session. During the test session, individuals participated in their regular 15-minute meditation period while a number of physiological indicators of relaxation were measured. One of these was heart rate. In general, the more relaxed an individual, the slower his or her heart rate will be. The heart rate measures for the 20 subjects appear in columns 1 and 3 of Table 10.5.

The null hypothesis is that individuals who practice TM will not differ in their mean heart rate from individuals who meditate using the word *One* in

TABLE 10.5 Data and Calculations for Meditation Technique and Heart Rate Experiment

MANTRA		"ONE"	
X	X^2	X	X^2
58	3,364	62	3,844
62	3,844	57	3,249
54	2,916	67	4,489
52	2,704	66	4,356
64	4,096	58	3,364
58	3,364	60	3,600
52	2,704	64	4,096
64	4,096	62	3,844
56	3,136	57	3,249
60	3,600	67	4,489
$\Sigma X_M = 580$	$\Sigma X_M^2 = 33{,}824$	$\Sigma X_O = 620$	$\Sigma X_O^2 = 38{,}580$
$\bar{X}_M = 58.00$		$\bar{X}_O = 62.00$	

$$SS_M = \Sigma X_M^2 - \frac{(\Sigma X_M)^2}{n_M} = 33{,}824 - \frac{580^2}{10} = 184$$

$$SS_O = \Sigma X_O^2 - \frac{(\Sigma X_O)^2}{n_O} = 38{,}580 - \frac{620^2}{10} = 140$$

place of a mantra, while the alternative hypothesis is that they will differ. These hypotheses can be formally stated as

$$H_0: \quad \mu_M = \mu_O$$
$$H_1: \quad \mu_M \neq \mu_O$$

where the subscript "M" denotes the mantra condition and the subscript "O" denotes the "One" condition. In other words, the null hypothesis states that there is no relationship between the type of meditation technique and relaxation level (as indicated by heart rate). The alternative hypothesis states that there is a relationship between these two variables.

Since the null hypothesis states that the population means are equivalent, $\mu_1 - \mu_2$ is equal to 0 for purpose of hypothesis testing. Other values necessary for the statistical analysis can be calculated using the information presented in Table 10.5:

$$\bar{X}_1 - \bar{X}_2 = 58.00 - 62.00 = -4.00$$

$$\hat{s}_{\bar{X}_1 - \bar{X}_2} = \sqrt{\left(\frac{SS_1 + SS_2}{n_1 + n_2 - 2}\right)\left(\frac{1}{n_1} + \frac{1}{n_2}\right)}$$

$$= \sqrt{\left(\frac{184 + 140}{10 + 10 - 2}\right)\left(\frac{1}{10} + \frac{1}{10}\right)} = 1.90$$

The observed value of t is thus

$$
\begin{aligned}
t &= \frac{(\bar{X}_1 - \bar{X}_2) - (\mu_1 - \mu_2)}{\hat{s}_{\bar{X}_1 - \bar{X}_2}} \\
&= \frac{-4.00 - 0}{1.90} = -2.11
\end{aligned}
$$

For an alpha level of .05, nondirectional test, and $n_1 + n_2 - 2 = 10 + 10 - 2 = 18$ degrees of freedom, the critical values of t are ± 2.101. Since the observed t value of -2.11 is less than the negative critical value of -2.101, we might be tempted to reject the null hypothesis and conclude that there is a relationship between the type of meditation technique and relaxation level (heart rate). However, since the two values are so similar to one another, we should first repeat our calculations to three (or more) decimal places to increase the precision of our answer. If we do so, we obtain an observed t value of -2.109. Since this is less than the negative critical value of -2.101, the null hypothesis is indeed rejected.

The strength of the relationship in the sample is indexed by eta-squared. Using Equation 10.11, we find that

$$
\begin{aligned}
\text{eta}^2 &= \frac{t^2}{t^2 + \text{df}} \\
&= \frac{(-2.11)^2}{(-2.11)^2 + 18} = .20
\end{aligned}
$$

The proportion of variability in heart rate that is associated with the type of meditation technique is .20. This represents a moderate effect.

The nature of the relationship is indicated by the mean scores for the two conditions. Since the mean heart rate for the mantra group (58.00) is lower than the mean heart rate for the "One" group (62.00), we conclude that the use of a mantra produces more relaxation during meditation than does the repetition of the word *One*.

You should think about these results in terms of the research design questions noted previously. Are there any potential confounding variables that have not been controlled? What has been the role of disturbance variables? What kinds of procedures could be used to reduce sampling error? What are the limitations of the experiment in terms of generalizability?

10.7 Planning an Investigation Using the Independent Groups *t* Test

When designing a study involving two independent groups, the investigator is faced with a number of important decisions. One set of decisions concerns the delineation of confounding and disturbance variables and how to control for these in the context of the study. Another decision concerns the number of subjects that should be sampled for each group. One consideration that influences the latter decision is practicality. One cannot sample more subjects than resources permit. Practical matters aside, there are also statistical considerations that must be taken into account. One major issue concerns the desired power of the statistical test that will be used to analyze the data. Recall from Chapter 8 that the power of a statistical test refers to the probability of rejecting the null hypothesis when it is false. When we say the power of a test is .70, this means that the chances are 7 in 10 we will correctly reject the null hypothesis when it is false.

The power of a statistical test is influenced by three major factors: (1) the strength of the relationship between the two variables *in the population,* (2) the sample sizes, and (3) the alpha level. Consider the first factor. If the strength of the relationship in the population is weak, then it is relatively unlikely that we will detect its presence, and hence, it is relatively unlikely that we will correctly reject the null hypothesis. The power of the test will be relatively low, everything else being equal. If the strength of the relationship is strong, however, then it is more likely that we will detect the relationship between the independent and dependent variables, and hence, it is more likely that we will correctly reject the null hypothesis. The power of the test will be relatively high. Thus, as the strength of the relationship in the population increases, the power of the statistical test also increases.

Next, consider sample size. As indicated in Chapters 8 and 9, when sample sizes are relatively large, it is more likely that we will detect an existing difference between population means (that is, the existence of a relationship). Again, the power of the statistical test is affected, with larger sample sizes leading to more powerful tests.

Finally, consider the case of the alpha level. When the alpha level is low, the likelihood that we will reject the null hypothesis is also low since we must be very certain a relationship exists before we are willing to say it does. Consequently, we will be relatively likely to overlook a relationship that does exist, especially if it is a weak one and we have small sample sizes. As the alpha level becomes larger, the power of a statistical test will increase.

If a Type I error is serious, then the investigator will want to minimize its chance of occurrence by setting a low alpha level. This means that the major factor the researcher can use to minimize a Type II error is the sample sizes used in the study. If a Type II error is also serious, then one will want to achieve high levels of power. Larger sample sizes can accomplish this.

Statisticians have developed procedures for estimating the sample sizes necessary to obtain a desired level of power, given the strength of the relationship in the population and the alpha level. The required sample sizes also differ depending on whether the test is directional or nondirectional. Appendix E.1 presents a set of tables that will be useful in this respect.* The portion of this

* These tables are based on power estimates given by Cohen (1977).

TABLE 10.6 Approximate Sample Sizes Necessary to Achieve Selected Levels of Power for Alpha = .05 as a Function of Population Values of Eta-Squared

Nondirectional Test, Alpha = .05

	POPULATION ETA-SQUARED									
POWER	.01	.03	.05	.07	.10	.15	.20	.25	.30	.35
.25	84	28	17	12	8	6	5	3	3	3
.50	193	63	38	27	18	12	9	7	5	5
.60	246	80	48	34	23	15	11	8	7	6
.67	287	93	55	39	27	17	12	10	8	6
.70	310	101	60	42	29	18	13	10	8	7
.75	348	113	67	47	32	21	15	11	9	7
.80	393	128	76	53	36	23	17	13	10	8
.85	450	146	86	61	41	26	19	14	11	9
.90	526	171	101	71	48	31	22	17	13	11
.95	651	211	125	87	60	38	27	21	16	13
.99	920	298	176	123	84	53	38	29	22	18

appendix for an alpha level of .05, nondirectional test, is shown in Table 10.6 for purpose of exposition. The first column of the table presents different levels of power. The values at the top of the table are population values of eta-squared. Table entries are required sample sizes *per group* to achieve the corresponding level of power.* For example, if the researcher suspects (either on the basis of past research, intuition, or theory) that the strength of the relationship he or she is trying to detect in the population is relatively weak, corresponding to an eta-squared of roughly .03, then the necessary sample size to achieve power of .95 for an alpha level of .05, nondirectional test, is 211 per group. If the researcher suspects that the strength of the relationship is much stronger, corresponding to an eta-squared of roughly .20, then the necessary sample size to achieve power of .95 is 27 per group. Examination of Table 10.6 and Appendix E.1 provides a general feel for the relationships between directionality, the alpha level, sample size, the strength of the relationship one is trying to detect in the population, and power for the independent groups t test.

It is not always practical to obtain high levels of power by increasing sample sizes, especially when the population relationship is judged to be weak. In the above example where eta-squared was .03, the required 211 subjects per group could be costly, especially if the study involved the use of nonhuman primates or other expensive laboratory animals. The seriousness of lack of power again depends on the seriousness of a Type II error. When testing for sex discrimination, what would be the consequences of saying discrimination does not exist when, in fact, it does? If the consequences are judged to be serious, one

* Given the imprecision of the procedures for estimating the necessary sample sizes, the values contained in this and the other power tables presented in this book are *approximate*.

would insist upon high levels of power. Investigators generally strive to achieve power of .80 to .95, as a rough guideline. However, the power level can be revised upward or downward, depending on the situation.

10.8 Method of Presentation

When reporting the results of an independent groups *t* test, all three questions outlined in Section 10.4 (whether there is a relationship between the two variables, the strength of the relationship, and the nature of the relationship) should be addressed. Of course, the last question is only relevant if the null hypothesis is rejected. In fact, if the null hypothesis is not rejected, discussion of the strength of the relationship is typically also excluded.

The results for the experiment on order of information and likeability discussed earlier in this chapter might be presented as follows:

Results

An independent groups *t* test was performed comparing the mean likeability rating for the pro–con condition (5.00) with that for the con–pro condition (3.00). This was found to be statistically significant, $t(18) = 3.57$, $p < .01$, indicating that strangers will be evaluated more favorably when positive information about them is followed by negative information than when the reverse is true. The strength of the relationship between order of information and likeability, as indexed by eta^2, was .41.

The first sentence states the type of test that was conducted and the values of the sample means. These should always be reported, either in the text itself or in an accompanying table. The statement that the *t* test was statistically significant signifies that the null hypothesis was rejected. If it had not been rejected, the terminology "statistically nonsignificant" would have been used instead. Paralleling the format for the one-sample *t* test, this is followed by the degrees of freedom associated with the relevant *t* distribution, the observed value of *t*, and the significance level. Since there is nothing to indicate otherwise, a nondirectional test is assumed. The remainder of this sentence specifies the nature of the relationship.

The final sentence contains the value of eta-squared. In practice, researchers often fail to present this or other information regarding the strength of the relationship. This can be problematic because sometimes investigators interpret their data as if a strong relationship exists when, in fact, it is so weak as to be almost trivial. Fortunately, however, eta-squared can be derived from the degrees of freedom and the observed value of t using Equation 10.11.

10.9

Examples from the Literature

Monetary Compensation and Enjoyableness of a Boring Task

Behavioral scientists have devoted a lot of work to identifying psychological factors that influence attitude change. In a classic experiment in this area (Festinger & Carlsmith, 1959), subjects individually came to a psychology laboratory and worked on a very boring task for 1 hour. After completing the task, they were informed that the experiment was over and that it had been a study of the effects of "expectancy" on performance—supposedly, half of the subjects had been led to believe that the task would be interesting while the other half had not been told anything about it so that the influence of the expectancy manipulation on task performance could be studied. In actuality, this was not the true purpose of the experiment, and none of the participants had actually been informed that the task would be interesting. Rather, as discussed below, it was a study of a psychological process known as cognitive dissonance.

Subjects were then told that the experimenter who usually informed waiting subjects that the task was interesting was sick, and were offered either $1 or $20 to take his place. All subjects complied with the request for help. Afterward, they were given a questionnaire about the original experiment and asked to rate, among other things, how enjoyable the task was on a scale from −5 ("extremely boring") to +5 ("extremely enjoyable").

At least two different principles could operate in this situation to influence the enjoyableness rating. The first is a reinforcement principle, which would state that subjects who were paid $20 for telling another subject that the task was interesting and enjoyable would receive greater reinforcement for doing so than subjects who were paid only $1 to do it. This greater reinforcement should generalize to their own perceptions of the task, and hence, subjects in the $20 condition should rate the task as being more enjoyable than those in the $1 condition.

An alternative explanation would predict just the opposite. According to this viewpoint, all subjects performed a behavior that was in contradiction to what they truly believed—they told someone that a boring task was enjoyable. Because of this contradiction, subjects should experience an unpleasant internal state known as *cognitive dissonance,* which they should then be motivated to reduce. Subjects in the $20 condition could easily justify their counterattitudinal behavior: They did it for the money. This was not the case, however, for subjects in the $1 condition, where the money was not very great. These subjects would have to reduce the dissonance in another way, and one possibility would be to rationalize that the task was really not all that boring. If this dissonance mechanism were operating, the reverse prediction would be made: Subjects in the \$1 condition should rate the task as being more enjoyable than those in the \$20 condition.

The results of the experiment were analyzed using an independent groups t test. The independent variable was the amount of money paid for telling another subject that the task was enjoyable (\$1 versus \$20) and the dependent variable was the enjoyableness of the task. The mean rating in the \$1 condition was 1.35 and the mean rating in the \$20 condition was −.05. Consistent with a dissonance interpretation, these means were found to be significantly different, $t(38) = 2.22$, $p < .05$. Festinger and Carlsmith did not report any statistics regarding the strength of the relationship between the amount of money paid and the enjoyableness ratings. However, using Equation 10.11, we find that eta-squared equals .11, which represents a moderate effect.

Students' Perceptions of Self-Monitoring of Good and Poor Teachers

An interesting question for teachers and students alike concerns the personality characteristics that distinguish good from poor teachers. One potentially relevant characteristic is self-monitoring. Among other things, self-monitoring concerns the extent to which individuals are sensitive to the behavior of others and able to modify their own social behavior accordingly. As self-monitoring increases, individuals become relatively more concerned with the situational appropriateness of their behavior and relatively less constrained by their own internal dispositions. Thus, for instance, high self-monitoring teachers might be expected to be more responsive to the needs of their students and more flexible in their classroom approach.

In an attempt to determine students' perception of the role teacher self-monitoring plays in the teaching process, Larkin (1987) asked 116 college undergraduates to think of either the best or the worst teacher that they had ever had. Students in each group then rated their "best" or "worst" teacher on a 13-item self-monitoring scale. A dependent variable score was derived for each subject by summing his or her responses across the 13 items. As predicted, an independent groups t test showed that the good teachers ($\bar{X} = 49.56$) were rated as significantly higher in self-monitoring than the poor teachers ($\bar{X} = 29.81$), $t(114) = 13.18$, $p < .001$. As indexed by eta-squared, the strength of the relationship was .60. This represents a strong effect. Are these results consistent with your own educational experience?

10.10

Summary

The independent groups t test is used to analyze the relationship between two variables when (1) the dependent variable is quantitative in nature and is measured on a level that approximates interval characteristics, (2) the independent variable is between-subjects in nature, and (3) the independent variable has two and only two levels. The existence of a relationship between the two variables is tested by converting the observed difference between sample means into a t value representing the number of estimated standard errors the difference is from the mean of the appropriate sampling distribution of the difference between two independent means (assuming the null hypothesis is true). This t value is compared with the critical values of t, and the decision to reject or not to reject the null hypothesis is made accordingly. The strength of the relationship is analyzed using the eta-squared statistic, which represents the proportion of variability in the dependent variable that is associated with the independent variable. Finally, the nature of the relationship is determined by examination of the mean scores in the two groups.

Tables are available for determining the sample sizes necessary for the independent groups t test to achieve a desired level of power. The required sample sizes are dependent on the strength of the relationship between the independent variable and the dependent variable in the population, the alpha level, and whether the test is directional or nondirectional.

Exercises

Answers to asterisked (*) exercises appear at the back of the book. Answers to exercises with two asterisks are also worked out step-by-step in the Study Guide.

Exercises to Review Concepts

1. Under what conditions is the independent groups t test typically used to analyze a bivariate relationship?

***2.** In general terms, what value will the mean of a sampling distribution of the difference between two independent means always be equal to?

***3.** Explain the rationale that allows us to combine the variance estimates from two samples to obtain a pooled variance estimate.

****4.** Compute the pooled variance estimate and estimated standard error of the difference for $n_1 = 10$, $n_2 = 13$, $\hat{s}_1^2 = 6.48$, and $\hat{s}_2^2 = 4.73$.

***5.** Two random samples were taken from their respective populations and the following summary statistics were observed:

Sample A	*Sample B*
$n_A = 49$	$n_B = 49$
$\bar{X}_A = 10.00$	$\bar{X}_B = 13.00$
$\hat{s}_A = 1.44$	$\hat{s}_B = 1.58$

a. Compute the estimated standard error of the mean for sample A.

b. Compute the estimated standard error of the mean for sample B.

c. Compute the estimated standard error of the difference between the two means.

d. Your answer for part **c** should be larger than your answer for part **a** or **b**. In fact, the estimated standard error of the difference between two independent means will always be larger than the respective estimated standard errors of the mean. Why do you think this is the case?

6. Match each symbol in the first column with the appropriate term in the second column:

Symbol	*Term*
1. $\hat{s}_{\bar{X}_1-\bar{X}_2}$	a. variance estimate
2. $\hat{s}^2_{\text{pooled}}$	b. null hypothesis
3. $\hat{s}_{\bar{X}}$	c. standard error of the mean
4. $\hat{s}^2$	d. estimated standard error of the mean
5. H_1	e. standard error of the difference between two independent means
6. H_0	f. alternative hypothesis
7. $\sigma_{\bar{X}}$	g. estimated standard error of the difference between two independent means
8. $\sigma_{\bar{X}_1-\bar{X}_2}$	h. pooled variance estimate

7. Summarize the five steps involved in hypothesis testing for the independent groups t test.

***8.** State the critical value(s) of t that would be used to reject the null hypothesis for an independent groups t test at an alpha level of .05 under each of the following conditions:

a. H_0: $\mu_1 = \mu_2$, H_1: $\mu_1 > \mu_2$, $n_1 = 10$, $n_2 = 10$
b. H_0: $\mu_1 = \mu_2$, H_1: $\mu_1 \neq \mu_2$, $n_1 = 10$, $n_2 = 10$
c. H_0: $\mu_1 = \mu_2$, H_1: $\mu_1 \neq \mu_2$, $n_1 = 15$, $n_2 = 15$
d. H_0: $\mu_1 = \mu_2$, H_1: $\mu_1 < \mu_2$, $n_1 = 15$, $n_2 = 15$
e. H_0: $\mu_1 = \mu_2$, H_1: $\mu_1 \neq \mu_2$, $n_1 = 10$, $n_2 = 20$

9. What are the assumptions underlying the independent groups t test?

Refer to the following information to answer Exercises 10–16.

An investigator was interested in the relationship between a person's gender and the tendency to hold discriminatory attitudes toward women. An attitude scale was administered to five men and five women. Scores could range from 1 to 10, with higher scores indicating more discriminatory attitudes. The relevant data are presented below:

Males	*Females*
8	5
7	4
7	4
7	4
6	3

***10.** Conduct a nondirectional independent groups t test on the data to test for a relationship between gender and discriminatory attitudes.

****11.** Compute the sum of squares total.

****12.** Compute the treatment effect for males and the treatment effect for females.

****13.** Based on your answers in Exercise 12, use the variance extraction procedures discussed in this chapter to nullify the effect of gender on discriminatory attitudes (that is, generate a set of scores on the dependent variable with the effect of gender removed).

****14.** Compute the sum of squares error and the sum of squares explained for the data derived in Exercise 13.

***15.** Compute the value of eta-squared for the given data using Equation 10.10. Recalculate this index using Equation 10.11. Compare the two sets of results. (*Note:* In practice, we would use only the approach of Equation 10.11.) Does the observed value represent a weak, moderate, or strong effect?

***16.** What is the nature of the relationship between gender and discriminatory attitudes for these data?

***17.** Explain the interrelationships among SS_{TOTAL}, $SS_{\text{EXPLAINED}}$, and SS_{ERROR}.

18. What would the value of eta-squared be if an independent groups t test yielded an observed t value of -1.73 for $n_1 = 16$ and $n_2 = 18$? Does this represent a weak, moderate, or strong effect?

***19.** Why is it inappropriate to estimate the strength of the relationship between two variables in the population from the sample value of eta-squared?

20. What is the rationale behind calculating eta-squared following nonsignificant statistical tests?

Refer to the following information to answer Exercises 21–27.

An investigator wanted to test the effect of alcohol on reaction time. Five subjects were given alcohol to consume until a certain level of intoxication was achieved (as indexed by physiological measures). Another group of subjects was not given any alcohol but instead consumed a placebo. All subjects then participated in a reaction time task. The reaction times, in seconds, for the two groups were as follows:

Alcohol	*Placebo*
2.00	1.50
2.50	1.00
2.00	1.00
1.50	1.00
2.00	.50

21. Conduct a nondirectional independent groups t test on the data to test for a relationship between alcohol consumption and reaction time.

22. Compute the sum of squares total.

23. Compute the treatment effect for alcohol and the treatment effect for the placebo.

24. Based on your answers in Exercise 23, use the variance extraction procedures discussed in this chapter to nullify the effect of alcohol consumption on reaction time (that is, generate a set of scores on the dependent variable with the effect of alcohol consumption removed).

25. Compute the sum of squares error and the sum of squares explained for the data derived in Exercise 24.

26. Compute the value of eta-squared for the given data using Equation 10.10. Recalculate this index using Equation 10.11. Compare the two sets of results. (*Note:* In practice, we would use only the approach of Equation 10.11.) Does the observed value represent a weak, moderate, or strong effect?

27. What is the nature of the relationship between alcohol consumption and reaction time for these data?

28. What are the three major factors that influence the power of a statistical test?

*29. If a researcher suspected that the strength of the relationship between two variables in the population was .07 as indexed by eta-squared, what sample size should she use in a study involving two independent groups and an alpha level of .05, nondirectional test, in order to achieve power of .95?

30. Suppose an investigator reported the results of a study using two independent groups with $n = 23$ per group. If the value of eta-squared in the population was .10, what would the power of his statistical test be at an alpha level of .05, nondirectional test?

Exercises to Apply Concepts

**31. Psychologists have studied extensively the effects of early experience on the development of individuals. It has long been recognized that a positive, challenging, and diverse environment (sometimes called an enriched environment) leads to the acquisition of more positive abilities and personality traits than an environment that is relatively impoverished and isolated. Bennett, Krech, and Rosenzweig (1964) suggested that the type of environment may even act to alter the physical characteristics of the brain and have reported a series of studies to investigate this possibility. In one study, laboratory rats from the same genetic strain were raised in one of two conditions. One-half of the rats were raised in an enriched environment, which involved being housed with 10 to 12 other animals in large cages that were equipped with a variety of "toys." Each day these rats were placed in a square field where they were allowed to explore a pattern of barriers that were changed daily. The other half of the rats were raised in an isolated environment. They were caged singly in a dimly lit room where they could not see or touch another animal (although they could hear and smell them). After 80 days all animals were killed for purposes of analyzing the structure of the brain as a function of the type of environment in which the animal was raised. One factor that was examined was the weight of the cortex (reported in milligrams). The hypothetical data presented below are representative of the results of the study. Analyze these data using a nondirectional test, draw a conclusion, and write up your results using the principles discussed in the Method of Presentation section.

Enriched environment	*Isolated environment*
685	660
690	642
675	640
660	626
645	612
630	610
635	592

32. McConnell (1966) reported a series of experiments that have been the subject of considerable controversy concerning the physiological bases of learning and memory. McConnell suggested that RNA and DNA protein molecules constitute the physiochemical substrate of learning, and that it may be possible to transfer memory, biochemically, between organisms by transferring the relevant RNA and DNA molecules. McConnell's initial experiments were conducted with planaria (small, worm-like organisms). Classical conditioning procedures were used to teach a group of planaria to contract in size

whenever they were exposed to a light. McConnell tried several different methods of transferring the relevant RNA and DNA molecules of these trained planaria to other planaria. The only practical method, however, was cannibalism. The trained planaria were chopped up into small pieces and fed to another group of planaria. The trained cannibals represented the experimental group. A control group of untrained cannibals was fed chopped planaria that had not been taught to contract when exposed to the light. The trained and untrained cannibals were then exposed to the light 25 times. McConnell counted the number of times each planarian contracted when exposed to the light. If the relevant RNA and DNA molecules had been transferred and appropriately used, then the trained cannibals should exhibit a greater number of contractions than the untrained cannibals. Such a demonstration would at least establish the *possibility* of transferring memory from one organism to another via biochemical means. The hypothetical data reported below are representative of the results of the study. Analyze these data using a nondirectional test, draw a conclusion, and write up your results using the principles discussed in the Method of Presentation section.

Trained cannibals					
4	8	10	13	15	21
6	8	10	14	15	22
7	9	11	14	15	
8	10	12	15	17	

Untrained cannibals					
1	4	5	6	10	16
1	4	6	7	10	19
3	5	6	7	11	
4	5	6	7	11	

ELEVEN

CORRELATED GROUPS t TEST

11.1

Use of the Correlated Groups *t* Test

The statistical technique developed in this chapter is called the **correlated groups *t* test.** It is typically used to analyze the relationship between two variables when

1. the dependent variable is quantitative in nature and is measured on approximately an interval level;
2. the independent variable is *within-subjects* in nature (it can be either qualitative or quantitative); and
3. the independent variable has two and only two levels.

The major difference between the correlated groups *t* test and the independent groups *t* test is that the former is used when the independent variable is within-subjects in nature and the latter is used when the independent variable is between-subjects in nature.*

An important advantage of the correlated groups *t* test over the independent groups *t* test relates to the control of disturbance variables. As discussed in Chapter 9, disturbance variables are variables that are unrelated to the independent variable but that influence the dependent variable. By thus contributing to variability in the dependent variable, disturbance variables create "noise" that makes it more difficult to detect a relationship between the independent and dependent variables.

One major source of "noise" is the difference in backgrounds and abilities of the individuals participating in an investigation. For example, the fact that some individuals are more intelligent than others might be expected to influence performance on a number of research tasks. If we knew the influence of individual differences on the dependent variable, then their influence could be

* Matched-subjects designs are analyzed as if the independent variable is within-subjects in nature.

separated from the effects of the independent variable. However, if their influence cannot be isolated, then these differences in background and ability remain as uncontrolled sources of variability and increase the error variance. As will be demonstrated shortly, within-subjects designs (but not between-subjects designs) allow us to estimate this source of variability and extract it from the dependent variable. Consequently, the correlated groups t test provides a more *sensitive* test of the relationship between the independent and dependent variables than does the independent groups t test. A test's sensitivity can be defined as its ability to detect a relationship between variables when a relationship exists in the population. Sensitivity is thus closely related to statistical power.

As an example of the type of study that meets the conditions for the correlated groups t test, suppose an investigator is studying the effects of two drugs, A and B, on learning. Five subjects are administered drug A and then work on a learning task. One month later, the same five subjects are administered drug B and work on the same type of learning task as before. The investigator has introduced a 1-month time interval between the two sessions in order to control for potential carry-over effects. The number of errors made following the administration of each drug is tabulated and serves as the dependent measure. These data are presented in the second and third columns of Table 11.1. The experimental design involves a within-subjects independent variable with two levels (drug A versus drug B) and a quantitative dependent variable. Hence, the correlated groups t test is the statistical technique that would typically be used to analyze the relationship between the variables.

TABLE 11.1 Data and Calculations for Drug and Learning Experiment

Subject	*X for drug A*	*X for drug B*	*Difference (D)*	D^2
1	3	7	−4	16
2	1	5	−4	16
3	4	4	0	0
4	2	8	−6	36
5	5	6	−1	1
	$\Sigma X_A = 15$	$\Sigma X_B = 30$	$\Sigma D = -15$	$\Sigma D^2 = 69$
	$\bar{X}_A = 3.00$	$\bar{X}_B = 6.00$	$\bar{D} = -3.00$	

$$SS_D = \Sigma D^2 - \frac{(\Sigma D)^2}{N} = 69 - \frac{(-15)^2}{5} = 24$$

$$\hat{s}_D^2 = \frac{SS_D}{N-1} = \frac{24}{4} = 6.00$$

$$\hat{s}_D = \sqrt{\hat{s}_D^2} = \sqrt{6.00} = 2.45$$

$$\hat{s}_{\bar{D}} = \frac{\hat{s}_D}{\sqrt{N}} = \frac{2.45}{\sqrt{5}} = 1.10$$

11.2

Inference of a Relationship Using the Correlated Groups t Test

Null and Alternative Hypotheses

The first question to be considered concerns the existence of a relationship between the type of drug administered and learning. We begin by formally phrasing this in the context of a null and an alternative hypothesis:

$$H_0: \quad \mu_A = \mu_B$$
$$H_1: \quad \mu_A \neq \mu_B$$

where the subscripts "A" and "B" respectively represent drug A and drug B. The null hypothesis states that there is no relationship between the type of drug and learning. If the effects of the two drugs on learning are the same, then one would expect the population means to be equal. The alternative hypothesis states that there is a relationship between the type of drug and learning; that is, which drug is used does matter, and the population means will, accordingly, not be equal.*

If the null hypothesis is true, and the type of drug does not make a difference, then we would expect performance while under the influence of drug A to be the same as performance while under the influence of drug B. Another way of stating this is that the difference between a subject's score in condition A and his score in condition B should be 0. Column 4 in Table 11.1 reports a difference score (D) for each individual in the experiment. The mean difference score across individuals ($\bar{D}$) is -3.00. Note that this is equal to the difference between the means for the two conditions ($3.00 - 6.00 = -3.00$). The question of interest is whether the mean difference of -3.00 is sufficiently different from 0 to reject the null hypothesis. To answer this question, it is necessary to specify the nature of the *sampling distribution of the mean of difference scores* for $N = 5$.

Sampling Distribution of the Mean of Difference Scores and the t Test

The **sampling distribution of the mean of difference scores** has properties similar to those of the sampling distribution of the mean described in Chapters 7 and 8 for testing a sample mean against a hypothesized population mean. It will be remembered that this latter distribution has a mean of μ and a standard error ($\sigma_{\bar{X}}$) that can be estimated using the formula $\hat{s}_{\bar{X}} = \hat{s}/\sqrt{N}$. Similarly, a sampling distribution of the mean of difference scores has a mean of $\mu_{\bar{D}}$ (the

* As with the tests discussed previously, if there is exclusive concern with a specific direction of mean differences, the alternative hypothesis can be phrased directionally rather than nondirectionally.

mean difference score in the population) and a standard error ($\sigma_{\bar{D}}$) that can be estimated as follows:

$$\hat{s}_{\bar{D}} = \frac{\hat{s}_D}{\sqrt{N}} \quad [11.1]$$

where $\hat{s}_{\bar{D}}$ is the estimated standard error of the mean of difference scores and $\hat{s}_D$ is the standard deviation estimate for the sample difference scores. The quantity $\hat{s}_D$ can be derived by applying the usual procedures for a standard deviation estimate to the D scores. As shown in Table 11.1, $\hat{s}_D$ in the drug and learning example is equal to 2.45, yielding an estimated standard error of $2.45/\sqrt{5} = 1.10$. Assuming the null hypothesis is true, $\mu_{\bar{D}}$ in this case is equal to 0.

The existence of a relationship between an independent variable and a dependent variable in the correlated groups case is tested by converting the observed mean difference in the sample into a t value and comparing this with the critical t values defining the rejection region. Since we are dealing with difference scores, the relevant t distribution has $N - 1$ degrees of freedom, where N represents the number of difference scores.* The observed value of t can be computed using the formula

$$t = \frac{\bar{D} - \mu_{\bar{D}}}{\hat{s}_{\bar{D}}} \quad [11.2]$$

When $\mu_{\bar{D}}$ is hypothesized to equal 0 (that is, when the population means for the two conditions are hypothesized to be identical), Equation 11.2 reduces to

$$t = \frac{\bar{D}}{\hat{s}_{\bar{D}}} \quad [11.3]$$

This is the most commonly used version of the correlated groups t test, as the null hypothesis will usually state H_0: $\mu_1 = \mu_2$.† However, for purpose of illustration, we will employ the more general formula represented by Equation 11.2 whenever a correlated groups t test is called for in the remainder of this chapter.

In the drug and learning example,

$$t = \frac{-3.00 - 0}{1.10} = -2.73$$

For an alpha level of .05, nondirectional test, the critical values of t for df $= N - 1 = 5 - 1 = 4$ are ± 2.776. Since -2.73 is not less than the negative critical value of -2.776, we fail to reject the null hypothesis of no relationship between the type of drug and learning.

---

* It will always be the case that $n_1 = n_2 = N$ in the type of design being considered in this chapter.

† As with the independent groups t test, other null hypotheses are also possible.

STUDY EXERCISE 11.1

Eight individuals indicated their attitudes toward socialized medicine before and after listening to a pro–socialized medicine lecture. Attitudes were assessed on a 7-point scale with higher numbers indicating more positive attitudes. The attitudes before and after listening to the lecture were as indicated below. Use a correlated groups t test to test for the existence of a relationship between the time of assessment and attitudes toward socialized medicine.

Individual	*Before speech*	*After speech*
1	3	6
2	4	6
3	3	3
4	5	7
5	2	4
6	5	6
7	3	7
8	4	6

Answer The null and alternative hypotheses are

$$H_0: \quad \mu_B = \mu_A$$

$$H_1: \quad \mu_B \neq \mu_A$$

where the subscripts "B" and "A" denote the before- and after-speech conditions, respectively. For an alpha level of .05, nondirectional test, and $N - 1 = 8 - 1 = 7$ degrees of freedom, the critical values of t are ± 2.365.

To obtain the observed value of t, we proceed as follows:

Individual	*X for before speech*	*X for after speech*	*Difference (D)*	D^2
1	3	6	−3	9
2	4	6	−2	4
3	3	3	0	0
4	5	7	−2	4
5	2	4	−2	4
6	5	6	−1	1
7	3	7	−4	16
8	4	6	−2	4
	$\Sigma X_B = 29$	$\Sigma X_A = 45$	$\Sigma D = -16$	$\Sigma D^2 = 42$
	$\bar{X}_B = 3.62$	$\bar{X}_A = 5.62$	$\bar{D} = -2.00$	

$$SS_D = 42 - \frac{(-16)^2}{8} = 10$$

$$\hat{s}_D^2 = \frac{10}{7} = 1.43$$

$$\hat{s}_D = \sqrt{1.43} = 1.20$$

$$\hat{s}_{\bar{D}} = \frac{1.20}{\sqrt{8}} = .42$$

Study Exercise 11.1 continued

Applying Equation 11.2, the observed t is thus

$$t = \frac{\bar{D} - \mu_{\bar{D}}}{\hat{s}_{\bar{D}}}$$

$$= \frac{-2.00 - 0}{.42} = -4.76$$

Since -4.76 is less than -2.365, we reject the null hypothesis and conclude that there is a relationship between the time of assessment and attitudes toward socialized medicine.

Assumptions of the *t* Test

The assumptions underlying the validity of the correlated groups t test parallel those for the one-sample t test, as discussed in Chapter 8. Specifically, it is assumed that

1. The sample is independently and randomly selected from the population of interest.
2. The population of difference scores is normally distributed.

In addition, the dependent variable should be quantitative in nature and measured on approximately an interval level. As with the one-sample t test, under certain conditions the correlated groups t test is robust with respect to violations of the normality assumption. This is particularly true when the sample size is relatively large. A detailed discussion of this issue can be found in Jaccard and Becker (1988).

11.3

Strength of the Relationship

The formula for computing eta-squared for the correlated groups t test is the same as that for the independent groups t test:

$$\text{eta}^2 = \frac{t^2}{t^2 + \text{df}} \qquad [11.4]$$

For the experiment on drugs and learning,

$$\text{eta}^2 = \frac{(-2.73)^2}{(-2.73)^2 + 4} = .65$$

TABLE 11.2 Raw and Nullified Scores for Drug and Learning Experiment

$Subject_i$	X for drug A	X for drug B	$\bar{X}_i$[a]	Nullified X for drug A	Nullified X for drug B
1	3	7	5.00	2.50	6.50
2	1	5	3.00	2.50	6.50
3	4	4	4.00	4.50	4.50
4	2	8	5.00	1.50	7.50
5	5	6	5.50	4.00	5.00
mean =	3.00	6.00	4.50	3.00	6.00

[a] $\bar{X}_i$ = Mean X score for subject i across conditions.

However, whereas eta-squared represents the proportion of variability in the dependent variable that is associated with the independent variable in the independent groups case, in the correlated groups case eta-squared represents the proportion of variability in the dependent variable that is associated with the independent variable *after variability due to individual differences has been removed.* In the present example, the proportion of variability in learning that is associated with the type of drug after the influence of individual differences has been removed is .65, which represents a strong effect. This interpretation of eta-squared can be best demonstrated by utilizing a variance extraction approach similar to that developed in Chapter 10.

Estimating and Extracting the Influence of Individual Differences

Consider an individual who tries to solve three math problems, each worth 5 points, on an exam. The individual obtains a score of 5 on the first problem, a score of 1 on the second problem, and a score of 3 on the third problem. Another individual works on the same three problems and scores 5 on the first one, 5 on the second one, and 2 on the third. Which individual has greater math ability? In the absence of any other information, the best approach to answering this question is to compare the average scores of the two individuals. The first individual's average score was $(5 + 1 + 3)/3 = 3.00$ and the second individual's average score was $(5 + 5 + 2)/3 = 4.00$. Using the average score across questions as an index of ability, you would conclude that the second individual has greater math ability than the first individual.

In within-subjects designs, estimates of the influence of individual difference variables are derived in a similar fashion. For each individual, the average score across conditions is computed. This has been done for the experiment on drugs and learning in column 4 of Table 11.2. For instance, the first subject had a score of 3 in condition A and a score of 7 in condition B for a mean of $(7 + 3)/2 = 5.00$. We can see by inspection of column 4 that some subjects had higher average scores than others. These differences reflect differences in the individuals' backgrounds (for example, intelligence, familiarity with the learning task, and so on).

We will now develop the logic of extracting this source of variability. In order to do so, we must first compute a grand mean (G). This involves summing *all* of the scores in columns 2 and 3 of Table 11.2 and dividing by the number of summed scores. We find that $G = (3 + 1 + 4 + 2 + 5 + 7 + 5 + 4 + 8 + 6)/10 = 4.50$. Thus, the average score across all subjects and across both experimental conditions is 4.50.

Consider the first subject. His average score across conditions is 5.00. The grand mean, which serves as a reference point for all subjects, is 4.50. If the average score across conditions is an index of the individual's background, then the effect of this background was to raise the individual's score one-half a unit above the grand mean ($5.00 - 4.50 = .50$). Using logic similar to that used to extract variability in scores in Chapter 10, we can nullify this effect by subtracting .50 unit from the scores of this individual. This has been done in columns 5 and 6 of Table 11.2. The score in condition A is 3, and this score is adjusted to 2.50 ($3 - .50 = 2.50$). The score in condition B is 7, and this score is adjusted to 6.50 ($7 - .50 = 6.50$). The new scores have had the effect of the individual's background removed, or nullified. This same approach is taken for each subject. For subject 2, the average response across conditions was 3.00. This is 1.50 units below the grand mean ($3.00 - 4.50 = -1.50$). The effect of this individual's background was to hold performance down 1.50 units from the overall average. To nullify the effect of this background, we simply add 1.50 to the two scores of the subject. The remaining subjects' scores are nullified in a similar fashion.

We now have an adjusted data set in which the effects of individual differences in background have been removed. Compare the nullified data in Table 11.2 with the original data. Note that there is less variability in the nullified scores because we have removed a source of variability—namely, individual abilities and background. If we were to compute the average nullified score across conditions for each subject, we would find that every subject would have the same average score. Again, this is because we have extracted variability due to individual differences.

Since there is a set of nullified scores for drug A and a set of nullified scores for drug B, the adjusted data can be analyzed with an independent groups t test. This test will tend to be more sensitive (and thus more powerful) than an independent groups t test applied to the raw scores since the "noise" created by individual differences has been eliminated. If we were to actually apply this test to the adjusted data, we would obtain the same value of t (-2.73) as that obtained earlier when the computational formula for the correlated groups t test was applied to difference scores.* The equivalence of the two approaches clarifies the nature of t in the correlated groups case—a correlated groups t test is analogous to an independent groups t test with the effects of

* Calculations for the nullified score approach can be found in Appendix 11.1.

individual differences extracted from the dependent variable. It follows that eta-squared in the correlated groups case is analogous to eta-squared in the independent groups case, but with variability due to individual differences removed from the dependent variable.

11.4 Nature of the Relationship

The analysis of the nature of the relationship is identical to that for the independent groups *t* test and involves examination of the mean scores in the two conditions. Because we failed to reject the null hypothesis, the question of the nature of the relationship is not meaningful for the present data. Had we rejected the null hypothesis, then the nature of the relationship would be that drug B produces more errors in learning ($\overline{X}_B = 6.00$) than drug A ($\overline{X}_A = 3.00$).

11.5 Methodological Considerations

The experiment on drugs and learning raises a number of methodological issues. First, consider the potential role of confounding variables. In this study, the investigator attempted to control for carry-over effects such as familiarity with the task, the persistence of the effects of drug A into the administration of drug B, and so on, by having a long time period between the administration of drug A and the administration of drug B. The introduction of the long time interval may, in fact, reduce such carry-over effects. But an additional problem arises. Perhaps during the time interval between the administration of the two drugs, the subjects experienced something that improved their performance on the learning task. For instance, if the subjects are introductory psychology students, they might have learned about certain memory aids in the context of their course work. A procedure that circumvents this problem, and that could have been used here, is **counterbalancing.** In this procedure, one-half of the subjects are given drug A first and then, at a later time when carry-over effects should be minimal, they are given drug B. The other half of the subjects are given drug B first and then drug A. If this is done, then the intervening events and familiarity effects should no longer be related to the administration of the two drugs. Any carry-over effects should influence learning performance under drug A (for the subjects who were given drug A second) and drug B (for subjects who were given drug B second) to an equal extent.* In essence, these confounding variables are turned into disturbance variables, which now create "noise" in the experiment. It is possible to remove the disturbance influence of these variables using certain advanced statistical techniques. These are, however, beyond the scope of the present text, and are discussed in Winer (1971).

Another problem with the experiment is the lack of a control condition in which subjects perform the learning task while not under the influence of any drugs. Although the effects of drug A and drug B relative to one another can be determined from the experiment, the effects of the drugs relative to baseline performance cannot. The addition of a control condition would be quite informative. Chapter 13 discusses statistical techniques that could be used to analyze the study if a control condition were included.

In Chapter 10, we saw that the role of disturbance variables in the independent groups case can be clarified by examining the magnitude of the estimated standard error of the difference and eta-squared. A similar approach can be taken with within-subjects designs. In the experiment on drugs and learning, eta-squared was equal to .65. An eta-squared of .65 indicates that, after variability due to individual differences has been removed, 65% of the variability in the dependent variable in the sample is associated with the independent variable and 35% is due to disturbance variables other than individual differences. This represents a strong relationship in the sample, yet the t test results were such that we could not conclude that a relationship exists in the population. Although this may appear contradictory, a consideration of the sample size will help to place this in proper perspective. The study was conducted using an extremely small sample size, namely, five individuals. With small sample sizes, very small magnitudes of sampling error are necessary before we can conclude that there is a relationship between the independent and dependent variables. Although the size of the estimated standard error (1.10) suggests that the role of sampling error (and thus the role of disturbance variables) was not of unreasonable magnitude, given the extremely small sample size, a relationship as strong as .65 in the sample is still not strong enough for us to be confident that sampling error is not producing the observed mean difference. The moral is to use large sample sizes when it is practical. As indicated in previous chapters, larger sample sizes will increase the power of the statistical test.

---

* In general, carry-over effects will influence the dependent variable equally in the two conditions when the nature of such effects is the same for the two levels of the independent variable. This will not be the case when the nature of carry-over effects is different for the two levels of the independent variable.

11.6

Power of Correlated Groups Versus Independent Groups *t* Tests

Since variability due to individual differences is extracted from the dependent variable as part of the correlated groups *t* test procedure, a correlated groups *t* test will generally be more powerful than a corresponding independent groups *t* test. However, this will not be the case when individual differences have only a minimal influence on the dependent variable. This stems from the fact that the number of degrees of freedom for the correlated groups test ($N - 1$) is smaller than the number of degrees of freedom for the independent groups test ($n_1 + n_2 - 2$). For instance, if there are two sets of 10 scores, df = 10 − 1 = 9 for the correlated groups test and df = 10 + 10 − 2 = 18 for the independent groups test. Since the *t* distribution requires more extreme values of *t* to reject the null hypothesis as the degrees of freedom become smaller, the statistical advantage of the correlated groups *t* test is offset when nullified scores and raw scores are similar because individual differences have only a minimal influence on the dependent variable. In fact, under this circumstance, a correlated groups test might actually be less powerful than an independent groups test. Thus, it is important that a researcher considering a within-subjects design accurately anticipate the role of individual differences. In general, however, the null hypothesis is more likely to be correctly rejected when a correlated groups analysis is used than when an independent groups analysis is used because individual differences will tend to have a sizable effect on the dependent variable. It is this greater sensitivity of the correlated groups analysis (along with the additional advantage of subject economy cited in Chapter 9) that recommends a within-subjects design over a between-subjects design when carry-over effects are not problematic.

11.7

Numerical Example

Developmental psychologists have attempted to specify the age period when infants tend to show signs of fearing strangers. At very early ages (1 to 2 months), infants generally will show positive reactions when approached or held by any adult. At some point, however, infants begin to discriminate adults

TABLE 11.3 Data and Calculations for Investigation on Age and Negative Responses to a Stranger

Infant	X for 3 months	X for 6 months	Difference (D)	D^2
1	0	3	−3	9
2	1	2	−1	1
3	2	4	−2	4
4	2	4	−2	4
5	2	4	−2	4
6	1	2	−1	1
7	0	3	−3	9
8	0	2	−2	4
	$\Sigma X_T = 8$	$\Sigma X_S = 24$	$\Sigma D = -16$	$\Sigma D^2 = 36$
	$\bar{X}_T = 1.00$	$\bar{X}_S = 3.00$	$\bar{D} = -2.00$	

$$SS_D = \Sigma D^2 - \frac{(\Sigma D)^2}{N} = 36 - \frac{(-16)^2}{8} = 4$$

$$\hat{s}_D^2 = \frac{SS_D}{N-1} = \frac{4}{7} = .57$$

$$\hat{s}_D = \sqrt{\hat{s}_D^2} = \sqrt{.57} = .75$$

$$\hat{s}_{\bar{D}} = \frac{\hat{s}_D}{\sqrt{N}} = \frac{.75}{\sqrt{8}} = .27$$

and will exhibit fear responses when a stranger, as opposed to a familiar person such as the mother or father, is present. In one investigation, a researcher wanted to compare negative responses to a stranger using infants at the age of 3 months and again at 6 months. Eight infants were involved one at a time in 10-minute interactions with a stranger in which the stranger attempted to engage them in playful behavior. The interactions were standardized as much as possible, and the number of minutes the infants spent crying was used as the measure of negative response. The scores on the dependent measure are presented in the second and third columns of Table 11.3.

In this study, the independent variable is age (3 months versus 6 months) and the dependent variable is the severity of the negative response to the stranger as indicated by the number of minutes spent crying. The independent variable has two levels and is within-subjects in nature since the same infants were involved in both the 3-month and 6-month conditions. The dependent variable is quantitative in nature. Given these conditions, the correlated groups t test is the statistical technique that would typically be used to analyze the relationship between the variables.

The null hypothesis posits that there is no relationship between infants' ages and their negative responses to a stranger, while the alternative hypothesis posits that these two variables are related. These hypotheses can be formally stated as

$$H_0: \quad \mu_T = \mu_S$$
$$H_1: \quad \mu_T \neq \mu_S$$

where the subscripts "T" and "S" denote the ages 3 months and 6 months, respectively.

The existence of a relationship is tested by converting the observed sample mean difference into a t value and comparing this with the values of t that define the rejection region. Since the null hypothesis states that the population means are equivalent, $\mu_{\bar{D}}$ can be set equal to 0. Other intermediate values necessary for the calculation of t are contained in Table 11.3. The observed t is found to equal

$$t = \frac{\bar{D} - \mu_{\bar{D}}}{\hat{s}_{\bar{D}}}$$
$$= \frac{-2.00 - 0}{.27} = -7.41$$

For an alpha level of .05, nondirectional test, and $N - 1 = 8 - 1 = 7$ degrees of freedom, the critical values of t are ± 2.365. Since -7.41 is less than -2.365, we reject the null hypothesis and conclude that a relationship exists between infants' ages and the severity of their negative responses to a stranger.

The strength of the relationship, as indexed by eta-squared, is

$$\text{eta}^2 = \frac{t^2}{t^2 + \text{df}}$$
$$= \frac{(-7.41)^2}{(-7.41)^2 + 7} = .89$$

The proportion of variability in the dependent variable (severity of negative response to a stranger) that is associated with the independent variable (age) after the influence of individual differences has been removed is .89. This represents a very strong effect.

The nature of the relationship is such that infants who are 6 months old respond more negatively to a stranger than infants who are 3 months old. This conclusion stems from the fact that the mean crying times in the two conditions are 3.00 and 1.00 minutes, respectively.

You should think about these results in terms of the research design questions discussed in Chapter 9. Are there any potential confounding variables that have not been controlled? What has been the role of disturbance variables? What kinds of procedures could be used to reduce sampling error? What are the limitations of the study in terms of generalizability?

11.8

Planning an Investigation Using the Correlated Groups *t* Test

As with the independent groups *t* test, the design of investigations that will use the correlated groups *t* test requires careful consideration of potential confounding and disturbance variables and the procedures one will use to control for them. Sample size selection is also a major issue relative to the power of the statistical test. Appendix E.1 can help select a sample size so as to obtain a desired level of power. Table 11.4 shows the portion of this appendix for an alpha level of .05, nondirectional test, for purpose of exposition. As in Chapter 10, different levels of power are listed in the first column, and different levels of the population value of eta-squared serve as column headings. It should be noted that these values of eta-squared are conceptualized as the proportion of variability in the dependent variable that is associated with the independent variable *after the effects of individual differences have been removed.* Thus, to the extent that the dependent variable is influenced by individual background, the population eta-squared will be greater in the correlated groups case than in the independent groups case.

As an example of the use of the table, if the desired level of power is .80 and the investigator suspects that the strength of the relationship in the population corresponds roughly to an eta-squared of .07, then the number of subjects he or she should use in a study for an alpha level of .05, nondirectional test, is 53. The level of power an investigator requires will, of course, depend on the seriousness of a Type II error. Examination of Table 11.4 and Appendix E.1 provides a general appreciation for the relationships between directionality, the alpha level, sample size, the strength of the relationship one is trying to detect in the population, and power for the correlated groups *t* test.

11.9

Method of Presentation

The method of presentation for the correlated groups *t* test is identical to that for the independent groups *t* test. This should include statements of the degrees of freedom, the observed value of *t*, the significance level, the sample means, and, if the analysis is statistically significant, the strength and nature of the relationship between the independent and dependent variables. For example, the results for the investigation on infants' ages and responses to strangers discussed earlier in this chapter might be presented as follows:

TABLE 11.4 Approximate Sample Sizes Necessary to Achieve Selected Levels of Power for Alpha = .05 as a Function of Population Values of Eta-Squared

Nondirectional Test, Alpha = .05

	POPULATION ETA-SQUARED									
POWER	.01	.03	.05	.07	.10	.15	.20	.25	.30	.35
.25	84	28	17	12	8	6	5	3	3	3
.50	193	63	38	27	18	12	9	7	5	5
.60	246	80	48	34	23	15	11	8	7	6
.67	287	93	55	39	27	17	12	10	8	6
.70	310	101	60	42	29	18	13	10	8	7
.75	348	113	67	47	32	21	15	11	9	7
.80	393	128	76	53	36	23	17	13	10	8
.85	450	146	86	61	41	26	19	14	11	9
.90	526	171	101	71	48	31	22	17	13	11
.95	651	211	125	87	60	38	27	21	16	13
.99	920	298	176	123	84	53	38	29	22	18

Results

A correlated groups t test compared the mean crying time for the infants when they were 3 months of age with the mean crying time when they were 6 months of age. This test was found to be statistically significant, $t(7) = -7.41$, $p < .001$, suggesting that infants react more negatively to strangers when they are 6 months old ($M = 3.00$) than when they are 3 months old ($M = 1.00$). The strength of the relationship between age and crying time was .89, as indexed by eta^2.

In this report, M is used to indicate the two sample means. As noted in Chapter 3, this symbol replaces $\bar{X}$ in American Psychological Association format.

11.10

Examples from the Literature

Number of Social Contacts with Same- Versus Opposite-Gender Individuals

Psychologists have studied interpersonal relations and friendship patterns in many contexts. One approach to this area has been to document the daily patterns of people's social behavior through the use of diaries. For instance, Nezlek (1978) asked subjects to keep daily diaries concerning the social contact they had with other people over four 2-week periods. One of the variables studied was the number of people that individuals reported they had met. Specifically, Nezlek was interested in comparing the number of same-gender contacts to the number of opposite-gender contacts that people made. Sixty-three individuals who kept diaries indicated that they contacted, on the average, 1.54 same-gender individuals per day and 1.01 opposite-gender individuals per day. A correlated groups *t* test indicated that this difference was statistically significant, $t(62) = 2.10$, $p < .05$, such that people tend to make more same-gender than opposite-gender contacts. The strength of the relationship, as indexed by eta-squared, was .07. This represents a weak effect and indicates, as one would expect, that many factors other than a person's gender influence social contact.

Self-Schema Similarity Before and After Discussion of a Hypothetical Person

The term "self-schema" refers to an individual's conception of who and what one is. According to a model of self-schema development proposed by Deutsch and Mackesy (1985), during the course of conversations about other people, individuals become aware of the person-description dimensions used by the other discussants and come to adopt these dimensions in their descriptions of others, and subsequently, themselves. According to this model, as reflected in their self-descriptions, individuals' self-schemas should become more similar after they have had the opportunity to share their opinions of another individual.

In a test of this hypothesis, Deutsch and Mackesy first instructed experimental participants to list 10 self-descriptive traits. Subjects then independently read a description of a hypothetical person and discussed their impressions of this person with a randomly assigned partner. In a final phase, participants were again instructed to list 10 self-descriptive traits.

The dependent variable was the number of overlapping self-traits reported by partners before versus after the discussion of the hypothetical person. Consistent with Deutsch and Mackesy's model, a correlated groups *t* test indicated that there was significantly more self-schema overlap after discussion ($\bar{X} = 2.05$) than before ($\bar{X} = 1.30$), $t(19) = 3.29$, $p < .01$. As indexed by eta-squared, the strength of the relationship was .36. This represents a strong effect.

11.11

Summary

The correlated groups *t* test involves the analysis of difference scores and is typically applied when (1) the dependent variable is quantitative in nature and is measured on approximately an interval level, (2) the independent variable is within-subjects in nature, and (3) the independent variable has two and only two levels. The existence of a relationship between the two variables is tested by converting the observed mean difference in the sample into a t value representing the number of estimated standard errors the difference is from the mean of the appropriate sampling distribution of the mean of difference scores (assuming the null hypothesis is true). This t value is compared with the critical values of t, and the decision to reject or not reject the null hypothesis is made accordingly. An important advantage of the correlated groups t test over the independent groups t test is that it extracts the influence of individual differences from the dependent variable, thereby providing a more sensitive test of the relationship between the independent and dependent variables. The strength of the relationship is analyzed using the eta-squared statistic, which in this case represents the proportion of variability in the dependent variable that is associated with the independent variable after the variability due to individual differences has been removed. Finally, the nature of the relationship is determined by examination of the mean scores in the two conditions.

Appendix **11.1** Computational Procedures for the Nullified Score Approach

The logic and computational procedures for the nullified score approach to correlated groups designs are similar to those developed in Chapter 10 for the independent groups t test, with one important exception: As noted earlier in this chapter, the relevant t distribution is distributed with $N - 1$ rather than $n_1 + n_2 - 2$ degrees of freedom. This has implications not only for the critical values of t that will be used in hypothesis testing, but also for the calculation of the estimated standard error of the difference.

We will demonstrate the relevant procedures by referring to the experiment on drugs and learning discussed earlier in this chapter. The nullified scores for this study from columns 5 and 6 of Table 11.2 have been reproduced in the table below. Also appearing in this table are calculations for the means and sums of squares for the two conditions.

The formula for the independent groups t test (see Equation 10.7) is

$$t = \frac{(\bar{X}_1 - \bar{X}_2) - (\mu_1 - \mu_2)}{\hat{s}_{\bar{X}_1 - \bar{X}_2}}$$

In the present example, $\bar{X}_1 - \bar{X}_2 = 3.00 - 6.00 = -3.00$ and $\mu_1 - \mu_2 = 0$, since μ_1 and μ_2 are hypothesized to be equal. The standard error of the difference can be estimated by the formula

$$\hat{s}_{\bar{X}_1 - \bar{X}_2} = \sqrt{\left(\frac{SS_1 + SS_2}{N - 1}\right)\left(\frac{1}{n_1} + \frac{1}{n_2}\right)}$$

which is the same formula as that presented in Chapter 10 (see Equation 10.6) with the exception that $N - 1$ (the degrees of freedom for the correlated groups case) replaces $n_1 + n_2 - 2$ (the degrees of freedom for the independent groups case). In the present example,

$$\hat{s}_{\bar{X}_1 - \bar{X}_2} = \sqrt{\left(\frac{6 + 6}{5 - 1}\right)\left(\frac{1}{5} + \frac{1}{5}\right)} = 1.10$$

The observed t is thus

$$t = \frac{-3.00 - 0}{1.10} = -2.73$$

This is the same value as was obtained in Section 11.2 using the difference score approach.

NULLIFIED DATA FOR DRUG A		NULLIFIED DATA FOR DRUG B	
X	X^2	X	X^2
2.50	6.25	6.50	42.25
2.50	6.25	6.50	42.25
4.50	20.25	4.50	20.25
1.50	2.25	7.50	56.25
4.00	16.00	5.00	25.00
$\Sigma X_A = 15.00$	$\Sigma X_A^2 = 51.00$	$\Sigma X_B = 30.00$	$\Sigma X_B^2 = 186.00$
$\bar{X}_A = 3.00$		$\bar{X}_B = 6.00$	

$$SS_A = \Sigma X_A^2 - \frac{(\Sigma X_A)^2}{N} = 51.00 - \frac{15.00^2}{5} = 6$$

$$SS_B = \Sigma X_B^2 - \frac{(\Sigma X_B)^2}{N} = 186.00 - \frac{30.00^2}{5} = 6$$

Exercises

Answers to asterisked (*) exercises appear at the back of the book. Answers to exercises with two asterisks are also worked out step-by-step in the Study Guide.

Exercises to Review Concepts

1. Under what conditions is the correlated groups t test typically used to analyze a bivariate relationship?

2. In general terms, what value will the mean of a sampling distribution of the mean of difference scores always be equal to?

***3.** State the critical value(s) of t that would be used to reject the null hypothesis for a correlated groups t test at an alpha level of .05 under each of the following conditions:

a. H_0: $\mu_1 = \mu_2$, H_1: $\mu_1 > \mu_2$, $N = 10$
b. H_0: $\mu_1 = \mu_2$, H_1: $\mu_1 \neq \mu_2$, $N = 10$
c. H_0: $\mu_1 = \mu_2$, H_1: $\mu_1 < \mu_2$, $N = 20$
d. H_0: $\mu_1 = \mu_2$, H_1: $\mu_1 \neq \mu_2$, $N = 20$

4. What are the assumptions underlying the correlated groups t test?

5. What does eta-squared represent in the correlated groups t test case?

***6.** Why is a correlated groups t test generally more powerful than a corresponding independent groups t test? When will this not be the case? Why?

Refer to the following information to answer Exercises 7–10.

Consider the following scores on a learning test that was administered to a group of subjects under two experimental conditions, A and B:

Subject	*Condition A*	*Condition B*
1	10	16
2	15	23
3	10	12
4	5	9
5	3	5

*7. Conduct a nondirectional correlated groups t test on the data to test for a relationship between experimental condition and learning scores.

*8. Compute the value of eta-squared for the given data. Does the observed value represent a weak, moderate, or strong effect?

*9. What is the nature of the relationship between the experimental condition one is exposed to and learning scores for these data?

*10. Analyze the given data as if the independent variable were between-subjects in nature. That is, conduct a nondirectional independent groups t test, compute the value of eta-squared, and determine the nature of the relationship between the experimental condition one is exposed to and learning test scores. Compare your findings with those for Exercises 7–9. How does this illustrate the advantage of within-subjects research designs?

**11. For the following data in which individuals' self-esteem was assessed at two different times, extract the effects of individual differences (that is, generate a set of nullified scores). Compute the mean for time 1 and the mean for time 2 for the original scores. Compute the corresponding means for the nullified scores. The respective mean values should be the same (for example, the mean for time 1 in the original data set should equal the mean for time 1 in the adjusted data set). Why do you think this is the case? (*Hint:* Remember that the individual differences whose effects have been extracted from the dependent variable serve as disturbance variables, as discussed in Chapter 9.)

Individual	*Time 1*	*Time 2*
1	10	12
2	11	13
3	12	14
4	13	17
5	14	14

12. Conduct a nondirectional correlated groups t test on the original scores from Exercise 11 to test for a relationship between time of assessment and self-esteem.

13. Compute the value of eta-squared for the data in Exercise 11. Does the observed value represent a weak, moderate, or strong effect?

14. What is the nature of the relationship between time of assessment and self-esteem for the data in Exercise 11?

**15. Using the procedures from Appendix 11.1, conduct a nondirectional independent groups t test on the nullified scores from Exercise 11. Compare the observed t with that obtained in Exercise 12.

For each of the studies described in Exercises 16–20, indicate whether an independent groups t test or a correlated groups t test should be used to analyze the relationship between the independent and dependent variables. Assume that the underlying assumptions of the tests have been satisfied.

*16. Frieze, Parsons, Johnson, Ruble, and Zellman (1978) examined the relationship between gender and mathematical ability as measured by the Graduate Record Exam. The mean score on the quantitative test was compared for males and females who took the exam during the year 1972.

*17. Smith, Phillipus, and Guard (1968) studied the effects of a drug known as ethosuximide on the verbal skills of children with learning problems. This research was based on previous research that had shown a similarity in EEG patterns between children with learning problems and those who have a mild form of epilepsy, called petit mal. Smith and colleagues speculated that since ethosuximide is effective in treating petit mal, it might also serve as a "learning facilitator" for children with learning problems. A group of children participated in the study for 6 weeks. During the first 3 weeks, some of the children were injected with ethosuximide and others were given a placebo. During the second 3 weeks, the procedures were reversed (that is, children who had previously been given a placebo were given ethosuximide and vice versa). Each child was given a verbal skills test after the first 3-week period and again after 6 weeks. For all children, the mean score on this test after they received the placebo was compared with the mean score after they received the ethosuximide.

18. A consumer psychologist was interested in the relative preferences of consumers for two types of headache remedies. Five hundred individuals were interviewed and rated each brand on a 10-point rating scale. The mean ratings for the two brands were compared.

19. Jensen (1973) reviewed research on the relationship between scores on an intelligence test and individuals' race. In one study, the mean scores of blacks and whites were compared.

20. Gallup (1976) studied how individuals' attitudes toward nuclear energy changed over a period of 10 years. The same individuals were interviewed in both 1965 and 1975, and during each interview, the respondents indicated their attitudes toward nuclear energy on a 5-point scale. The mean attitude scores were compared across interviews.

21. If a researcher suspected that the strength of the relationship between two variables in the population was .25 as indexed by eta-squared, what sample size should he use in a study involving two correlated groups and an alpha level of .05, directional test, in order to achieve power of .90?

***22.** Suppose an investigator reported the results of a study using two correlated groups with $N = 7$. If the value of eta-squared in the population was .30, what would the power of her statistical test be at an alpha level of .05, nondirectional test?

Exercises to Apply Concepts

****23.** The process of decision making has been studied extensively by psychologists. One area of inquiry has been the effects of making a decision on the subsequent evaluation of the decision alternatives. Brehm (1956) has suggested that when an individual is forced to choose between two equally attractive alternatives, the individual will justify his or her decision, after the choice has been made, by downgrading the unchosen alternative and upgrading the chosen one. In one experiment, subjects rated the desirability of two household products. The products were pretested so that the experimenter knew they would be rated with about equal desirability. Subjects then read marketing reports on each product and were asked to choose one as payment for participation in the study. Finally, subjects were told that their first rating should be considered a first impression and that the experimenters wanted to get a second rating since the subject now had more time to think about the products. The ratings were made on an 8-point scale, where 1 indicated low desirability, and 8 indicated high desirability. We will consider only the results for the unchosen alternative. The hypothetical data presented below are representative of the results of this study. Analyze these data using a nondirectional test, draw a conclusion, and write up your results using the principles developed in the Method of Presentation section.

Subject	*Before choice*	*After choice*
1	8	4
2	7	3
3	7	5
4	4	6
5	8	4
6	6	4
7	6	2
8	5	5
9	5	3
10	4	4

24. As noted in Exercise 30 in Chapter 8, investigators have studied the effects of marijuana on human physiology. One common belief held by laypersons is that marijuana affects pupil size. Weil, Zinberg, and Nelson (1968) studied 10 subjects. Each was administered a high dose of marijuana by smoking a potent marijuana cigarette. The subjects were males, 21 to 26 years of age, all of whom smoked tobacco cigarettes regularly but had never tried marijuana. In the study, pupil size was measured with a millimeter rule under constant illumination with eyes focused on an object at a constant distance. Pupil size was measured before and after smoking marijuana. The data are presented below. Analyze these data using a nondirectional test, draw a conclusion, and write up your results using the principles developed in the Method of Presentation section.

Subject	Before marijuana	After marijuana
1	5	7
2	7	5
3	6	8
4	7	5
5	6	6
6	5	7
7	3	9
8	3	5
9	5	9
10	3	9

TWELVE

ONE-WAY BETWEEN-SUBJECTS ANALYSIS OF VARIANCE

12.1

Use of One-Way Between-Subjects Analysis of Variance

One-way between-subjects analysis of variance, more simply known as *one-way analysis of variance* (abbreviated as "one-way ANOVA"), is typically used to analyze the relationship between two variables when

1. the dependent variable is quantitative in nature and is measured on approximately an interval level;
2. the independent variable is *between-subjects* in nature (it can be either qualitative or quantitative); and
3. the independent variable has three or more levels.

In short, one-way analysis of variance is used under the same circumstances as the independent groups t test except that the independent variable has more than two levels.* Let us consider an example of an investigation that meets the conditions for this technique.

An investigator was interested in the relationship between individuals' religion and what they consider the ideal family size. Twenty-one people were interviewed—seven Catholics, seven Protestants, and seven Jews. Each was asked what he or she considered to be the ideal number of children to have in a family. Their responses are presented in Table 12.1.

In this investigation, religion is the independent variable. It is between-subjects in nature and has three levels. Ideal family size is the dependent vari-

---

* For ease of presentation, we will restrict our discussion of one-way analysis of variance to the situation where sample sizes in the groups being considered are all equivalent. While the same general logic can be extended to unequal sample sizes, some of the computational procedures may require minor modification. More generally, our discussion of analysis of variance in this and subsequent chapters will deal only with the type most commonly encountered in the behavioral sciences, referred to as *fixed-effects analysis*. For a discussion of other possibilities, see Hays (1981).

TABLE 12.1 Data for Religion and Ideal Family Size Investigation

Catholic	Protestant	Jewish
4	1	1
3	2	1
2	0	2
3	2	0
3	4	2
4	3	0
2	2	1
$\Sigma X_C = 21$	$\Sigma X_P = 14$	$\Sigma X_J = 7$
$\bar{X}_C = 3.00$	$\bar{X}_P = 2.00$	$\bar{X}_J = 1.00$

able and is quantitative in nature. Given these conditions, one-way between-subjects analysis of variance is the statistical technique that would typically be used to analyze the relationship between the variables.

12.2

Inference of a Relationship Using One-Way Between-Subjects Analysis of Variance

Null and Alternative Hypotheses

The first question concerns whether a relationship exists between religion and ideal family size. The null hypothesis states that there is no relationship between religion and the ideal number of children. It is expressed in terms of population mean scores, as we have done in previous chapters:

$$H_0\colon \quad \mu_C = \mu_P = \mu_J$$

where the subscript "C" represents the Catholic group, the subscript "P" represents the Protestant group, and the subscript "J" represents the Jewish group. The alternative hypothesis states that there is a relationship between religion and the ideal number of children and that the three population means are not all equal to one another. Unlike previous tests, however, the alternative hypothesis for one-way analysis of variance cannot be summarized in a single mathematical statement. This is because there are a number of different ways in which the three population means can pattern themselves so that they are not all equal to one another. Some examples are

$$\mu_C = \mu_P > \mu_J$$

$$\mu_C > \mu_P > \mu_J$$

$$\mu_C > \mu_P = \mu_J$$

In all, there are 12 possible patterns. The alternative hypothesis does not distinguish among these different possibilities. It simply states that a relationship exists such that the three population means are *not* all equal to one another. The question of the exact patterning of means is addressed in the context of the nature of the relationship. Thus, the alternative hypothesis is

H_1: The three population means are not all equal

The problem is to choose between the null and alternative hypotheses. If we look at the mean scores for the samples, we find that they are all different. The Catholic mean is 3.00, the Protestant mean is 2.00, and the Jewish mean is 1.00. This would seem to support the alternative hypothesis. However, we know that the nonequivalence of sample means could reflect sampling error. We therefore want to test the viability of a sampling error interpretation.

The logic for testing this interpretation is similar to the logic discussed in prior chapters. First, we will assume that the null hypothesis is true. We will then state an expected result based on this assumption. Next, we will compute a sample statistic (in this instance, the statistic is a *variance ratio,* which we will discuss shortly). The statistic will be treated as a score in a sampling distribution where the null hypothesis is assumed to be true. If the score falls within the rejection region (as defined by the alpha level), then we will reject the null hypothesis and conclude that there is a relationship between the two variables. Otherwise, we will fail to reject the null hypothesis.

Between- and Within-Group Variability

For the data in Table 12.1, we can distinguish two types of variability. The first concerns the differences between the mean scores in the three groups, or in more formal terms, **between-group variability.** If the mean scores in the three groups were all equal, then there would be no between-group variability. The more different the three means are from one another, the more between-group variability there is.

Why do mean differences exist? There are two factors that contribute to between-group variability. The first is sampling error: Even if the population means were, in fact, equal, we might observe between-group variability in the samples because of sampling error. A second factor that contributes to between-group variability is the effect of the independent variable on the dependent variable: If religion influences the ideal number of children, then this will tend to make the sample means different from one another. Thus, between-group variability in mean scores on the dependent variable reflects two things, (1) sampling error and (2) the effect of the independent variable on the dependent variable.

The second type of variability concerns the variability of scores *within* each of the three groups, or **within-group variability.** Examine the seven scores for Catholics in Table 12.1. Note that there is variability in the scores; some Catholics have a higher ideal number of children than other Catholics. The same is true for Protestants and Jews. Since religion is held constant for each group, the variability in the scores cannot be attributed to religious affiliation.

Other factors are operating to cause the variability in scores (such as how religious the person is, and other such disturbance variables). The fact that scores vary within a group suggests that there is variability in scores in the *population* represented by that group. Recall from Chapter 7 that when the variability of scores in a population is large, we expect more sampling error than when the population variability is small. As such, greater variability of scores within a group is indicative of greater variability of scores within the corresponding population and, thus, a greater amount of sampling error. In short, then, within-group variability reflects sampling error. Note that within-group variability is *not* influenced by the effect of the independent variable on the dependent variable. Since the variability of scores is considered within each group *separately,* the values of the independent variable are held constant and cannot be a source of variability within groups.

Since between-group variability reflects both sampling error *and* the effect of the independent variable, and within-group variability reflects just sampling error, the ratio of these two sources of variability can be used to test the sampling error interpretation discussed in the previous section. This ratio, known as the **variance ratio,** can be represented as

$$\frac{\text{Between-group variability}}{\text{Within-group variability}} = \frac{\text{Sampling error + Effect of the independent variable}}{\text{Sampling error}} \qquad [\mathbf{12.1}]$$

Consider the case where the null hypothesis is true. In this case, the independent variable has no effect on the dependent variable, so the between-group variability reflects only sampling error. Since within-group variability also reflects sampling error, the ratio in Equation 12.1 is simply dividing one estimate of sampling error by another estimate of sampling error. The result will be a value which, over the long run, will approach 1.00 (since any number divided by itself is 1.00). In contrast, if the null hypothesis is not true, between-group variability reflects both sampling error and the effect of the independent variable. In this case, we would expect the variance ratio, over the long run, to be greater than 1.00.

Partitioning of Variability

In order to compute the variance ratio, it is necessary first to derive numerical estimates of between- and within-group variability. While this is most readily accomplished through the computational formulas presented later in this chapter, it is important to have a conceptional understanding of what these formulas entail. We will demonstrate this in the present section by drawing a parallel between the way in which we can split up, or *partition,* an individual's score on the dependent variable into the mean score of the group of which he or she is a member and how far his or her score deviates from this mean, and the way in which we can partition the total variability in a set of scores into two similar components.

Consider the first individual in the Catholic group. Her score of 4 deviated $4 - 3 = 1$ unit from the group mean of 3.00. Her score can thus be represented as

TABLE 12.2 Breakdown of Scores on the Dependent Variable

Individual	*Religion*	*X*	=	$\bar{X}_j$	+	d^a
1	C	4	=	3	+	1
2	C	3	=	3	+	0
3	C	2	=	3	+	−1
4	C	3	=	3	+	0
5	C	3	=	3	+	0
6	C	4	=	3	+	1
7	C	2	=	3	+	−1
8	P	1	=	2	+	−1
9	P	2	=	2	+	0
10	P	0	=	2	+	−2
11	P	2	=	2	+	0
12	P	4	=	2	+	2
13	P	3	=	2	+	1
14	P	2	=	2	+	0
15	J	1	=	1	+	0
16	J	1	=	1	+	0
17	J	2	=	1	+	1
18	J	0	=	1	+	−1
19	J	2	=	1	+	1
20	J	0	=	1	+	−1
21	J	1	=	1	+	0

[a] $d = X - \bar{X}_j$.

$$4 = 3 + 1$$

We can write a general equation that reflects this relationship:

$$X = \bar{X}_j + d \qquad [12.2]$$

where X represents an individual's score, $\bar{X}_j$ is the mean of the group in which the individual is a member (group j), and d is the deviation of the individual's score from the group mean $(X - \bar{X}_j)$. It is possible to express the score of every individual in the study in this manner, as demonstrated in Table 12.2.

The column labeled X contains the score of each individual on the dependent variable. We can compute an index of how much variability there is in these scores by computing a sum of squares for this column of numbers using the standard formula for a sum of squares. This sum of squares is called the **sum of squares total** (symbolized SS_{TOTAL}) since it reflects the total variability in the dependent variable across all individuals. It is identical to the sum of squares total discussed in Chapter 10. If we were to perform the required calculations, we would find that $SS_{TOTAL} = 32$ in this instance.

Examine the scores in the column labeled $\bar{X}_j$ in Table 12.2. The source of variability in these scores is the differences between the group means. If the means of all three groups had been equal, every score in this column would be the same. The more the three means differ, the more these scores will tend to be different from one another. A sum of squares for this column would therefore reflect between-group variability. This sum of squares, called the **sum of squares**

between (symbolized $SS_{BETWEEN}$), is identical to the sum of squares explained discussed in Chapter 10 and can be calculated by applying the standard formula for a sum of squares to the $\bar{X}_j$ scores. In this instance, $SS_{BETWEEN} = 14$.

Examine the scores in the column labeled *d*. Each of these scores represents a deviation from the group mean, or how far an individual's score deviates from the average score *within the group*. If there were little variability within groups, all of these scores would be close to 0 and would tend to equal one another. If there were considerable variability within groups, there would be considerable variability in these scores. A sum of squares for this column of numbers would therefore reflect how much within-group variability there is. This sum of squares is called the **sum of squares within** (symbolized SS_{WITHIN}) and can be calculated by applying the standard formula for a sum of squares to the *d* scores. In this instance, $SS_{WITHIN} = 18$. Another name given to the sum of squares within is the sum of squares error because it reflects the effects of disturbance variables and, hence, sampling error. This is the term we used when we first introduced this concept in Chapter 10.

In Chapter 10 we also noted that the total variability in the dependent variable can be partitioned into two components, one reflecting the influence of the independent variable and one reflecting the influence of disturbance variables. In the present context, the partitioning of variability can be represented as

$$SS_{TOTAL} = SS_{BETWEEN} + SS_{WITHIN} \quad [12.3]$$

This partitioning is consistent with the idea of expressing each individual's score in terms of (1) the group mean and (2) the score's deviation from the group mean, as we did in Table 12.2. Thus, just as each score can be expressed in terms of these two components, the total *variability* in scores can be expressed in terms of (1) the *variability* between the group means (between-group variability) and (2) the *variability* of deviations from the group means (within-group variability). This is clearly demonstrated in our example (where $SS_{TOTAL} = 32$, $SS_{BETWEEN} = 14$, and $SS_{WITHIN} = 18$), as $32 = 14 + 18$.

Mean Squares and the *F* Ratio

The variance ratio that is computed to test the null hypothesis is the between-group variability divided by the within-group variability. The particular variance ratio that is employed does not utilize measures of sums of squares, but rather measures of variance, or *mean squares*. As noted in Chapter 7, a mean square is simply a sum of squares divided by its corresponding degrees of freedom. The reason why measures of mean squares rather than measures of sums of squares are used will be made explicit shortly. We will first consider the computation of the relevant quantities. The formulas are

$$MS_{BETWEEN} = \frac{SS_{BETWEEN}}{df_{BETWEEN}} \quad [12.4]$$

$$MS_{WITHIN} = \frac{SS_{WITHIN}}{df_{WITHIN}} \quad [12.5]$$

where $MS_{BETWEEN}$, referred to as the **mean square between,** represents the mean square for between-group variability; MS_{WITHIN}, referred to as the **mean square within,** represents the mean square for within-group variability; $df_{BETWEEN}$ is the degrees of freedom associated with the sum of squares between; and df_{WITHIN} is the degrees of freedom associated with the sum of squares within. These degrees of freedom can be represented as follows:

$$df_{BETWEEN} = k - 1 \qquad [12.6]$$

$$df_{WITHIN} = N - k \qquad [12.7]$$

where k is the number of levels of the independent variable (that is, the number of groups) and N is the total number of subjects in the study.* As with the sums of squares, the degrees of freedom are additive; if we sum the degrees of freedom between and the degrees of freedom within, we will obtain the degrees of freedom associated with the sum of squares total:

$$df_{TOTAL} = df_{BETWEEN} + df_{WITHIN} \qquad [12.8]$$

Since $df_{BETWEEN} = k - 1$ and $df_{WITHIN} = N - k$, and $(k - 1) + (N - k) = N - 1$, the degrees of freedom total can be calculated directly as

$$df_{TOTAL} = N - 1 \qquad [12.9]$$

In our example, $k = 3$ and $N = 21$. Thus,

$$df_{BETWEEN} = 3 - 1 = 2$$
$$df_{WITHIN} = 21 - 3 = 18$$
$$df_{TOTAL} = 21 - 1 = 20$$

Using Equations 12.4 and 12.5, we obtain the mean squares:

$$MS_{BETWEEN} = \frac{14}{2} = 7.00$$

$$MS_{WITHIN} = \frac{18}{18} = 1.00$$

The mean square between reflects variability between group means. As such, it bears a relationship to the quantity we would obtain if we treated the sample means like any other set of scores and applied the usual computations for a variance estimate. In fact, when sample sizes are equal, the mean square between is exactly equivalent to the variance estimate of the sample means multiplied by n to reflect the number of scores on which each mean is based. In our example, the sample means are 3.00, 2.00, and 1.00. The variance estimate for these three "scores" is 1.00. Multiplying this by 7, the per-group sample

* The rationale for the equivalence of the degrees of freedom to the indicated quantities is discussed in Appendix 12.1.

size, we obtain a value of 7.00, the same quantity as was obtained above using Equation 12.4.

The terminology "mean square within" is merely another name for the pooled variance estimate discussed in Chapter 10. It will be remembered that a pooled variance estimate is derived by multiplying the variance estimate for each group by its degrees of freedom, summing the products across groups, and then dividing the resulting quantity by the total number of degrees of freedom. In our example, there are three groups—Catholics, Protestants, and Jews. If we were to individually compute the variance estimate for each, we would find that the estimated variance for Catholics is .67; for Protestants, 1.67; and for Jews, .67. Since each sample contains seven subjects, each variance estimate has $n - 1 = 7 - 1 = 6$ degrees of freedom associated with it and the total degrees of freedom are $6 + 6 + 6 = 18$. The pooled variance estimate is thus

$$\hat{s}^2_{\text{pooled}} = \frac{(6)(.67) + (6)(1.67) + (6)(.67)}{18} = 1.00$$

This is the same quantity as was obtained above using Equation 12.5.

The variance ratio, formally referred to as the **F ratio** (in honor of the statistician Sir Ronald Fisher, who developed this approach), is

$$F = \frac{\text{MS}_{\text{BETWEEN}}}{\text{MS}_{\text{WITHIN}}} \qquad [12.10]$$

which in our example equals

$$F = \frac{7.00}{1.00} = 7.00$$

To recapitulate, the F ratio is a ratio of two variances. In the context of one-way analysis of variance, it is the mean square between (an index of between-group variability) divided by the mean square within (an index of within-group variability). The mean square between and mean square within are derived by computing the sum of squares between and the sum of squares within and dividing each by its corresponding degrees of freedom. If the null hypothesis is true, we would expect the F ratio, over the long run, to approach 1.00. If the null hypothesis is not true, we would expect the F ratio, over the long run, to be greater than 1.00. Let us now consider this issue in more detail.

Sampling Distribution of the *F* Ratio

Consider three populations, A, B, and C, where $\mu_A = \mu_B = \mu_C$. Suppose we select a random sample of 30 individuals from population A, a random sample of 30 individuals from population B, and a random sample of 30 individuals from population C. Using the procedures described previously, we could compute an F ratio. Although we would expect the value of this ratio to be near 1.00 (because the null hypothesis is true), it might not equal exactly 1.00 because $\text{MS}_{\text{BETWEEN}}$ and $\text{MS}_{\text{WITHIN}}$ are independent estimates of variability in the populations, based on sample data. Suppose the F ratio were found to equal

1.52. Further suppose that we repeated this procedure by randomly selecting new samples of the same sizes from the same populations and this time observed an F ratio of 1.32. We could repeat this procedure a large number of times (in principle, for all possible samples of size 30 for each population), and the result would be a large number of F ratios. These F ratios can be treated as "scores" in a distribution and constitute a **sampling distribution of the *F* ratio** analogous to the types of sampling distributions we have discussed in previous chapters.

As illustrated in Figure 12.1, the sampling distribution of the F ratio takes on different shapes depending on the particular values of $df_{BETWEEN}$ and df_{WITHIN} associated with it. In all of the distributions, the lowest possible value for F is 0. Also, the average (median) value for F in all of the distributions is 1.00. This should not be surprising since the F ratios are calculated from the case where the null hypothesis is true. Values of F become less frequent as they become increasingly greater than 1.00, and large values of F are highly unlikely. The question then becomes similar to the one posed in Chapter 8: How large must the F ratio computed from the sample information be before we reject the null hypothesis of no relationship? It turns out that the sampling distribution of the F ratio, under certain conditions to be discussed shortly, corresponds to an important theoretical distribution called the ***F* distribution.** The reason why measures of mean squares rather than measures of sums of squares are used to define the F ratio is that when the relevant conditions are met, measures of mean squares yield a sampling distribution that closely approximates an F distribution, whereas measures of sums of squares do not.

Just as probability statements can be made with respect to scores in the normal and t distributions, so can probability statements be made with respect to scores in the F distribution. It is therefore possible to set an alpha level and define a critical value such that if the null hypothesis is true, the probability of obtaining an F ratio larger than that critical value is less than alpha. Since all departures from the null hypothesis are reflected in the upper tail of the F distribution (as defined by the critical value), the ***F* test** is, by its nature, nondirectional.

Before conducting the F test, we will summarize all of our previous calculations in a **summary table:**

Source	*SS*	*df*	*MS*	*F*
Between	14.00	2	7.00	7.00
Within	18.00	18	1.00	
Total	32.00	20		

The first column of the summary table indicates the source of the information in the other columns. The second column reports the sums of squares between, within, and total. The next column provides the associated degrees of freedom for each sum of squares, and column 4 presents the mean squares for the two sources of variability. The last column contains the F ratio.

FIGURE 12.1 Sampling Distributions for *F* Ratios with Different Degrees of Freedom

a Sampling Distribution for $df_{BETWEEN} = 3$ and $df_{WITHIN} = 12$

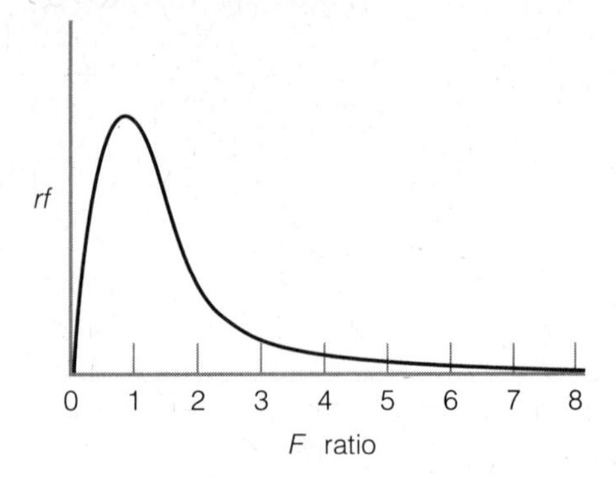

b Sampling Distribution for $df_{BETWEEN} = 2$ and $df_{WITHIN} = 9$

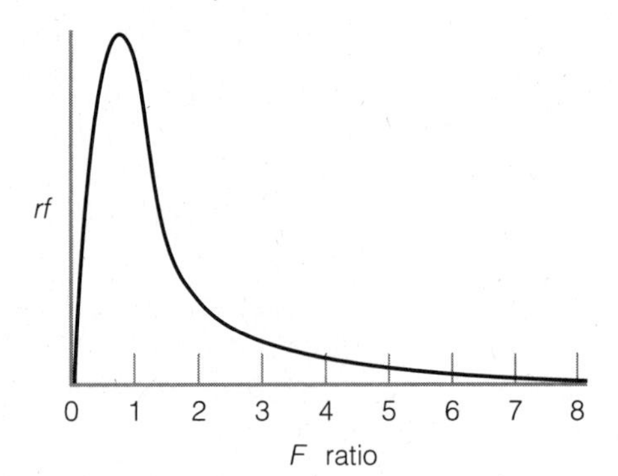

c Sampling Distribution for $df_{BETWEEN} = 3$ and $df_{WITHIN} = 8$

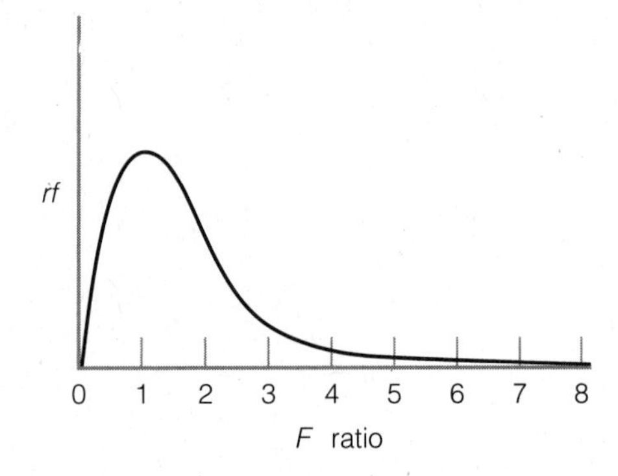

The test of the null hypothesis involves comparing the observed value of *F* with the appropriate critical value of *F* from the *F* table in Appendix F. Instructions for using this table are also contained in the appendix. When reporting an *F* value, it is conventional to report the degrees of freedom between followed by the degrees of freedom within. For an alpha level of .05 and $df_{BETWEEN} = 2$ and

$df_{WITHIN} = 18$, the critical value of F is 3.55. In order to be statistically significant, the observed value of F must exceed the critical value of F. Since the observed F value of 7.00 is greater than 3.55, we reject the null hypothesis and conclude that a relationship exists between religion and ideal family size. We have answered our first question about the existence of a relationship.

Computational Formulas

There are several different ways in which the information necessary for the calculation of the F ratio can be derived. We have already seen how the sum of squares between and the sum of squares within can be conceptualized in terms of variability between the group means and variability of deviations from the group means, respectively. We have also seen the equivalence between the mean square between and the variance estimate of the sample means scaled by per-group sample size, on the one hand, and the mean square within and the pooled variance estimate, on the other.

While both of these approaches help to clarify the conceptual foundation of one-way analysis of variance, they are computationally inefficient. In practice, the F ratio is most commonly (and most efficiently) obtained by calculating the sum of squares between and the sum of squares within with computational formulas available for this purpose and then dividing these by their corresponding degrees of freedom. We will now consider these formulas.

We begin by examining the computational formula for the sum of squares total. This is simply the standard computational formula for a sum of squares applied to the entire data set. More specifically, to obtain the sum of squares total using the computational formula, square each score in the data set, add these squared scores, and subtract the square of the sum of the scores divided by the total sample size. Symbolically,

$$SS_{TOTAL} = \Sigma X^2 - \frac{(\Sigma X)^2}{N} \qquad [12.11]$$

where ΣX^2 is the *sum of the squared scores,* $(\Sigma X)^2$ is the *square of the summed scores,* and N is the total sample size.

The data for the example on religion and ideal family size are reproduced in Table 12.3. To obtain ΣX^2 we must first square each score in the data set and then add these squared scores. For each group, the scores have been squared in the column labeled X^2. Adding all 21 of these squared scores together, we find that

$$\Sigma X^2 = 16 + 9 + \cdots + 0 + 1 = 116$$

The notation "· · ·" is used to symbolize the fact that not all of the X^2 scores contributing to the observed sum of 116 are physically represented. Rather, as a means of saving space, we have chosen to write out only the first two and the last two X^2 scores. When the "· · ·" notation is used, the inclusion in the summation of *all* relevant scores falling between the written values is implicit. We will make extensive use of this notation in the remainder of the book.

TABLE 12.3 Computation of Sums of Squares for Religion and Ideal Family Size Investigation

CATHOLIC		PROTESTANT		JEWISH	
X	X^2	X	X^2	X	X^2
4	16	1	1	1	1
3	9	2	4	1	1
2	4	0	0	2	4
3	9	2	4	0	0
3	9	4	16	2	4
4	16	3	9	0	0
2	4	2	4	1	1
$T_C = \Sigma X_C = 21$	$\Sigma X_C^2 = 67$	$T_P = \Sigma X_P = 14$	$\Sigma X_P^2 = 38$	$T_J = \Sigma X_J = 7$	$\Sigma X_J^2 = 11$
$T_C^2 = 441$		$T_P^2 = 196$		$T_J^2 = 49$	
$SS_C = 67 - \frac{21^2}{7} = 4$		$SS_P = 38 - \frac{14^2}{7} = 10$		$SS_J = 11 - \frac{7^2}{7} = 4$	

$$\Sigma X^2 = 16 + 9 + \cdots + 0 + 1 = 116$$

$$\Sigma X = 4 + 3 + \cdots + 0 + 1 = 42$$

$$\frac{(\Sigma X)^2}{N} = \frac{42^2}{21} = 84$$

$$\frac{\Sigma T_j^2}{n} = \frac{441 + 196 + 49}{7} = 98$$

$$SS_{WITHIN} = SS_C + SS_P + SS_J = 4 + 10 + 4 = 18$$

The sum of the X scores is

$$\Sigma X = 4 + 3 + \cdots + 0 + 1 = 42$$

where the "· · ·" notation again indicates that *all* relevant scores (in this case, X scores) are included in the summation. Squaring the observed sum of 42 and dividing by N, we obtain

$$\frac{(\Sigma X)^2}{N} = \frac{42^2}{21} = 84$$

Thus, the sum of squares total in this example is

$$SS_{TOTAL} = \Sigma X^2 - \frac{(\Sigma X)^2}{N}$$
$$= 116 - 84 = 32$$

which is the same value as reported previously.

The sum of squares within is equal to the total of the sums of squares within the individual groups. For any given group, this is equal to the sum of that group's squared scores minus that group's squared sum after it has been divided by the relevant sample size. These procedures are mathematically equivalent to the computational formula for the sum of squares within:

$$SS_{WITHIN} = \Sigma X^2 - \frac{\Sigma T_j^2}{n} \quad [12.12]$$

where ΣX^2 is as defined above, T_j^2 is the square of the sum of the scores in group j, and n is the per-group sample size. In our example, T_j^2 equals 441 for Catholics, 196 for Protestants, and 49 for Jews. Thus,

$$\frac{\Sigma T_j^2}{n} = \frac{441 + 196 + 49}{7} = 98$$

so

$$SS_{WITHIN} = \Sigma X^2 - \frac{\Sigma T_j^2}{n}$$
$$= 116 - 98 = 18$$

This is the same value as reported previously for the sum of squares within and is also equal to the total of the individual groups' sums of squares as reported in Table 12.3.

The computational formula for the sum of squares between is

$$SS_{BETWEEN} = \frac{\Sigma T_j^2}{n} - \frac{(\Sigma X)^2}{N} \quad [12.13]$$

where all terms are as defined above. In our example,

$$SS_{BETWEEN} = 98 - 84 = 14$$

as previously reported.

Of course, given that the sum of squares total is equal to the sum of squares between plus the sum of squares within, if two of these quantities have already been calculated, the third can be determined through simple algebraic manipulation. For instance, we could have obtained the sum of squares between as follows:

$$SS_{BETWEEN} = SS_{TOTAL} - SS_{WITHIN}$$
$$= 32 - 18 = 14$$

STUDY EXERCISE 12.1

An investigator was interested in the effects of various types of performance feedback on self-esteem. In order to examine the relationship between these two variables, she had 15 subjects take a general knowledge test. Five participants were randomly assigned to a positive feedback condition where, irrespective of actual performance, they were informed that they had scored at a very high level. An additional five subjects were randomly assigned to a negative feedback condition and informed that they had performed very poorly. A final group of five subjects constituting a control condition were not provided with any feedback regarding their test scores. All participants then responded to a measure of self-esteem. Scores on this measure could range from 0 to 10, with higher scores indicating greater self-esteem. The data for the experiment can be found in the accompanying table. Conduct a one-way analysis of variance to test for a relationship between the type of feedback and self-esteem.

Answer The null and alternative hypotheses can be phrased as follows:

H_0: $\mu_P = \mu_N = \mu_C$

H_1: The three population means are not all equal

where the subscripts "P," "N," and "C" respectively denote the positive feedback, negative feedback, and control conditions.

Intermediate statistics necessary for the calculation of the sums of squares are contained in the table. The sum of squares between is computed using Equation 12.13:

$$SS_{BETWEEN} = \frac{\Sigma T_j^2}{n} - \frac{(\Sigma X)^2}{N}$$

$$= 525.00 - 481.67 = 43.33$$

The sum of squares within is computed using Equation 12.12:

$$SS_{WITHIN} = \Sigma X^2 - \frac{\Sigma T_j^2}{n}$$

$$= 555 - 525.00 = 30.00$$

The sum of squares total is computed using Equation 12.11:

$$SS_{TOTAL} = \Sigma X^2 - \frac{(\Sigma X)^2}{N}$$

$$= 555 - 481.67 = 73.33$$

The degrees of freedom are computed using Equations 12.6, 12.7, and 12.9:

$$df_{BETWEEN} = k - 1 = 3 - 1 = 2$$

$$df_{WITHIN} = N - k = 15 - 3 = 12$$

$$df_{TOTAL} = N - 1 = 15 - 1 = 14$$

The mean square between and the mean square within are then computed by dividing the corresponding sums of squares by their degrees of freedom:

$$MS_{BETWEEN} = \frac{SS_{BETWEEN}}{df_{BETWEEN}} = \frac{43.33}{2} = 21.67$$

$$MS_{WITHIN} = \frac{SS_{WITHIN}}{df_{WITHIN}} = \frac{30.00}{12} = 2.50$$

Finally, the F ratio is derived by dividing $MS_{BETWEEN}$ by MS_{WITHIN}:

$$F = \frac{MS_{BETWEEN}}{MS_{WITHIN}} = \frac{21.67}{2.50} = 8.67$$

These calculations yield the following summary table:

Source	*SS*	*df*	*MS*	*F*
Between	43.33	2	21.67	8.67
Within	30.00	12	2.50	
Total	73.33	14		

The critical value of F from Appendix F for an alpha level of .05 and 2 and 12 degrees of freedom is 3.88. Since the observed F value of 8.67 is greater than 3.88, we reject the null hypothesis and conclude that a relationship exists between the type of feedback and self-esteem.

POSITIVE FEEDBACK		NEGATIVE FEEDBACK		CONTROL	
X	X^2	X	X^2	X	X^2
8	64	5	25	2	4
7	49	6	36	4	16
9	81	7	49	5	25
10	100	4	16	3	9
6	36	3	9	6	36
$T_P = 40$		$T_N = 25$		$T_C = 20$	
$\bar{X}_P = 8.00$		$\bar{X}_N = 5.00$		$\bar{X}_C = 4.00$	
$T_P^2 = 1{,}600$		$T_N^2 = 625$		$T_C^2 = 400$	

$$\Sigma X^2 = 64 + 49 + \cdots + 9 + 36 = 555$$

$$\Sigma X = 8 + 7 + \cdots + 3 + 6 = 85$$

$$\frac{(\Sigma X)^2}{N} = \frac{85^2}{15} = 481.67$$

$$\frac{\Sigma T_j^2}{n} = \frac{1{,}600 + 625 + 400}{5} = 525.00$$

Assumptions of the *F* Test

The F test for one-way between-subjects analysis of variance is appropriate when the dependent variable is quantitative in nature and measured on a level that approximates interval characteristics. The test also requires that the sampling distribution of the F ratio be distributed in accord with the relevant F distribution. This is true when the following assumptions are met:

1. The samples are independently and randomly selected from their respective populations.
2. The scores in each population are normally distributed.
3. The scores in each population have homogeneous variances.

Note that these assumptions parallel those for the independent groups t test, as discussed in Chapter 10.

For the test to be valid, it is essential that the assumption of independent and random selection be met. However, under certain conditions, the sampling distribution of the F ratio will approximate the relevant F distribution even when the normality or homogeneity of variance assumption has been violated. In general, the test is more robust when sample sizes are equal and relatively large than when they are discrepant and/or small. The implication is that a researcher should try, when possible, to employ an equal number of subjects in each group, with sample sizes being as large as circumstances reasonably permit. A detailed discussion of the robustness of one-way analysis of variance can be found in Jaccard and Becker (1988).

12.3 Relationship of the *F* Test to the *t* Test

The F test that we have been considering is closely related to the independent groups t test presented in Chapter 10. The distinction between the two approaches is more a matter of historical than practical significance. For the two-group between-subjects situation, the F distribution bears a mathematical relationship to the t distribution such that

$$F = t^2 \qquad [12.14]$$

One could apply the procedures developed in this chapter to the problems in Chapter 10 and exactly the same conclusions would be drawn. The values of F would be equal to the squares of the corresponding t scores.

12.4 Strength of the Relationship

Earlier, we noted that the sum of squares between is identical to the sum of squares explained discussed in Chapter 10. Accordingly, the strength of the relationship between the independent and dependent variables can be indexed by substituting $SS_{BETWEEN}$ for $SS_{EXPLAINED}$ in the defining formula for eta-squared (Equation 10.10):

$$\text{eta}^2 = \frac{\text{SS}_{\text{EXPLAINED}}}{\text{SS}_{\text{TOTAL}}}$$

This yields the following formula for computing eta-squared for one-way analysis of variance:

$$\text{eta}^2 = \frac{\text{SS}_{\text{BETWEEN}}}{\text{SS}_{\text{TOTAL}}} \qquad [12.15]$$

For the religion and ideal family size investigation,

$$\text{eta}^2 = \frac{14}{32} = .44$$

In this investigation, 44% of the variability in ideal family size is associated with religion. This represents a strong effect.

An alternative formula for eta-squared based on the degrees of freedom and the observed value of F is

$$\text{eta}^2 = \frac{(\text{df}_{\text{BETWEEN}})F}{(\text{df}_{\text{BETWEEN}})F + \text{df}_{\text{WITHIN}}} \qquad [12.16]$$

For the religion and ideal family size investigation,

$$\text{eta}^2 = \frac{(2)(7.00)}{(2)(7.00) + 18} = .44$$

which is the same result as obtained above.

12.5

Nature of the Relationship

As noted at the beginning of this chapter, the F test considers the null hypothesis against all possible alternatives. For example, possible patterns of unequal population means for three groups include $\mu_1 > \mu_2 > \mu_3$, $\mu_1 = \mu_2 > \mu_3$, and $\mu_1 < \mu_2 = \mu_3$. If any one of the alternatives holds, the null hypothesis will be rejected, unless a Type II error occurs. However, the alternative hypothesis states only that the population means are not all equal. Given this state of affairs, it becomes necessary to conduct additional analyses to determine the exact nature of the relationship between the two variables when three or more groups are involved.

Which procedure to use for determining the nature of the relationship after the null hypothesis has been rejected is controversial among statisticians. Among the most common **multiple comparison procedures** are the Scheffé test, the Newman–Keuls test, Duncan's multiple range test, Tukey's honest significant difference (HSD) test, and Fisher's least significant difference (LSD) test.

Each of these is discussed in Kirk (1968). The most general technique is the one proposed by Scheffé. However, the Scheffé procedure tends to produce a high incidence of Type II errors and, for this reason, is often not the test of choice for researchers. Because of its ease of presentation and desirable statistical properties (see Jaccard, Becker & Wood, 1984), we will focus on the test proposed by Tukey.

The **Tukey HSD test** discerns the nature of the relationship by testing a null hypothesis for each possible pair of group means. In the study on religion and ideal family size, there are three groups: Catholic, Protestant, and Jewish. The questions of interest for the HSD test are: (1) Is the population mean for Catholics different from the population mean for Protestants? (2) Is the population mean for Catholics different from the population mean for Jews? and (3) Is the population mean for Protestants different from the population mean for Jews? These three questions can be cast in terms of three sets of null and alternative hypotheses:

$$H_0\colon\ \mu_C = \mu_P$$
$$H_1\colon\ \mu_C \neq \mu_P$$

$$H_0\colon\ \mu_C = \mu_J$$
$$H_1\colon\ \mu_C \neq \mu_J$$

$$H_0\colon\ \mu_P = \mu_J$$
$$H_1\colon\ \mu_P \neq \mu_J$$

These are similar to the statements of null and alternative hypotheses for the independent groups t test discussed in Chapter 10. Could we simply test each pair of group means by conducting three independent groups t tests? The answer is no. The problem with doing this is that multiple t tests increase the probability of making a Type I error beyond that specified by the alpha level. Witte (1980) has presented a coin tossing analogy that illustrates the problem:

> When a fair coin is tossed only once, the probability of heads equals .50—just as when a single t test is to be conducted at the .05 level of significance, the probability of a Type I error equals .05. When a fair coin is tossed three times, however, heads can appear not only on the first toss but on the second or third toss as well, and hence the probability of heads on at least one of the three tosses exceeds .50. By the same token, when a series of three t tests are conducted at the .05 level of significance, a Type I error can be committed not only on the first test but on the second or third test as well, and hence the probability of committing a Type I error on at least one of the three tests exceeds .05. In fact, the cumulative probability of at least one Type I error could be as large as .15 for this series of three t tests.

The Tukey HSD test circumvents this problem and maintains the probability of making one or more Type I errors in the set of multiple comparisons at the specified alpha level. In most applications, researchers adopt an overall alpha level of .05. We will follow this practice in the remainder of the book. The logic underlying the HSD test is rather complex and is presented in Tukey (1953). We will focus on the mechanics of applying the approach.

In order to facilitate presentation, the summary table for the study on religion and ideal family size is reproduced below:

Source	*SS*	*df*	*MS*	*F*
Between	14.00	2	7.00	7.00
Within	18.00	18	1.00	
Total	32.00	20		

The HSD test involves the computation of a **critical difference,** which is defined as follows:

$$CD = q\sqrt{\frac{MS_{WITHIN}}{n}} \quad [12.17]$$

In this formula, CD is the abbreviation for the critical difference, MS_{WITHIN} is the mean square within from the summary table, n is the per-group sample size, and q is a *Studentized range value* obtainable from the *Studentized range table* located in Appendix G. The value of q is determined with reference to the overall alpha level (in this case, .05), the degrees of freedom within (in this case, 18), and the number of groups, k (in this case, 3). Using the instructions provided in Appendix G, we find that $q = 3.61$. We calculate the critical difference as follows:

$$CD = 3.61\sqrt{\frac{1.00}{7}} = 1.36$$

Consider the first set of hypotheses specified above:

$$H_0: \mu_C = \mu_P$$
$$H_1: \mu_C \neq \mu_P$$

We want to choose between these two competing hypotheses. The rule for doing so for the HSD test is the following: If the absolute difference between sample means for the two groups involved in the comparison exceeds the critical difference, then reject the null hypothesis. Otherwise, fail to reject the null hypothesis. If the null hypothesis is rejected, conclude that the group with the larger sample mean also has a larger mean in the population. In this case, the absolute difference between the sample means is $|\bar{X}_C - \bar{X}_P| = |3.00 - 2.00| = 1.00$. Since 1.00 is not greater than 1.36, we fail to reject the null hypothesis stated above.

The same logic is applied to the other sets of hypotheses. We can summarize the results of the entire analysis in a table:

Null hypothesis tested	Absolute difference between sample means	Value of CD	Null hypothesis rejected?
$\mu_C = \mu_P$	$\|3.00 - 2.00\| = 1.00$	1.36	No
$\mu_C = \mu_J$	$\|3.00 - 1.00\| = 2.00$	1.36	Yes
$\mu_P = \mu_J$	$\|2.00 - 1.00\| = 1.00$	1.36	No

The nature of the relationship is now apparent: Catholics ($\bar{X}_C = 3.00$) have a larger ideal family size than Jews ($\bar{X}_J = 1.00$). However, we cannot confidently conclude that the ideal family size for either of these groups differs from the ideal family size for Protestants ($\bar{X}_P = 2.00$). The HSD test provides a clear picture of just how religion is related to the ideal number of children.

STUDY EXERCISE 12.2

Calculate eta-squared and use the HSD test to analyze the nature of the relationship between the two variables for the data in Study Exercise 12.1.

Answer Eta-squared is most easily calculated using Equation 12.15:

$$\text{eta}^2 = \frac{SS_{BETWEEN}}{SS_{TOTAL}} = \frac{43.33}{73.33} = .59$$

This represents a strong effect.

The first step in applying the HSD test is to calculate the critical difference. Referring to Appendix G, we find that for an overall alpha level of .05, $df_{WITHIN} = 12$, and $k = 3$, q is equal to 3.77. Thus,

$$CD = q\sqrt{\frac{MS_{WITHIN}}{n}} = 3.77\sqrt{\frac{2.50}{5}} = 2.67$$

The HSD procedure can now be applied as follows:

Null hypothesis tested	Absolute difference between sample means	Value of CD	Null hypothesis rejected?
$\mu_P = \mu_N$	$\|8.00 - 5.00\| = 3.00$	2.67	Yes
$\mu_P = \mu_C$	$\|8.00 - 4.00\| = 4.00$	2.67	Yes
$\mu_N = \mu_C$	$\|5.00 - 4.00\| = 1.00$	2.67	No

The nature of the relationship is such that self-esteem will be greater when people receive positive feedback ($\bar{X}_P = 8.00$) than when they receive negative feedback ($\bar{X}_N = 5.00$) or no feedback ($\bar{X}_C = 4.00$). However, we cannot confidently conclude that negative feedback affects self-esteem relative to no feedback.

12.6 Methodological Considerations

Although the results of the investigation of religion and ideal family size suggest that a relationship exists between the two variables, the data must be interpreted in the context of certain methodological constraints. Consider first the role of confounding variables. Because religion is not an experimental manipulation, subjects in the study could not be randomly assigned to groups. Consequently, all variables that are naturally related to religion will serve as confounding variables. These will include such variables as social class, size of the family in which one grew up, and education, to name a few. It is impossible to conclude unambiguously that a causal relationship exists between religion and ideal family size. Religion per se might have no effect on ideal family size, and the observed relationship might simply be a function of the causal influence of one or more confounding variables.

In terms of disturbance variables, a large number of such variables were uncontrolled in the study. For instance, research has shown that ideal family size is influenced by one's religiosity (that is, how religious one is). Subjects within a given religion in this study probably differed in religiosity, and this would, in turn, create error variance. One minus eta-squared provides an index of the extent to which disturbance variables have influenced the dependent measure. Specifically, 1.00 minus eta-squared represents the proportion of variability in the dependent variable that can be attributed to disturbance variables. For the religion and ideal family size data, this equals $1.00 - .44 = .56$. Thus, more than half of the variability in ideal family size is due to disturbance variables. Certainly, this could have been reduced through additional control procedures.

The results of the study must also be considered in terms of their generalizability. No details were given about who the subjects were. If they were all college students, this would suggest limitations on the generalizability of the findings. Research reports should provide reasonable descriptions of the nature of the samples studied.

12.7 Numerical Example

An important issue in court cases is the validity of eyewitness testimony. Several behavioral scientists have suggested that the nature of eyewitness accounts can be influenced by many psychological factors. One of these is the way in which a question is phrased to an eyewitness. If subtle wording of questions can influence the type of answer an eyewitness gives, then such reports must be interpreted with considerable caution. Consider the following hypothetical experiment.

Twenty individuals watched a film of a car accident in which car A ran a stop sign and hit car B. Car A was traveling at a speed of 20 miles per hour. The entire incident was filmed, including the arrival of a police officer and the citation of the motorist who was in the wrong. After watching the film, each individual was asked to estimate the speed of car A at the moment of impact (subjects did not know the actual speed of the car). The question was phrased in four different ways. The first five individuals were asked, "How fast was car A going at the time of the accident with car B?" The second five individuals were asked, "How fast was car A going when it hit car B?" An additional five individuals were asked, "How fast was car A going when it crashed into car B?" Finally, the last five individuals were asked, "How fast was car A going when it smashed into car B?" The question of interest is whether estimates of the car's speed vary as a function of the specific wording used in asking about the accident. The speed estimates are provided in Table 12.4.

The null hypothesis is that the phrasing of the question does not influence speed estimates. The alternative hypothesis is that the phrasing of the question does influence speed estimates. These hypotheses can be formally stated as

$$H_0: \quad \mu_A = \mu_H = \mu_C = \mu_S$$

H_1: The four population means are not all equal.

TABLE 12.4 Data and Computation of Sums of Squares for Phraseology and Speed Estimation Experiment

ACCIDENT		HIT		CRASHED		SMASHED	
X	X^2	X	X^2	X	X^2	X	X^2
18	324	23	529	25	625	29	841
20	400	20	400	27	729	28	784
17	289	22	484	26	676	30	900
19	361	19	361	23	529	27	729
21	441	21	441	24	576	31	961
$T_A = 95$		$T_H = 105$		$T_C = 125$		$T_S = 145$	
$\bar{X}_A = 19.00$		$\bar{X}_H = 21.00$		$\bar{X}_C = 25.00$		$\bar{X}_S = 29.00$	
$T_A^2 = 9{,}025$		$T_H^2 = 11{,}025$		$T_C^2 = 15{,}625$		$T_S^2 = 21{,}025$	

$$\Sigma X^2 = 324 + 400 + \cdots + 729 + 961 = 11{,}380$$

$$\Sigma X = 18 + 20 + \cdots + 27 + 31 = 470$$

$$\frac{(\Sigma X)^2}{N} = \frac{470^2}{20} = 11{,}045$$

$$\frac{\Sigma T_j^2}{n} = \frac{9{,}025 + 11{,}025 + 15{,}625 + 21{,}025}{5} = 11{,}340$$

where the subscript "A" represents the "accident" phraseology, the subscript "H" represents the "hit" phraseology, the subscript "C" represents the "crashed" phraseology, and the subscript "S" represents the "smashed" phraseology.

Intermediate statistics necessary for the calculation of the sums of squares are contained in Table 12.4. The sum of squares between is

$$SS_{BETWEEN} = \frac{\Sigma T_j^2}{n} - \frac{(\Sigma X)^2}{N}$$

$$= 11{,}340 - 11{,}045 = 295$$

The sum of squares within is

$$SS_{WITHIN} = \Sigma X^2 - \frac{\Sigma T_j^2}{n}$$

$$= 11{,}380 - 11{,}340 = 40$$

Finally, the sum of squares total is

$$SS_{TOTAL} = \Sigma X^2 - \frac{(\Sigma X)^2}{N}$$

$$= 11{,}380 - 11{,}045 = 335$$

The degrees of freedom are

$$df_{BETWEEN} = k - 1 = 4 - 1 = 3$$

$$df_{WITHIN} = N - k = 20 - 4 = 16$$

$$df_{TOTAL} = N - 1 = 20 - 1 = 19$$

The relevant mean squares are

$$MS_{BETWEEN} = \frac{SS_{BETWEEN}}{df_{BETWEEN}} = \frac{295}{3} = 98.33$$

$$MS_{WITHIN} = \frac{SS_{WITHIN}}{df_{WITHIN}} = \frac{40}{16} = 2.50$$

Thus, the F ratio is

$$F = \frac{MS_{BETWEEN}}{MS_{WITHIN}} = \frac{98.33}{2.50} = 39.33$$

These calculations yield the following summary table:

Source	*SS*	*df*	*MS*	*F*
Between	295.00	3	98.33	39.33
Within	40.00	16	2.50	
Total	335.00	19		

The first question is whether there is a relationship between the type of wording used and the speed estimates. For an alpha level of .05, and 3 and 16 degrees of freedom, the critical value of F is 3.24. The observed value of F is 39.33. This exceeds the critical value, so we reject the null hypothesis and conclude that a relationship exists between the two variables.

The second question concerns the strength of the relationship. This is indexed by eta-squared and is computed as follows:

$$\text{eta}^2 = \frac{SS_{\text{BETWEEN}}}{SS_{\text{TOTAL}}} = \frac{295}{335} = .88$$

This represents a very strong effect and indicates that 88% of the variability in speed estimates is associated with the way in which the question was phrased.

The final question concerns the nature of the relationship. This requires application of the HSD test. The critical difference is derived using Equation 12.17. From Appendix G, the value of q for an overall alpha level of .05, $df_{\text{WITHIN}} = 16$, and $k = 4$ is 4.05. Thus,

$$\text{CD} = q\sqrt{\frac{MS_{\text{WITHIN}}}{n}}$$

$$= 4.05\sqrt{\frac{2.50}{5}} = 2.86$$

The HSD procedure can now be applied as follows:

Null hypothesis tested	*Absolute difference between sample means*	*Value of CD*	*Null hypothesis rejected?*
$\mu_A = \mu_H$	$\lvert 19.00 - 21.00\rvert = 2.00$	2.86	No
$\mu_A = \mu_C$	$\lvert 19.00 - 25.00\rvert = 6.00$	2.86	Yes
$\mu_A = \mu_S$	$\lvert 19.00 - 29.00\rvert = 10.00$	2.86	Yes
$\mu_H = \mu_C$	$\lvert 21.00 - 25.00\rvert = 4.00$	2.86	Yes
$\mu_H = \mu_S$	$\lvert 21.00 - 29.00\rvert = 8.00$	2.86	Yes
$\mu_C = \mu_S$	$\lvert 25.00 - 29.00\rvert = 4.00$	2.86	Yes

Inspection of the table suggests the following conclusions: Speed estimates obtained when the question is phrased in terms of car A smashing into car B ($\bar{X}_S = 29.00$) will be higher than speed estimates obtained when the question is phrased in terms of the two cars being involved in an accident ($\bar{X}_A = 19.00$), car A hitting car B ($\bar{X}_H = 21.00$), or car A crashing into car B ($\bar{X}_C = 25.00$). Furthermore, speed estimates obtained using the "crash" phraseology will be higher than such estimates obtained using either the "accident" or the "hit" phraseology. However, we cannot confidently conclude that the "accident" phraseology affects speed estimates relative to the "hit" phraseology.

You should think about the experiment in terms of basic research design questions: What potential confounding variables might be operating? What has been the role of disturbance variables? What kinds of procedures could be used to reduce sampling error? What are the potential limitations of the experiment in terms of generalizability? All these questions are critical to drawing appropriate conclusions from the study.

12.8

Planning an Investigation Using One-Way Between-Subjects Analysis of Variance

Appendix E.2 contains tables of the per-group sample sizes necessary to achieve various levels of power for one-way between-subjects analysis of variance. A set of tables is presented for each research design involving a different $df_{BETWEEN}$. Table 12.5 presents the portion of this appendix for $df_{BETWEEN} = 2$ (that is, for three groups) and an alpha level of .05 for illustration. The first column presents different levels of power and the column headings are values of eta-squared in the population. As an example of the use of this table, if the desired power level is .80 and the researcher suspects that the strength of the relationship in the population corresponds to an eta-squared of roughly .15, then the number of subjects that should be sampled in *each group* is 19. Inspection of Table 12.5 and Appendix E.2 provides a general appreciation for the relation-

TABLE 12.5 Approximate Sample Sizes Necessary to Achieve Selected Levels of Power for $df_{BETWEEN} = 2$ and Alpha = .05 as a Function of Population Values of Eta-Squared

Degrees of Freedom Between = 2
Alpha = .05

	POPULATION ETA-SQUARED									
POWER	*.01*	*.03*	*.05*	*.07*	*.10*	*.15*	*.20*	*.25*	*.30*	*.35*
.10	22	8	5	4	3	2	2	2	—	—
.50	165	55	32	23	16	10	8	6	5	4
.70	255	84	50	35	24	16	11	9	7	6
.80	319	105	62	44	30	19	14	11	9	7
.90	417	137	81	57	39	25	18	14	11	9
.95	511	168	99	69	47	30	22	16	13	11
.99	708	232	137	96	65	41	29	22	18	14

ships between $df_{BETWEEN}$, the alpha level, sample size, the strength of the relationship one is trying to detect in the population, and power for one-way between-subjects analysis of variance.

12.9

Method of Presentation

Reports of a one-way between-subjects analysis of variance should include statements of the degrees of freedom between, the degrees of freedom within, the observed value of *F*, the significance level, and the sample means. In addition, if the analysis is statistically significant, the strength and nature of the relationship between the independent and dependent variables should also be addressed.

The results for the study on religion and ideal family size discussed earlier in this chapter might be presented as follows:

Results

A one-way analysis of variance compared the mean ideal family sizes of Catholics, Jews, and Protestants. This was found to be statistically significant, F(2, 18) = 7.00, p < .01. The strength of the relationship, as indexed by eta^2, was .44. A Tukey HSD test indicated that the mean for Catholics (3.00) was significantly greater than the mean for Jews (1.00). The mean for Protestants (2.00) did not significantly differ from the mean for either of these groups.

The first sentence identifies the statistical test that was used. The term *between-subjects* is not used to describe the analysis of variance because, unless stated otherwise, it is assumed that a between-subjects analysis was performed.

The second sentence states the results of the test for the existence of a relationship. The numbers within the parentheses are the degrees of freedom between and the degrees of freedom within, respectively. This is followed by the observed value of *F* and the significance level.

The third sentence indicates the strength of the relationship. In practice, this is rarely reported. Fortunately, however, eta-squared can be derived from the degrees of freedom and the observed value of F using Equation 12.16.

The last two sentences present the results of the HSD test, including specification of the sample means. The intermediate statistics computed when applying a multiple comparison procedure are not reported.

12.10 Examples from the Literature

The Effectiveness of Different Incentives for Learning

Educational psychologists have studied the effects of rewards and punishment on learning. In the early 1900s, many educators and parents assumed that the threat of punishment or actual punishment itself was the most effective means for motivating children to learn. More recently, the dominant philosophy seems to be rewards as an incentive for learning.

Hurlock (1925) reported a study relevant to this issue in which fourth- and sixth-grade students were divided into four groups. Each group took addition tests in class on 5 successive days. Students in the first group were separated from the other students and told to work on these tests as usual. This constituted the control condition. Students in another group were brought to the front of the room each day before the test was given and praised for their good work. Students in the third group were also brought to the front of the room, but these students were reproved (reprimanded) for their poor work. The students in the fourth group were ignored. Of interest to Hurlock was whether the different incentives had different effects on performance as indicated by scores on the final addition test. Since students were randomly assigned to the four conditions, group means would be expected to reflect the effects of the type of incentive on learning.

Hurlock's investigation was conducted before the technique of analysis of variance had been developed. However, Kerlinger (1973) analyzed the Hurlock data using one-way analysis of variance and found the null hypothesis to be rejected, $F(3, 102) = 10.08$, $p < .001$, indicating that the type of incentive is related to performance. The strength of the relationship, as indexed by eta-squared, was .23. This represents a moderate effect. An HSD test indicated that the mean for the praised group ($\bar{X} = 20.22$) was significantly larger than the mean for each of the other three groups ($\bar{X} = 11.35$ for the control group; $\bar{X} =$ 12.38 for the ignored group; $\bar{X} = 14.19$ for the reproved group). The means for these latter groups, however, did not significantly differ. Thus, consistent with a reward philosophy, the most effective incentive in this experiment was that of praise.

Body Posture of Message Recipients and Susceptibility to Influence

Research in nonverbal behavior has shown that the body posture of a communicator can affect his or her ability to influence the attitudes of the message recipients. A related issue, but one that has received far less attention, concerns how susceptibility to influence is affected by the recipient's body posture. In one of the few studies on this topic, Petty, Wells, Heesacker, Brock, and Cacioppo (1983) had 78 college students listen to a persuasive message advocating a 20% tuition increase at their university while either standing, sitting, reclining on a cushioned table, or reclining on an uncushioned table. Following exposure to this message, subjects responded to the question, "In general, to what extent do you agree that the tuition should be increased?" Responses were made on a 12-point scale, with higher scores indicating greater agreement.

A one-way analysis of variance applied to these data was statistically significant, $F(3, 74) = 3.33$, $p < .025$. As indexed by eta-squared, the strength of the relationship was .12. This represents a moderate effect. Multiple comparisons showed that subjects who had heard the persuasive communication while reclining on the cushioned table ($\bar{X} = 7.60$) reported significantly greater positive attitudes than did subjects who had heard this message while standing ($\bar{X} = 5.63$). The means for the sitting ($\bar{X} = 6.00$) and uncushioned reclining ($\bar{X} = 6.95$) conditions did not significantly differ from one another nor from the means reported above.

A follow-up study by Petty and colleagues suggests that "a reclining posture facilitates message-relevant thinking over a standing posture and thereby enhances the importance of message content in producing persuasion" (p. 219). According to this perspective, reclining individuals are better able to differentiate strong from weak arguments than are standing individuals, and are thus more susceptible to influence when persuasive messages are compelling but less susceptible to influence when persuasive messages are specious. The exact reasons for this, however, are unclear.

12.11 Summary

One-way between-subjects analysis of variance is typically used to analyze the relationship between two variables when (1) the dependent variable is quantitative in nature and is measured on approximately an interval level, (2) the independent variable is between-subjects in nature, and (3) the independent variable has three or more levels. The essence of one-way analysis of variance is the partitioning of variability into between-group and within-group components. Between-group variability reflects both sampling error and the effect of the independent variable on the dependent variable. Within-group variability reflects just sampling error. The ratio of variance measures (mean squares) based on these sources of variability is called the F ratio. The F ratio has a sampling distribution that closely approximates an F distribution and that can be used to test for the existence of a relationship between the independent and dependent variables. The strength of the relationship is analyzed using eta-squared, and the nature of the relationship is analyzed using the Tukey HSD test. The HSD test ensures that the probability of making one or more Type I errors in the set of multiple comparisons will be equal to alpha.

Appendix 12.1 Rationale for the Degrees of Freedom

Unlike the sum of squares discussed in Chapter 7, the sum of squares between does not have $N - 1$ degrees of freedom associated with it. This is because the sum of squares between is based on deviations of the k group means from the grand mean. Since the sum of deviations about a mean (in this case, the grand mean) will always equal 0 (see Chapter 3 for the logic underlying this property), if all but one of the group means are known, the last one is not free to vary. Thus, the degrees of freedom between is $k - 1$.

The sum of squares within is based on the deviation of scores about their respective group means. Within each group, there will be $n - 1$ degrees of freedom. Given k groups, this yields $k(n - 1)$ degrees of freedom, which can be rewritten as $kn - k$. Since $kn = N$, the degrees of freedom within is $N - k$.

Exercises

Answers to asterisked (*) exercises appear at the back of the book. Answers to exercises with two asterisks are also worked out step-by-step in the Study Guide.

Exercises to Review Concepts

1. Under what conditions is one-way between-subjects analysis of variance typically used to analyze a bivariate relationship?

*2. What general form does the alternative hypothesis for one-way analysis of variance take? Why can it not be summarized in a single mathematical statement?

*3. Distinguish between between-group variability and within-group variability.

4. What does between-group variability reflect? Why?

5. What does within-group variability reflect? Why?

*6. Under what circumstance will the F ratio, over the long run, approach 1.00? Under what circumstance will the F ratio, over the long run, be greater than 1.00?

7. Distinguish between the sum of squares total, the sum of squares between, and the sum of squares within. How are they interrelated?

*8. Consider the following scores in an experiment involving three groups, A, B, and C, with five subjects in each group:

A	B	C
3	5	7
3	5	7
3	5	7
3	5	7
3	5	7

Without actually computing the sum of squares within, what must its value be? Why?

*9. What does a mean square represent in analysis of variance?

10. What is the relationship between the mean square within and the pooled variance estimate?

11. What are the assumptions underlying one-way between-subjects analysis of variance?

*12. State the critical value of F that would be used to reject the null hypothesis for a one-way between-subjects analysis of variance at an alpha level of .05 under each of the following conditions:

a. $k = 3, N = 21$
b. $k = 4, N = 20$
c. $k = 3, N = 30$
d. $k = 5, N = 75$

*13. Complete the missing entries in the summary table for a one-way analysis of variance having three levels of the independent variable and $n = 20$ per group:

Source	SS	df	MS	F
Between	—	—	—	—
Within	152.00	—	—	
Total	182.00	—		

14. Complete the missing entries in the summary table for a one-way analysis of variance having four levels of the independent variable:

Source	SS	df	MS	F
Between	—	—	18.00	3.60
Within	—	—	—	
Total	—	23		

15. What is the advantage, following rejection of the null hypothesis, of determining the nature of the relationship between an independent variable and a dependent variable with the Tukey HSD test rather than with multiple t tests?

**16. An investigator wanted to test the effects of marital status on attitudes toward divorce. A scale measuring attitudes on this issue was administered to 10 single, 10 married, and 10 divorced individuals. Scores could range from 1 to 12, with higher scores representing more positive attitudes. The means and summary table are as follows:

$\bar{X}_S = 6.00 \qquad \bar{X}_M = 8.00 \qquad \bar{X}_D = 10.00$

Source	SS	df	MS	F
Between	80.00	2	40.00	8.00
Within	135.00	27	5.00	
Total	215.00	29		

Analyze the nature of the relationship between marital status and attitudes toward divorce using the HSD test.

*17. Suppose one of "Nader's Raiders" was interested in comparing the performance of three types of cars, X, Y, and Z. Random samples of five owners were drawn from the list of owners of each model. These owners were asked how many times their cars had undergone major repairs in the past 2 years. Conduct a one-way analysis of variance on the following data to test for a relationship between the type of car and repair records:

X	Y	Z
2	5	9
1	4	6
2	3	3
3	4	7
2	4	5

***18.** Compute the value of eta-squared for the data in Exercise 17 using Equation 12.15. Recalculate eta-squared using Equation 12.16. Compare the two sets of results. Does the observed value represent a weak, moderate, or strong effect?

***19.** Analyze the nature of the relationship for the data in Exercise 17 using the HSD test.

20. An investigator was interested in the relationship between supposed task difficulty and task performance. Twenty-four subjects worked on the identical spatial ability task, but six of these individuals were led to believe that the task was of low difficulty, six were led to believe that the task was of moderate difficulty, six were led to believe that the task was of high difficulty, and six were not given any information about the task's difficulty. Scores could range from 0 to 10, with higher scores indicating better task performance. Conduct a one-way analysis of variance on the following data to test for a relationship between supposed task difficulty and task performance:

Low	Moderate	High	No information
8	6	4	4
7	7	1	5
5	4	2	5
8	5	4	6
9	4	6	8
7	6	3	6

21. Compute the value of eta-squared for the data in Exercise 20. Does the observed value represent a weak, moderate, or strong effect?

22. Analyze the nature of the relationship for the data in Exercise 20 using the HSD test.

***23.** Consider the following scores for two groups of subjects who were tested under either condition A or condition B:

A	B
3	9
4	10
5	11
6	12
7	13

Test for a relationship between the independent and dependent variables using the one-way analysis of variance procedures developed in the present chapter. (*Note:* Even though one-way analysis of variance is typically applied in cases where there are three or more groups, for this example, proceed to use it for analyzing two groups.)

***24.** Test for a relationship between the independent and dependent variables for the data in Exercise 23 using the procedures developed in Chapter 10 for a nondirectional independent groups t test. Square the observed value of t. Compare this with the observed value of F from Exercise 23. Square the critical value of t. Compare this with the critical value of F from Exercise 23. What does this indicate about the relationship between one-way analysis of variance and the independent groups t test in the two-group case?

For each of the studies described in Exercises 25–27, indicate the appropriate statistical test for analyzing the relationship between the independent and dependent variables. Assume that the underlying assumptions of the tests have been satisfied.

***25.** Barron (1965) administered an intelligence test to creative individuals in four different occupations: mathematicians, writers, psychologists, and architects. The mean intelligence scores were compared for the four groups.

26. A researcher was interested in the ability of people to process information presented to their right versus their left ears. Twelve subjects participated in the experiment. Headphones were used to present simultaneously one list of 20 words to participants' right ears and a different list of 20 words to participants' left ears. After the lists were presented, subjects were asked to recall as many words from each list as they could. The mean number of words recalled from the list presented to the right ear was compared with the mean number of words recalled from the list presented to the left ear.

27. An investigator was interested in the effects of practice on problem-solving performance. A total of 150 individuals participated in the study. Subjects were randomly assigned to two conditions. Seventy-five subjects tried to solve 30 problems after having a 10-minute practice session on a similar set of problems. The other 75 subjects tried to solve the same 30 problems, but with no practice session. The average number of correct solutions was compared for people in the practice condition with those in the no-practice condition.

***28.** If a researcher suspected that the strength of the relationship between two variables in the population was .07 as indexed by eta-squared, what sample size should she use in a study involving four independent groups and an alpha level of .05 in order to achieve power of .80?

29. Suppose an investigator reported the results of a study using five independent groups with $n = 5$ per group. If the value of eta-squared in the population were .25, what would the power of his statistical test be at an alpha level of .05?

Exercises to Apply Concepts

****30.** Of recent interest to psychologists has been the study of jury decision making and factors influencing judgments of guilt on the part of jurors. Stephen (1975) reviewed a number of studies that examine the effect of a defendant's race on guilty verdicts. In one experiment, subjects were presented a transcript of a trial and asked to indicate the probability that the defendant was guilty. Judgments were made on a 10-point scale where 1 = "definitely not guilty" through 10 = "definitely guilty." The higher the number, the higher the perceived probability of guilt. All subjects read the same transcript. However, one-third of the subjects were told that the defendant was white, another third were told that the defendant was black, and the last third were told that the defendant was Mexican-American. The hypothetical data presented below are representative of the results of this experiment. Analyze these data, draw a conclusion, and write up your results using the principles discussed in the Method of Presentation section.

White defendant	*Black defendant*	*Mexican-American defendant*
6	10	10
7	10	6
2	9	10
3	4	5
5	4	10
0	10	5
1	10	2
0	10	10
6	3	2
0	10	10

31. Ainsworth, Blehar, Waters, and Wall (1978) studied how infants become psychologically attached to their mothers. They delineated three qualitatively different types of attachment, which Sroufe and Waters (1977) labeled "security," "avoidance," and "ambivalence." Infants who are securely attached, in general, seek to be near their mother and have contact with her. These infants, when separated from their mother, may or may not exhibit distress and are generally not anxious about being left alone. Infants who are "avoidant" exhibit few tendencies toward maintaining proximity and contact with the mother. Finally, infants who are "ambivalent" display both contact- and proximity-seeking behaviors as well as contact- and interaction-resisting behaviors (hence the term "ambivalent"). When separated from the mother, these infants tend to exhibit anger or become conspicuously passive until the mother returns.

In one study, Ainsworth and colleagues were interested in examining the relationship between the type of attachment exhibited by infants and maternal behavior. Trained observers watched interactions between mothers and infants and rated these interactions on a number of dimensions. One of these was the extent to which the mother was sensitive or insensitive to the infant's signals and communications. Ratings were made on a scale from 1 to 9, with higher numbers indicating greater sensitivity. The hypothesis was that the type of attachment exhibited by an infant would be associated with different levels of sensitivity on the part of the mother. Hypothetical data for this study representative of the results of Ainsworth and colleagues are presented below. Analyze these data, draw a conclusion, and write up your results using the principles discussed in the Method of Presentation section.

Security	*Avoidance*	*Ambivalence*
5	1	3
9	5	3
9	1	1
5	1	1
9	5	3
5	5	1

THIRTEEN

ONE-WAY REPEATED MEASURES ANALYSIS OF VARIANCE

13.1

Use of One-Way Repeated Measures Analysis of Variance

One-way repeated measures analysis of variance (abbreviated as "one-way repeated measures ANOVA") is typically used to analyze the relationship between two variables when

1. the dependent variable is quantitative in nature and is measured on approximately an interval level;
2. the independent variable is *within-subjects* in nature (it can be either qualitative or quantitative); and
3. the independent variable has three or more levels.

In short, one-way repeated measures analysis of variance is used under the same circumstances as one-way between-subjects analysis of variance except that the independent variable is within-subjects in nature rather than between-subjects in nature. Just as one-way between-subjects analysis of variance is an extension of the independent groups t test for instances when the independent variable has more than two levels, one-way repeated measures analysis of variance is an extension of the correlated groups t test. Let us consider an example of an investigation that meets the conditions for this technique.

Suppose a consumer psychologist is interested in the effect of label information on the perceived quality of wine. She designs an experiment with three conditions. In each condition, an individual tastes a wine, and then rates its taste. The ratings are made on a 20-point scale, with higher scores indicating that the wine tasted better. In all three conditions the wine is identical. However, in one condition the label indicates that it is a French wine, in a second condition the label indicates that it is an Italian wine, and in a third condition the label indicates that it is an American wine. The experiment is conducted as a within-subjects design in that each of the six participants successively tastes and rates the wine in the three different conditions. In order to counteract pos-

TABLE 13.1 Data for Wine Label and Perceived Quality Experiment

Subject	*French*	*Italian*	*American*	$\bar{X}_i$
1	14	10	9	11.00
2	16	12	12	13.33
3	17	13	14	14.67
4	16	14	16	15.33
5	15	12	10	12.33
6	12	11	8	10.33
	$\Sigma X_F = 90$	$\Sigma X_I = 72$	$\Sigma X_A = 69$	
	$\bar{X}_F = 15.00$	$\bar{X}_I = 12.00$	$\bar{X}_A = 11.50$	

sible carry-over effects, the type of label presented first, second, and third is randomized for each subject.* The taste ratings are presented in the second, third, and fourth columns of Table 13.1.

In this experiment, the independent variable is the type of label, and it has three levels. It is within-subjects in nature since the same subjects participate in all three conditions. The dependent variable is the perceived quality of the wine as reflected in the taste ratings. It represents a quantitative variable. Given these conditions, one-way repeated measures analysis of variance is the statistical technique that would typically be used to analyze the relationship between the variables.

The mean score across conditions for each subject has been calculated in column 5 of Table 13.1. Note that there is variability in the mean scores; some subjects, on the average, rated the wine higher than other subjects. This reflects the influence of individual differences. If we could remove this influence (which serves as a disturbance variable), we could increase the sensitivity of the statistical test of the relationship between the independent and dependent variables.

When discussing the correlated groups *t* test in Chapter 11, we noted that an advantage of within-subjects designs is their ability to systematically remove variability due to individual differences from the dependent variable and developed the logic of extracting this source of variability in the two-group case. The same procedures can be used to generate a set of nullified scores when a within-subjects independent variable has more than two levels. The nullified data can then be analyzed to determine whether a relationship exists between the independent variable and the dependent variable after the effects of individual background have been removed. The results of this analysis would be identical to the results that would be obtained using the more computationally efficient procedure for one-way repeated measures analysis of variance described in the following section. As with the correlated groups *t* test, the computational approach utilizes raw rather than nullified scores. For purpose of illustration, we will return to the wine label and perceived quality experiment.

---

* The rationale for randomly ordering the sequence of conditions across subjects as a way to deal with carry-over effects is discussed in Section 13.5.

13.2

Inference of a Relationship Using One-Way Repeated Measures Analysis of Variance

Null and Alternative Hypotheses

The null hypothesis states that the type of label does not influence the perceived quality of the wine. The alternative hypothesis states that there is a relationship between the two variables. Stated formally, the hypotheses are

H_0: $\mu_F = \mu_I = \mu_A$

H_1: The three population means are not all equal

where the subscripts "F," "I," and "A" denote the French, Italian, and American labels, respectively.

Partitioning of Variability

In Chapter 12, we demonstrated that the total variability in the dependent variable in a between-subjects design (the sum of squares total) can be partitioned into two components, one reflecting the influence of the independent variable (the sum of squares between) and one reflecting the influence of disturbance variables (the sum of squares within). We can also identify and partition the total variability when a repeated measures design is used. As before, the total variability in the dependent variable across all individuals and all conditions can be represented by the sum of squares total. This can be partitioned into three components, one reflecting the influence of the independent variable (the **sum of squares treatments**), one reflecting the influence of individual differences (the **sum of squares subjects**), and one reflecting the influence of disturbance variables other than individual differences (the **sum of squares error**). Symbolically,

$$SS_{TOTAL} = SS_{TREATMENTS} + SS_{SUBJECTS} + SS_{ERROR} \quad [13.1]$$

The sum of squares treatments is conceptually equivalent to the sum of squares between in a between-subjects design. In that they both reflect the influence of disturbance variables, the sum of squares error and the sum of squares within are also conceptually equivalent.

The sum of squares subjects has no counterpart in a between-subjects design. In fact, it is this component that differentiates the two approaches. When a between-subjects design is used, each score is provided by a different subject, so it is not possible to identify the effects of individual differences. Differences in background (including such things as affinity for wine) thus contribute to sampling error as reflected in the sum of squares within. However, when a repeated measures design is used, the influence of individual differences can be identified and statistically removed from the dependent variable in the form of the sum of squares subjects. This explains the equivalence between the computational approach to repeated measures analysis of variance and the variance extraction approach referred to earlier. It also explains why a repeated measures analysis of variance is a more sensitive test of the relationship between the independent and dependent variables than is a between-subjects analysis of variance: Since the sum of squares within in a between-subjects design includes the effects of individual differences while the sum of squares error in a repeated measures design does not, the mean square error, which forms the denominator in the F test of a relationship between the independent and dependent variables in the repeated measures case, will tend to be smaller than the mean square within forming the denominator of the F test in the between-subjects case. To the extent that the mean square error is in fact smaller than the mean square within, a larger F ratio and a greater likelihood of rejecting the null hypothesis should result.*

Computation of the Sums of Squares

The formula for calculating the sum of squares total is

$$SS_{TOTAL} = \Sigma X^2 - \frac{(\Sigma X)^2}{kN} \qquad [13.2]$$

where ΣX^2 is the *sum of the squared scores* constituting the data set, $(\Sigma X)^2$ is the *square of the summed scores*, k is the number of levels of the independent variable (that is, the number of conditions), and N is the sample size.

The data for the wine label and perceived quality experiment have been reproduced in Table 13.2, where it can be seen that the sum of the 18 X^2 scores is

$$\Sigma X^2 = 196 + 256 + \cdots + 100 + 64 = 3{,}081$$

and the sum of the 18 X scores is

$$\Sigma X = 14 + 16 + \cdots + 10 + 8 = 231$$

Squaring the latter value and dividing by kN, we obtain

$$\frac{(\Sigma X)^2}{kN} = \frac{231^2}{(3)(6)} = 2{,}964.50$$

Thus, the sum of squares total in this example is

$$SS_{TOTAL} = \Sigma X^2 - \frac{(\Sigma X)^2}{kN}$$
$$= 3{,}081 - 2{,}964.50 = 116.50$$

The formula for calculating the sum of squares treatments is

$$SS_{TREATMENTS} = \frac{\Sigma T_j^2}{N} - \frac{(\Sigma X)^2}{kN} \qquad [13.3]$$

---

* We will return to the issue of the sensitivity of repeated measures versus between-subjects analysis of variance later in this chapter.

TABLE 13.2 Computation of Sums of Squares for Wine Label and Perceived Quality Experiment

	FRENCH		ITALIAN		AMERICAN			
SUBJECT	X	X^2	X	X^2	X	X^2	s_i	s_i^2
1	14	196	10	100	9	81	33	1,089
2	16	256	12	144	12	144	40	1,600
3	17	289	13	169	14	196	44	1,936
4	16	256	14	196	16	256	46	2,116
5	15	225	12	144	10	100	37	1,369
6	12	144	11	121	8	64	31	961
	$T_F = 90$		$T_I = 72$		$T_A = 69$			
	$T_F^2 = 8{,}100$		$T_I^2 = 5{,}184$		$T_A^2 = 4{,}761$			

$$\Sigma X^2 = 196 + 256 + \cdots + 100 + 64 = 3{,}081$$

$$\Sigma X = 14 + 16 + \cdots + 10 + 8 = 231$$

$$\frac{(\Sigma X)^2}{kN} = \frac{231^2}{(3)(6)} = 2{,}964.50$$

$$\frac{\Sigma T_j^2}{N} = \frac{8{,}100 + 5{,}184 + 4{,}761}{6} = 3{,}007.50$$

$$\frac{\Sigma s_i^2}{k} = \frac{1{,}089 + 1{,}600 + 1{,}936 + 2{,}116 + 1{,}369 + 961}{3} = 3{,}023.67$$

where T_j^2 is the square of the sum of the scores in condition j and all other terms are as defined previously. In our example, T_j^2 equals 8,100 for the French label, 5,184 for the Italian label, and 4,761 for the American label. Thus,

$$\frac{\Sigma T_j^2}{N} = \frac{8{,}100 + 5{,}184 + 4{,}761}{6} = 3{,}007.50$$

so

$$SS_{\text{TREATMENTS}} = \frac{\Sigma T_j^2}{N} - \frac{(\Sigma X)^2}{kN}$$
$$= 3{,}007.50 - 2{,}964.50 = 43.00$$

The formula for calculating the sum of squares subjects is

$$SS_{\text{SUBJECTS}} = \frac{\Sigma s_i^2}{k} - \frac{(\Sigma X)^2}{kN} \quad [13.4]$$

where s_i^2 is the square of the sum of the scores of subject i and all other terms are as defined previously. For instance, the first subject in Table 13.2 provided a score of 14 in the French condition, 10 in the Italian condition, and 9 in the American condition. The value of s_i^2 for this subject is thus $(14 + 10 + 9)^2 =$ 1,089. The s_i^2 values for the other subjects in our example have been calculated in the last two columns of Table 13.2. We find that s_i^2 equals 1,600 for the second subject, 1,936 for the third, 2,116 for the fourth, 1,369 for the fifth, and 961 for the sixth. Thus,

$$\frac{\Sigma s_i^2}{k} = \frac{1{,}089 + 1{,}600 + 1{,}936 + 2{,}116 + 1{,}369 + 961}{3} = 3{,}023.67$$

so

$$\begin{aligned} SS_{\text{SUBJECTS}} &= \frac{\Sigma s_i^2}{k} - \frac{(\Sigma X)^2}{kN} \\ &= 3{,}023.67 - 2{,}964.50 = 59.17 \end{aligned}$$

The formula for calculating the sum of squares error is

$$SS_{\text{ERROR}} = \Sigma X^2 + \frac{(\Sigma X)^2}{kN} - \frac{\Sigma T_j^2}{N} - \frac{\Sigma s_i^2}{k} \qquad [13.5]$$

where all terms are as defined above. In our example,

$$SS_{\text{ERROR}} = 3{,}081 + 2{,}964.50 - 3{,}007.50 - 3{,}023.67 = 14.33$$

Of course, given that the sum of squares total is equal to the sum of squares treatments plus the sum of squares subjects plus the sum of squares error, if three of these quantities have already been calculated, the fourth can be determined through simple algebraic manipulation. For instance, we could have obtained the sum of squares error as follows:

$$\begin{aligned} SS_{\text{ERROR}} &= SS_{\text{TOTAL}} - SS_{\text{TREATMENTS}} - SS_{\text{SUBJECTS}} \\ &= 116.50 - 43.00 - 59.17 = 14.33 \end{aligned}$$

This relationship reflects the fact that, as noted above, the sum of squares error represents the influence of disturbance variables other than individual differences—that is, variability in the dependent variable remaining after the effects of the independent variable and individual differences have been partitioned out.

Derivation of the Summary Table

Each of the above sums of squares has a certain number of degrees of freedom associated with it. These can be represented as follows: *

$$df_{\text{TREATMENTS}} = k - 1 \qquad [13.6]$$

$$df_{\text{SUBJECTS}} = N - 1 \qquad [13.7]$$

$$df_{\text{ERROR}} = (k - 1)(N - 1) \qquad [13.8]$$

As with the sums of squares, the degrees of freedom are additive:

$$df_{\text{TOTAL}} = df_{\text{TREATMENTS}} + df_{\text{SUBJECTS}} + df_{\text{ERROR}} \qquad [13.9]$$

* The rationale for the equivalence of the degrees of freedom to the indicated quantities is conceptually similar to that discussed in Appendix 12.1 for the degrees of freedom in the between-subjects case.

Since $df_{TREATMENTS} = k - 1$, $df_{SUBJECTS} = N - 1$, and $df_{ERROR} = (k - 1)(N - 1)$, and $(k - 1) + (N - 1) + [(k - 1)(N - 1)] = k - 1 + N - 1 + (kN - k - N + 1) = kN - 1$, the degrees of freedom total can be calculated directly as

$$df_{TOTAL} = kN - 1 \qquad [13.10]$$

In our example,

$$df_{TREATMENTS} = 3 - 1 = 2$$
$$df_{SUBJECTS} = 6 - 1 = 5$$
$$df_{ERROR} = (3 - 1)(6 - 1) = 10$$
$$df_{TOTAL} = (3)(6) - 1 = 17$$

In order to test the null hypothesis of equivalent population means, it is necessary to calculate mean squares for the treatments and error components. These are obtained by dividing the relevant sums of squares by their degrees of freedom:

$$MS_{TREATMENTS} = \frac{SS_{TREATMENTS}}{df_{TREATMENTS}} = \frac{43.00}{2} = 21.50 \qquad [13.11]$$

$$MS_{ERROR} = \frac{SS_{ERROR}}{df_{ERROR}} = \frac{14.33}{10} = 1.43 \qquad [13.12]$$

The mean square subjects, on the other hand, does not directly figure in the F test constituting the hypothesis testing procedure and, thus, is not typically calculated. This is consistent with the fact that the sole function of the sum of squares subjects is to remove variability due to individual differences from the dependent variable so that a more sensitive test of the relationship between the independent and dependent variables can be performed.

The F test for one-way repeated measures analysis of variance follows the same general logic as outlined in Chapter 12 for one-way between-subjects analysis of variance. First, we form an F ratio by dividing the mean square treatments by the mean square error. If the null hypothesis is true, both of these components reflect only sampling error. Thus, we would expect the F ratio, over the long run, to approach 1.00. If the null hypothesis is not true, the mean square error again reflects sampling error but the mean square treatments reflects both sampling error and the effect of the independent variable on the dependent variable. We would thus expect the F ratio, over the long run, to be greater than 1.00. Next, we compare the observed value of F with the appropriate critical value of F from Appendix F. If the observed value of F exceeds the critical value of F, we reject the null hypothesis and conclude that there is a relationship between the independent and dependent variables. Otherwise, we fail to reject the null hypothesis.

The F ratio for one-way repeated measures analysis of variance is *

$$F = \frac{MS_{TREATMENTS}}{MS_{ERROR}} \qquad [13.13]$$

which in our example equals

$$F = \frac{21.50}{1.43} = 15.03$$

As we did in Chapter 12, we can summarize our calculations in a summary table. This would appear as follows:

Source	*SS*	*df*	*MS*	*F*
Treatments	43.00	2	21.50	15.03
Subjects	59.17	5		
Error	14.33	10	1.43	
Total	116.50	17		

When reporting an F value, it is conventional to report the degrees of freedom treatments followed by the degrees of freedom error. For an alpha level of .05 and 2 and 10 degrees of freedom, the critical value of F from Appendix F is 4.10. Since the observed F value of 15.03 is greater than 4.10, we reject the null hypothesis and conclude that a relationship exists between the type of label and the perceived quality of the wine.

STUDY EXERCISE 13.1

An investigator was interested in the effect of exercise on psychological well-being. In order to examine the relationship between these two variables, she placed five volunteers on an aerobic exercise regimen. All of these individuals were of normal weight and health, and none was presently involved in a physical fitness program. Participants responded to a measure of psychological well-being on four occasions: before beginning the exercise regimen and 2, 4, and 6 weeks later. Scores on this measure could range from 1 to 70, with higher scores indicating greater psychological well-being. The data for the study can be found in the accompanying table. Conduct a one-way repeated measures analysis of variance to test for a relationship between the amount of exercise and psychological well-being.

* Analogous to the between-subjects case, the F distribution in the two-group within-subjects situation bears a mathematical relationship to the t distribution such that the values of F obtained using Equation 13.13 will be equal to the squares of the t scores that would be obtained if a correlated groups t test were applied to the same data.

Answer The null and alternative hypotheses can be phrased as follows:

H_0: $\mu_Z = \mu_T = \mu_F = \mu_S$

H_1: The four population means are not all equal

where the subscripts "Z," "T," "F," and "S" represent 0, 2, 4, and 6 weeks of exercise, respectively.

Intermediate statistics necessary for the calculation of the sums of squares are contained in the table. Applying Equations 13.2–13.5, we obtain the following values:

$$SS_{TREATMENTS} = \frac{\Sigma T_j^2}{N} - \frac{(\Sigma X)^2}{kN}$$

$$= 70{,}726.00 - 70{,}567.20 = 158.80$$

$$SS_{SUBJECTS} = \frac{\Sigma s_i^2}{k} - \frac{(\Sigma X)^2}{kN}$$

$$= 71{,}524.00 - 70{,}567.20 = 956.80$$

$$SS_{ERROR} = \Sigma X^2 + \frac{(\Sigma X)^2}{kN} - \frac{\Sigma T_j^2}{N} - \frac{\Sigma s_i^2}{k}$$

$$= 71{,}714 + 70{,}567.20 - 70{,}726.00 - 71{,}524.00 = 31.20$$

$$SS_{TOTAL} = \Sigma X^2 - \frac{(\Sigma X)^2}{kN}$$

$$= 71{,}714 - 70{,}567.20 = 1{,}146.80$$

The degrees of freedom are computed using Equations 13.6, 13.7, 13.8, and 13.10:

$$df_{TREATMENTS} = k - 1 = 4 - 1 = 3$$

$$df_{SUBJECTS} = N - 1 = 5 - 1 = 4$$

$$df_{ERROR} = (k - 1)(N - 1) = (4 - 1)(5 - 1) = 12$$

$$df_{TOTAL} = kN - 1 = (4)(5) - 1 = 19$$

The mean square treatments and the mean square error are then computed by dividing the corresponding sums of squares by their degrees of freedom:

$$MS_{TREATMENTS} = \frac{SS_{TREATMENTS}}{df_{TREATMENTS}} = \frac{158.80}{3} = 52.93$$

$$MS_{ERROR} = \frac{SS_{ERROR}}{df_{ERROR}} = \frac{31.20}{12} = 2.60$$

Finally, the F ratio is derived by dividing $MS_{TREATMENTS}$ by MS_{ERROR}:

$$F = \frac{MS_{TREATMENTS}}{MS_{ERROR}} = \frac{52.93}{2.60} = 20.36$$

These calculations yield the following summary table:

Source	*SS*	*df*	*MS*	*F*
Treatments	158.80	3	52.93	20.36
Subjects	956.80	4		
Error	31.20	12	2.60	
Total	1,146.80	19		

The critical value of F from Appendix F for an alpha level of .05 and 3 and 12 degrees of freedom is 3.49. Since the observed F value of 20.36 is greater than 3.49, we reject the null hypothesis and conclude that a relationship exists between the amount of exercise and psychological well-being.

	0 WEEKS		2 WEEKS		4 WEEKS		6 WEEKS			
INDIVIDUAL	X	X^2	X	X^2	X	X^2	X	X^2	s_i	s_i^2
1	60	3,600	59	3,481	63	3,969	68	4,624	250	62,500
2	52	2,704	53	2,809	58	3,364	61	3,721	224	50,176
3	61	3,721	67	4,489	69	4,761	69	4,761	266	70,756
4	44	1,936	46	2,116	50	2,500	50	2,500	190	36,100
5	63	3,969	62	3,844	66	4,356	67	4,489	258	66,564

$T_Z = 280$, $\bar{X}_Z = 56.00$, $T_Z^2 = 78,400$

$T_T = 287$, $\bar{X}_T = 57.40$, $T_T^2 = 82,369$

$T_F = 306$, $\bar{X}_F = 61.20$, $T_F^2 = 93,636$

$T_S = 315$, $\bar{X}_S = 63.00$, $T_S^2 = 99,225$

$$\Sigma X^2 = 3,600 + 2,704 + \cdots + 2,500 + 4,489 = 71,714$$

$$\Sigma X = 60 + 52 + \cdots + 50 + 67 = 1,188$$

$$\frac{(\Sigma X)^2}{kN} = \frac{1,188^2}{(4)(5)} = 70,567.20$$

$$\frac{\Sigma T_j^2}{N} = \frac{78,400 + 82,369 + 93,636 + 99,225}{5} = 70,726.00$$

$$\frac{\Sigma s_i^2}{k} = \frac{62,500 + 50,176 + 70,756 + 36,100 + 66,564}{4} = 71,524.00$$

Assumptions of the *F* Test

The F test for one-way repeated measures analysis of variance is appropriate when the dependent variable is quantitative in nature and measured on approximately an interval level. Its validity rests on the following assumptions:

1. The sample is independently and randomly selected from the population of interest.
2. Each population of scores is normally distributed.
3. The k populations of scores have homogeneous variances.

4. The variance of the population difference scores for any two conditions is the same as the variance of the population difference scores for any other two conditions. This is known as the **sphericity** assumption. For example, the variance of the scores we would obtain across the population by subtracting individuals' scores in the French condition of the wine label and perceived quality experiment from individuals' scores in the Italian condition is assumed to be the same as the variance of the scores we would obtain by subtracting individuals' scores in the French or Italian conditions from their scores in the American condition.

Under certain conditions, the F test is robust to violations of the normality and homogeneity of variance assumptions. This is particularly true when the sample size is relatively large. However, it is essential that the assumption of independent and random selection be met. In that they tend to increase the Type I error rate beyond the level specified by alpha, violations of the sphericity assumption also tend to be serious. In fact, many statisticians recommend that modifications be made to the traditional F test when there is suspicion that the sphericity assumption is violated. These modifications take the form of multiplying the degrees of freedom treatments and the degrees of freedom error by *adjustment factors* to obtain new degrees of freedom to be used in assessing the significance of the observed F ratio. Two of the most frequently encountered adjustment factors are the *Huynh–Feldt epsilon* and the *Greenhouse–Geisser epsilon*. Both approaches result in adjusted degrees of freedom that are less than or equal to the usual degrees of freedom of $df_{TREATMENTS} = k - 1$ and $df_{ERROR} = (k - 1)(N - 1)$. The use of these adjusted degrees of freedom serves to decrease the Type I error rate and, thus, to increase the robustness of the statistical test to violations of the sphericity assumption. Unfortunately, the formulas for calculating the adjustment factors are complex, and difficult to apply by hand. However, many statistical computer programs provide both the Huynh–Feldt epsilon and the Greenhouse–Geisser epsilon as part of the normal statistical output for repeated measures analysis of variance. Because of certain desirable statistical properties, the Huynh–Feldt procedure is the preferred alternative for most applications.

13.3 Strength of the Relationship

The formula for computing eta-squared for one-way repeated measures analysis of variance is

$$\text{eta}^2 = \frac{SS_{TREATMENTS}}{SS_{TREATMENTS} + SS_{ERROR}} \qquad [13.14]$$

As with the correlated groups t test, eta-squared in the context of one-way repeated measures analysis of variance represents the proportion of variability in the dependent variable that is associated with the independent variable *after variability due to individual differences has been removed.*

The strength of the relationship for the wine label and perceived quality experiment is

$$\text{eta}^2 = \frac{43.00}{43.00 + 14.33} = .75$$

In this experiment, 75% of the variability in the dependent variable (the perceived quality of the wine) is associated with the independent variable (the type of label) after the influence of individual differences has been removed. This represents a very strong effect.

An alternative formula for eta-squared based on the degrees of freedom and the observed value of F is

$$\text{eta}^2 = \frac{(\text{df}_{\text{TREATMENTS}})F}{(\text{df}_{\text{TREATMENTS}})F + \text{df}_{\text{ERROR}}} \quad [13.15]$$

For the wine label and perceived quality experiment,

$$\text{eta}^2 = \frac{(2)(15.03)}{(2)(15.03) + 10} = .75$$

which is the same result as obtained above.

13.4 Nature of the Relationship

The nature of the relationship following a statistically significant one-way repeated measures analysis of variance is addressed using an HSD procedure conceptually identical to the one discussed in Chapter 12. The only differences are that N and MS_{ERROR} replace n and $\text{MS}_{\text{WITHIN}}$ in the formula for the critical difference, and that the value of q is determined with reference to the overall alpha level, the degrees of freedom error, and k (the number of conditions) rather than the overall alpha level, the degrees of freedom within, and k. Specifically, the critical difference is defined as

$$\text{CD} = q\sqrt{\frac{\text{MS}_{\text{ERROR}}}{N}} \quad [13.16]$$

Referring to Appendix G, we find that for an overall alpha level of .05, $df_{ERROR} = 10$, and $k = 3$, q is equal to 3.88. Thus, the critical difference for the wine label and perceived quality experiment is

$$CD = 3.88\sqrt{\frac{1.43}{6}} = 1.89$$

The HSD procedure can now be applied as follows:

Null hypothesis tested	*Absolute difference between sample means*	*Value of CD*	*Null hypothesis rejected?*
$\mu_F = \mu_I$	$\lvert 15.00 - 12.00 \rvert = 3.00$	1.89	Yes
$\mu_F = \mu_A$	$\lvert 15.00 - 11.50 \rvert = 3.50$	1.89	Yes
$\mu_I = \mu_A$	$\lvert 12.00 - 11.50 \rvert = .50$	1.89	No

The nature of the relationship is such that the wine will be rated as tasting better when it is labeled as French ($\bar{X}_F = 15.00$) than when it is labeled as either Italian ($\bar{X}_I = 12.00$) or American ($\bar{X}_A = 11.50$). However, we cannot confidently conclude that the wine will be rated differently as a function of being labeled Italian versus American.

STUDY EXERCISE 13.2

Calculate eta-squared and use the Tukey HSD test to analyze the nature of the relationship for the data in Study Exercise 13.1.

Answer Eta-squared is most easily calculated using Equation 13.14:

$$\text{eta}^2 = \frac{SS_{TREATMENTS}}{SS_{TREATMENTS} + SS_{ERROR}}$$

$$= \frac{158.80}{158.80 + 31.20} = .84$$

This represents a very strong effect.

The first step in applying the HSD test is to calculate the critical difference. Referring to Appendix G, we find that for an overall alpha level of .05, $df_{ERROR} = 12$, and $k = 4$, q is equal to 4.20. Thus,

$$CD = 4.20\sqrt{\frac{2.60}{5}} = 3.03$$

The HSD procedure can now be applied as follows:

Study Exercise 13.2 continued

Null hypothesis tested	*Absolute difference between sample means*	*Value of CD*	*Null hypothesis rejected?*
$\mu_Z = \mu_T$	$\lvert 56.00 - 57.40 \rvert = 1.40$	3.03	No
$\mu_Z = \mu_F$	$\lvert 56.00 - 61.20 \rvert = 5.20$	3.03	Yes
$\mu_Z = \mu_S$	$\lvert 56.00 - 63.00 \rvert = 7.00$	3.03	Yes
$\mu_T = \mu_F$	$\lvert 57.40 - 61.20 \rvert = 3.80$	3.03	Yes
$\mu_T = \mu_S$	$\lvert 57.40 - 63.00 \rvert = 5.60$	3.03	Yes
$\mu_F = \mu_S$	$\lvert 61.20 - 63.00 \rvert = 1.80$	3.03	No

The nature of the relationship is such that psychological well-being will be greater after 4 ($\bar{X}_F = 61.20$) or 6 ($\bar{X}_S = 63.00$) weeks of aerobic exercise than before beginning the exercise regimen ($\bar{X}_Z = 56.00$) or after exercising for only 2 weeks ($\bar{X}_T = 57.40$). However, we cannot confidently conclude that 2 weeks of aerobic exercise affects psychological well-being relative to no exercise, or that 6 weeks of aerobic exercise affects psychological well-being relative to 4 weeks of exercise.

13.5 Methodological Considerations

The sum of squares subjects in the wine label and perceived quality experiment was equal to 59.17. This constitutes a relatively large portion of the total sum of squares (116.50), thus indicating that individual differences played an important role in the taste ratings. One way to examine the extent of this influence is to compare the observed F ratio with the F ratio that would have been observed if the experiment had been conducted using a between-subjects rather than a repeated measures design (that is, if each wine had been rated by six *different* subjects) and the same data hypothetically obtained.

As discussed earlier, since it is not possible to identify the effects of individual differences when a between-subjects analysis is performed, variability due to subjects contributes to the sum of squares within along with all other disturbance variables that are operating. Not surprisingly, then, it turns out that the sum of squares within from a one-way between-subjects analysis of variance is mathematically equal to the total of the sum of squares subjects (which reflects the influence of individual differences) and the sum of squares error (which reflects the influence of disturbance variables other than individual differences) from a one-way repeated measures analysis of variance applied to the same scores. Furthermore, assuming a common data set, it also turns out that the degrees of freedom within in the between-subjects case is mathematically equal to the total of the degrees of freedom subjects and the degrees of freedom error in the repeated measures case, and that the sum of squares and degrees of freedom between are mathematically equal to the sum of squares and degrees of freedom treatments, respectively.

Thus, if the wine label and perceived quality data had been obtained using a between-subjects design, the summary table would take the following form:

Source	*SS*	*df*	*MS*	*F*
Between	43.00	2	21.50	4.39
Within	73.50	15	4.90	
Total	116.50	17		

Note that the *F* ratio is 4.39 as compared with the *F* ratio of 15.03 computed earlier when a repeated measures analysis was used. Since the mean square treatments and the mean square between are identical, the difference in *F* ratios must be due to differences in the denominator of the *F* test. Indeed, a comparison of the two summary tables shows that the denominator of the *F* ratio increased from 1.43 (the mean square error) to 4.90 (the mean square within) when variability due to individual differences was not separately removed from the dependent variable. Although the *F* test in this instance is still statistically significant (since the critical value of *F* for an alpha level of .05 and 2 and 15 degrees of freedom is 3.68), there will be instances where the *F* ratio yielded by a repeated measures analysis of variance will be large enough to lead to rejection of a false null hypothesis, whereas the *F* ratio yielded by the corresponding between-subjects analysis will not be.

On the other hand, the degrees of freedom for the denominator of the *F* test will always be less in the repeated measures case than in the between-subjects case. Since the value of *F* required to reject the null hypothesis becomes more extreme as the degrees of freedom become smaller, repeated measures analysis of variance might actually be less powerful than between-subjects analysis of variance when individual differences have only a minimal influence on the dependent variable. This is parallel to the situation that exists with independent and correlated groups *t* tests, as discussed in Section 11.6. Thus, as noted in that section, it is important that a researcher considering a within-subjects (repeated measures) design accurately anticipate the role of individual differences. In general, however, the null hypothesis is more likely to be correctly rejected when a repeated measures analysis is used than when a between-subjects analysis is used.

In Chapter 11 we saw that one way to deal with carry-over effects is through *counterbalancing*. This involves exposing subjects to *predetermined* sequences of the independent variable such that carry-over effects are evenly distributed across conditions. An alternative to counterbalancing is to *randomly* order the conditions for each subject. This is the approach that was taken in the wine label and perceived quality experiment. The rationale is that when the sequence of conditions across subjects is randomly determined, chance will ensure that each condition occurs in each position an approximately equal number of times and, thus, that carry-over effects are evenly distributed across conditions. This approach is particularly valuable when the number of conditions is so great that counterbalancing is impractical. For instance, there are 24 different ways that four conditions can be ordered and 120 different ways that five conditions can be ordered!* Can you think of some potential carry-over effects that might have been counteracted in the wine label and perceived quality experiment by using the randomization procedure?

13.6

Numerical Example

One area of interest to cognitive psychologists is memory. In order to examine the role that the amount of exposure to a stimulus plays in the memory process, suppose a cognitive psychologist conducts an experiment in which the same list of 10 nonsense syllables (for instance, "blux," "gonk," and "delp") is presented to subjects four different times. The list is presented for 30 seconds on each occasion, and subjects are given 60 seconds to recall as many of the syllables as they can after the list is removed. Test sessions are separated by 10 minutes, during which time subjects work on a math task designed to occupy their thoughts so that they will not be able to practice the nonsense syllables. Hypothetical data for the seven experimental participants are presented in Table 13.3.

The null hypothesis is that there is no relationship between the amount of exposure and recall, while the alternative hypothesis is that there is a relationship. These hypotheses can be formally stated as

$$H_0: \mu_1 = \mu_2 = \mu_3 = \mu_4$$

H_1: The four population means are not all equal

where the subscripts "1," "2," "3," and "4" represent the four test times.

Intermediate statistics necessary for the calculation of the sums of squares are contained in Table 13.3. The sum of squares treatments is

$$SS_{\text{TREATMENTS}} = \frac{\Sigma T_j^2}{N} - \frac{(\Sigma X)^2}{kN}$$
$$= 1{,}306.71 - 1{,}302.89 = 3.82$$

* This can be determined using the formula for permutations (Equation 6.10) presented in Chapter 6: ${}_nP_r = n!/(n - r)!$ For instance, the number of permutations of five conditions taken five at a time is $5!/(5 - 5)! = [(5)(4)(3)(2)(1)]/1 = 120$.

TABLE 13.3 Data and Computation of Sums of Squares for Amount of Exposure and Recall Experiment

SUBJECT	TIME 1 X	TIME 1 X^2	TIME 2 X	TIME 2 X^2	TIME 3 X	TIME 3 X^2	TIME 4 X	TIME 4 X^2	s_i	s_i^2
1	5	25	6	36	6	36	5	25	22	484
2	7	49	6	36	7	49	8	64	28	784
3	8	64	9	81	9	81	10	100	36	1,296
4	3	9	4	16	4	16	6	36	17	289
5	9	81	8	64	9	81	7	49	33	1,089
6	5	25	4	16	6	36	6	36	21	441
7	7	49	10	100	8	64	9	81	34	1,156
	$T_1 = 44$		$T_2 = 47$		$T_3 = 49$		$T_4 = 51$			
	$\bar{X}_1 = 6.29$		$\bar{X}_2 = 6.71$		$\bar{X}_3 = 7.00$		$\bar{X}_4 = 7.29$			
	$T_1^2 = 1{,}936$		$T_2^2 = 2{,}209$		$T_3^2 = 2{,}401$		$T_4^2 = 2{,}601$			

$$\Sigma X^2 = 25 + 49 + \cdots + 36 + 81 = 1{,}405$$
$$\Sigma X = 5 + 7 + \cdots + 6 + 9 = 191$$
$$\frac{(\Sigma X)^2}{kN} = \frac{191^2}{(4)(7)} = 1{,}302.89$$
$$\frac{\Sigma T_j^2}{N} = \frac{1{,}936 + 2{,}209 + 2{,}401 + 2{,}601}{7} = 1{,}306.71$$
$$\frac{\Sigma s_i^2}{k} = \frac{484 + 784 + 1{,}296 + 289 + 1{,}089 + 441 + 1{,}156}{4} = 1{,}384.75$$

The sum of squares subjects is

$$SS_{\text{SUBJECTS}} = \frac{\Sigma s_i^2}{k} - \frac{(\Sigma X)^2}{kN}$$
$$= 1{,}384.75 - 1{,}302.89 = 81.86$$

The sum of squares error is

$$SS_{\text{ERROR}} = \Sigma X^2 + \frac{(\Sigma X)^2}{kN} - \frac{\Sigma T_j^2}{N} - \frac{\Sigma s_i^2}{k}$$
$$= 1{,}405 + 1{,}302.89 - 1{,}306.71 - 1{,}384.75 = 16.43$$

Finally, the sum of squares total is

$$SS_{\text{TOTAL}} = \Sigma X^2 - \frac{(\Sigma X)^2}{kN}$$
$$= 1{,}405 - 1{,}302.89 = 102.11$$

The degrees of freedom are

$$df_{\text{TREATMENTS}} = k - 1 = 4 - 1 = 3$$
$$df_{\text{SUBJECTS}} = N - 1 = 7 - 1 = 6$$

$$df_{ERROR} = (k - 1)(N - 1) = (4 - 1)(7 - 1) = 18$$
$$df_{TOTAL} = kN - 1 = (4)(7) - 1 = 27$$

The relevant mean squares are

$$MS_{TREATMENTS} = \frac{SS_{TREATMENTS}}{df_{TREATMENTS}} = \frac{3.82}{3} = 1.27$$
$$MS_{ERROR} = \frac{SS_{ERROR}}{df_{ERROR}} = \frac{16.43}{18} = .91$$

Thus, the F ratio is

$$F = \frac{MS_{TREATMENTS}}{MS_{ERROR}} = \frac{1.27}{.91} = 1.40$$

These calculations yield the following summary table:

Source	*SS*	*df*	*MS*	*F*
Treatments	3.82	3	1.27	1.40
Subjects	81.86	6		
Error	16.43	18	.91	
Total	102.11	27		

The critical value of F for an alpha level of .05 and 3 and 18 degrees of freedom is 3.16. Since the observed F value of 1.40 is not greater than 3.16, we fail to reject the null hypothesis of equivalent recall during the four test sessions.

As discussed in Section 10.3, it is worthwhile to calculate eta-squared even when the inferential statistical test is nonsignificant. In our example,

$$\text{eta}^2 = \frac{SS_{TREATMENTS}}{SS_{TREATMENTS} + SS_{ERROR}}$$
$$= \frac{3.82}{3.82 + 16.43} = .19$$

Thus, the proportion of variability in recall scores that is associated with the amount of exposure to the nonsense syllables after the effects of individual differences have been removed is .19. Although this indicates a moderate relationship between the independent variable and the dependent variable in the sample, we are unable to conclude that the two variables are related in the population. If the null hypothesis had been rejected, we could have determined the nature of the relationship using the HSD procedure discussed in Section 13.4. As it is, it is interesting to note the high variability due to individual differences extracted from the dependent variable by the sum of squares subjects and to contemplate whether additional disturbance and/or confounding variables might have been operating in the experiment. Given that a sample eta-squared of .19 is indicative of a moderate effect, one possibility is that the amount of exposure to nonsense syllables and recall are actually related in the population but that the sample size ($N = 7$) was not large enough to yield a statistically significant result with the desired degree of consistency.

13.7

Planning an Investigation Using One-Way Repeated Measures Analysis of Variance

Appendix E.3 contains tables of the sample sizes necessary to achieve various levels of power for one-way repeated measures analysis of variance. A set of tables is presented for each research design involving a different $df_{TREATMENTS}$. Table 13.4 presents the portion of this appendix for $df_{TREATMENTS} = 2$ (that is, for three conditions) and an alpha level of .05 for illustration. The first column presents different levels of power and the column headings are values of eta-squared in the population. It should be noted that these values of eta-squared are conceptualized as the proportion of variability in the dependent variable that is associated with the independent variable *after the effects of individual differences have been removed.* Thus, to the extent that the dependent variable is influenced by individual background, the population eta-squared will be greater in the within-subjects case than in the between-subjects case.

As an example of the use of the table, if the desired power level is .80 and the researcher suspects that the strength of the relationship in the population corresponds to an eta-squared of roughly .15, then the number of subjects that should be used in the study is 28. Inspection of Table 13.4 and Appendix E.3 provides a general appreciation for the relationships between $df_{TREATMENTS}$, the

TABLE 13.4 Approximate Sample Sizes Necessary to Achieve Selected Levels of Power for $df_{TREATMENTS} = 2$ and Alpha $= .05$ as a Function of Population Values of Eta-Squared

Degrees of Freedom Treatments = 2
Alpha = .05

	POPULATION ETA-SQUARED									
POWER	.01	.03	.05	.07	.10	.15	.20	.25	.30	.35
.10	32	11	7	5	4	3	2	2	2	2
.50	247	81	48	34	23	15	11	8	7	6
.70	382	125	74	52	36	23	16	13	10	8
.80	478	157	93	65	44	28	20	15	12	10
.90	627	206	121	85	58	37	26	20	16	13
.95	765	251	148	104	70	45	32	24	19	15
.99	1060	347	204	143	97	62	44	33	26	21

alpha level, sample size, the strength of the relationship one is trying to detect in the population, and power for one-way repeated measures analysis of variance.

13.8

Method of Presentation

The method of presentation for a one-way repeated measures analysis of variance parallels that for a one-way between-subjects analysis of variance, the only difference being that the degrees of freedom treatments and the degrees of freedom error are reported instead of the degrees of freedom between and the degrees of freedom within. For example, the results for the wine label and perceived quality experiment discussed earlier in this chapter might be reported as follows:

Results

A one-way repeated measures analysis of variance was performed relating the type of label (French, Italian, or American) to the perceived quality of the wine. The obtained _F_ ratio was statistically significant, _F_(2, 10) = 15.03, _p_ < .01, and the strength of the relationship, as indexed by eta^2, was .75. A Tukey HSD test revealed that the wine was rated significantly higher when it was labeled as French (_M_ = 15.00) than when it was labeled as either Italian (_M_ = 12.00) or American (_M_ = 11.50), but that the means for the latter two conditions did not significantly differ.

13.9

Examples from the Literature

Age Regression and the Magnitude of the Poggendorff Illusion

The use of hypnosis to "return" individuals to an earlier chronological age is known as *age regression*. The technique is based on the assumption that individuals regressed to a particular age will behave as if they were actually that age. Among the types of age-regressed behavior that have been studied is perception. For instance, Parrish, Lundy, and Leibowitz (1969) investigated the effect of age regression on the magnitude of the Poggendorff illusion. This illusion refers to the fact that the right half of a diagonal white bar running upward from left to right through a solid black bar appears to be higher than it actually is.

The magnitude of the Poggendorff illusion decreases with age from age 5 to approximately age 10 and then stabilizes. If age regression affects perceptual behavior, this same pattern of results should thus be observed with age-regressed individuals. In an attempt to determine whether this is indeed the case, Parrish and colleagues exposed a group of 10 college students to each of four conditions: no hypnosis, hypnosis but no age regression, hypnosis and regression to age 5, and hypnosis and regression to age 9. The order of these conditions was randomized such that subjects participated in one condition at a time at weekly intervals. The experimental task involved adjusting the right half of the diagonal bar on a metal Poggendorff figure to make it straight with the stationary left half. The magnitude of the illusion was conceptualized as the difference in inches between where subjects set the right half of the bar and the point at which this portion was actually aligned with the left half. Six trials were undertaken during each experimental session with the average of the six responses constituting the dependent variable.

A one-way repeated measures analysis of variance applied to these data was statistically significant, $F(3, 27) = 4.10$, $p < .05$. The strength of the relationship, as indexed by eta-squared, was .31. This represents a strong effect. Multiple comparisons showed that the Poggendorff illusion was stronger for the hypnosis/regressed-to-age-5 condition ($\bar{X} = 1.41$ inches) than for either the no-hypnosis ($\bar{X} = .96$ inch) or the hypnosis/no-regression ($\bar{X} = 1.03$ inches) condition. (The mean for the hypnosis/regressed-to-age-9 condition was 1.26 inches.) None of the other comparisons were statistically significant. When considered in conjunction with similar findings for a second illusion (the Ponzo illusion), these results suggest that age regression affects perceptual behavior in the expected manner.

Sex-Role Orientation and Perceived Sexual Attractiveness

Sex-role orientation refers to the tendency to exhibit traditionally masculine (for instance, aggression, risk-taking, ambition) versus traditionally feminine (for instance, warmth, sympathy, sensitivity) behaviors. Based on their responses to a self-report inventory, individuals can be classified as falling into one of four sex-role groups: (1) *Masculine* individuals are characteristically high in traditionally masculine behaviors and low in traditionally feminine behaviors. (2) *Feminine* individuals are characteristically high in traditionally feminine behaviors and low in traditionally masculine behaviors. (3) *Androgynous* individuals characteristically exhibit both traditionally masculine and traditionally feminine behaviors. (4) *Undifferentiated* individuals characteristically exhibit neither traditionally masculine nor traditionally feminine behaviors.

Among the results of research on sex-role orientation is the finding that an individual's sex-role standing affects the ways in which he or she is perceived by others. For instance, Becker and Gaeddert (1988) presented 18 male college students with descriptions of masculine, feminine, androgynous, and undifferentiated female targets. For each of the four descriptions, participants were asked to indicate how sexually attractive they thought the target person was. Responses were made on a 9-point scale ranging from "not at all" to "very much." The order of presentation was randomized for each subject.

A one-way repeated measures analysis of variance comparing the mean sexual attractiveness ratings for the four targets was found to be statistically significant, $F(3, 51) = 29.42$, $p < .01$. As indexed by eta-squared, the strength of the relationship was .63. This represents a very strong effect. Application of the HSD test showed that the feminine ($\bar{X} = 5.33$) and the androgynous ($\bar{X} = 6.17$) targets were rated as significantly more sexually attractive than were the masculine ($\bar{X} = 2.78$) and the undifferentiated ($\bar{X} = 2.22$) targets. The means for the feminine versus the androgynous target and the masculine versus the undifferentiated target did not significantly differ.

13.10

Summary

One-way repeated measures analysis of variance is typically used to analyze the relationship between two variables when (1) the dependent variable is quantitative in nature and is measured on approximately an interval level, (2) the independent variable is within-subjects in nature, and (3) the independent variable has three or more levels. One-way repeated measures analysis of variance differs from one-way between-subjects analysis of variance in that it removes the influence of individual differences from the dependent variable. The repeated measures analysis thus yields a more sensitive test of the relationship between the independent and dependent variables. Otherwise, the logic underlying the two techniques is identical. The strength of the relationship is analyzed using eta-squared, and the nature of the relationship is analyzed using the Tukey HSD test.

Exercises

Answers to asterisked (*) exercises appear at the back of the book. Answers to exercises with two asterisks are also worked out step-by-step in the Study Guide.

Exercises to Review Concepts

1. Under what conditions is one-way repeated measures analysis of variance typically used to analyze a bivariate relationship?

2. Distinguish between the sum of squares total, the sum of squares treatments, the sum of squares subjects, and the sum of squares error. How are they interrelated?

***3.** Why is a repeated measures analysis of variance a more sensitive test of the relationship between the independent and dependent variables than a between-subjects analysis of variance?

***4.** Complete the missing entries in the summary table for a one-way repeated measures analysis having five levels of the independent variable and $N = 12$:

Source	*SS*	*df*	*MS*	*F*
Treatments	20.00	—	—	—
Subjects	—	—		
Error	132.00	—	—	
Total	198.00	—		

5. Complete the missing entries in the summary table for a one-way repeated measures analysis of variance having three levels of the independent variable:

Source	*SS*	*df*	*MS*	*F*
Treatments	—	—	20.00	10.00
Subjects	—	9		
Error	—	18	—	
Total	106.00	—		

***6.** For a repeated measures analysis of variance with $N = 21$ and five conditions, what would the degrees of freedom treatments, degrees of freedom subjects, degrees of freedom error, and degrees of freedom total be?

***7.** State the critical value of F that would be used to reject the null hypothesis for a one-way repeated measures analysis of variance at an alpha level of .05 under each of the following conditions:

a. $k = 3, N = 20$
b. $k = 4, N = 21$
c. $k = 3, N = 15$
d. $k = 5, N = 12$

8. What are the assumptions underlying one-way repeated measures analysis of variance?

***9.** How do the Huynh–Feldt epsilon and the Greenhouse–Geisser epsilon increase the robustness of one-way repeated measures analysis of variance to violations of the sphericity assumption?

10. How is the interpretation of eta-squared in one-way repeated measures analysis of variance different from that in one-way between-subjects analysis of variance?

11. Why is a one-way repeated measures analysis of variance generally more powerful than a corresponding one-way between-subjects analysis of variance? When will this not be the case? Why?

12. What is the rationale behind randomly ordering the sequence of conditions across subjects in repeated measures designs? How does this differ from counterbalancing?

***13.** Consider the following scores on an anxiety test that was administered to a group of 5 individuals at three different times:

Individual	*Time 1*	*Time 2*	*Time 3*
1	2	3	4
2	3	5	8
3	4	5	6
4	5	7	9
5	5	6	4

Conduct a one-way repeated measures analysis of variance to test for a relationship between the time of assessment and anxiety scores.

***14.** Compute the value of eta-squared for the data in Exercise 13. Does the observed value represent a weak, moderate, or strong effect?

***15.** Analyze the nature of the relationship for the data in Exercise 13 using the HSD test.

***16.** Analyze the data in Exercise 13 as if the independent variable were between-subjects in nature. That is, conduct a one-way between-subjects analysis of variance, compute the value of eta-squared, and use the HSD test to analyze the nature of the relationship. Compare your findings with those in Exercises 13–15. How does this illustrate the advantage of within-subjects research designs?

17. Conduct a one-way repeated measures analysis of variance for the following data:

Subject	Condition A	Condition B	Condition C
1	5	7	2
2	9	7	5
3	8	5	6
4	6	5	4
5	9	6	6

18. Compute the value of eta-squared for the data in Exercise 17 using Equation 13.14. Recalculate this index using Equation 13.15. Compare the two sets of results. Does the observed value represent a weak, moderate, or strong effect?

19. Analyze the nature of the relationship for the data in Exercise 17 using the HSD test.

For each of the studies described in Exercises 20–23, indicate the appropriate statistical test for analyzing the relationship between the independent and dependent variables. Assume that the underlying assumptions of the tests have been satisfied.

***20.** Kelman and Hovland (1953) studied the effects of the source of a persuasive communication on attitude change. Three groups of subjects listened to a persuasive message on the treatment of juvenile delinquents. For one group of subjects, the message was attributed to a trustworthy and well-informed source, for another group it was attributed to an untrustworthy and poorly informed source, and for the third group it was attributed to a "neutral" source. The amount of attitude change was computed for each individual by subtracting indices of attitude after hearing the message from indices of attitude before hearing the message. The mean change scores for the three groups were compared.

***21.** A researcher wanted to test the effects of age on infants' memory capacities. Forty infants were given a memory test at the age of 5 months and again at the age of 7 months. Scores on this test could range from 1 to 15. The mean test scores were compared for the two ages.

22. An investigation was undertaken to determine the effects of different types of drugs on driving skills. Thirty subjects participated in a driving simulation task under each of three conditions: (1) while under the influence of small amounts of alcohol, (2) while under the influence of small amounts of marijuana, and (3) while not under the influence of any drug. The simulation task yielded "driving scores" that could range from 1 to 100. The mean driving scores were compared for the three conditions.

23. Advertisements often present results of surveys about medical products in the following format: "8 out of 10 doctors recommend . . . ." A researcher was interested in whether the base in which such information was presented influences its impact on consumer preferences. Two groups of individuals were presented with identical descriptions of a brand of aspirin. However, in one group, subjects were told that "8 out of 10 doctors recommended the aspirin," whereas in the other group subjects were told that "80 out of 100 doctors recommended the aspirin." Although the proportion of doctors recommending the aspirin was identical for both groups, the number of doctors differed. For each subject a measure of attitude toward the aspirin was obtained on a scale of 1 to 10. The mean attitude scores were compared for the two groups.

24. If a researcher suspected that the strength of the relationship between two variables in the population was .10 as indexed by eta-squared, what sample size should he use in a study involving a within-subjects independent variable with five levels and an alpha level of .05 in order to achieve power of .80?

***25.** Suppose an investigator reported the results of a study using a within-subjects independent variable with four levels with $N = 12$. If the value of eta-squared in the population was .15, what would the power of her statistical test be at an alpha level of .05?

Exercises to Apply Concepts

****26.** Population growth trends have a dramatic impact on society, both economically and politically. Behavioral scientists have studied extensively factors that influence population growth. One area of research has been the individual's decision to use birth control. Although numerous studies have examined the factors that females take into consideration when choosing a birth control method, relatively few studies have examined birth control from the standpoint of males. Male oral contraceptives will probably be available to the general public in the next 10 years. Jaccard (1980) conducted a study to discover those factors that males would be most concerned about when evaluating male oral contraceptives. In this investigation, subjects rated how important each of four factors would be to them in deciding whether to use an oral contraceptive. The ratings were made on a 21-point scale, with higher numbers indicating greater degrees of importance. The data presented below are representative of those obtained by Jaccard. Analyze these data, draw a conclusion, and write up your results using the principles given in the Method of Presentation section.

Individual	*Health risks*	*Effectiveness*	*Cost*	*Convenience*
1	20	12	8	8
2	19	15	11	11
3	18	14	10	10
4	17	13	9	9
5	16	16	12	12

27. Evidence suggests that many rape victims fail to report their victimization to the police or other public authorities. For instance, a survey conducted in five large metropolitan areas found that only about 50% of most crimes—including rape—are ever reported to the police. With this in mind, Feldman-Summers and Ashworth (1980) conducted an experiment to identify factors that influence intentions to report sexual victimization to various agencies and individuals. As part of this study, each participant was asked the likelihood that she would report a rape to her husband/boyfriend, the police, her parents, and a female friend. A separate probability judgment was obtained for each source. The judgments were made on a 7-point scale, with 1 labeled "very unlikely I would report the crime to the source" and 7 labeled "very likely I would report the crime to the source." Hypothetical data for this study representative of the results of Feldman-Summers and Ashworth are presented below. Analyze these data, draw a conclusion, and write up your results using the principles given in the Method of Presentation section.

Individual	*Husband/ boyfriend*	*Police*	*Parents*	*Female friend*
1	7	6	3	4
2	6	5	4	5
3	7	6	1	2
4	6	5	6	7
5	7	6	3	4
6	6	5	4	5

FOURTEEN

PEARSON CORRELATION AND REGRESSION: INFERENTIAL ASPECTS

14.1

Use of Pearson Correlation

In Chapter 5, we noted the conditions under which Pearson correlation is used to determine the extent of linear relationship between two variables. Formally stated, Pearson correlation is typically used to analyze the relationship between two variables when

1. both variables are quantitative in nature and are measured on approximately an interval level;
2. the two variables have been measured on the same individuals; and
3. the observations on each variable are *between-subjects* in nature.

Our discussion of correlation in Chapter 5 focused on the description of the relationship between two variables. In practice, however, the most common use of correlation procedures involves making inferences about correlation coefficients in populations based on sample data. For instance, the correlation between traditionalism and ideal family size for the 10 individuals discussed in Chapter 5 was .66. The question of interest is whether we can conclude that a linear relationship exists between these two variables in the *population,* given a correlation of this magnitude in the observed *sample.* The present chapter considers procedures for drawing inferences about population correlation coefficients from sample correlation coefficients in situations of this type.

14.2

Inference of a Relationship Using Pearson Correlation

As noted in previous chapters, even if a relationship is observed between two variables in a set of sample data, this does not necessarily mean that a relationship exists between the variables in the corresponding population. A relationship might exist in a sample even though it does not exist in the population, because of sampling error. The task at hand, then, is to test the viability of a sampling error interpretation.

Null and Alternative Hypotheses

Recall that a correlation coefficient can range from -1.00 to $+1.00$. A coefficient of 0 means that there is no linear relationship between the variables. In contrast, nonzero correlation coefficients indicate some approximation to a linear relationship. We can therefore state the null and alternative hypotheses in the nondirectional case as follows:

$$H_0: \quad \rho = 0$$
$$H_1: \quad \rho \neq 0$$

where ρ (lowercase Greek *r*, called "rho") represents the true correlation in the population. The null hypothesis states that there is no linear relationship between the two variables—that is, that the population correlation between the two variables is 0. The alternative hypothesis states that the correlation between the two variables is not 0 in the population. If we were interested in a directional rather than a nondirectional relationship, the alternative hypothesis would state either that rho is greater than 0 (that is, that the two variables are positively correlated in the population) or that rho is less than 0 (that is, that the two variables are negatively correlated in the population). Following the logic developed when discussing the power of nondirectional and directional *t* tests, directional hypotheses should be used only when there is clear reason for doing so.

Based on the data collected in an investigation, we want to choose between the two hypotheses. We will again use the logic developed in Chapter 8. In the present context, this involves converting the value of *r* in the sample into a *t* value under the assumption that the null hypothesis is true and determining whether this value of *t* falls in the rejection region. If it does, then we will reject the null hypothesis and conclude that there is a linear relationship between the two variables. Otherwise, we will fail to reject the null hypothesis.

Sampling Distribution of a Correlation Coefficient

Consider the case of a population in which the correlation between two variables is 0. From the population, we select a random sample of 10 individuals and compute the correlation coefficient. We might find that $r = .15$. The fact that *r* does not equal 0 is due to sampling error, as discussed in previous chapters. Now suppose we randomly select another sample of 10 individuals from the population. For this sample, the correlation might be .03. In principle, we could continue this process a large number of times until we have calculated *r* for all possible samples of size 10. The resulting distribution of correlation coefficients based on all random samples of size 10 would constitute a **sampling distribution of the correlation coefficient** and would have many of the same properties as the sampling distributions we have discussed in prior chapters.

The mean of a sampling distribution of the correlation coefficient is approximately ρ, the true population correlation, and the standard deviation of the sampling distribution, or *standard error of r*, is $\sigma_r = \sqrt{(1 - \rho^2)/(N - 2)}$.

FIGURE 14.1 Sampling Distributions of r for Three Values of ρ for $N = 10$ (From Minium, 1970)

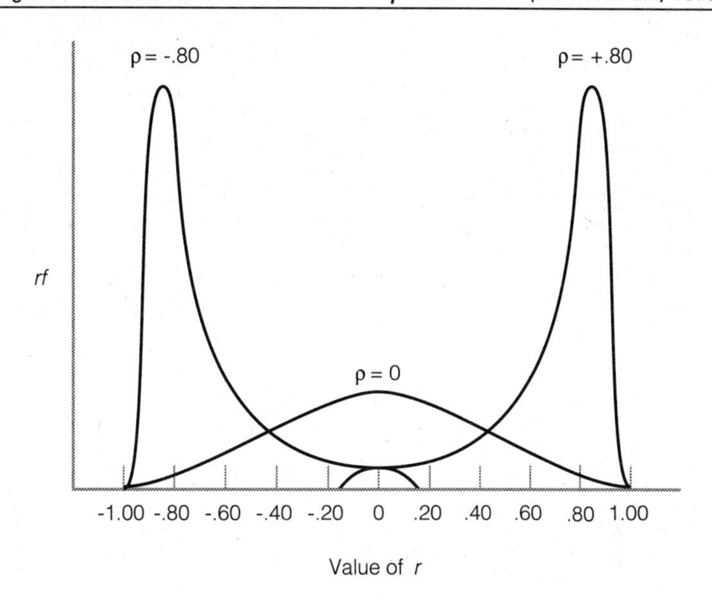

When $\rho = 0$, the sampling distribution is symmetrical and approximately normal in shape. When $\rho \neq 0$, the sampling distribution is skewed. Figure 14.1 illustrates sampling distributions for $\rho = -.80$, $\rho = 0$, and $\rho = +.80$, for $N = 10$.

To estimate σ_r when ρ is unknown, we can use the formula

$$\hat{s}_r = \sqrt{\frac{1 - r^2}{N - 2}} \qquad \textbf{[14.1]}$$

where $\hat{s}_r$ is the estimated standard error of r, r is the observed sample correlation, and N is the sample size. Applying this equation to the example on traditionalism and ideal family size from Chapter 5 (where $r = .66$ and $N = 10$), we find that the estimated standard error of r is

$$\hat{s}_r = \sqrt{\frac{1 - .66^2}{10 - 2}} = .27$$

Testing the Significance of a Correlation Coefficient

The values yielded by the following formula for testing the null hypothesis that $\rho = 0$ have been shown to be approximately distributed as a t distribution with $N - 2$ degrees of freedom: *

$$t = \frac{r - \rho}{\hat{s}_r} \qquad \textbf{[14.2]}$$

* Null hypotheses other than $\rho = 0$ can be tested using the procedure discussed in Appendix 14.1.

In this formula, r and $\hat{s}_r$ are as defined above, and ρ is the hypothesized value of the correlation in the population, which in the present context is always equal to 0. For the traditionalism and ideal family size investigation,

$$t = \frac{.66 - 0}{.27} = 2.44$$

For an alpha level of .05, nondirectional test, and $N - 2 = 10 - 2 = 8$ degrees of freedom, the critical values of t from Appendix D are ± 2.306. Since the observed t value is greater than $+2.306$, we reject the null hypothesis. A sample correlation coefficient of .66 based on 8 degrees of freedom is too large to attribute to sampling error. We therefore conclude that the population correlation is nonzero—that is, that there is a relationship between traditionalism and ideal family size.

Tabled Values of *r*

When Equation 14.2 is used to test the null hypothesis that $\rho = 0$, the numerator will be equal to $r - 0 = r$. Since the denominator, $\hat{s}_r = \sqrt{(1 - r^2)/(N - 2)}$, depends only on the values of r and N, it is possible to determine critical values of r directly by substituting the value of N, which will always be known, and the corresponding critical values of t into Equation 14.2 and solving for r. This has been done in Appendix H, which contains a table of values of r that lead to rejection of the null hypothesis that $\rho = 0$ at various levels of alpha for selected degrees of freedom of $\text{df} = N - 2$. If the observed value of r is greater than the positive critical value of r or less than the negative critical value of r, we will reject the null hypothesis. Otherwise, we will fail to reject the null hypothesis. For instance, based on $N - 2 = 10 - 2 = 8$ degrees of freedom and an alpha level of .05, nondirectional test, the critical values of r for the traditionalism and ideal family size example are $\pm .632$. Since the observed correlation of .66 is greater than $+.632$, we reject the null hypothesis and conclude that there is a relationship between traditionalism and ideal family size. This is the same conclusion we reached using the t test procedure. Given its greater computational ease, we will follow the strategy of referring to Appendix H when testing the significance of correlation coefficients in the remainder of this chapter.

STUDY EXERCISE 14.1

In Study Exercise 5.2, the correlation between voters' perceptions that a candidate supported labor unions and their willingness to vote for her was calculated to be .50 for a sample of nine individuals. Test the viability of the hypothesis that there is no relationship between the two variables in the population.

Answer The null and alternative hypotheses are

$$H_0: \quad \rho = 0$$
$$H_1: \quad \rho \neq 0$$

Study Exercise 14.1 continued

For an alpha level of .05, nondirectional test, and $N - 2 = 9 - 2 = 7$ degrees of freedom, the critical values of r from Appendix H are $\pm.666$. Since .50 is neither greater than $+.666$ nor less than $-.666$, we fail to reject the null hypothesis of no linear relationship between the two variables.

Assumptions of Pearson Correlation

The test of the null hypothesis that $\rho = 0$ is based on the following assumptions:

1. The sample is independently and randomly selected from the population of interest.
2. The population distributions of X and Y are such that their joint distribution (that is, their scatterplot) represents a **bivariate normal distribution.** This means that the distribution of Y scores at any value of X is normal in the population (see Figure 14.2), and vice versa.
3. The variability of Y scores is the same at each value of X, and vice versa.

In addition, both variables should be quantitative in nature and measured on approximately an interval level. While it is essential that the first assumption not be violated, under certain conditions Pearson correlation is robust with respect to violations of the last two. As discussed in detail in Jaccard and Becker (1988), this is particularly true when the sample size is relatively large.

FIGURE 14.2 Distribution of Y Scores at Given Values of X for a Bivariate Normal Distribution (Adapted from Glass & Stanley, 1970)

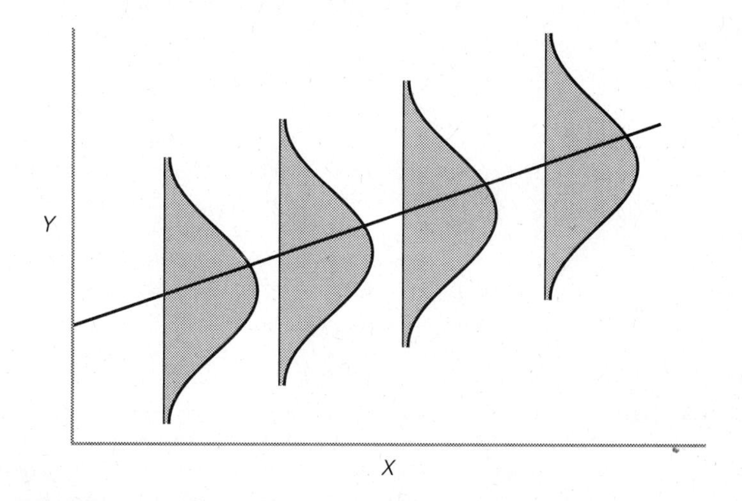

14.3

Strength of the Relationship

The strength of the relationship between two variables in a correlational analysis can be represented by eta-squared in the form of the ratio of explained variance ($SS_{EXPLAINED}$) to total variance (SS_{TOTAL}). However, it is not necessary to actually calculate $SS_{EXPLAINED}$ and SS_{TOTAL} since eta-squared bears a direct relationship to the correlation coefficient. Specifically,

$$r^2 = \text{eta}^2 = \frac{SS_{EXPLAINED}}{SS_{TOTAL}} \qquad [\mathbf{14.3}]$$

Given the nature of correlational analysis, eta-squared typically has a somewhat different interpretation in the context of Pearson correlation than with the previous tests that we have considered. Rather than representing the proportion of variability in the dependent variable that is associated with the independent variable, r^2 represents the proportion of variability that is *shared by the two variables*. This interpretation reflects the fact that the independent variable–dependent variable distinction is not relevant to most applications of correlational analysis. As we have seen, the usual focus of Pearson correlation is the extent to which two variables *share a linear relationship*, not whether one variable influences a second variable. Of course, if one variable can be identified as the independent variable and one variable identified as the dependent variable in a given correlation problem, the previous interpretation of eta-squared (r^2) as the proportion of variability in the dependent variable that is associated with the independent variable will apply. We will consider such an instance later in this chapter.

In the example on traditionalism and ideal family size, r was found to equal .66. Thus, r^2, formally known as the **coefficient of determination,** is equal to

$$r^2 = .66^2 = .44$$

The proportion of variability in this investigation that is shared by traditionalism and ideal family size is .44, as indexed by a linear model. This represents a strong effect.

14.4

Nature of the Relationship

The nature of the relationship between two correlated variables is determined through examination of the sign of the correlation coefficient observed in the sample. If the null hypothesis is rejected and the sample correlation coefficient is positive, the appropriate conclusion is that the population correlation coefficient is also positive. On the other hand, if the null hypothesis is rejected and the sample correlation coefficient is negative, the appropriate conclusion is that the two variables are inversely related in the population. Aside from this, no additional analyses are performed to discern the nature of the relationship. This is because the correlation coefficient assesses the extent to which two variables approximate a *linear* relationship and, hence, the nature of the relationship is built in. Given an observed correlation of .66 in the traditionalism and ideal family size example, we conclude that a direct linear relationship exists between these two variables.

As discussed in Chapters 5 and 9, it is important to remember that the fact that two variables are related does not necessarily imply that one variable *causes* the other to vary as it does. Since the correlation between variable X and variable Y might be due to X causing Y, Y causing X, or some additional variable(s) causing both X and Y, caution is warranted when drawing causal inferences from correlational analyses.

14.5 Planning an Investigation Using Pearson Correlation

Appendix E.4 contains tables of the sample sizes necessary to achieve various levels of power for Pearson correlation for tests of the null hypothesis that $\rho = 0$. Table 14.1 presents the portion of this appendix for an alpha level of .05, nondirectional test, for purpose of exposition. The first column presents different

TABLE 14.1 Approximate Sample Sizes Necessary to Achieve Selected Levels of Power for Alpha = .05 as a Function of the Population Correlation Coefficient Squared

Nondirectional Test, Alpha = .05

	POPULATION CORRELATION COEFFICIENT SQUARED									
POWER	.01	.03	.05	.07	.10	.15	.20	.25	.30	.35
.25	166	56	34	25	17	12	9	8	6	6
.50	384	127	76	54	38	25	19	15	12	10
.60	489	162	97	69	48	31	23	18	15	13
.67	570	188	112	80	55	36	27	21	17	14
.70	616	203	121	86	59	39	29	23	18	15
.75	692	228	136	96	67	43	32	25	20	17
.80	783	258	153	109	75	49	36	28	23	19
.85	895	294	175	124	85	56	41	32	26	21
.90	1046	344	204	144	100	65	47	37	30	25
.95	1308	429	255	180	124	80	58	46	37	30
.99	1828	599	355	251	172	111	81	63	50	42

levels of power and the column headings are population values of the correlation coefficient squared (that is, rho-squared). As an example of the use of this table, if the desired level of power is .80 and the researcher suspects that the strength of the relationship in the population corresponds to a rho-squared of roughly .15, then the number of subjects that should be sampled is 49. Inspection of Table 14.1 and Appendix E.4 provides a general appreciation for the relationship between directionality, the alpha level, sample size, the strength of the relationship one is trying to detect in the population, and power for Pearson correlation.

14.6 Method of Presentation for Pearson Correlation

Reports of a correlational analysis should include statements of the degrees of freedom, the observed value of *r*, and the significance level. It is conventional to also indicate the mean of each variable. Although it is acceptable to do so, it is not necessary to explicitly state the strength and nature of the relationship between the variables as these are indicated by the square and sign of the correlation coefficient, respectively. The results for the investigation on traditionalism and ideal family size discussed earlier might be reported as follows: *

Results

A Pearson correlation addressed the relationship between traditionalism ($\underline{M}$ = 5.00) and ideal family size ($\underline{M}$ = 5.00). This was found to be statistically significant, $\underline{r}(8) = .66$, $\underline{p} < .05$, indicating that these two variables are positively related.

14.7 Examples from the Literature

Personality Characteristics and Creativity

Psychologists have studied extensively the process of creativity and what distinguishes creative individuals from noncreative individuals. For instance, Barron (1965) reported the results of an investigation designed to illuminate person-

* The reported sample means are as derived in Chapter 5.

ality differences between certain creative and noncreative persons. One part of this investigation focused on groups of professionals, including 44 female mathematicians who were invited to attend a 3-day interview session at the Institute of Personality Assessment and Research (IPAR) at the University of California at Berkeley. These individuals were chosen on the basis of a "nomination" technique and consisted of 16 women considered by a panel of mathematics experts to be "the most original and important women mathematicians in the United States and Canada," and 28 women who were nationally recognized but not distinctively creative in the mathematics field.

While at IPAR, the mathematicians interacted in both formal and informal settings with the staff of the institute, who were kept unaware of who the creative versus noncreative women were. The staff rated each woman on a number of personality characteristics and these ratings were correlated with ratings of the women's creativity as previously generated by the above-mentioned panel of mathematics experts. The strongest correlations, all of which are statistically significant, were as shown in the accompanying table. In each case, these represent moderate or strong effects.

POSITIVE CORRELATIONS

Personality characteristic	r	r^2
Thinks and associates to ideas in unusual ways; has unconventional thought processes.	.64	.41
Is an interesting, arresting person.	.55	.30
Tends to be rebellious and nonconforming.	.51	.26
Genuinely values intellectual and cognitive matters.	.49	.24
Appears to have a high degree of intellectual capacity.	.46	.21
Is self-dramatizing; histrionic.	.42	.18
Has fluctuating moods.	.40	.16

NEGATIVE CORRELATIONS

Personality characteristic	r	r^2
Judges self and others in conventional terms like "popularity," the "correct thing to do," "social pressures," and so forth.	−.62	.38
Is a genuinely dependable and responsible person.	−.45	.20
Behaves in a sympathetic or considerate manner.	−.43	.18
Favors conservative values in a variety of areas.	−.40	.16
Is moralistic.	−.40	.16

Barron summarized the results as follows: "The emphasis is upon genuine unconventionality, high intellectual ability, vividness or even flamboyance of character, moodiness and preoccupation, courage, and self-centeredness. These are people who stand out, and who probably are willing to strike out if impelled to do so."

Desire to Control and Gambling Frequency

Among the factors that have been hypothesized to influence gambling behavior is the belief that one can exert personal control over outcomes that are largely determined by chance. This belief, referred to as *illusion of control,* is supposedly induced by elements of gambling situations that hint at potential control (for instance, the observation of horses during prerace warm-ups or the decision to hold or draw cards at a poker game).

Burger and Smith (1985) proposed that one's characteristic desire to control events in one's life should relate to gambling behavior on games that hold an element of illusion of control but not on games that do not. To test this hypothesis, they asked 18 members of Gamblers Anonymous to complete a scale assessing characteristic desire to control and to indicate the frequency with which they had bet on a number of gambling games. Some of these games (casino games, lotteries) were conceptualized as not inducing an illusion of control, whereas others (poker and card games, horse racing, sports events) were conceptualized as capable of inducing such an illusion. For each type of game, a total frequency score was created by adding the appropriate frequency ratings. As predicted, desire-to-control scores were found to correlate significantly with gambling frequency on illusion-of-control games, $r(16) = .46$, $p < .05$, but not on the other games, $r(16) = .04$, *ns*. As indexed by r^2, the strength of the relationships were .21 (representative of a moderate effect) and .002 (representative of a weak effect), respectively.

14.8 Regression

In Chapter 5, we noted that a regression equation can be used to identify the value of Y that is predicted to be paired with an individual's score on X. Our discussion at that time focused on the scores of individuals who were members of the group on which the regression equation was based. An important characteristic of regression is that prediction procedures can be extended to individuals who were not included in the original data set. This is accomplished as follows. Scores on X and Y are determined for a sample of individuals. The procedures presented in Chapter 5 are then used to derive a regression equation based on the sample data. This equation can then be applied to individuals outside the original sample to make predictions about their scores on Y from their scores on X. This is accomplished by substituting an individual's X score into the regression equation. The resulting value of $\hat{Y}$ is that individual's predicted score on Y.

In the context of regression, prediction merely refers to the fact that we are making inferences about one variable from a second variable and does not imply that the latter variable necessarily precedes the former. For instance, individuals might formulate their opinions about ideal family size (variable Y) several years before they develop their traditionalism orientation (variable X). Nevertheless, we can use regression procedures to get an idea of what a person's ideal family size might be given knowledge of her traditionalism score. The variable being predicted, Y, is formally known as the *dependent* or **crite-**

rion variable. The variable from which predictions are made, X, is formally known as the *independent* or **predictor variable.***

Consider the following example. The personnel director of an insurance company has rated every underwriter who has worked for the company during the past 5 years on a scale of 1 to 10, indicating how successful the underwriter was during his or her tenure. The higher the rating, the more successful the underwriter. Each underwriter was also administered a test designed to measure his or her potential on the job. The test scores could range from 0 to 100, with higher scores indicating greater potential. Suppose correlational analysis yielded a Pearson correlation of .80 between test scores (X) and success ratings (Y). A correlation coefficient of this magnitude indicates a very strong linear relationship between test scores and job success. The personnel director might therefore construct a regression equation to help him screen future applicants for underwriting jobs by predicting who is likely to be successful. Suppose this equation took the form

$$\hat{Y} = .50 + .10X$$

Further suppose an applicant took the test and obtained a score of 85. If we substitute this score into the regression equation, we find a predicted success rating of

$$\hat{Y} = .50 + (.10)(85) = 9.00$$

A success rating of 9.00 is very high; thus, based on this criterion, this applicant stands a good chance of being hired. On the other hand, the predicted success rating for an applicant who obtained a test score of 40 is

$$\hat{Y} = .50 + (.10)(40) = 4.50$$

A success rating of 4.50 indicates low job success, so this individual will probably not be considered further for a position with the company. The success ratings for other applicants can be similarly estimated by substituting their test scores into the regression equation.

The Estimated Standard Error of Estimate

Unless the two variables are perfectly correlated, the use of a regression equation to predict scores on Y from scores on X will have some degree of error associated with it. For instance, the regression equation for predicting ideal family size from traditionalism was found in Chapter 5 to be $\hat{Y} = 2.10 + .58X$. The predicted ideal family size for a traditionalism score of 8 according to this equation is $2.10 + (.58)(8) = 6.74$, or, rounded to the nearest whole number, 7. Surely, though, not every woman obtaining a score of 8 on the traditionalism questionnaire desires a family this large. What is needed in order to gain insight

* While the use of X to represent the independent (predictor) variable in the case of regression differs from the use of X to represent the dependent variable as presented in previous chapters, this notation is consistent with conventional practice in both instances.

into the predictive utility of a regression equation is an index of how much error will occur when predicting Y from X. Such an index is provided by the **estimated standard error of estimate.**

When we introduced the concept of the standard error of estimate in Chapter 5, it was defined as

$$s_{YX} = \sqrt{\frac{\Sigma(Y - \hat{Y})^2}{N}}$$

where s_{YX} is the standard error of estimate, Y is an individual's actual score on Y, $\hat{Y}$ is an individual's predicted score on Y based on the regression equation, and N is the sample size (see Equation 5.12). While this formula represents the average amount of predictive error *for the original data set,* it is not an appropriate measure of how much error will occur when predicting the scores on Y of other individuals. When regression is used to make predictions about the Y scores of people outside the original sample, the population value of the standard error of estimate (σ_{YX}) can be estimated from sample data as follows:

$$\hat{s}_{YX} = \sqrt{\frac{\Sigma(Y - \hat{Y})^2}{N - 2}} \qquad [14.4]$$

where $\hat{s}_{YX}$ is the estimated standard error of estimate and all other terms are as previously defined. The only difference between this formula and the formula for the sample standard error of estimate from above is that the denominator reflects the associated degrees of freedom ($N - 2$) rather than the sample size (N). This is equal to $N - 2$ because both the slope and the intercept of the regression line must be estimated from sample data. The estimated standard error of estimate thus estimates the average error that will be made across individuals when predicting scores on Y from the regression equation.

Analogous to the computational formulas for the sample standard error of estimate presented in Chapter 5 (Equations 5.13 and 5.14), in practice the estimated standard error of estimate is calculated using the equation

$$\hat{s}_{YX} = \hat{s}_Y \sqrt{\left(\frac{N - 1}{N - 2}\right)(1 - r^2)} \qquad [14.5]$$

if one is working with the standard deviation estimate for Y, and the equation

$$\hat{s}_{YX} = \sqrt{\frac{SS_Y(1 - r^2)}{N - 2}} \qquad [14.6]$$

if one is working with the sum of squares for Y.

The interpretation of $\hat{s}_{YX}$ directly follows that for s_{YX}. First, its absolute magnitude provides a measure of the amount of predictive error that will occur. Second, $\hat{s}_{YX}$ can be compared with the estimated standard deviation of Y. The estimated standard deviation of Y estimates what the average error in prediction would be if one were to predict everyone had a Y score equal to the mean of Y. If X helps to predict Y, then the estimated standard error of estimate will be smaller than the estimated standard deviation of Y, and the better the predictor X is, the smaller the estimated standard error of estimate will be.

14.9 Method of Presentation for Regression

There are several ways in which the results of a regression analysis might be presented. One approach is to include the regression equation along with the statistics for the correlation. For the example on voters' perceptions that a candidate supported labor unions and willingness to vote for her referred to in Study Exercise 14.1, this might appear as follows: *

Results

A Pearson correlation between voters' perceptions that a candidate supported labor unions (M = 8.00) and their willingness to vote for her (M = 5.00) was statistically nonsignificant, r(7) = .50, ns. The regression equation for predicting willingness to vote for the candidate from perceptions that she supported labor unions was found to be Ŷ = 1.00 + .50X.

On occasion, written reports of regression results are supplemented by a scatterplot showing the observed data points and the regression line.

In practice, regression analyses involving only one predictor variable are rarely encountered in the behavioral sciences. Far more common is the practice of predicting a criterion variable from two or more predictor variables. This technique, known as *multiple regression,* is reviewed in Chapter 18.

* The reported sample means and quantities in the regression equation are as derived in Study Exercise 5.3.

14.10

Issues Associated with the Use of Correlation and Regression

Nonlinear Relationships

There are many ways in which two variables might be related. For instance, Figure 14.3 illustrates a *curvilinear relationship* between two variables. Such a relationship has been shown to represent, for example, the association between motivation and task performance—performance is best at moderate levels of motivation and increasingly deteriorates as motivation decreases or increases past this point.

Two variables might be related, but if they are related in a fashion that is *nonlinear,* Pearson correlation will not be sensitive to this. Thus, a Pearson correlation coefficient computed on the data in Figure 14.3 would be near 0 even though there is clearly a strong curvilinear relationship. This is because Pearson correlation is only concerned with linear relationships. In a similar vein, the utility of using the type of regression analysis presented in this chapter for predicting scores on variable Y from scores on variable X decreases as the relationship between X and Y deviates from linearity. The extent of nonlinear relationship can, however, be determined using a technique known as *curvilinear* or *polynomial regression,* and this same technique can also be used in a predictive fashion (see Pedhazur, 1982).

Predicting X from Y

A second consideration to bear in mind when contemplating the application of regression procedures is that *the regression equation for predicting X from Y is not the same as the regression equation for predicting Y from X.* This is be-

FIGURE 14.3 Scatterplot of a Curvilinear Relationship Between Two Variables

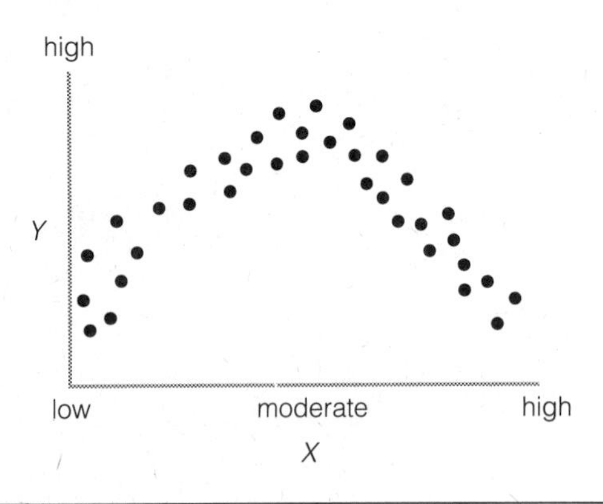

cause the least squares criterion defines the regression line such that the sum of the squared discrepancies between actual Y scores and predicted Y scores is minimized. If we wanted to use our knowledge of Y to predict scores on X, we would have to identify the line that minimizes the sum of the squared discrepancies between actual and predicted X scores. Unless the relationship between the two variables is perfect ($r = -1.00$ or $+1.00$), this would be a different line than the line for predicting Y from X. In contrast, r and r^2 have the same interpretation regardless of whether we are predicting X from Y or Y from X, because these indices are concerned with the extent to which two variables *share* a linear relationship.

From a *statistical* perspective, the designation of one variable as the predictor variable and one variable as the criterion variable is arbitrary—it is as easy to derive the line for predicting the first variable from the second as it is to derive the line for predicting the second variable from the first. This involves merely reversing which variable is labeled X and which variable is labeled Y and applying the usual formulas for the slope and intercept. From a *conceptual* perspective, however, the designation of one variable as the predictor and one variable as the criterion has important implications. After all, it is one thing to predict individuals' job success from their aptitude test scores and another thing to predict individuals' aptitude test scores from their job success. Chances are that the insurance company personnel director referred to earlier would himself be looking for a job if he took the latter focus. The use of regression presupposes an underlying rationale for making predictions about variable Y from variable X rather than vice versa. If interest is merely in whether a given variable is *related* to another variable, Pearson correlation can be applied.

Restricted Range

A third issue of note is the effect on the observed correlation of examining only a portion of the range of a variable. Depending on the particular circumstances, the linear approximation when a limited portion of this range is considered might be either lesser or greater than if the range had not been so restricted. Suppose, for instance, that we were interested in the correlation between motivation and task performance. As indicated above, research has shown that these variables generally tend to be curvilinearly related. However, suppose the sample were selected such that the research participants were all moderately to highly motivated. Such might be the case, for instance, if the sample were recruited from an organization of self-employed entrepreneurs, as these individuals are characteristically highly motivated to achieve. Figure 14.3 suggests that the correlation between motivation and task performance for this range of motivation should be a relatively strong negative one even though the linear approximation across the whole range of motivation scores is extremely poor. In general, when two variables are related in a curvilinear fashion, the Pearson correlation coefficient will tend to be greater in magnitude the smaller the range of variable X that is examined.

On the other hand, if two variables are linearly related, the effect of restricting the range of one of them is often to reduce the magnitude of the

FIGURE 14.4 Illustration of the Effect of a Restricted Range on the Correlation Coefficient When Two Variables Are Linearly Related (Adapted from Howell, 1985)

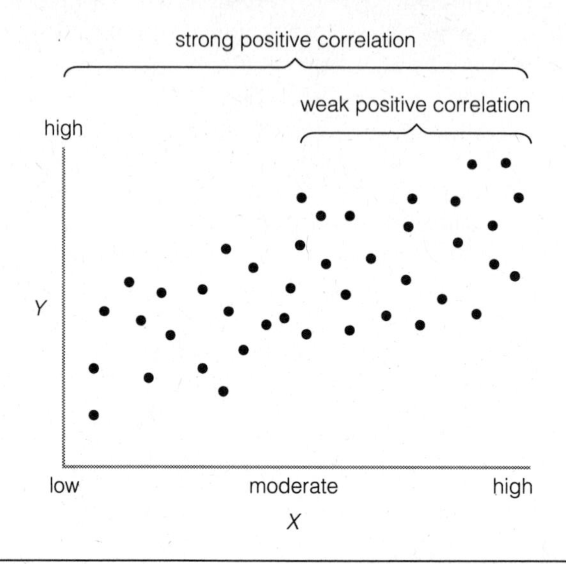

observed correlation coefficient. For instance, Figure 14.4 illustrates a strong positive correlation between two variables. Note, though, that if only individuals having moderate to high scores on X had been included in the study, the correlation observed between X and Y would be substantially reduced. In general, the effect of restricting the range of a variable that is linearly related with another variable is to reduce the magnitude of the observed correlation. This might explain the relatively weak correlations that have been reported between Scholastic Aptitude Test (SAT) scores and college grade point averages. Typically, only students who are interested in pursuing a college education take the SAT. To the extent that this group of students is more academically inclined than the population of high school students in general (and, thus, more likely to obtain high SAT scores), the correlations that are observed between SAT scores and grade point averages are based on a restricted range of SAT scores. It might in fact be the case that if *all* graduating high school students took the SAT and went on to college, the correlation between SAT scores and grade point averages would reflect a strong positive association.

Since the effects of a restricted range on the Pearson correlation coefficient can never be known with certainty, we must be careful to select our sample such that the entire range of values of interest is studied. We must also be careful not to extend our interpretation of correlational results *outside* the range of the original data set. The conclusions drawn from a correlational analysis apply only to the range of the variables on which the correlation was based. This is because the Pearson correlation coefficient represents the extent to which two variables approximate a linear relationship *for the range of the variables included in its calculation*. The possibility always exists that the relationship between X and Y outside this range is not the same as the relationship between X and Y within this range. For instance, it would be meaningless to try to generalize about the relationship between motivation and task performance at low to moderate levels of motivation from the relationship between these variables at moderate to high levels of motivation.

As you might expect, a similar caution holds for regression analysis—prediction of Y from X is meaningful only for the range of X values that formed the basis for the calculation of the regression equation. If, in fact, the regression equation does not accurately represent the linear relationship between X and Y for the values of X from which one is making predictions, the results yielded by application of this equation will be inaccurate.

Suppose, for instance, that a researcher is interested in predicting the vividness with which people are able to recall their dreams from the length of time for which they sleep. One hundred volunteers spend the night in a sleep laboratory. In the morning, the vividness with which they are able to recall their night's dreams is measured on a scale of 1 to 9, where higher scores indicate more vivid dream recall. The range of sleep duration for the subjects participating in this study is from 5.30 to 8.60 hours, and the regression equation for predicting the vividness of dream recall from the amount of sleep is found to be $\hat{Y} = 8.71 - .68X$. This equation indicates, for instance, that the vividness of dream recall for an individual who sleeps for 6 hours is predicted to be $8.71 - (.68)(6) = 4.63$ and the vividness of dream recall for an individual who sleeps for 8 hours is predicted to be $8.71 - (.68)(8) = 3.27$. This equation also indicates that the vividness of dream recall for an individual who fails to fall asleep is predicted to be a very vivid $8.71 - (0)(.68) = 8.71$ even though a person cannot dream if he or she is not sleeping! This inconsistency reflects the fact that the regression equation was established on an X value range of 5.30 to 8.60 and does not represent the relationship between sleep duration and vividness of dream recall for an X score of 0.

14.11 Numerical Example

Generally, a participant in a computer dating program completes a questionnaire indicating his or her interests and habits. Responses to the questions are then analyzed and the person is matched with a member of the opposite gender who has similar interests. Underlying the matching of individuals is an important assumption: People with similar interests will be attracted to one another.

In order to test the validity of this assumption, a researcher surveyed 15 women who had participated in a computer dating program. Each of these

TABLE 14.2 Data and Calculation of Intermediate Statistics for Similarity and Attraction Experiment

Individual	X	Y	X^2	Y^2	XY
1	10	8	100	64	80
2	8	6	64	36	48
3	6	4	36	16	24
4	4	2	16	4	8
5	2	3	4	9	6
6	10	6	100	36	60
7	8	8	64	64	64
8	6	5	36	25	30
9	7	9	49	81	63
10	4	5	16	25	20
11	1	3	1	9	3
12	3	1	9	1	3
13	5	3	25	9	15
14	7	5	49	25	35
15	9	7	81	49	63
	$\Sigma X = 90$	$\Sigma Y = 75$	$\Sigma X^2 = 650$	$\Sigma Y^2 = 453$	$\Sigma XY = 522$
	$\bar{X} = 6.00$	$\bar{Y} = 5.00$			

individuals had already been on one date with her partner, and each was asked to indicate how much she was attracted to him on a scale of 1 to 10. A score of 1 meant she was not at all attracted to him, while a score of 10 meant she was very attracted to him. Before being matched, subjects had responded to a 10-item questionnaire about their interests and habits, as had their eventual partners. For purposes of the study, the similarity between assigned partners differed. Some individuals were assigned partners who responded the same way on all 10 questions, some were assigned partners who responded the same way on only 9 of the questions, and so on, all the way down to 1 similar questionnaire response. Thus, similarity between partners could range from 1 to 10, with low scores indicating little similarity and high scores indicating a great deal of similarity. The data for the experiment appear in columns 2 and 3 of Table 14.2.

As noted previously, the usual focus of Pearson correlation is the extent to which two variables share a linear relationship. The identification of one variable (the independent variable) as a presumed influence and one variable (the dependent variable) as a presumed effect is not relevant in situations of this type. Note, however, that in the present study the researcher explicitly manipulated one variable (similarity with one's partner) in order to determine how it affects a second variable (attraction to the partner). The independent variable–dependent variable distinction is thus applicable in this instance. The independent variable is the similarity with one's partner, and the dependent variable is the attraction to the partner. The null hypothesis of no linear relationship states that the population correlation between these two variables is 0. The alternative hypothesis states that the population correlation is nonzero. These hypotheses can be formally stated as

$$H_0: \ \rho = 0$$
$$H_1: \ \rho \neq 0$$

The correlation coefficient can be represented as (see Equation 5.4)

$$r = \frac{\text{SCP}}{\sqrt{\text{SS}_X \text{SS}_Y}}$$

and calculated as (see Equation 5.8)

$$r = \frac{\Sigma XY - \dfrac{(\Sigma X)(\Sigma Y)}{N}}{\sqrt{\left[\Sigma X^2 - \dfrac{(\Sigma X)^2}{N}\right]\left[\Sigma Y^2 - \dfrac{(\Sigma Y)^2}{N}\right]}}$$

Intermediate statistics necessary for the application of this formula have been computed in Table 14.2. From these we can derive the correlation coefficient:

$$r = \frac{522 - \dfrac{(90)(75)}{15}}{\sqrt{\left(650 - \dfrac{90^2}{15}\right)\left(453 - \dfrac{75^2}{15}\right)}}$$

$$= \frac{72}{\sqrt{(110)(78)}} = .78$$

For an alpha level of .05, nondirectional test, and $N - 2 = 15 - 2 = 13$ degrees of freedom, the critical values of r from Appendix H are $\pm.514$. Since .78 is greater than $+.514$, we reject the null hypothesis and conclude that there is a relationship between similarity with one's partner and attraction.

The strength of the relationship is indexed by the square of the correlation coefficient:

$$r^2 = .78^2 = .61$$

The proportion of variability in attraction ratings that is associated with partner similarity in this experiment is .61, as indexed by a linear model. This represents a strong effect.

The nature of the relationship is indicated by the sign of the correlation coefficient. In this case, the correlation is positive, indicating that as similarity between partners increases, so does attraction.

If the researcher was interested in identifying the regression equation for predicting attraction (Y) from similarity scores (X), he would first determine the slope using Equation 5.10:

$$b = \frac{\text{SCP}}{\text{SS}_X}$$

$$= \frac{72}{110} = .65$$

The obtained value of b can then be substituted into Equation 5.11 along with the means of the two variables to yield the intercept:

$$a = \overline{Y} - b\overline{X}$$

$$= 5.00 - (.65)(6.00) = 1.10$$

The regression equation is thus

$$\hat{Y} = a + bX$$

$$= 1.10 + .65X$$

This equation can now be used to make predictions about the outcome of dating situations. For instance, a woman is unlikely to enjoy a computer date with a man who agrees with her on only 3 questionnaire items, since the predicted attraction level for a similarity score of 3 is $1.10 + (.65)(3) = 3.05$. In contrast, the predicted attraction score is $1.10 + (.65)(10) = 7.60$ when the two individuals agree on all 10 questions.

These predictions are, of course, subject to error. Specifically, the standard error of estimate in the present example is estimated to be

$$\hat{s}_{YX} = \sqrt{\frac{\text{SS}_Y(1 - r^2)}{N - 2}}$$

$$= \sqrt{\frac{78(1 - .78^2)}{15 - 2}} = 1.53$$

indicating that on the average, predicted Y (attraction) scores are estimated to deviate from actual Y scores by 1.53 units. This represents a relatively small degree of error. To gain further insight into the predictive utility of the regression equation, we can compare the estimated standard error of estimate with the estimated standard deviation of Y. Since $\text{SS}_Y = 78$ and $N = 15$, $\hat{s}_Y = \sqrt{\text{SS}_Y/(N - 1)} = \sqrt{78/(15 - 1)} = 2.36$. Thus, predicting scores on Y from scores on X leads to considerably less error (an estimated average prediction error of 1.53 units) than if all Y scores were predicted to be equal to $\overline{Y}$ (an estimated average prediction error of 2.36 units).

As noted in Chapter 9, the ability to make a causal inference between two variables is a function of one's research design, not the statistical technique used to analyze the data that are yielded by that research design. Thus, as discussed previously, correlation does not necessarily imply causation. Note, however, that the present study used an experimental research strategy—individuals were *randomly assigned* partners ranging in similarity from 1 to 10. To the extent that confounding variables were controlled, we can thus infer that in this experiment similarity between partners *caused* attraction to vary as it did. As with all statistical techniques, causality can similarly be inferred following a statistically significant correlation analysis any time an experimental, as opposed to an observational, research strategy is employed (assuming, of course, that confounding variables are controlled). We should note, however, that in practice most applications of correlation and regression utilize an observational approach.

The above results relate only to the linear relationship between the two variables. If we were interested in the extent of nonlinear relationship, we could use curvilinear regression techniques. The use of regression to predict attraction scores from similarity scores is predicated on the fact that the researcher explicitly manipulated similarity so that he could determine how it affects attraction. Thus, there would be no justification for deriving the regression equation for predicting X (similarity) from Y (attraction) in this example. Since the range of similarity scores included in the sample was from 1 to 10 similar questionnaire responses, our conclusions apply only across these values. For instance, we are unable to infer the effect of *no* similarity (0 similar questionnaire responses) on attraction ratings or what attraction ratings would be if assigned dating partners shared more than 10 interests and habits.

14.12

Summary

Pearson correlation is typically used to analyze the relationship between two variables when (1) both variables are quantitative in nature and are measured on approximately an interval level, (2) the two variables have been measured on the same individuals, and (3) the observations on each variable are between-subjects in nature. The test from sample data of the hypothesis that the population correlation is 0 is based on a sampling distribution of the correlation coefficient. This distribution permits an application of basic hypothesis testing principles. The strength of the linear relationship is indicated by the square of the correlation coefficient. The nature of the relationship is indicated by the sign of the correlation coefficient.

Regression equations can be applied to individuals outside of the original data set to make predictions about their scores on variable Y (the criterion variable) from their scores on variable X (the predictor variable). An estimate of the average error that will be made across individuals when making such predictions is provided by the estimated standard error of estimate.

Several issues are associated with the use of correlation and regression. First, Pearson correlation assesses only linear relationships. If two variables are related in a nonlinear fashion, this relationship will not be reflected in the correlation coefficient and scores on variable Y cannot be accurately predicted from scores on variable X. A second consideration is that the regression equation for predicting X from Y is not the same as the regression equation for predicting Y from X. A third issue concerns the effects of a restricted range on the observed correlation. In general, when two variables are related in a curvilinear fashion, the correlation coefficient will tend to be greater in magnitude the smaller the range of variable X that is examined. On the other hand, if two variables are linearly related, the effect of restricting the range of one of them is to reduce the magnitude of the observed correlation coefficient. Lastly, we must be careful not to extend our interpretation of correlational results or our predictions from the regression equation outside the range of the original data set.

Appendix 14.1 Testing Null Hypotheses Other Than $\rho = 0$

Occasionally, an investigator will want to test a null hypothesis that ρ is equal to a value other than 0. As noted in Section 14.2, when ρ does not equal 0, the sampling distribution of the correlation coefficient is skewed. Fisher (1950) derived a logarithmic transformation of r, which we will symbolize r' (read "*r prime*"), that has two desirable properties: (1) the sampling distribution of r' is approximately normally distributed irrespective of the value of ρ, and (2) the standard error of r' is essentially independent of ρ.* Because of these properties, it is possible to convert a sample r into r' and then use the normal distribution to test the null hypothesis that ρ is equal to some value other than 0.

The formula for r' is

$$r' = .50[\log_e(1 + r) - \log_e(1 - r)] \quad [14.7]$$

where $\log_e$ is the symbol for the natural logarithm of a number and r is the sample correlation coefficient.

* Fisher referred to the r transformation using the symbol Z. Thus, you might see reference in the literature to "Fisher's r to Z transform." We will use the symbol r' to avoid confusing the transformation with a z score.

All terms in Equation 14.7 are constants except for r. Thus, the value of r' is completely dependent on r, and r' is simply a rescaling of r. It is not necessary to actually calculate r' when testing a given null hypothesis since this has been done in Appendix I, which presents a table of values of r' for selected values of r.

Figure 14.5 presents sampling distributions of r' for $\rho = -.80$, $\rho = 0$, and $\rho = +.80$, for $N = 10$. Compare these distributions with the comparable sampling distributions of r illustrated in Figure 14.1. Unlike the sampling distributions of r, the sampling distributions of r' are similar in shape and variability. This reflects the fact that, as noted above, the standard error of r' is independent of ρ. Specifically, the standard error of r' (symbolized $\sigma_{r'}$) is equal to

$$\sigma_{r'} = \frac{1}{\sqrt{N - 3}} \quad [14.8]$$

The test of a null hypothesis that ρ is equal to some value other than 0 is based on the following equation:

FIGURE 14.5 Sampling Distributions of r' for Three Values of ρ for $N = 10$

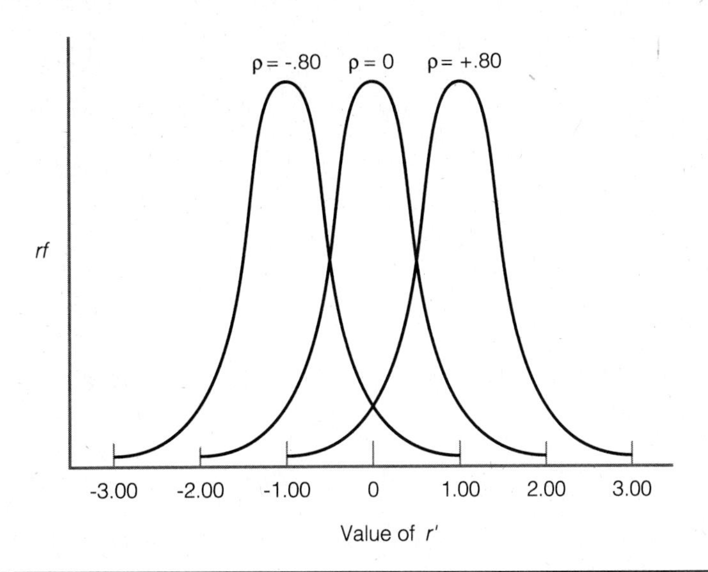

$$z = \frac{r' - \rho'}{\sigma_{r'}} \qquad [14.9]$$

where ρ' is the log-transformed value corresponding to the hypothesized value of ρ, and r' and $\sigma_{r'}$ are as defined above. The observed value of z is then compared with the appropriate critical values of z from the table of the normal distribution in Appendix B.

As an example of the above procedures, consider the following null and alternative hypotheses:

$$H_0: \ \rho = .25$$
$$H_1: \ \rho \neq .25$$

Suppose an r of .45 is observed for $N = 100$. The standard error of r' in this case equals

$$\sigma_{r'} = \frac{1}{\sqrt{N - 3}}$$
$$= \frac{1}{\sqrt{100 - 3}} = .10$$

From Appendix I, the transformed value (r') corresponding to an r of .45 is .485, and the transformed value (ρ') corresponding to a ρ of .25 is .255. The observed z score is thus

$$z = \frac{r' - \rho'}{\sigma_{r'}}$$
$$= \frac{.485 - .255}{.10} = 2.30$$

For an alpha level of .05, nondirectional test, the critical values of z are ± 1.96. The observed z is greater than $+1.96$ and the null hypothesis is therefore rejected. Since the sample correlation of .45 is greater than the hypothesized population correlation of .25, we conclude that the population correlation between the two variables under study is greater than .25.

151

Exercises

Answers to asterisked (*) exercises appear at the back of the book. Answers to exercises with two asterisks are also worked out step-by-step in the Study Guide.

Exercises to Review Concepts

1. Under what conditions is Pearson correlation typically used to analyze a bivariate relationship?

***2.** Explain the equivalence between the t test procedure for testing the significance of a correlation coefficient using Equation 14.2 and the procedure for testing the significance of a correlation coefficient by comparing the observed value of r with critical values of r.

***3.** State the critical value(s) of r that would be used to reject the null hypothesis for Pearson correlation at an alpha level of .05 under each of the following conditions:

a. $H_0: \rho = 0, H_1: \rho \neq 0, N = 32$
b. $H_0: \rho = 0, H_1: \rho \neq 0, N = 22$
c. $H_0: \rho = 0, H_1: \rho > 0, N = 22$
d. $H_0: \rho = 0, H_1: \rho < 0, N = 15$
e. $H_0: \rho = 0, H_1: \rho \neq 0, N = 67$

4. What are the assumptions underlying Pearson correlations?

5. What is the relationship between eta-squared and the correlation coefficient?

***6.** What interpretation does eta-squared have in the context of Pearson correlation? How does this interpretation differ from that of previous tests that we have considered?

***7.** Compute the Pearson correlation coefficient for the following data and test the viability of the hypothesis that the population correlation between X and Y is 0 using a nondirectional test:

Individual	*X*	*Y*
1	4	9
2	7	11
3	9	14
4	9	10
5	2	8
6	12	14
7	4	8
8	5	9
9	13	10
10	3	7

***8.** Compute the value of eta-squared for the data in Exercise 7. Does the observed value represent a weak, moderate, or strong effect?

***9.** What is the nature of the relationship between the two variables for the data in Exercise 7?

10. Compute the Pearson correlation coefficient for the following data and test the viability of the hypothesis that the population correlation between X and Y is 0 using a nondirectional test:

Individual	*X*	*Y*
1	4	5
2	8	2
3	3	4
4	9	10
5	2	4
6	1	2
7	7	8
8	4	8
9	1	5
10	7	9

11. Compute the value of eta-squared for the data in Exercise 10. Does the observed value represent a weak, moderate, or strong effect?

12. What is the nature of the relationship between the two variables for the data in Exercise 10?

For each of the studies described in Exercises 13–17, indicate the appropriate statistical test for analyzing the relationship between the variables. Assume that the underlying assumptions of the tests have been satisfied.

***13.** An investigator wanted to test the relative effects of two drugs on learning. Thirty subjects were randomly assigned to take the first drug and a different 30 subjects were randomly assigned to take the second drug. The mean scores on a learning task were compared for the two groups.

*14. An investigator wanted to determine preference for three brands of ice cream. One hundred individuals rated each of the three brands on a scale ranging from 1 to 15. The mean ratings for the three brands were compared.

15. An investigator wanted to test the effects of the nuclear accident at Three Mile Island on attitudes toward nuclear energy. One hundred people had been interviewed prior to the accident and their attitudes measured on a 10-point scale. Five days after the accident they were reinterviewed and their attitudes measured again. The mean attitude scores were compared across interviews.

16. An investigator tested the relationship between age and blood pressure in a sample of 250 adults.

17. An investigator wanted to test the relationship between driving conditions and amount of gas used. Twenty midsize cars were driven on smooth roads, 20 were driven on hilly roads, and 20 were driven on mountainous roads. The amount of gas used over a 50-mile trip was measured and the mean gas consumption compared for the three conditions.

*18. If a researcher suspected that the strength of the relationship between two variables in the population was .15 as indexed by the correlation coefficient squared, what sample size should she use in a study in order to achieve power of .90 for an alpha level of .05, directional test?

19. Suppose an investigator reported the results of a correlational analysis with $N = 32$. If the value of the correlation coefficient squared in the population was .20, what would the power of his statistical test be at an alpha level of .05, nondirectional test?

20. What is the meaning of prediction in the context of regression?

*21. What information is conveyed by the estimated standard error of estimate?

22. Under what circumstance will the regression equation for predicting X from Y be the same as the regression equation for predicting Y from X? Under what circumstance will the regression equation for predicting X from Y be different from the regression equation for predicting Y from X? Why?

*23. What is the effect on the correlation coefficient of restricting the range of one of the variables?

*24. Compute the regression equation for predicting Y from X for the data in Exercise 7. What is the predicted Y score for an X score of 3? What is the predicted Y score for an X score of 7? What is the predicted Y score for an X score of 11?

*25. Compute the estimated standard error of estimate for the data in Exercise 7.

26. Compute the regression equation for predicting Y from X for the data in Exercise 10. What is the predicted Y score for an X score of 2? What is the predicted Y score for an X score of 8? What is the predicted Y score for an X score of 8.50?

27. Compute the estimated standard error of estimate for the data in Exercise 10.

**28. Compute the regression equation for predicting X from Y for the data in Exercise 7. Compare this with the regression equation for predicting Y from X that you computed in Exercise 24. What accounts for the difference between the two equations?

Exercises to Apply Concepts

**29. Behavioral scientists have studied extensively factors that determine an effective group leader in small-group problem-solving situations. Much of this work has used a scale developed by Fiedler (1967) to measure different leadership styles. The scale involves having individuals think of all the people with whom they have ever worked and singling out their least preferred coworker (LPC). They then rate this LPC on a series of dimensions, such as the extent to which the person was pleasant, friendly, cooperative, etc. A total LPC score is obtained by summing these ratings. According to Fiedler, individuals with high LPC scores tend to see even a poor coworker in a relatively favorable light. These leaders tend to behave in a manner described as compliant, nondirective, and generally relaxed. In contrast, individuals with low LPC scores tend to be demanding, controlling, and managing in their group interactions.

It is generally recognized that the effectiveness of any leader depends not only on his or her leadership style, but also on the type of problem being addressed and the situation in which the group finds itself. In one investigation, Fiedler examined the relationship between a leader's LPC score and group problem-solving performance (minutes until solution of the problem). In this investigation, the task was relatively unstructured, there were good leader–member relations, and the leader had relatively little control over the rewards given to each group member. The hypothetical data presented below are representative of the results of the study. Analyze the data using a nondirectional test, draw a conclusion, and write up your results using principles discussed in the Method of Presentation section.

Group	*X* (*leader's LPC score*)	*Y* (*minutes until solution*)
1	10	11
2	15	12
3	18	15
4	12	10
5	8	6
6	22	19
7	11	9
8	10	7
9	17	13
10	20	7

30. Borden (1978) investigated factors that differentiate individuals who are ecologically concerned from those who tend to exhibit little concern for the environment. He hypothesized that one difference between these individuals might be the extent to which they value technology: People who value technological innovation and who believe that technology can solve most world problems should be less likely to perform environmental-conserving behaviors than those who question technology as a panacea to environmental problems. In order to test this hypothesis, Borden administered two scales to a group of individuals. One scale measured the extent to which participants performed ecological-related behaviors (for example, saving energy), and the other scale measured belief in technology. Scores on each scale could range from 0 to 20, with higher scores representing the performance of more ecological-related behaviors and greater belief in technology, respectively. The hypothetical data presented below are representative of the outcome of this investigation. Analyze the data using a directional test, draw a conclusion, and write up your results using principles discussed in the Method of Presentation section.

Individual	*X* (*ecological-related behaviors*)	*Y* (*belief in technology*)
1	13	12
2	7	17
3	14	13
4	10	15
5	16	10
6	12	14
7	8	16
8	5	19
9	10	16
10	11	7

****31.** Greater numbers of students are attending graduate school than ever before in academic history. Very few graduate programs can admit all applicants, and accordingly, most have a screening committee whose task is to select those applicants with the most promise. What kinds of criteria discriminate good graduate students from poor ones? Willingham (1974) reviewed 43 studies from 1952 to 1972 that addressed this question. In his review, Willingham noted that the most commonly used selection criteria were the following: (1) scores on the Graduate Record Exam, which consist of a quantitative ability score (GRE-Q), a verbal ability score (GRE-V), and an advanced score (GRE-A) that measures mastery and comprehension of material basic to graduate study in a specific major field; (2) grade point average (GPA) during undergraduate education; and (3) letters of recommendation. Some of these criteria were found to be reasonable discriminators of the performance of graduate students, while others were quite poor in their prediction of graduate school success.

The hypothetical data in Set I below examine the relationship between GRE-A scores and graduate students' grade point averages after 2 years of graduate study in a particular program. Compute the re-

	SET I			SET II
Student	*X* (*GRE-A*)	*Y* (*GPA*)	Applicant	X (*GRE-A*)
1	533	3.11	1	532
2	497	2.89	2	478
3	612	3.66	3	589
4	564	3.50	4	483
5	582	3.29	5	527
6	476	3.34	6	493
7	607	3.61	7	546
8	621	3.74		
9	590	3.42		
10	512	2.61		

gression equation for predicting graduate grade point average from GRE-A scores. If only students who are likely to maintain a 2-year grade point average of 3.00 (B) or better are to be admitted to the program, which of the applicants in Set II should be accepted for admission?

FIFTEEN

CHI-SQUARE TEST

15.1

Use of the Chi-Square Test

The **chi-square test** is typically used to analyze the relationship between two variables when

1. both variables are qualitative in nature (that is, measured on a nominal level);
2. the two variables have been measured on the same individuals; and
3. the observations on each variable are *between-subjects* in nature.

As an example of an investigation that meets these criteria, suppose that a researcher interested in the relationship between gender and political party identification in a particular geographic area conducts a survey in which a total of 170 residents indicate whether they identify as Democrats, Republicans, or Independents. This investigation concerns the relationship between two qualitative variables: (1) gender (male or female) and (2) political party identification (Democrat, Republican, or Independent). The two variables are measured on the same individuals, and the observations on each dimension are between-subjects in nature. Given these conditions, the chi-square test is the statistical technique that would typically be used to analyze the relationship between the variables.

15.2

Two-Way Contingency Tables

Unlike the previous statistical tests that we have considered, the chi-square test analyzes relationships between variables using frequency information. The use of frequency information is predicated on the fact that chi-square test is designed for use with qualitative variables, and it is not appropriate to compute means for variables of this type.

TABLE 15.1 Contingency Table for Gender and Political Party Identification Investigation

GENDER	PARTY IDENTIFICATION			TOTALS
	Democrat	*Republican*	*Independent*	
Male	63	17	10	90
Female	35	20	25	80
TOTALS	98	37	35	170

The basis of analysis for the chi-square test is a **contingency table** (also called a *frequency* or *crosstabulation table*). A *two-way* contingency table is illustrated in Table 15.1, which examines the relationship between gender (male or female) and political party identification (Democrat, Republican, or Independent) for the sample of 170 individuals referred to above. This table is called a *two-way* table because it examines two variables. We can also refer to this as a 2 × 3 (read "two by three") table, where the first number indicates the number of rows (excluding the "Totals" row) and the second number indicates the number of columns (excluding the "Totals" column). These are equivalent to the numbers of levels of the two variables. The assignment of one variable as the row variable and the other variable as the column variable is arbitrary. If gender had been represented in columns and political party identification in rows, we would have a 3 × 2 rather than a 2 × 3 table, but would apply the identical analytical procedures described below.

Each unique combination of variables in a contingency table is referred to as a **cell.** The entries within the cells represent the number of individuals in the sample who are characterized by the corresponding levels of the variables. In the present example, there are 63 male Democrats, 17 male Republicans, 10 male Independents, 35 female Democrats, 20 female Republicans, and 25 female Independents. The numbers in the last column and bottom row of Table 15.1 indicate how many individuals have each separate characteristic. For example, there are 98 Democrats, 37 Republicans, and 35 Independents. Also, there are 90 males and 80 females. These frequencies are referred to as **marginal frequencies** and are the sum of the frequencies in the corresponding row (for example, 63 + 17 + 10 = 90) or column (for example, 63 + 35 = 98). Finally, the number in the lower right-hand corner indicates the overall sample size (in this case, 170).

15.3

Chi-Square Tests of Independence and Homogeneity

Since subjects in the gender and political party identification study were selected for participation without regard to gender or political party identification, the marginal frequencies for both of these variables were free to vary, or **random.**

As noted above, the actual sample contained 98 Democrats, 37 Republicans, and 35 Independents, of whom 90 were males and 80 were females. However, assuming that the researcher wanted to recruit exactly 170 subjects, the sampling procedures could have hypothetically produced other combinations of marginal frequencies, such as 72 males and 98 females for the gender variable, or 81 Democrats, 44 Republicans, and 45 Independents for the party identification variable. In contrast, suppose the researcher had explicitly selected the sample so that it would consist of a specified number of individuals of each gender (for instance, 100 males and 100 females) and had then classified each subject into one of the three party identification categories. In this instance, the marginal frequencies for the party identification variable would still be random, but the marginal frequencies for the gender variable would be predetermined, or **fixed.** Alternatively, the researcher might have designed the study such that a predetermined number of subjects from each party identification category would be included (for instance, 50 Democrats, 50 Republicans, and 50 Independents) and then classified these individuals as being either male or female. In this case, the marginal frequencies for gender would be random but the marginal frequencies for party identification would be fixed.

When the marginal frequencies of both variables under study are random, the chi-square test described in this chapter is known as the **chi-square test of independence.** When the marginal frequencies are random for one variable and fixed for the other, the analytical procedures are identical but the test is referred to as the **chi-square test of homogeneity.** Although the two names are often used interchangeably in the literature, they refer to different situations and the reader is encouraged to apply them appropriately. Since the two tests are identical in all other respects, the general terminology *chi-square test* will be used in the remainder of this book to encompass both situations.

15.4 Inference of a Relationship Using the Chi-Square Test

The logic underlying the chi-square test focuses on the concept of **expected frequencies.** This concept can be illustrated using the coin flipping example discussed in Chapter 8. Suppose you are given a coin and asked to determine whether it is fair. If you assume that the coin is fair, your best guess about the outcome of 100 flips would be 50 heads and 50 tails (since the probability of each is .50). These expected frequencies can then be compared with the **observed frequencies** when you actually flip the coin. Suppose that, after 100 flips, you had observed 53 heads and 47 tails. Although the observed frequencies are not exactly 50/50, you would be hesitant to conclude that the coin is biased because of your knowledge of sampling error. But suppose the observed frequencies had been 65/35. These are much more discrepant from the expected frequencies. At some point the observed frequencies will become so discrepant from the expected frequencies that you will conclude that the coin is biased. This is analogous to the logic of the chi-square test whereby we (1) assume a null hypothesis is true, (2) derive a set of expected frequencies based on this assumption, (3) compare the expected frequencies with the frequencies observed in the investigation, and (4) reject the null hypothesis if the overall difference between observed frequencies and expected frequencies is large (as defined by a given alpha level). We will use the example on gender and political party identification to illustrate the chi-square test.

Null and Alternative Hypotheses

The null hypothesis for a chi-square test states that the variables of interest are unrelated in the population, while the alternative hypothesis states that there is a population relationship between the two variables.* Since this relationship can take a number of different forms, the alternative hypothesis for the chi-square test, as with the F test in analysis of variance, is nondirectional in nature. The null and alternative hypotheses can be phrased as follows for the study on gender and political party identification:

H_0: Gender and political party identification are unrelated in the population.

H_1: Gender and political party identification are related in the population.

Expected Frequencies and the Chi-Square Statistic

Application of the chi-square test requires computation of an expected frequency for each cell under the assumption of no relationship between the two variables. Consider the Democrat category for the party identification variable. If party identification and gender are unrelated, we would expect the proportion of males who are Democrats to be the same as the proportion of females who are Democrats. In our example, 98/170 or 57.65% of the *total* sample are Democrats. This represents an estimate of the percentage of Democrats in the population. If gender and party identification are unrelated, we would expect 57.65% of the males to be Democrats and 57.65% of the females to be Democrats. In the example, there are 90 males, and 57.65% of 90 is 51.88. This is the expected frequency for Democratic males. There are 80 females in the example, and 57.65% of 80 is 46.12. This is the expected frequency for Democratic females.

Now consider the Republican category for the party identification variable. In our example, 37/170 or 21.76% of the total sample are Republicans. If gender and party identification are unrelated, then 21.76% of the 90 males

* If we let A represent a given outcome on gender and B represent a given outcome on political party identification, then the null hypothesis can be stated formally using Equation 6.6: H_0: $p(A, B) = p(A)p(B)$ for all values of A and B. The corresponding alternative hypothesis states H_1: $p(A, B) \neq p(A)p(B)$ for all values of A and B.

TABLE 15.2 Computation of Chi-Square for Gender and Political Party Identification Investigation

Cell	*Observed frequency (O)*	*Expected frequency (E)*	*O − E*	*(O − E)²*	*(O − E)²/E*
Democratic males	63	51.88	11.12	123.65	2.38
Democratic females	35	46.12	−11.12	123.65	2.68
Republican males	17	19.59	− 2.59	6.71	.34
Republican females	20	17.41	2.59	6.71	.39
Independent males	10	18.53	− 8.53	72.76	3.93
Independent females	25	16.47	8.53	72.76	4.42
					$\chi^2 = 14.14$

should be Republicans (yielding an expected frequency of 19.59) and 21.76% of the 80 females should be Republicans (yielding an expected frequency of 17.41). The same procedure can be applied to the Independent category.

The foregoing steps involved in the calculation of expected frequencies can be summarized as follows:

1. For the cell in question, divide the total of the column (the column marginal frequency) in which it appears by the total number of observations.
2. Multiply this value by the total of the row (the row marginal frequency) in which the cell appears.

This can be represented symbolically as

$$E_j = \left(\frac{CMF_j}{N}\right)(RMF_j) \qquad [15.1]$$

where E_j is the expected frequency for cell j, CMF_j is the column marginal frequency associated with the cell in question, N is the overall sample size, and RMF_j is the row marginal frequency associated with the cell in question.* For instance, the expected frequency for Independent males is $(35/170)(90) = 18.53$.

The third column of Table 15.2 presents the expected frequency for each cell in our example. In order to determine how discrepant an observed frequency (column 2) is from the corresponding expected frequency, we can compute the difference between the two. This has been done for each cell in column 4. It is now necessary to combine the discrepancies into an index that reflects the overall difference between the observed and the expected frequencies. We will accomplish this using the **chi-square statistic,** which is defined as follows:

$$\chi^2 = \sum \frac{(O_j - E_j)^2}{E_j} \qquad [15.2]$$

* For the sake of simplicity, the j subscript will be excluded wherever possible from these and the other terms that we will encounter in this chapter.

where χ^2 is the chi-square statistic, O_j is the observed frequency for cell j, E_j is the expected frequency for cell j derived using Equation 15.1, and the summation is across all cells.

The calculation of chi-square for our example is worked through in columns 4–6 of Table 15.2. First, the difference between the observed and expected frequencies is determined for each cell (column 4). Second, each of these difference scores is squared (column 5). The squared difference for each cell is then divided by the corresponding expected frequency (column 6). Finally, column 6 is summed, yielding a chi-square value of 14.14.

Sampling Distribution of the Chi-Square Statistic

If two variables are unrelated in the population, the population value of chi-square will equal 0. However, because of sampling error, a chi-square computed from sample data might be greater than 0 even when the null hypothesis is true. Consider a population where there is no relationship between the two variables. A sample of size N is randomly selected and the chi-square statistic is computed. The value of chi-square might be .50. Now suppose we repeat this procedure with a second random sample of size N and observe a chi-square of 1.17. In principle, we could continue to do this a large number of times with the result being a large number of chi-square statistics. The sample chi-square statistics, if computed for all possible random samples of a given size, would form a **sampling distribution of the chi-square statistic.** This distribution closely approximates, under certain conditions to be discussed shortly, a theoretical distribution called the **chi-square distribution.**

There are different chi-square distributions depending on the degrees of freedom associated with them. For the chi-square test that we are considering, $\text{df} = (r - 1)(c - 1)$, r being the number of levels of the row variable and c being the number of levels of the column variable. In the gender and party identification example, there are two rows and three columns, so there are $(2 - 1)(3 - 1) = 2$ degrees of freedom.

Figure 15.1 presents examples of chi-square distributions for 1, 2, and 8 degrees of freedom. Analogous to the previous sampling distributions that we have considered, probability statements can be made with respect to scores in the chi-square distribution. It is therefore possible to set an alpha level and define a critical value such that if the null hypothesis is true, the probability of obtaining a chi-square value larger than that critical value is less than alpha. Since all discrepancies from the expected frequencies are reflected in the upper tail of the chi-square distribution (as defined by the critical value), the chi-square test is, by its nature, nondirectional.

Appendix J presents a table of critical values for the chi-square distribution. For the gender and political party identification investigation, the critical value of chi-square for an alpha level of .05 and 2 degrees of freedom is 5.991. The observed chi-square of 14.14 is greater than this value, and the null hypothesis is therefore rejected. We conclude that gender and political party identification are related in the population.

FIGURE 15.1 Chi-Square Distributions for Various Degrees of Freedom

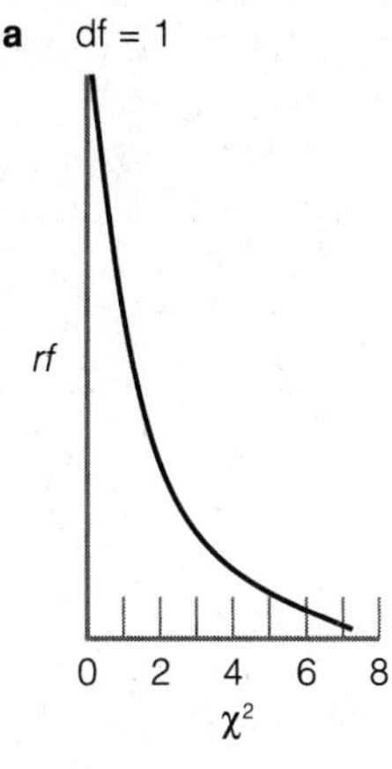

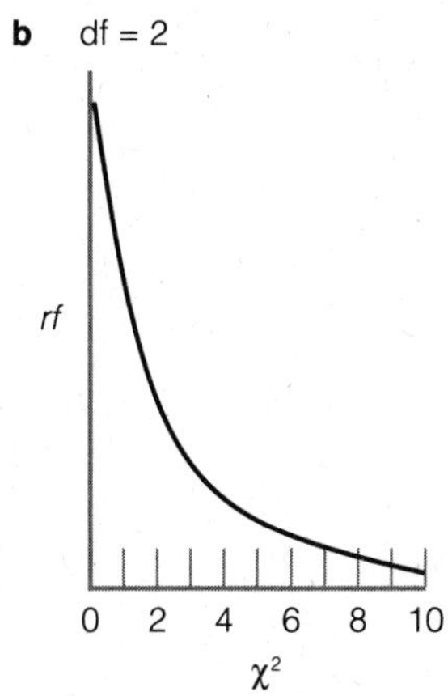

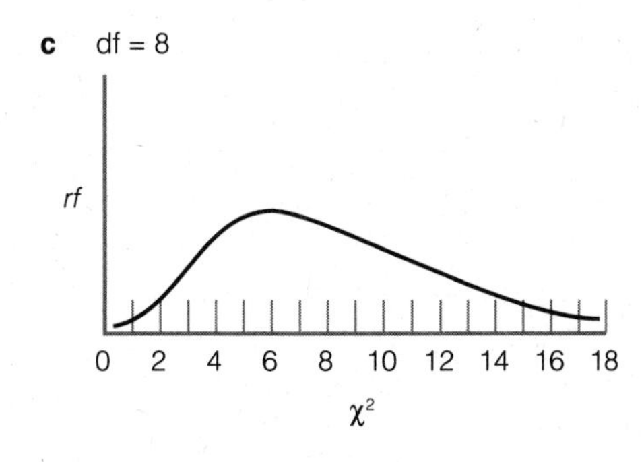

STUDY EXERCISE 15.1

A researcher was interested in the relationship between the social class of parents and the discipline style they use with their children. Groups of working-class, middle-class, and upper-class parents were studied. Each participant was classified as using primarily physical discipline (emphasizing physical punishment), primarily psychological discipline (emphasizing psychological punishment such as scolding and withdrawal of affection), or a mixture of physical and psychological discipline. The following contingency table was observed:

	DISCIPLINE STYLE			
SOCIAL CLASS	*Physical*	*Psychological*	*Mixed*	TOTALS
Working class	60	29	48	137
Middle class	24	49	25	98
Upper class	18	31	16	65
TOTALS	102	109	89	300

Test for the existence of a relationship between the two variables.

Answer The null hypothesis states that social class and discipline style are unrelated in the population, while the alternative hypothesis states that there is a population relationship between these two variables. The calculations for the chi-square statistic are shown in the table below. We find that $\chi^2 = 25.71$. For an alpha level of .05 and $(r - 1)(c - 1) = (3 - 1)(3 - 1) = 4$ degrees of freedom, the critical value of chi-square from Appendix J is 9.488. Since the observed chi-square value of 25.71 is greater than 9.488, we reject the null hypothesis and conclude that a relationship exists between social class and discipline style.

Cell	*O*	*E*	*O − E*	$(O - E)^2$	$(O - E)^2/E$
Physical–working class	60	46.58	13.42	180.10	3.87
Physical–middle class	24	33.32	− 9.32	86.86	2.61
Physical–upper class	18	22.10	− 4.10	16.81	.76
Psychological–working class	29	49.78	−20.78	431.81	8.67
Psychological–middle class	49	35.61	13.39	179.29	5.03
Psychological–upper class	31	23.62	7.38	54.46	2.31
Mixed–working class	48	40.64	7.36	54.17	1.33
Mixed–middle class	25	29.07	− 4.07	16.56	.57
Mixed–upper class	16	19.28	− 3.28	10.76	.56
					$\chi^2 = 25.71$

Assumptions of the Chi-Square Test

The chi-square test is based on several assumptions. These ensure that the sampling distribution of the chi-square statistic approximates a chi-square distribution. Specifically, the following is assumed:

1. The observations are independently and randomly sampled from the population of all possible observations.
2. The *expected* frequency of each cell is nonzero. In fact, while the issue is controversial, statisticians generally recommend that the lowest expected frequency one should have in order to use the chi-square test is somewhere around 5, with the required frequency decreasing as the dimensions of the contingency table increase.

   Note that this assumption is concerned with *expected* frequencies, not observed frequencies. Thus, as long as the expected frequency assumption is met, observed frequencies can be as low as 0. As you might anticipate, expected frequencies tend to increase as does the size of the overall sample.

15.5

2 × 2 Tables

The analysis of 2 × 2 contingency tables raises several unique issues. These will now be considered.

Yates' Correction for Continuity

When both of the variables under study have only two levels, the sampling distribution of the chi-square statistic corresponds less closely to a chi-square distribution than when one or both variables have more than two levels. In this case, a correction factor to be incorporated into the formula for computing the chi-square statistic has been suggested that supposedly yields a sampling distribution similar to the chi-square distribution. This correction factor is known as **Yates' correction for continuity** and involves subtracting .5 from the *absolute value* of $O_j - E_j$ before these quantities are squared, divided by E_j, and summed across cells. Symbolically, the formula for computing the chi-square statistic incorporating Yates' correction is

$$\chi^2 = \sum \frac{(|O_j - E_j| - .5)^2}{E_j} \qquad [15.3]$$

Whether one should apply Yates' correction for continuity has been the subject of considerable debate among statisticians (see Camilli & Hopkins, 1978; Conover, 1974a, 1974b; Mantel, 1974). In practice, the correction is applied more often than not. However, a recent review of the literature on this topic (Jaccard & Becker, 1988) clearly suggests that the correction should *not* be used, as it tends to reduce the power of the statistical test below what it would otherwise be. It is presented here merely for the convenience of the reader who might encounter it in other sources.

Computational Formula for 2 × 2 Tables

In cases where both variables have only two levels, an efficient computational formula can be used to derive the chi-square statistic. This formula is

$$\chi^2 = \frac{N(ad - bc)^2}{(a + b)(c + d)(a + c)(b + d)} \qquad [15.4]$$

where the letters a, b, c, and d represent observed frequencies for the four cells as follows:

	Variable A	
Variable B	a	b
	c	d

Consider the following contingency table:

	Variable A		*Totals*
Variable B	25	16	41
	15	23	38
Totals	40	39	79

According to Equation 15.4, the observed value of chi-square is

$$\chi^2 = \frac{79[(25)(23) - (16)(15)]^2}{(25 + 16)(15 + 23)(25 + 15)(16 + 23)}$$

$$= \frac{(79)(335)^2}{2{,}430{,}480} = 3.65$$

This is the same value of chi-square that we would obtain if we were to analyze the data using the more general Equation 15.2.

15.6

Strength of the Relationship

A number of different indices have been proposed for measuring the strength of the relationship between two variables in a contingency table. Among the more popular are Pearson's index of mean square contingency, the phi coefficient, gamma, the coefficient of contingency, and the Goodman–Kruskal index of predictive association. Statisticians do not agree on which of these statistics is most appropriate. Probably the most common index of the strength of association is a measure known as the **fourfold point correlation coefficient** (as it is called when applied to the relationship between variables with two levels each) or **Cramér's statistic** (as it is called when one or both variables have more than two levels). The strength of association using this measure is defined as

$$V = \sqrt{\frac{\chi^2}{N(L - 1)}} \qquad [15.5]$$

where V is the fourfold point correlation coefficient/Cramér's statistic, χ^2 is the observed chi-square, N is the overall sample size, and L is the number of levels of the variable that has the fewer values. For the study on gender and political party identification, $\chi^2 = 14.14$; $N = 170$; and $L = 2$ because the gender variable has two levels, the party identification variable has three levels, and 2 is less than 3. Thus,

$$V = \sqrt{\frac{14.14}{170(2 - 1)}} = .29$$

The fourfold point correlation coefficient/Cramér's statistic can range from 0 to 1.00, where a value of 0 indicates no relationship and a value of 1.00 indicates a perfect relationship. The magnitude of V is interpreted like that of the Pearson correlation coefficient. To illustrate the relationship between the Pearson correlation coefficient and the present approach, consider the case of a contingency table with two levels of variable A and two levels of variable B. For each participant in the study, a score of 1 is assigned if the individual is characterized by level 1 of variable A and a score of 2 is assigned if the individual is characterized by level 2. The same is done for variable B. If the Pearson correlation between the variables were computed, the absolute value of the result would be equivalent to the fourfold point correlation. Conceptually, a large value of V indicates a tendency for particular categories of one variable to be associated with particular categories of the other variable.

15.7 Nature of the Relationship

If the null hypothesis of no relationship is rejected, then additional steps are required to discern more fully the nature of the relationship. The test of the chi-square statistic applies to the data taken as a whole and provides no information as to which cells are responsible for rejecting the null hypothesis. Just as the HSD test can be applied to break down the overall relationship following a statistically significant analysis of variance, comparable tests can be applied following a statistically significant chi-square test. Procedures have been suggested by Cohen (1967), Goodman (1964), Ryan (1959), and Wike (1971).

The procedure recommended by Goodman is the most popular among statisticians. The mathematics of the approach, however, are complex, and interested readers are referred to Appendix 15.1.

From an intuitive perspective, insight into the nature of the relationship can be gained by examining the $O - E$ and $(O - E)^2/E$ values for each cell. For the gender and political party identification investigation, these correspond to columns 4 and 6 of Table 15.2. The numbers in the latter column are summed to yield the overall chi-square statistic. The smaller numbers within this column contribute less to the rejection of the null hypothesis than the larger numbers. The two largest values in our example are for Independent females (4.42) and Independent males (3.93). Examination of the observed minus expected frequencies $(O - E)$ indicates that males are less likely to be Independents than expected (−8.53) and females are more likely to be Independents than expected (8.53), given the assumption of no relationship between gender and party identification. The $(O - E)^2/E$ values for Democratic males (2.38) and Democratic females (2.68) are also relatively large, and the corresponding values of $O - E$ indicate that males are more likely to be Democrats than expected (11.12) and females are less likely to be Democrats than expected (−11.12). The values of $(O - E)^2/E$ for Republican males (.34) and Republican females (.39) are relatively small and suggest that these groups contribute little to the rejection of the null hypothesis.

STUDY EXERCISE **15.2**

Determine the strength and nature of the relationship between social class and discipline style for the contingency table in Study Exercise 15.1.

Answer The strength of the relationship can be determined using Cramér's statistic. Since both variables have three levels, L in this case is equal to 3. Applying Equation 15.5,

$$V = \sqrt{\frac{\chi^2}{N(L - 1)}} = \sqrt{\frac{25.71}{300(3 - 1)}} = .21$$

The nature of the relationship can be discerned intuitively by examining columns 4 and 6 of the calculations table in Study Exercise 15.1. The largest $(O - E)^2/E$ values are for the psychological–working class, psychological–middle class, and physical–working class cells. Examination of the observed minus expected frequencies indicates that working-class parents are less likely to use psychological discipline than expected (−20.78) and more likely to use physical discipline (13.42). In contrast, the frequency of psychological discipline used by middle-class parents was greater than expected (13.39).

15.8 Methodological Considerations

In Section 5.4, we noted three reasons why two variables might be related: (1) the first variable might cause the second variable, (2) the second variable might cause the first variable, or (3) some additional variable(s) might cause both variables. A fourth reason why two variables might be related is that one of them bears a relationship to some additional variable(s) that cause(s) the second variable. This is probably the case in the gender and political party identification investigation. Even though the chi-square test was statistically significant, this does not necessarily mean that gender *causes* individuals to identify with one political party or another. It is more likely that the relationship between the two variables is due to the different experiences that males and females have in terms of their socialization, learning, and personal development. The fact that the strength of the relationship, as indexed by Cramér's statistic, was only .29 suggests that in addition to gender and associated factors, disturbance variables also have an important influence on political party identification. We can speculate that these include such characteristics as parents' political party identification, race, and social class.

15.9 Numerical Example

Social psychologists have studied variables that influence altruistic behavior. In one experiment, a researcher investigated the effects of a role model on people's willingness to donate money to charity. Three hundred individuals were approached in a shopping center and asked to make a charitable donation. One hundred of these individuals were assigned to a positive-model condition such that they were approached just after they had seen an assistant of the experimenter make a donation. Another 100 individuals were assigned to a negative-model condition such that they were approached just after they had seen an assistant of the experimenter refuse to make a donation. Finally, the last 100 individuals were approached without having observed a model. The numbers of people who did and did not donate money in each condition are presented in Table 15.3.

The null hypothesis is that model status and outcome are unrelated in the population, while the alternative hypothesis is that there is a population rela-

TABLE 15.3 Contingency Table for Model Status and Outcome Experiment

OUTCOME	MODEL STATUS			TOTALS
	Positive	*Negative*	*No model*	
Donated	63	34	49	146
Did not donate	37	66	51	154
TOTALS	100	100	100	300

TABLE 15.4 Computation of Chi-Square for Model Status and Outcome Experiment

Cell	*O*	*E*	*O − E*	$(O - E)^2$	$(O - E)^2/E$
Positive–donated	63	48.67	14.33	205.35	4.22
Positive–did not donate	37	51.33	−14.33	205.35	4.00
Negative–donated	34	48.67	−14.67	215.21	4.42
Negative–did not donate	66	51.33	14.67	215.21	4.19
No model–donated	49	48.67	.33	.11	.002
No model–did not donate	51	51.33	− .33	.11	.002
					$\chi^2 = 16.83$

tionship between these two variables. The calculations for the chi-square statistic are shown in Table 15.4. We find that $\chi^2 = 16.83$. The critical value of chi-square for an alpha level of .05 and $(r - 1)(c - 1) = (2 - 1)(3 - 1) = 2$ degrees of freedom is 5.991. Since the observed chi-square value of 16.83 is greater than 5.991, we reject the null hypothesis and conclude that a relationship exists between model status and outcome.

The strength of the relationship is indicated by Cramér's statistic. Since the model status variable has three levels and the outcome variable has two levels, L in this case is equal to 2. Using Equation 15.5, we find that

$$V = \sqrt{\frac{\chi^2}{N(L - 1)}}$$

$$= \sqrt{\frac{16.83}{300(2 - 1)}} = .24$$

The nature of the relationship can be discerned intuitively by examining columns 4 and 6 of Table 15.4. It can be seen that the no-model condition contributed very little to the overall chi-square [$(O - E)^2/E$ values of .002]. The other $(O - E)^2/E$ and $O - E$ values indicate that a positive model increases donating behavior, whereas a negative model decreases donating behavior. These results should, of course, be interpreted in the context of relevant methodological considerations. Do any of these come to mind?

15.10

Use of Quantitative Variables in the Chi-Square Test

While the chi-square test is typically used to analyze the relationship between two qualitative variables, it can also be applied when one or both variables are quantitative. For instance, an investigator concerned with the relationship between the number of siblings one has (a quantitative variable) and preference for group versus individual activity (a qualitative variable) might use this approach. A common procedure involves classifying scores on a quantitative variable into a small number of groups before applying the chi-square test. The following table for analyzing the relationship between men's ages at the time of marriage and the number of years that their marriages lasted illustrates this strategy:

	DURATION OF MARRIAGE			
AGE AT MARRIAGE	*<5*	*5–9*	*10–14*	*≥15*
<19	42	28	16	18
19–24	33	26	23	23
25–34	12	9	17	15
>34	14	14	10	11

When possible, it is usually preferable to analyze quantitative variables with the parametric tests in prior chapters rather than with the chi-square test. For instance, an alternative way to approach the problem of the relationship between age at marriage and marriage duration would be to perform a Pearson correlation between individuals' marriage-age and marriage-duration scores. Parametric tests are usually preferred because they tend to be more powerful. The power of the chi-square test is further reduced when quantitative variables are collapsed into categories because considerable information is likely to be lost by placing individuals with different scores (for instance, an age at marriage of 25 and an age at marriage of 34 in the above example) into the same group. While this approach implicitly assumes that all individuals assigned to a given category are equivalent on the underlying dimension, this may not in fact be the case. An advantage of the chi-square approach, however, is that a quantitative variable need be measured on only an ordinal level as opposed to the approximately interval level required for parametric tests.

15.11

Planning an Investigation Using the Chi-Square Test

Appendix E.5 contains tables of the sample sizes necessary to achieve various levels of power for the chi-square test. A set of tables is presented for each research design involving contingency tables of different dimensions. Table 15.5 reproduces the portion of this appendix for a 3 × 3 contingency table and an alpha level of .05 for illustration. The first column presents different levels of power and the column headings are population values for Cramér's statistic. As an example of the use of this table, if the desired level of power is .80 and if the researcher suspects that the strength of the relationship in the population corresponds to a Cramér's statistic of .30, then the number of subjects that should be sampled is 66. Inspection of Table 15.5 and Appendix E.5 provides a general appreciation for the relationship between the dimensions of the table, the alpha level, sample size, the strength of the relationship one is trying to detect in the population, and power for the chi-square test.

TABLE 15.5 Approximate Sample Sizes Necessary to Achieve Selected Levels of Power for 3 × 3 Table and Alpha = .05 as a Function of Population Values of Cramér's Statistic

Type of Table: 3 × 3
Alpha = .05

	POPULATION VALUE OF CRAMÉR'S STATISTIC								
POWER	*.10*	*.20*	*.30*	*.40*	*.50*	*.60*	*.70*	*.80*	*.90*
.25	154	39	17	10	6	4	3	2	2
.50	321	80	36	20	13	9	7	5	4
.60	396	99	44	25	16	11	8	6	5
.70	484	121	54	30	19	13	10	8	6
.75	536	134	60	34	21	15	11	8	7
.80	597	149	66	37	24	17	12	9	7
.85	671	168	75	42	27	19	14	10	8
.90	770	193	86	48	31	21	16	12	10
.95	929	232	103	58	37	26	19	15	11
.99	1262	316	140	79	50	35	26	20	16

15.12

Method of Presentation

When presenting the results of a chi-square analysis, the researcher should report the degrees of freedom for the chi-square test, the sample size on which it is based, the observed value of chi-square, and the significance level. If the analysis is statistically significant, the strength and nature of the relationship should also be reported.

Textual presentation of a chi-square test is typically supplemented by a two-way contingency table of the observed frequencies. Such a table would have a format similar to that of Table 15.1. In unpublished manuscripts, tables are placed at the very end of the research report and are referred to in the text, where appropriate, by number, starting with Table 1. Of course, there should be a clear tie-in between the content of a table and the content of the text. Quoting from pages 84 and 86 of the *Publication Manual of the American Psychological Association* (American Psychological Association, 1983), "An informative table supplements—it does not duplicate—the text. In the text, refer to every table and its data. Although in text you should tell the reader what to look for in the table, discuss only the table's highlights. If you discuss every item of the table in text, the table is unnecessary."

The location of a table is indicated in the text by the instructions "Insert Table (number) about here" set off by lines above and below and centered. In published form, the actual table will replace its reference.

The results for the study on gender and political party identification discussed earlier in this chapter might be reported as follows:

Results

A chi-square test was applied to the relationship between gender and political party identification and found to be statistically significant, $\chi^2(2, \underline{N} = 170) = 14.14$, $\underline{p} < .01$. The observed frequencies for the six cells can be found in Table 1.

Insert Table 1 about here

As indexed by Cramér's statistic, the strength of the relationship was .29. This reflects the fact that males are more likely than females to be Democrats, and females

```
are more likely than males to be Independents. Gender dif-
ferences in indentification with the Republican party were
found to be minimal.
```

The first sentence specifies the statistical test and the decision with respect to the null hypothesis. The first number in the parentheses is the degrees of freedom, and the number indicated by N is the overall sample size. This is followed by the observed chi-square value, the significance level, and reference to a table of the observed frequencies. As noted above, this table would take a form similar to Table 15.1 in the present book.

Cramér's statistic indicates the strength of the relationship. If the variables had had two levels each, this would have been called a fourfold point correlation. The last two sentences report the nature of the relationship. This should be determined using the procedure developed by Goodman or a similar technique. However, in practice, few behavioral scientists break down the overall relationship using formal tests. The more common approach is to examine the nature of the relationship via inspection of the frequency values.

15.13

Examples from the Literature

Religion and College Attendance

Several sociologists have argued that American Catholics have not contributed to the scientific and intellectual development of the United States relative to their numbers in the population. Considerable documentation for this assertion exists. This deficiency is most commonly considered to result from the cultural values of American Catholics, which are believed to impede intellectual achievement. Warkov and Greeley (1966), however, have argued that the deficiency may be due to the social conditions surrounding early American Catholic immigrants. According to Warkov and Greeley, Catholic immigrants were too poor to be concerned with much beyond sheer economic survival. They argue that immigrants had little time to concern themselves with matters such as intellectualism and going to college since all of their effort had to be directed toward establishing an adequate economic base. This should not, however, be true for more recent generations who have had the opportunity to work from the economic base established by their ancestors.

Based on this logic, Warkov and Greeley hypothesized that there should be a relationship between religion (Catholic versus Protestant) and college attendance for older individuals but not for younger individuals, since Catholics in the latter group have had the economic freedom to concern themselves with education while Catholics in the former group have not. To examine this issue, they collected data on religion and college attendance for each of the two age groups. The older group consisted of individuals between 50 and 59 years of age. The contingency table for this group can be represented as follows:

	RELIGION	
COLLEGE ATTENDANCE	*Catholic*	*Protestant*
Attended	42	24
Did not attend	310	73

The younger group consisted of individuals between 23 and 29 years of age. The contingency table for this group can be represented as follows:

	RELIGION	
COLLEGE ATTENDANCE	*Catholic*	*Protestant*
Attended	99	265
Did not attend	253	642

A chi-square test for the older group indicated a statistically significant relationship between religion and college attendance, $\chi^2(1, N = 449) = 9.95$, $p < .01$. The nature of the relationship was such that Protestants were more likely to have attended college than Catholics. The strength of the relationship, as indexed by the fourfold point correlation coefficient, was .15. In contrast, a chi-square test for the younger group was statistically nonsignificant, $\chi^2(1, N = 1{,}259) = .15$, *ns*.* The fourfold point correlation coefficient in this instance was equal to .01. These findings are consistent with the reasoning of Warkov and Greeley. However, there are other possible interpretations as well. Can you think of them?

Gender Differences in the Importance of Romantic Love for Marriage

Do men and women have different feelings about the importance of romantic love as a prerequisite for establishing a marital relationship? In one attempt to address this issue, Simpson, Campbell, and Berscheid (1986) asked male and female college students to respond to the question, "If a man (woman) had all the other qualities you desired, would you marry this person if you were not in love with him (her)?" Response categories were "no," "yes," or "undecided." The obtained data were as follows:

	GENDER	
RESPONSE	*Male*	*Female*
No	148	141
Yes	3	6
Undecided	22	19

* An alternative to performing separate chi-square tests for the two age groups would be to simultaneously examine the age, religion, and college attendance variables in the context of a three-way contingency table. Simultaneous analysis of three or more between-subjects qualitative variables is most readily accomplished using a statistical technique known as *log-linear analysis*.

A chi-square test applied to this contingency table was not statistically significant, $\chi^2(2, N = 339) = 1.24$, *ns*. Thus, the importance of romantic love as a prerequisite for marriage appears to be independent of gender. The strength of the relationship, as indexed by Cramér's statistic, was .06. Examination of the observed frequencies shows a clear consensus among both genders about the importance of romantic love; no matter how ideal a prospective partner might otherwise be, marriage is not believed to be a viable option unless one is in love.

Additional questions assessed the importance of romantic love as a prerequisite for *maintaining* a marital relationship. Using the response categories "agree," "disagree," and "neutral," participants were asked to respond to the statements, "If love has completely disappeared from a marriage, I think it is probably best for a couple to make a clean break and start new lives" and "In my opinion, the disappearance of love is not a sufficient reason for ending a marriage, and should not be viewed as such." As before, chi-square tests involving the two sets of responses were statistically nonsignificant, with a sizeable portion of both genders indicating that remaining in love is an important condition for the continuation of a marriage.

15.14

Chi-Square Goodness-of-Fit Test

In addition to its role in the chi-square test discussed in the previous sections, the chi-square statistic is also involved in the application of the **goodness-of-fit test.** This test involves determining whether a distribution of frequencies for a variable in a sample is representative of a specified population distribution.* For example, suppose the relative frequencies of marital status for the population of adult American females under 40 years of age are as follows:

Marital status	*rf*
Married	.60
Single	.23
Separated	.04
Divorced	.12
Widowed	.01

Further suppose an investigator wanted to know whether a particular sample of 200 adult females under age 40 was drawn from a population that is repre-

* Since the goodness-of-fit test is concerned with the distribution of frequencies for a single variable, unlike the other techniques discussed in Part 2 of this book, it is not a bivariate test.

sentative of the general population of adult American females under age 40 in terms of marital status. The observed frequencies for the sample were:

Marital status	*O*
Married	100
Single	44
Separated	16
Divorced	36
Widowed	4
Total	200

The null hypothesis is that the population represented by the sample does not differ from the general population in the distribution of marital status. The alternative hypothesis is that the population represented by the sample does differ from the general population in the distribution of marital status. Based on the relative frequencies in the general population noted above, we can compute the frequencies that would be expected in the sample under the null hypothesis. The expected frequencies are equal to the respective relative frequencies multiplied by the overall sample size. Symbolically,

$$E_j = (rf_j)(N) \quad [15.6]$$

where E_j is the expected frequency for category j, rf_j is the relative frequency in the population for category j, and N is the overall sample size.

In our example, the expected frequency for the married category is

$$E_j = (.60)(200) = 120.00$$

Similarly, the expected frequency for the single category is

$$E_j = (.23)(200) = 46.00$$

The results of this procedure for all five categories have been placed in column 3 of Table 15.6. The chi-square statistic is derived by applying Equation 15.2 to the observed and expected frequencies in the same manner as we did previously. The computations are shown in Table 15.6, where it can be seen that $\chi^2 = 19.42$.

TABLE 15.6 Computation of Chi-Square for Marital Status Investigation

Marital status	*O*	*E*	*O − E*	*(O − E)²*	*(O − E)²/E*
Married	100	120.00	−20.00	400.00	3.33
Single	44	46.00	− 2.00	4.00	.09
Separated	16	8.00	8.00	64.00	8.00
Divorced	36	24.00	12.00	144.00	6.00
Widowed	4	2.00	2.00	4.00	2.00
					$\chi^2 = 19.42$

The degrees of freedom for the goodness-of-fit test are $k - 1$, k being the number of categories. From Appendix J, the critical value of chi-square for an alpha level of .05 and $5 - 1 = 4$ degrees of freedom is 9.488. The observed chi-square of 19.42 is greater than 9.488, so we reject the null hypothesis and conclude that the sample was drawn from a population with a marital status distribution different from that of the general population of adult American females under age 40.

If we examine the last column in Table 15.6, we can gain an appreciation for which categories are most responsible for the rejection of the null hypothesis. The largest $(O - E)^2/E$ values are for the separated and divorced categories. Examination of the observed minus expected frequencies for these categories indicates that in the population represented by this sample, women are more likely than expected to be separated or divorced.

A common use of the goodness-of-fit test is to determine whether frequencies in the population are evenly distributed across the categories under study. For example, a consumer psychologist might ask each of 60 subjects to report a preference for oblong, round, or irregularly shaped potato chips. Under the null hypothesis that no differentiation is made between the three shapes in the population, 20 subjects should indicate a preference for oblong potato chips, 20 subjects should indicate a preference for round potato chips, and 20 subjects should prefer irregularly shaped potato chips. Thus, the expected frequency is 20.00 in each instance. If the observed frequencies were 17 for oblong chips, 24 for round chips, and 19 for irregularly shaped chips, the observed value of chi-square would be as computed in Table 15.7. For an alpha level of .05 and $k - 1 = 3 - 1 = 2$ degrees of freedom, the critical value of chi-square is 5.991. Since the observed chi-square of 1.30 does not exceed the critical value, we fail to reject the null hypothesis that preference between the three shapes of potato chips is evenly distributed in the population.

The chi-square goodness-of-fit test is based on several assumptions. First, the observations must be independently and randomly selected from the population of all possible observations. In addition, the overall sample size should be large enough so that the *expected* frequency of any category is at least 10 if only two categories are involved and at least 5 if more than two categories are involved.

TABLE 15.7 Computation of Chi-Square for Potato Chip Preference Investigation

Preference	*O*	*E*	*O − E*	*(O − E)²*	*(O − E)²/E*
Oblong	17	20.00	−3.00	9.00	.45
Round	24	20.00	4.00	16.00	.80
Irregular	19	20.00	−1.00	1.00	.05
					$\chi^2 = 1.30$

15.15 Summary

The chi-square test is typically applied to the analysis of the relationship between two variables when (1) both variables are qualitative in nature, (2) the two variables have been measured on the same individuals, and (3) the observations on each variable are between-subjects in nature. Although the analytical procedures are identical, the test is referred to as a test of independence when the marginal frequencies of both variables are random and a test of homogeneity when the marginal frequencies are random for one variable and fixed for the other. In both instances, the test is based on the analysis of discrepancies between observed frequencies and frequencies expected under the null hypothesis.

The chi-square statistic has a sampling distribution that approximates a distribution called the chi-square distribution. The test for the existence of a relationship using this distribution requires that the expected frequency be nonzero. The strength of the relationship between the two variables is indicated by an index referred to as the fourfold point correlation coefficient when both variables have two levels and Cramér's statistic when one or both variables have more than two levels. The nature of the relationship can be analyzed by examining the $O - E$ and $(O - E)^2/E$ values for the individual cells.

In addition to its role in the chi-square test of independence/homogeneity, the chi-square statistic is also involved in the application of the goodness-of-fit test. This test involves determining whether a distribution of frequencies for a variable in a sample is representative of a specified population distribution. The analytical procedures parallel those for the two-variable case.

Appendix 15.1 Goodman's Simultaneous Confidence Interval Procedure

The procedure recommended by Goodman (1964) for analyzing the nature of the relationship following a statistically significant chi-square test is called the *simultaneous confidence interval procedure*. We will develop the logic of this approach using the data in Table 15.1. The technique considers all possible 2 × 2 contingency tables. In the case of a 2 × 3 analysis, there are three such tables:

	PARTY IDENTIFICATION	
GENDER	*Democrat*	*Republican*
Male	63	17
Female	35	20

GENDER	PARTY IDENTIFICATION: *Democrat*	*Independent*
Male	63	10
Female	35	25

GENDER	PARTY IDENTIFICATION: *Republican*	*Independent*
Male	17	10
Female	20	25

For each 2 × 2 table, we must first perform the following calculation:

$$g = \frac{ad}{bc} \qquad [15.7]$$

where the letters *a, b, c,* and *d* represent observed frequencies for the four cells as follows:

	Variable A	
Variable B	a	b
	c	d

For the first 2 × 2 table,

$$g = \frac{(63)(20)}{(17)(35)} = 2.12$$

The value of *g* for each table is then substituted into the following equation:

$$\hat{y} = (2.3026)(\log_{10} g) \qquad [15.8]$$

where $\hat{y}$ is the sample estimate of *y*, a measure of association between the two variables in the population. A $\hat{y}$ value of 0 indicates that the variables are unrelated, while nonzero $\hat{y}$ values indicate the presence of a relationship. The standard error of $\hat{y}$ can be estimated as follows:

$$\hat{s}_{\hat{y}} = \sqrt{\frac{1}{a} + \frac{1}{b} + \frac{1}{c} + \frac{1}{d}} \qquad [15.9]$$

For the first 2 × 2 table,

$$\hat{y} = (2.3026)[\log_{10}(2.12)]$$
$$= (2.3026)(.33) = .76$$

and

$$\hat{s}_{\hat{y}} = \sqrt{\frac{1}{63} + \frac{1}{17} + \frac{1}{35} + \frac{1}{20}} = .39$$

For large samples, the sampling distribution of $\hat{y}$ is normally distributed. Under the null hypothesis, the observed $\hat{y}$ can be converted into a *z* score as follows:

$$z = \frac{\hat{y} - 0}{\hat{s}_{\hat{y}}} \qquad [15.10]$$

For the first 2 × 2 table,

$$z = \frac{.76 - 0}{.39} = 1.95$$

These procedures can also be applied to the remaining 2 × 2 tables, yielding the following values of *z:*

$$z \text{ for table } 1 = 1.95$$
$$z \text{ for table } 2 = 3.49$$
$$z \text{ for table } 3 = 1.52$$

Each *z* score is then compared against positive and negative critical values, which are defined as follows:

$$z_{\text{critical}} = \pm\sqrt{\chi^2_{\text{critical}}} \qquad [15.11]$$

where z_{critical} represents the critical values of *z* and χ^2_{critical} is the critical value of chi-square associated with the original contingency table. In this case, for an alpha level of .05, $\chi^2_{\text{critical}} = 5.991$, since the original contingency table has 2 degrees of freedom. Thus, the critical values of *z* are

$$z_{\text{critical}} = \pm\sqrt{5.991} = \pm 2.45$$

The *z* score for the second 2 × 2 table is greater than the positive critical value and, consequently, indicates the source of rejection of the original null hypothesis. Males are more likely than females to be Democrats, and females are more likely than males to be Independents. This is consistent with the conclusions drawn from the intuitive procedure developed in Section 15.7.

Exercises

Answers to asterisked (*) exercises appear at the back of the book. Answers to exercises with two asterisks are also worked out step-by-step in the Study Guide.

Exercises to Review Concepts

1. Under what conditions is the chi-square test typically used to analyze a bivariate relationship?

2. How many rows are there in a 3×4 contingency table? How many columns?

3. Differentiate between the chi-square test of independence and the chi-square test of homogeneity.

***4.** Differentiate between observed frequencies and expected frequencies.

Exercises 5–8 refer to the following contingency table summarizing movie attendance and movie preference for a sample of individuals:

MOVIE ATTENDANCE	MOVIE PREFERENCE		
	Romance	*Comedy*	*Drama*
Infrequent	10	20	30
Moderate	15	30	40
Frequent	20	40	50

***5.** What is the marginal frequency for frequent moviegoers?

6. What is the marginal frequency for people who prefer comedies?

7. How many observations are represented?

***8.** Compute the expected frequency for each cell under the assumption of no relationship between movie attendance and movie preference.

***9.** State the critical value of chi-square that would be used to reject the null hypothesis for the chi-square test at an alpha level of .05 under each of the following conditions:

a. $r = 2, c = 2$
b. $r = 3, c = 3$
c. $r = 2, c = 3$
d. $r = 4, c = 4$

10. What are the assumptions underlying the chi-square test?

***11.** What is the disadvantage of using Yates' correction for continuity when analyzing 2×2 contingency tables?

12. What is the relationship between the fourfold point correlation coefficient and Cramér's statistic?

***13.** What is the advantage of analyzing quantitative variables with the chi-square test? What is the disadvantage?

***14.** A researcher investigated the relationship between race and voting behavior during a presidential election and observed the following sample data. Conduct a chi-square test to test for a relationship between the two variables.

RACE	VOTING BEHAVIOR	
	Voted	*Did not vote*
Black	60	40
White	70	30
Other	75	25

***15.** Compute the value of Cramér's statistic for the data in Exercise 14.

***16.** Based on the intuitive approach discussed in the text, what is the nature of the relationship between race and voting behavior for the data in Exercise 14?

17. A study was undertaken relating political party identification to residential area. The following sample data were observed. Conduct a chi-square test to test for a relationship between the two variables.

PARTY IDENTIFICATION	RESIDENTIAL AREA		
	City	*Suburbs*	*Country*
Democrat	40	60	30
Republican	40	20	10
Independent	20	20	10

18. Compute the value of Cramér's statistic for the data in Exercise 17.

19. Based on the intuitive approach discussed in the text, what is the nature of the relationship between party identification and residential area for the data in Exercise 17?

***20.** The relationship between smoking behavior and cause of death was studied for a sample of people who recently died. The data are presented below. Conduct a chi-square test to test for a relationship between the two variables using Equation 15.2. Recalculate the value of chi-square using Equation 15.4. Compare the two sets of results.

	SMOKING BEHAVIOR	
CAUSE OF DEATH	*Smoker*	*Nonsmoker*
Cancer	46	25
Other	34	45

***21.** Compute the value of the fourfold point correlation coefficient for the data in Exercise 20.

***22.** Based on the intuitive approach discussed in the text, what is the nature of the relationship between smoking behavior and cause of death for the data in Exercise 20?

For each of the studies described in Exercises 23–27, indicate the appropriate statistical test for analyzing the relationship between the variables. Assume that the assumptions underlying the tests have been satisfied.

***23.** A researcher tested the relationship between college students' need for achievement as assessed on a 20-item test and their grade point averages.

***24.** An investigator tested the relationship between individuals' gender and their preference for Ford, Chevrolet, or Chrysler cars. One hundred males and 100 females were interviewed and asked which make they preferred.

25. Thirty psychology majors, 30 business majors, and 30 biology majors responded to a social anxiety scale. The mean anxiety scores for the three groups were compared.

26. A researcher tested the effect of a special summer school learning program on reading skills. A test of reading skills was administered to the same group of students before and after the program and the mean performances at the two times were compared.

27. An investigator tested the relationship between individuals' marital status and whether they were for or against tax reform. Four hundred individuals were interviewed and their marital status and tax reform attitudes (pro or con) noted.

28. If a researcher suspected that the strength of the relationship between two variables in the population was .50 as indexed by Cramér's statistic, what sample size should he use in a study involving a 2×4 contingency table and an alpha level of .05 in order to achieve power of .95?

***29.** Suppose an investigator reported the results of a study involving a 2×2 contingency table with $N = 100$. If the value of the fourfold point correlation coefficient in the population was .30, what would the power of her statistical test be at an alpha level of .05?

****30.** Suppose the relative frequencies of preference for the ABC, NBC, and CBS evening news programs as determined by national ratings are as follows:

Preference	*rf*
ABC	.20
NBC	.23
CBS	.21
No preference	.36

Further suppose that a survey of the preferences of 1,000 college students yielded the following observed frequencies:

Preference	*O*
ABC	177
NBC	252
CBS	240
No preference	331

Conduct a goodness-of-fit test to test the hypothesis that the evening news program preference for the population of college students represented by this sample does not differ from that reflected in the national ratings.

31. Conduct a goodness-of-fit test on the following sample data to test the hypothesis that blue, brown, and green eyes are equally common in the population of interest:

Eye color	O
Blue	23
Brown	52
Green	45

Exercises to Apply Concepts

**32. Since the inception of television, behavioral scientists have been concerned with the influence of TV on the social development of children. Violence in programming and its effects on aggressive behavior have been the major source of investigation. However, behavioral scientists have also studied how television affects sex stereotyping in children. For instance, an investigation by McArthur and Eisen (1976) sought to document the amount of sex stereotyping that occurs on television. In one analysis, McArthur and Eisen classified commercials shown on Saturday morning children's television programs in terms of whether the central character was male or female and whether he or she was portrayed as an authority (that is, an expert about the product) or simply as a user of the product. (Barcus, 1971, has estimated that by age 17 the average viewer has seen approximately 350,000 commercials on television.) The data for the investigation are presented below. Analyze the relationship between gender of the central character and the role portrayed by that character, draw a conclusion, and write up your results using principles discussed in the Method of Presentation section.

	ROLE	
GENDER	*Authority*	*User*
Male	138	114
Female	20	43

33. It is commonly believed that the "winner" of a televised presidential debate will benefit greatly at the polls. Research has indicated, however, that the assessment of who wins a debate is not entirely objective. In one study, a sample of individuals were asked whether they thought the Democratic or the Republican candidate won a presidential debate. These individuals had been interviewed prior to the debate to determine the political party they identified with. The data for the study are presented below. Analyze the relationship between political party identification and judged winner of the debate, draw a conclusion, and write up your results using principles discussed in the Method of Presentation section.

	JUDGED WINNER	
PARTY IDENTIFICATION	*Democratic candidate*	*Republican candidate*
Democrat	71	10
Republican	17	58
Independent	35	32

SIXTEEN

NONPARAMETRIC STATISTICAL TESTS

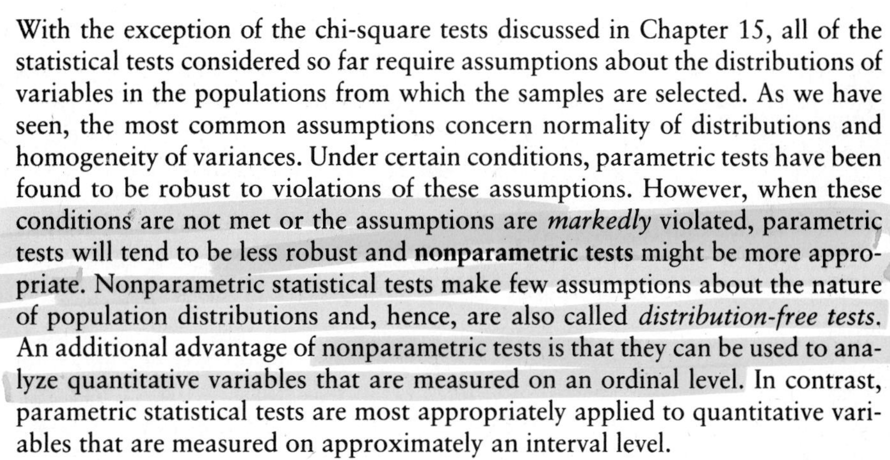

With the exception of the chi-square tests discussed in Chapter 15, all of the statistical tests considered so far require assumptions about the distributions of variables in the populations from which the samples are selected. As we have seen, the most common assumptions concern normality of distributions and homogeneity of variances. Under certain conditions, parametric tests have been found to be robust to violations of these assumptions. However, when these conditions are not met or the assumptions are *markedly* violated, parametric tests will tend to be less robust and **nonparametric tests** might be more appropriate. Nonparametric statistical tests make few assumptions about the nature of population distributions and, hence, are also called *distribution-free tests.* An additional advantage of nonparametric tests is that they can be used to analyze quantitative variables that are measured on an ordinal level. In contrast, parametric statistical tests are most appropriately applied to quantitative variables that are measured on approximately an interval level.

If nonparametric tests require fewer assumptions about population distributions and the level of measurement, then why are parametric tests more popular? The major reason is that parametric tests tend to be more powerful, in a statistical sense, than their nonparametric counterparts. Furthermore, parametric tests are often relatively robust to violations of assumptions. These factors have contributed greatly to the general preference for parametric procedures. The development of parametric statistics has tended to receive more attention from statisticians than the development of nonparametric statistics. This historical bias has yielded a situation whereby certain important statistical questions have been addressed from a parametric perspective but not from a nonparametric perspective. As advances are made in this latter respect, the use of nonparametric statistics should increase.

Most of the parametric tests that we have considered have a nonparametric counterpart. Actually, there are two types of nonparametric counterparts—*rank tests* and *sign tests*. We will consider only tests based on ranks, since they are the more popular of the two. Rank tests also have the advantage of being readily applicable to data that naturally occur in the form of ranks, as when students are ranked from least to most mature by their teacher. The relevant sign tests are discussed in Marascuilo and McSweeney (1977).

Column 2 of Table 16.1 presents the nonparametric tests that will be considered in this chapter. Each of these is a counterpart to the parametric test listed

TABLE 16.1 Parametric Tests and Their Nonparametric Counterparts

Parametric test	*Nonparametric counterpart*
Independent groups *t* test	Wilcoxon rank sum test/ Mann–Whitney *U* test
Correlated groups *t* test	Wilcoxon signed-rank test
One-way between-subjects analysis of variance	Kruskal–Wallis test
One-way repeated measures analysis of variance	Friedman analysis of variance by ranks
Pearson correlation	Spearman rank-order correlation

differences between distribution of scores

in the first column. In contrast to the focus on differences between population means taken by the first four parametric tests, the corresponding nonparametric tests are designed to detect differences between *distributions of scores*. Analogous to Pearson correlation, Spearman correlation tests whether rank scores for the two variables under study are linearly related in the population.

Our focus will be on the application and calculation of the nonparametric tests that we will be considering. A more detailed explanation of their underlying rationale can be found in advanced nonparametric statistics books. For all of the tests to be described, we will assume that the assumptions of the parametric tests have been violated to the extent that nonparametric tests are required. We will consider the issue of selecting parametric versus nonparametric tests in Chapter 18.

16.1 Ranked Scores

All of the tests discussed in this chapter use ranked (ordinal-level) scores. To illustrate ranking procedures, consider the following scores on an intelligence test for five individuals:

Individual	*Score*	*Rank*
1	142	5
2	139	4
3	138	3
4	130	2
5	126	1

We can rank order the scores from least intelligent to most intelligent by assigning the number 1 to the least intelligent individual and then assigning successive integers to individuals in increasing order of intelligence test scores, as has been done in the table. Alternatively, we could rank the scores from most intelligent to least intelligent, if we wished. In this instance, the intelligence test score of 142 would be assigned a rank of 1, the intelligence test score of 139 would be assigned a rank of 2, and so forth. While the computational procedures described in this chapter will be the same regardless of whether scores are ranked from low to high or from high to low, the *interpretation* of results is facilitated if scores are ranked such that lower scores receive the lower ranks and higher scores receive the higher ranks. This is the strategy that we will follow in the remainder of this chapter.

In some instances, we will encounter tied score values. For example, consider the following distribution:

Individual	*Score*	*Rank*
1	142	5
2	135	3.5
3	135	3.5
4	130	2
5	126	1

In cases where a tie occurs, the scores involved are assigned the average of the ranks involved. In the above example, a tie occurred where ranks 3 and 4 were to be assigned. The average of these ranks is $(3 + 4)/2 = 3.5$. Thus, a rank of 3.5 is assigned to the two scores. The next highest score is assigned a rank of 5 because ranks 3 and 4 have already been "used."

Rank-order tests assume that quantitative variables are continuous in nature. Given this assumption, tied scores are viewed as being the result of imprecision of measurement since, in principle, more precise measures would eliminate ties. Nevertheless, tied scores often occur in practice and must be dealt with when applying a nonparametric test. The most common approach is to rank the scores using the preceding procedure (assigning the average rank for ties) and then to introduce a *correction term* into the formula for the test statistic to adjust for the presence of ties. Unfortunately, statisticians do not agree on the optimal correction terms for the various tests. The statistical tests in this chapter will be developed in pure form where no ties exist. As long as the number of ties is minimal, the test statistics will not be affected greatly by tied ranks. Appendix 16.1 presents formulas based on the correction terms most frequently used when a large number of ties occur.

16.2 Rank Test for Two Independent Groups

Use of the Rank Test for Two Independent Groups

The **Wilcoxon rank sum test** and the **Mann–Whitney *U* test** are the nonparametric counterparts of the independent groups t test. They are typically used to analyze the relationship between two variables when (1) scores on the

TABLE 16.2 Data for Wilcoxon Rank Sum Test/Mann–Whitney *U* Test Example

Individual	*Gender*	*Prejudice score*	*Rank*
1	M	48	20
2	M	45	19
3	M	35	16
4	M	33	15
5	M	32	14
6	M	27	9
7	M	31	13
8	M	20	4
9	M	24	6
10	M	25	7
11	F	43	18
12	F	28	10
13	F	29	11
14	F	30	12
15	F	40	17
16	F	19	3
17	F	15	1
18	F	17	2
19	F	26	8
20	F	23	5

$R_1 = 20 + 19 + 16 + 15 + 14 + 9 + 13 + 4 + 6 + 7 = 123$

$R_2 = 18 + 10 + 11 + 12 + 17 + 3 + 1 + 2 + 8 + 5 = 87$

dependent variable are in the form of ranks, (2) the independent variable is between-subjects in nature (it can be either qualitative or quantitative), and (3) the independent variable has two and only two levels. Although the computational procedures are somewhat different, the two tests yield the same results.

Analysis of the Relationship

Consider an investigation in which 10 males and 10 females were administered a questionnaire designed to measure prejudice against women. Scores on this questionnaire could range from 0 to 50, with higher scores indicating greater prejudice. The scores for each individual are presented in column 3 of Table 16.2. The first step of the analysis is to rank order the scores from lowest (rank = 1) to highest (rank = 20) across all individuals. This has been done in column 4 of Table 16.2. Next, the sum of the ranks is separately computed for each group. This is symbolized R_j and referred to as the ***R* statistic.** Let R_1 represent the sum of the ranks for group 1 (males) and R_2 represent the sum of the ranks for group 2 (females). From Table 16.2 we find that $R_1 = 123$ and $R_2 = 87$.

Given a null hypothesis of no relationship between gender and prejudice against women, any difference between R_1 and R_2 would reflect sampling error. There is no reason, under the assumption of the null hypothesis, to expect high scores to concentrate themselves in one group or the other, so we should find that both high- and low-ranked scores are intermingled among the two groups; that is, R_1 should equal R_2 except for sampling error. Specifically, we would expect to find that R_1 and R_2 are both approximately equal to the sum of ranks 1 through 20 divided by 2, or $(1 + 2 + \cdots + 19 + 20)/2 = 105$. In practice, we would not actually perform the above calculations, as we can determine the expected summed rank score for group j through a simpler computational formula:

$$E_j = \frac{n_j(n_1 + n_2 + 1)}{2} \qquad [16.1]$$

where E_j is the expected rank sum for group j, and n_j is the sample size for group j. Since both groups in our example are based on the same number of cases, E_j will be the same for males and females:

$$E_j = E_1 = E_2 = \frac{10(10 + 10 + 1)}{2} = 105.00$$

Note that R_1 and R_2 are equidistant from E_j ($R_1 - E_j = 123 - 105.00 = 18.00$; $R_2 - E_j = 87 - 105.00 = -18.00$). This will occur whenever $n_1 = n_2$.

When $n_1 \neq n_2$, the values of E_j will differ for the two groups. Either value can be used in the statistical test to be presented below as long as the value of R_j also to be included is calculated on the same group. In either instance, the sampling distribution of R_j will have a mean equal to E_j and a standard deviation equal to

$$\sigma_R = \sqrt{\frac{n_1 n_2(n_1 + n_2 + 1)}{12}} \qquad [16.2]$$

In our example,

$$\sigma_R = \sqrt{\frac{(10)(10)(10 + 10 + 1)}{12}} = 13.23$$

When the sample sizes of both groups are 10 or greater, the shape of the sampling distribution of R_j approximates a normal distribution, so we can convert the R statistic into a z score using the following formula (Ferguson, 1976):

$$z = \frac{|R_j - E_j| - 1}{\sigma_R} \qquad [16.3]$$

The same value of z will be obtained regardless of whether R_j and E_j are derived from group 1 or group 2. Focusing on the female scores,

$$z = \frac{|87 - 105| - 1}{13.23} = 1.28$$

in our example. For an alpha level of .05, nondirectional test, the critical values of z from Appendix B are ± 1.96. Since 1.28 is neither less than -1.96 nor greater than $+1.96$, we fail to reject the null hypothesis of no relationship between gender and prejudice against women.

The above test, known as the *Wilcoxon rank sum test,* is applicable when the sample sizes of both groups are 10 or greater. When one or both sample sizes are smaller than 10, the sampling distribution of the R statistic does not approximate a normal distribution, so application of the Wilcoxon rank sum test is not appropriate. Under this circumstance, the data can be analyzed using the *Mann–Whitney U test.* This involves the determination of the **U statistic,** which is the smaller of the two values of U_1 and U_2 as calculated by the following formulas:

$$U_1 = n_1 n_2 + \frac{n_1(n_1 + 1)}{2} - R_1 \qquad [16.4]$$

$$U_2 = n_1 n_2 + \frac{n_2(n_2 + 1)}{2} - R_2 \qquad [16.5]$$

The hypothesis testing procedure consists of comparing the observed value of U against the relevant critical value of U from Appendix K. The observed U is statistically significant if it is *equal to or less than* the critical U.

As noted above, when one or both sample sizes are smaller than 10, the Wilcoxon rank sum test is not appropriate and the Mann–Whitney U test must be used instead. When the sample sizes of both groups are 10 or greater, either test can be applied as long as a critical value for U can be obtained from the U table in Appendix K. Application of the Mann–Whitney U test to our example would proceed as follows:

$$U_1 = (10)(10) + \frac{(10)(10 + 1)}{2} - 123 = 32$$

and

$$U_2 = (10)(10) + \frac{(10)(10 + 1)}{2} - 87 = 68$$

Since 32 is less than 68, U in this case is equal to 32. Referring to Appendix K, we find that the critical value of U for an alpha level of .05, nondirectional test, and $n_1 = n_2 = 10$ is 23. Since 32 is not equal to or less than 23, as we did with the Wilcoxon rank sum test, we fail to reject the null hypothesis of no relationship between gender and prejudice against women.

Regardless of whether the Wilcoxon rank sum test or the Mann–Whitney U test is used, the strength of the relationship between the two variables can be measured using the **Glass rank biserial correlation coefficient** (symbolized r_g). This coefficient is derived using the following formula:

$$r_g = \frac{2(\bar{R}_2 - \bar{R}_1)}{N} \qquad [16.6]$$

where $\bar{R}_1$ is the mean rank for group 1, $\bar{R}_2$ is the mean rank for group 2, and N is the total number of individuals in the study. The value of r_g can range from -1.00 to $+1.00$, and its magnitude can be interpreted like that of the Pearson correlation coefficient. In our example, $\bar{R}_1 = R_1/n_1 = 123/10 = 12.30$ and $\bar{R}_2 = R_2/n_2 = 87/10 = 8.70$. Thus,

$$r_g = \frac{2(8.70 - 12.30)}{20} = -.36$$

The nature of the relationship between the two variables is ascertained by inspection of $\bar{R}_1$ and $\bar{R}_2$. If we had rejected the null hypothesis in the present example, the appropriate conclusion would be that males are more prejudiced against women than are females (since $\bar{R}_1 = 12.30$ and $\bar{R}_2 = 8.70$, and higher rank scores represent greater degrees of prejudice).

Method of Presentation

The results of a rank test for two independent groups are reported in terms of the Wilcoxon rank sum test when the *R* statistic is converted into a *z* score and the Mann–Whitney *U* test when the *U* statistic is calculated. Reports typically include information about the value of *z* or *U* (whichever is appropriate), the sample sizes, the significance level, and, if the test is statistically significant, the strength and nature of the relationship. Unfortunately, the strength of the relationship is seldom reported. This can be accomplished by including the value of the Glass rank biserial correlation coefficient. The results for the study on gender and prejudice against women might be presented as follows:

Results

A Wilcoxon rank sum test compared males' (n = 10) and females' (n = 10) prejudice against women. The mean ranks were found to be nonsignificantly different, z = 1.28, ns.

If the null hypothesis had been rejected, additional sentences would have stated that the Glass rank biserial correlation coefficient was −.36 and that males are more prejudiced against women than are females.

16.3

Rank Test for Two Correlated Groups

Use of the Rank Test for Two Correlated Groups

The **Wilcoxon signed-rank test** is the nonparametric counterpart of the correlated groups *t* test. It is typically used to analyze the relationship between two variables when (1) scores on the dependent variable are in the form of ranked differences, (2) the independent variable is within-subjects in nature (it can be either qualitative or quantitative), and (3) the independent variable has two and only two levels.

TABLE 16.3 Data for Wilcoxon Signed-Rank Test Example

Individual	*Before accident*	*After accident*	*Difference*	*Rank*
1	27	15	12	9
2	26	16	10	8
3	24	16	8	7
4	20	26	− 6	6
5	19	15	4	4
6	18	20	− 2	2
7	15	14	1	1
8	13	16	− 3	3
9	11	16	− 5	5
10	9	9	0	—

$R_p = 9 + 8 + 7 + 4 + 1 = 29$

$R_n = 6 + 5 + 3 + 2 = 16$

Analysis of the Relationship

Consider an investigation in which the attitudes toward nuclear energy for each of 10 individuals were measured before a nuclear accident and again just after it. The attitude scale was such that scores could range from 0 to 30, with higher scores indicating more favorable attitudes. The investigator was interested in comparing attitudes before and after the accident. The scores for each individual are presented in columns 2 and 3 of Table 16.3.

The analysis begins with the computation of the difference between scores in the two conditions for each individual. This has been done in column 4 of Table 16.3. The differences are then rank ordered from smallest to largest, disregarding the sign of the difference and ignoring differences of 0. This has been done in column 5 of Table 16.3. We then separately sum the rank scores for the positive differences (R_p) and the rank scores for the negative differences (R_n). This has been done in Table 16.3, and we find that $R_p = 29$ and $R_n = 16$. If the null hypothesis of no relationship between time of attitude assessment and attitudes toward nuclear energy is true, we would expect the scores for a given individual to be approximately the same in the two conditions. Since unfavorable and favorable attitude changes should be equally likely to occur and since the average size of the changes in each direction should be approximately equal, we would expect the sum of the ranks of the positive differences (R_p) to equal the sum of the ranks of the negative differences (R_n), except for sampling error. In our example, we would expect both sums to equal $(29 + 16)/2$, or 22.5. The symbol E will represent the expected value of the sums of ranks. The question then becomes the following: Under the assumption that the null hypothesis is true, how likely is it that we would obtain rank sums of 29 and 16 when we expected to find rank sums of 22.5? The answer to this question depends on the value of the ***T* statistic,** which is equal to the smaller of R_p and R_n. In this instance, $T = R_n = 16$.

It is possible to construct sampling distributions of the T statistic under the assumption of the null hypothesis and to define critical values of T for purpose of hypothesis testing. This has been done by Wilcoxon (1949), and a table of critical values is presented in Appendix L for sample sizes (N) of 50 or less.

In the present instance, we are interested in the critical value for $N = 9$ rather than $N = 10$ because we disregarded the individual whose difference score was 0. For an alpha level of .05, nondirectional test, the critical value of T is 5. In order for the null hypothesis to be rejected, the observed value of T must be *equal to or less than* the critical value. Since the observed T value of 16 is not equal to or less than 5, the appropriate decision in this instance is to fail to reject the null hypothesis of no relationship between time of assessment and attitudes toward nuclear energy.

When the sample size is greater than 40, the sampling distribution of T is normally distributed with a standard deviation equal to

$$\sigma_T = \sqrt{\frac{N(N + 1)(2N + 1)}{24}} \qquad [16.7]$$

When the sample size requirement is met, we can thus convert the T statistic into a z score and test the null hypothesis using the appropriate critical values from Appendix B. The formula for the z score transformation is

$$z = \frac{T - E}{\sigma_T} \qquad [16.8]$$

The strength of the relationship for the Wilcoxon signed-rank test can be measured using the **matched-pairs rank biserial correlation coefficient** (symbolized r_c). This coefficient can be computed using the following formula:

$$r_c = \frac{4(T - E)}{N(N + 1)} \qquad [16.9]$$

where all terms are as previously defined. The value of r_c can range from -1.00 to $+1.00$, and its magnitude can be interpreted in the same manner as the Pearson correlation coefficient. In our example,

$$r_c = \frac{4(16 - 22.5)}{9(9 + 1)} = -.29$$

The nature of the relationship between the two variables is addressed by inspection of the rank sums. In our example, if we had rejected the null hypothesis, we would have concluded that attitudes toward nuclear energy are less favorable after a nuclear accident than before since the rank sum for the negative differences ($R_n = 16$) was smaller than the rank sum for the positive differences ($R_p = 29$).

Method of Presentation

When presenting the results of a Wilcoxon signed-rank test, one should include information about the sample size, the value of the T statistic, and the significance level. If the test is statistically significant, the strength and nature of the relationship should also be addressed. This can be accomplished in terms of the matched-pairs rank biserial correlation coefficient and discussion of the magnitude of the rank sums, respectively. The results for the study on time of assessment and attitudes toward nuclear energy might be reported as follows:

Results

A Wilcoxon signed-rank test was applied to individuals' attitudes toward nuclear energy before versus after the nuclear accident. The rank sums were found to be nonsignificantly different, $N = 9$, $T = 16$, *ns*.

16.4 Rank Test for Three or More Independent Groups

Use of the Rank Test for Three or More Independent Groups

The **Kruskal–Wallis test** is the nonparametric counterpart of one-way between-subjects analysis of variance. It is typically used to analyze the relationship between two variables when (1) scores on the dependent variable are in the form of ranks, (2) the independent variable is between-subjects in nature (it can be either qualitative or quantitative), and (3) the independent variable has three or more levels.

Analysis of the Relationship

Consider an experiment on the effect of number of roommates on attitudes toward living in dormitories. Eighteen students were assigned dormitory rooms with either one, two, or three roommates. After living with their roommates for one semester, the students were administered a questionnaire measuring their attitudes toward living in dormitories. Scores on this questionnaire could range from 0 to 30, with higher scores indicating more favorable attitudes. The scores for each individual are presented in column 3 of Table 16.4, and these have been rank ordered from lowest (rank = 1) to highest (rank = 18) in column 4.

The rank scores have been retabled in Table 16.5 for purpose of calculation of the Kruskal–Wallis test statistic. If the null hypothesis of no relationship between number of roommates and attitudes toward living in dormitories is true, we would expect the average ranks for the three conditions to be the same, within the constraints of sampling error. To the extent that the average ranks are different, the null hypothesis is questionable. The logic underlying the measurement of differences in average (mean) ranks using the Kruskal–Wallis test can be illustrated by analogy (Friedman, 1972). For instance, consider the following sets of two numbers (5, 5; 4, 6; 3, 7; and 2, 8) that each sum

TABLE 16.4 Data for Kruskal–Wallis Test Example

Individual	*Number of roommates*	*Attitude score*	*Rank*
1	1	29	17
2	1	24	12
3	1	27	15
4	1	22	10
5	1	30	18
6	1	25	13
7	2	23	11
8	2	26	14
9	2	21	9
10	2	17	5
11	2	28	16
12	2	19	7
13	3	18	6
14	3	15	3
15	3	16	4
16	3	20	8
17	3	14	2
18	3	13	1

TABLE 16.5 Ranked Data for Kruskal–Wallis Test Example

One roommate	*Two roommates*	*Three roommates*
17	11	6
12	14	3
15	9	4
10	5	8
18	16	2
13	7	1
$R_1 = 85$	$R_2 = 62$	$R_3 = 24$

to 10 and note what happens when each number is squared and the squares are summed for each set:

$$5^2 + 5^2 = 25 + 25 = 50$$
$$4^2 + 6^2 = 16 + 36 = 52$$
$$3^2 + 7^2 = 9 + 49 = 58$$
$$2^2 + 8^2 = 4 + 64 = 68$$

Note that as the difference between the two numbers increases, the sum of the squared values also increases.

The foregoing property was used by Kruskal and Wallis (1952) in proposing the following test statistic (the ***H* statistic**) to reflect differences between groups:

$$H = \left[\frac{12}{N(N + 1)}\right]\left(\sum \frac{R_j^2}{n_j}\right) - 3(N + 1) \qquad [16.10]$$

where N is the total number of individuals in the study, R_j is the sum of the ranks for group j, n_j is the sample size for group j, and the summation of R_j^2/n_j is across all groups. According to this formula, the sum of the ranks in each group is to be squared and then divided by the number of subjects in that group. The sum of these values combines with the other terms in the equation to yield a statistic, H, which has a sampling distribution that approximates a chi-square distribution with $k - 1$ degrees of freedom, where k represents the number of groups.

For the data in Table 16.5, the value of H is

$$H = \left[\frac{12}{18(18 + 1)}\right]\left(\frac{85^2}{6} + \frac{62^2}{6} + \frac{24^2}{6}\right) - 3(18 + 1)$$
$$= (.035)(1{,}940.83) - 57 = 10.93$$

For an alpha level of .05 and $k - 1 = 3 - 1 = 2$ degrees of freedom, the critical value of H from the chi-square table in Appendix J is 5.991. Since 10.93 is greater than 5.991, we reject the null hypothesis and conclude that number of roommates and attitudes toward living in dormitories are related.

The strength of the relationship for the Kruskal–Wallis test can be measured using an index known as **epsilon-squared.** Epsilon-squared can range from 0 to 1.00, where a value of 0 indicates no relationship and a value of 1.00 indicates a perfect relationship.The formula for epsilon-squared in the between-subjects case is

$$E_R^2 = \frac{H - k + 1}{N - k} \qquad [16.11]$$

where all terms are as previously defined. In our example,

$$E_R^2 = \frac{10.93 - 3 + 1}{18 - 3} = .60$$

Analogous to analysis of variance, rejection of the null hypothesis for a nonparametric test applied to more than two groups tells us only that at least two of the k groups differ in ranks, but does not indicate the nature of these differences. Several procedures have been proposed for analyzing the nature of the relationship following a statistically significant Kruskal–Wallis test (for instance, see Dunn, 1964; Miller, 1966; Ryan, 1959; Steel, 1960; Wike, 1971). Which one to use, however, is controversial among statisticians. We suggest the procedure recommended by Dunn (1964). The **Dunn procedure** involves using the Wilcoxon rank sum test or the Mann–Whitney U test to compare the mean ranks for all possible pairs of conditions, just as the HSD test was applied to all possible pairs of means following a statistically significant analysis of variance. In order to maintain the overall Type I error rate at the desired level of alpha, however, Dunn recommends that the critical value for each comparison be set based on a revised alpha level equal to alpha/C, where C represents the number of pairs of mean ranks to be tested.*

Since the desired alpha level in our example is .05 and there are three pairs of mean ranks (those for the one roommate versus the two roommates condition; those for the one roommate versus the three roommates condition; those for the two roommates versus the three roommates condition), the revised alpha level to be used with the Dunn procedure in this instance is $.05/3 = .017$. If we were to apply this procedure, we would find that the mean rank for the three roommates condition is significantly lower than the mean rank for the one roommate condition, suggesting that the attitudes toward living in dormitories of dormitory students who live with three roommates are less favorable than the attitudes of dormitory students who live with one roommate.

Method of Presentation

Reports of a Kruskal–Wallis test should include information about the degrees of freedom, the overall sample size, the value of the H statistic, and the significance level. Following the format for the chi-square test, the first two indices are placed in parentheses following the symbol for the test statistic. If the test is statistically significant, the strength of the relationship can be addressed by epsilon-squared and the nature of the relationship can be addressed in terms of the Dunn procedure. The results for the experiment on number of roommates and attitudes toward living in dormitories might be presented as follows:

Results

A Kruskal–Wallis test was applied to the ranked data relating number of roommates to attitudes toward living in dormitories. The resulting value of H was statistically significant, H(2, N = 18) = 10.93, p < .01. The strength of the relationship, as indexed by epsilon-squared, was .60. A follow-up procedure suggested by Dunn (1964) indicated that the attitudes of dormitory students who live with three roommates are less favorable than the attitudes of dormitory students who live with one roommate.

Notice that a reference has been provided for the Dunn procedure. This should be done anytime a procedure is not widely known, as doing so will allow interested readers to locate information about the technique.

* Critical values for use with the Dunn procedure corresponding to values of alpha/C can be estimated based on the critical values reported in the appropriate appendix for tabled levels of alpha.

16.5

Rank Test for Three or More Correlated Groups

Use of the Rank Test for Three or More Correlated Groups

Friedman analysis of variance by ranks is the nonparametric counterpart of one-way repeated measures analysis of variance. It is typically used to analyze the relationship between two variables when (1) scores on the dependent variable are in the form of ranks across conditions for each subject, (2) the independent variable is within-subjects in nature (it can be either qualitative or quantitative), and (3) the independent variable has three or more levels.

Analysis of the Relationship

Consider an experiment in which each of 10 individuals attempted to solve a different complex problem under quiet, slightly noisy, and noisy conditions. Performance under each condition was scored 0 to 20, with higher scores indicating better performance. The three problems had been pretested such that they were of equal difficulty and the order of conditions was counterbalanced. Table 16.6 presents the experimental data.

Friedman analysis of variance by ranks involves first rank ordering the scores *for each individual* across the research conditions, as has been done in Table 16.7. For example, the first individual had a score of 10 in the quiet condition, a score of 6 in the slightly noisy condition, and a score of 5 in the noisy condition. These correspond to ranks of 3, 2, and 1, respectively.

If the null hypothesis that the level of noise is unrelated to problem solving is true, then differences in the rank scores are merely a function of sampling error, and we would expect the sum of the ranks in each condition to be approximately equal. Friedman (1972) suggested the following statistic for ascertaining the magnitude of differences in rank sums:

$$\chi_r^2 = \frac{12\Sigma R_j^2}{Nk(k+1)} - 3N(k+1) \qquad [16.12]$$

where R_j is the sum of the rank scores in condition j, N is the sample size, k is the number of conditions, and the summation of R_j^2 is across all conditions. The χ_r^2 statistic has a sampling distribution that approximates a chi-square distribution with $k - 1$ degrees of freedom.

In our example,

$$\chi_r^2 = \frac{(12)(25^2 + 20^2 + 15^2)}{(10)(3)(3+1)} - (3)(10)(3+1)$$

$$= 125.00 - 120 = 5.00$$

For an alpha level of .05 and $k - 1 = 3 - 1 = 2$ degrees of freedom, the critical value of χ_r^2 from the chi-square table in Appendix J is 5.991. Since 5.00 does not exceed 5.991, we fail to reject the null hypothesis of no relationship between noise level and problem solving performance.

TABLE 16.6 Data for Friedman Analysis of Variance by Ranks Example

Individual	*Quiet*	*Slightly noisy*	*Noisy*
1	10	6	5
2	15	10	9
3	15	14	8
4	15	13	17
5	8	5	3
6	6	5	2
7	8	7	3
8	9	8	6
9	9	14	15
10	8	14	9

TABLE 16.7 Ranked Data for Friedman Analysis of Variance by Ranks Example

Individual	*Quiet*	*Slightly noisy*	*Noisy*
1	3	2	1
2	3	2	1
3	3	2	1
4	2	1	3
5	3	2	1
6	3	2	1
7	3	2	1
8	3	2	1
9	1	2	3
10	1	3	2
	$R_1 = 25$	$R_2 = 20$	$R_3 = 15$

The strength of the relationship for Friedman analysis of variance by ranks can be measured using an index of epsilon-squared applicable to the within-subjects situation:

$$E_R^2 = \frac{\chi_r^2 - (k + 1)}{Nk} \qquad [16.13]$$

where all terms are as previously defined. As with the measure of epsilon-squared presented in Section 16.4, this index can range from 0 to 1.00, with higher values indicating a stronger relationship between the independent and dependent variables. In our example,

$$E_R^2 = \frac{5.00 - (3 + 1)}{(10)(3)} = .03$$

If we had rejected the null hypothesis, it would have been necessary to conduct additional analyses to discern the nature of the relationship. The procedure recommended by Dunn (1964) could be used in such an instance. This would involve comparing the rank sums for all possible pairs of conditions using the Wilcoxon signed-rank test, but setting the alpha level for defining the critical value for each comparison at alpha/C (where C represents the number of pairs of rank sums to be tested) as a means of maintaining the overall Type I error rate at the desired level of alpha.

Method of Presentation The method of presentation for Friedman analysis of variance by ranks parallels that for the Kruskal–Wallis test as discussed in Section 16.4, except that the value of χ_r^2 is reported instead of H and the nature of the relationship is discussed in terms of the application of the Dunn procedure to the Wilcoxon signed-rank test rather than to the Wilcoxon rank sum test/Mann–Whitney U test.

16.6 Rank Test for Correlation

Use of the Rank Test for Correlation **Spearman rank-order correlation,** more simply known as *Spearman correlation,* is a nonparametric counterpart of Pearson correlation. It is typically used to analyze the relationship between two variables when (1) scores on both variables are in the form of ranks, (2) the two variables have been measured on the same individuals, and (3) the observations on each variable are between-subjects in nature.

Analysis of the Relationship Consider a study in which an investigator wanted to examine the relationship between pollution and cancer mortality rates. A pollution index was developed and applied to 20 cities in the United States. Scores on this index could range from 0 to 4, with higher scores indicating greater pollution. For each city, a measure of cancer mortality per 100,000 people was also obtained. The data for each city are presented in columns 2 and 3 of Table 16.8.

The first step of the analysis is to rank order the scores separately for each variable such that the lower scores receive the lower ranks and the higher scores receive the higher ranks. The ranked scores for the first variable (pollution) in our example are presented in column 4 of Table 16.8 under the heading R_1 and the ranked scores for the second variable (cancer mortality) are presented in column 5 under the heading R_2. The Spearman rank-order correlation procedure involves computing a correlation coefficient with respect to the ranked data. This coefficient can range from -1.00 through 0 to $+1.00$ and is interpreted in the same manner as the Pearson correlation coefficient. In fact, the Spearman correlation coefficient can be derived by applying the formula for the Pearson correlation coefficient directly to the ranked data using the proce-

TABLE 16.8 Data for Spearman Rank-Order Correlation Example

City	Pollution	Cancer mortality	R_1	R_2	Difference (D)	D^2
1	1.0	124	1	1	0	0
2	1.5	131	4	7	− 3	9
3	2.0	135	8	10	− 2	4
4	2.5	139	11	14	− 3	9
5	3.0	145	15	18	− 3	9
6	1.6	125	5	2	3	9
7	2.8	130	14	6	8	64
8	1.2	136	2	11	− 9	81
9	1.7	140	6	15	− 9	81
10	2.2	146	9	19	−10	100
11	2.6	126	12	3	9	81
12	3.2	129	16	5	11	121
13	3.7	137	18	12	6	36
14	4.0	141	20	16	4	16
15	1.3	133	3	9	− 6	36
16	1.9	127	7	4	3	9
17	2.4	132	10	8	2	4
18	2.7	138	13	13	0	0
19	3.4	142	17	17	0	0
20	3.8	148	19	20	− 1	1
						$\Sigma D^2 = 670$

dures discussed in Chapter 5. In practice, however, the Spearman correlation between two variables is most commonly calculated using a special formula available for this purpose.

The first step in deriving the Spearman correlation coefficient using this formula is to compute the difference, D, between the ranked scores for each individual. This has been done in column 6 of Table 16.8. The differences are then squared, as has been done in column 7. Lastly, the Spearman correlation coefficient (symbolized r_s) is calculated as follows:

$$r_s = 1 - \frac{6\Sigma D^2}{N(N^2 - 1)} \qquad [16.14]$$

where N is the sample size and D is as defined above. For our data, the Spearman correlation between pollution and cancer mortality is

$$r_s = 1 - \frac{(6)(670)}{(20)(20^2 - 1)}$$

$$= 1 - \frac{4{,}020}{7{,}980} = .50$$

When the sample size is greater than 10, the Spearman correlation coefficient can be converted into a score in a t distribution with $N - 2$ degrees of freedom. The following estimate of the standard error of r_s constitutes the denominator of the t test:

$$\hat{s}_{r_s} = \sqrt{\frac{1 - r_s^2}{N - 2}} \qquad [16.15]$$

Except for the fact that r_s has replaced r in the numerator, this formula is identical to the formula for the estimated standard error of Pearson r presented in Equation 14.1. In our example,

$$\hat{s}_{r_s} = \sqrt{\frac{1 - .50^2}{20 - 2}} = .20$$

The null hypothesis for Spearman correlation states that there is no linear relationship between the two rank variables, that is, that the population correlation, ρ_s, between the two rank variables is 0. Under the assumption that this hypothesis is true, the sample correlation coefficient can be converted into a t score as follows:

$$t = \frac{r_s - \rho_s}{\hat{s}_{r_s}} \qquad [16.16]$$

Note that this formula parallels the formula for converting r into t presented in Equation 14.2. In our example,

$$t = \frac{.50 - 0}{.20} = 2.50$$

The critical values of t for an alpha level of .05, nondirectional test, and $N - 2 = 20 - 2 = 18$ degrees of freedom are ± 2.101. Since the observed t value of 2.50 is greater than 2.101, we reject the null hypothesis and conclude that there is a relationship between pollution and cancer mortality.

Appendix M presents critical values of r_s that can be used to test for linear relationships between variables for selected sample sizes when $N \leq 30$. This approach is easier to use than the t test approach and *must* be used when $N \leq 10$, as application of the t test is not appropriate under this circumstance.

The strength of the relationship for Spearman correlation is indicated by the magnitude of the correlation coefficient and the nature of the relationship is indicated by the sign of the correlation coefficient. In our example, $r_s = .50$, so the relationship between the variables is direct; as one variable increases, so does the other. This interpretation was facilitated by the fact that the original scores for both variables were ranked from low to high. If one variable had been ranked from low to high and the other had been ranked from high to low, the magnitude of r_s would remain the same but the sign would be negative, indicating that as rank scores on one variable increase (high ranks representing high levels of the variable), rank scores on the second variable decrease (low ranks representing high levels of the variable). To avoid the possible misinterpretation that might arise from situations of this type, it is important that lower scores consistently receive the lower ranks and higher scores consistently receive the higher ranks when Spearman correlation is applied.

An alternative index of correlation for ranked scores is a statistic proposed by Kendall called *tau*. The computation of tau is considerably more complex than that of r_s, but it has certain advantageous statistical properties. For a comparison of the two techniques, see Glass and Stanley (1970).

Method of Presentation

Reports of an analysis using Spearman correlation should include statements of the sample size, the observed value of r_s, and the significance level. It is not necessary to explicitly state the strength of the relationship between the variables since this is indicated by the magnitude of the correlation coefficient. Although the nature of the relationship is similarly indicated by the sign of the correlation coefficient, researchers often choose to state this verbally. The results for the study on pollution and cancer mortality might be reported as follows:

Results

A Spearman rank-order correlation addressed the relationship between rank scores on the pollution and cancer mortality indices for the 20 cities. The observed correlation was found to be statistically significant, $\underline{r}_s = .50$, $\underline{p} < .05$, suggesting that as pollution in cities increases, so does mortality due to cancer.

16.7 Examples from the Literature

Closeness of Relationships and Intimacy of Touching Behavior

One factor that has been hypothesized to affect nonverbal intimacy is the nature of the relationship between the persons involved in the interaction. In a study of this issue, Heslin and Boss (1980) observed 103 travelers at an airport and a randomly selected member from the party that had come to greet or send off each one. It was proposed that the closeness of the relationship between the traveler and the other person would be positively related to the intimacy of their touching.

The nature of each relationship was determined after observation of the interaction by asking a nonobserved member of the greeting or send-off party. Each relationship was subsequently classified into one of seven ranked categories reflecting familiarity, involvement, and having influence over one another.

Nonverbal intimacy was assessed by classifying each interaction into one of six ranked touch categories ranging from "none" to "extended kissing and embracing." Consistent with the research hypothesis, a Spearman rank-order correlation indicated a statistically significant direct relationship between closeness of the relationship and intimacy of touching behavior, $r_s = .54$, $p < .01$.

Severity of the Final Verdict as a Function of the Order of Consideration of Lenient Versus Harsh Verdicts

Consider the following scenario (Greenberg, Williams, & O'Brien, 1986, p. 41):

> A defendant in a criminal case is charged with first degree murder in the stabbing death of a woman. Before deliberation, a judge instructs the jury first to consider whether or not the defendant is guilty of first degree murder. If they cannot agree to this verdict, they should next consider whether the defendant is guilty of second degree murder. Then, if the jurors cannot agree to this verdict, they are instructed to consider the verdict of voluntary manslaughter. If they still do not agree they are to consider involuntary manslaughter. Finally, if the jurors still have not reached agreement, they are instructed to find the defendant not guilty.

This procedure of having juries consider a series of increasingly less severe verdicts against a defendant when making their deliberations is a common component of criminal trials. However, it is possible that certain cognitive biases associated with this process will lead to harsher verdicts than if the jury were to reverse the order of deliberation and consider the verdicts from least severe to most severe instead.

To study this issue, Greenberg, Williams, and O'Brien (1986) had 16 college students read a condensed version of evidence presented at an actual murder trial. Subjects in the harsh-to-lenient condition were instructed to consider the verdict of "guilty of murder in the first degree" first and then to proceed progressively down the list of less severe verdicts noted above until a final verdict was reached. In contrast, subjects in the lenient-to-harsh condition were instructed to reach a final verdict by considering the verdict of "not guilty" first and then proceeding progressively up the list of more severe verdicts as necessary. This was done on an individual basis by each experimental participant.

A Mann–Whitney U test showed that subjects' verdicts were significantly harsher in the harsh-to-lenient than in the lenient-to-harsh condition, $U = 7$, $p < .01$. To the extent that this finding is generalizable to actual jury trials, it raises serious questions about the fairness of the judicial process. Greenberg and colleagues did not report any information regarding the strength of the relationship.

Effect of Exposure to an Aggressive Model on Observers' Aggressive Behavior

An important question concerns the effect of exposure to an aggressive model on observers' aggressive behavior. Bandura, Ross, and Ross (1963) studied this issue by assigning 96 nursery school children to either a no-model control condition or to conditions where they observed either an aggressive adult model on film, an aggressive adult model live, or an aggressive cartoon character. All models performed identical aggressive acts of repeatedly punching, kicking, hitting, and throwing an inflated Bobo doll. Subjects in the four conditions were matched on the basis of their teacher's ratings of their characteristic levels of aggressive behavior. Following the model manipulation, children in all conditions spent 20 minutes alone in a room containing a variety of toys, including a Bobo doll. The number of aggressive acts directed toward the Bobo doll and the other toys during this time span served as the measure of aggressive behavior.

In Chapter 11 it was noted that matched-subjects designs are analyzed as if they represent within-subjects variables. Given this and the fact that the researchers were concerned with violations of both the normality and the homogeneity of variance assumptions, the aggressiveness scores were analyzed using Friedman analysis of variance by ranks. This test was found to be statistically significant, $\chi_r^2(3, N = 24) = 9.06, p < .05$.* The strength of the relationship, as indexed by epsilon-squared, was .04. Follow-up tests showed that the level of aggressive behavior exhibited by children in the three model conditions did not significantly differ, but that children in all three model conditions behaved significantly more aggressively than children in the no-model condition. This basic finding has been replicated in a number of other studies.

16.8 Summary

When the assumptions of a parametric test are markedly violated, a nonparametric test can be used to analyze the relationship between two variables. Nonparametric statistics have the advantage of making few assumptions about population distributions but tend to be less powerful than their parametric counterparts. The major rank-based nonparametric tests for analyzing bivariate relationships are the Wilcoxon rank sum test/Mann–Whitney U test, the Wilcoxon signed-rank test, the Kruskal–Wallis test, Friedman analysis of variance by ranks, and Spearman rank-order correlation. Each of these is a counterpart to a parametric test, as indicated in Table 16.1. When scores are ranked for a given variable, the interpretation of results will be facilitated if ranks are assigned such that lower scores receive the lower ranks and higher scores receive the higher ranks.

---

* The overall sample size for purpose of the statistical test is 24 rather than 96 because the 96 children were matched and assigned to conditions such that 24 scores were provided in each one.

Appendix 16.1 Corrections for Ties for Nonparametric Rank Tests

All of the tests we have considered in this chapter deal with ties by assigning the average rank that the tied observations occupy. However, some of the procedures require modifications in the derivation of the test statistics in order to maintain a close approximation to the relevant sampling distribution when a large number of ties occur. Two tests where this is the case are the Wilcoxon rank sum test and the Kruskal–Wallis test.

For the Wilcoxon rank sum test, the formula for z when numerous ties occur is

$$z = \frac{|R_i - E_i| - 1}{\hat{s}_R} \qquad [16.17]$$

where

$$\hat{s}_R = \sqrt{\left[\frac{n_1 n_2}{N(N-1)}\right]\left(\frac{N^3 - N}{12} - \Sigma T\right)} \qquad [16.18]$$

In this formula, T equals $(t^3 - t)/12$, where t is the number of scores tied at a particular rank. The summation of T extends across all groups of ties. All other terms in Equations 16.17 and 16.18 are as defined in Section 16.2.

For the Kruskal–Wallis test, when numerous ties occur the value of H is computed by the following formula:

$$H = \frac{\left[\frac{12}{N(N+1)}\right]\left(\sum \frac{R_j^2}{n_j}\right) - 3(N+1)}{1 - \frac{\Sigma T}{N^3 - N}} \qquad [16.19]$$

where again T equals $(t^3 - t)/12$ and t is the number of scores tied at a particular rank. As before, the summation of T extends across all groups of ties. All other terms in Equation 16.19 are as defined in Section 16.4.

Exercises

Answers to asterisked (*) exercises appear at the back of the book. Answers to exercises with two asterisks are also worked out step-by-step in the study guide.

Exercises to Review Concepts

1. What are the advantages and disadvantages of nonparametric versus parametric statistical tests?

2. Match each parametric test in the first column with its nonparametric counterpart in the second column:

Parametric test	*Nonparametric test*
1. correlated groups *t* test	a. Wilcoxon rank sum test/Mann–Whitney *U* test
2. Pearson correlation	b. Friedman analysis of variance by ranks
3. one-way between-subjects analysis of variance	c. Spearman rank-order correlation
4. independent groups *t* test	d. Wilcoxon signed-rank test
5. one-way repeated measures analysis of variance	e. Kruskal–Wallis test

(continued)

***3.** Rank order each of the following sets of scores such that the lowest score in each set receives a rank of 1:

Set I	*Set II*	*Set III*	*Set IV*
135	131	130	104
132	131	136	102
136	136	135	102
131	135	130	102
134	132	135	105
134	134	134	106

4. Rank order each of the following sets of scores such that the lowest score in each set receives a rank of 1:

Set I	Set II	Set III	Set IV
12	6	4	10
16	3	1	6
9	4	1	13
16	7	2	12
9	11	8	6
13	10	3	10
8	2	7	5

5. What is the relationship between the Wilcoxon rank sum test and the Mann–Whitney U test?

****6.** A high school counselor wanted to test the relationship between car ownership and performance in school. The grade point averages of 15 students who own cars and 15 students who do not own cars were obtained. The data for the two groups are presented below. Conduct a nondirectional Wilcoxon rank sum test on these data and specify the nature of the relationship between the two variables.

Owns car			Does not own car		
3.10	2.10	2.90	2.20	3.00	2.00
2.50	2.30	3.90	2.40	2.19	3.40
2.75	2.15	2.70	2.60	2.98	3.60
2.96	3.30	2.64	2.80	2.77	4.00
3.50	3.70	1.90	3.20	2.66	3.80

***7.** Compute the value of the Glass rank biserial correlation coefficient for the data in Exercise 6.

****8.** A researcher was interested in comparing the attitudes of males and females toward an experimental male birth control pill. Eight males and eight females were presented a description of the characteristics of the pill and asked how favorably they would feel about its use (either by themselves or by their partner). Ratings were made on a scale of 0 to 100, where higher numbers indicated more favorable attitudes. The data for the study are presented below. Conduct a nondirectional Mann–Whitney U test on these data and specify the nature of the relationship between the two variables.

Males		Females	
50	59	52	61
48	65	50	66
74	67	76	86
56	82	63	84

***9.** Compute the value of the Glass rank biserial correlation coefficient for the data in Exercise 8.

10. A researcher was interested in the effect of marijuana on recognition of auditory stimuli. Eight subjects were presented an auditory stimulus that was increased in intensity until it could be heard (as signaled by the subject). The amount of time, in seconds, it took each subject to recognize the stimulus was measured. This task was performed in each of two conditions: (1) while under the influence of marijuana, and (2) while not under the influence of marijuana. The data for the study are presented below. Conduct a nondirectional Wilcoxon signed-rank test on these data and specify the nature of the relationship between the two variables.

Subject	Under influence of marijuana	Not under influence of marijuana
1	.73	.63
2	.83	.65
3	.69	.64
4	.67	.41
5	.72	.71
6	.43	.44
7	.84	.83
8	1.02	.86

11. Compute the value of the matched-pairs rank biserial correlation coefficient for the data in Exercise 10.

12. A sociologist wanted to compare the relative amount of violence on the three major television networks during prime time. During a week, eight randomly selected television shows for each network were watched and rated in terms of their violence. The ratings were made on a scale from 0 ("no violence") to 50 ("considerable violence"). The data for the three networks are presented below. Conduct a Kruskal–Wallis test on these data.

Network A		Network B		Network C	
20	38	21	39	17	43
28	40	29	41	27	26
32	42	33	45	31	50
35	15	34	13	30	16

13. Compute the value of epsilon-squared for the data in Exercise 12.

****14.** A consumer psychologist wanted to compare the quality ratings for picture tubes of three brands of television, A, B, C. Ten repairmen rated each brand on a scale from 0 to 50, where higher scores indicated higher quality. The data for the study are presented below. Conduct a Friedman analysis of variance by ranks on these data.

Repairman	*Brand A*	*Brand B*	*Brand C*
1	15	17	19
2	27	28	29
3	31	33	32
4	18	20	22
5	41	39	36
6	44	35	37
7	13	14	16
8	43	44	45
9	24	26	30
10	20	23	25

***15.** Compute the value of epsilon-squared for the data in Exercise 14.

****16.** A researcher was interested in the relationship between crime rates in cities and the size of a city's police force. For 18 large cities, the rank order of the crime rate (1 = lowest, 18 = highest) and the rank order of the size of the police force (1 = smallest, 18 = largest) were determined. The data for the study are presented below. Compute the Spearman rank-order correlation for these data, test for a relationship between the two variables using a nondirectional test, and specify the strength and nature of the relationship.

City	*Crime rate*	*Size of police force*
1	3	4
2	15	16
3	18	5
4	1	2
5	8	12
6	14	15
7	13	6
8	2	1
9	7	11
10	12	7
11	16	18
12	6	10
13	11	9
14	17	14
15	10	8
16	4	17
17	9	13
18	5	3

Exercises to Apply Concepts

****17.** A psychologist tested the effect of a group encounter session on individuals' self-esteem. A scale measuring self-esteem was administered to 10 individuals just before they participated in the encounter session and again just after. Scores on this scale could range from 10 to 80, with higher scores indicating higher self-esteem. The data for the study are presented below. Analyze these data using a nondirectional Wilcoxon signed-rank test, draw a conclusion, and write up your results using the principles given in the relevant Method of Presentation section.

Individual	*Before session*	*After session*
1	40	42
2	38	39
3	36	37
4	39	41
5	28	26
6	30	32
7	65	63
8	69	70
9	72	71
10	71	82

****18.** A psychologist was interested in the effect of the affective content of words on learning. Fifteen subjects were read a list of 25 words and asked to recall as many as they could. For five subjects, the words were all positive (for example, "intelligent"), for another five subjects the words were all negative (for example, "conceited"), and for the remaining five subjects, the words were all neutral (for example, "door"). The words comprising the three lists were equated on potentially relevant dimensions such as the frequency with which they occur in literature, etc. Learning scores were derived by scoring the number of words correctly recalled. The data for the study are presented below. Analyze these data using the Kruskal–Wallis test, draw a conclusion, and write up your results using the principles given in the relevant Method of Presentation section.

Positive words	*Negative words*	*Neutral words*
20	15	16
18	10	21
22	9	14
17	7	13
19	11	12

19. An educator wanted to test the effect of a film on students' knowledge about health. A health knowledge test was administered to 10 students before the film, just after the film, and again 1 week later. Scores on this test could range from 0 to 100, with higher scores indicating greater knowledge. The data for the study are presented below. Analyze these data using Friedman analysis of variance by ranks, draw a conclusion, and write up your results using the principles given in the relevant Method of Presentation section.

Individual	*Before film*	*Immediately after film*	*1 week after film*
1	28	31	29
2	32	39	37
3	14	23	21
4	20	17	15
5	36	35	33
6	40	43	41
7	24	27	25
8	16	19	18
9	10	13	11
10	44	47	45

20. A researcher wanted to test the relationship between conservatism and support for gun control. Twenty individuals were administered a scale measuring political conservatism and another scale measuring attitudes toward gun control. Scores on the first scale could range from 0 to 100, and scores on the second scale could range from 0 to 50, with higher scores indicating greater conservatism and more positive attitudes, respectively. The data for the study are presented below. Analyze these data using Spearman correlation, nondirectional test, draw a conclusion, and write up your results using the principles given in the relevant Method of Presentation section.

Individual	*Conservatism score*	*Attitude toward gun control*
1	74	10
2	67	48
3	75	28
4	56	14
5	64	24
6	66	26
7	85	47
8	79	35
9	87	46
10	60	23
11	68	32
12	52	11
13	54	15
14	58	16
15	83	44
16	81	45
17	77	34
18	73	30
19	62	22
20	70	31

3

ADDITIONAL TOPICS

SEVENTEEN

TWO-WAY BETWEEN-SUBJECTS ANALYSIS OF VARIANCE

Our focus to this point has been on the application of statistical techniques to the analysis of the relationship between two variables. For instance, we have seen how one-way between-subjects analysis of variance can be used to examine the relationship between a qualitative or quantitative independent variable and a quantitative dependent variable. However, in real life, it is rare that a given dependent variable is influenced by only one independent variable. For instance, a person's attitude toward abortion might be affected by the religion she was raised in, how religious she is, the type of upbringing she had, and so on. Thus, statistical techniques have been developed to analyze the relationship between a dependent variable and two or more independent variables. This chapter will consider one such technique, called **two-way between-subjects analysis of variance,** or more simply, *two-way analysis of variance* (abbreviated "two-way ANOVA").*

17.1 Factorial Designs

Suppose an investigator interested in factors that influence the number of children people want to have in their completed families posits that one important factor is an individual's religion. To examine this issue, he might conduct a study in which 50 Catholics and 50 Protestants are asked what they consider to be the ideal number of children. Suppose this is done, that the data are analyzed using the independent groups t test discussed in Chapter 10 (the independent variable being religion and the dependent variable being ideal family size), and that the results indicate that Catholics want more children than Protestants.

Suppose the same investigator is also interested in studying the effect of a second independent variable—how religious an individual is—on ideal family

* Our discussion of two-way analysis of variance will focus on the situation where the sample sizes in the individual groups being considered are all equal. The case of unequal sample sizes will be considered in Section 17.9.

TABLE 17.1 Example of a Two-Way Factorial Design

	RELIGIOSITY	
RELIGION	*Religious*	*Nonreligious*
Catholic	Religious Catholics	Nonreligious Catholics
Protestant	Religious Protestants	Nonreligious Protestants

size. Suppose further that the investigator has a measure of religiosity that reliably categorizes individuals into one of two categories, religious versus nonreligious. Again, a study could be conducted in which 50 religious and 50 nonreligious individuals are asked what they consider to be the ideal number of children. Suppose this is done, the independent groups t test is applied, and the results indicate that religious individuals want more children than nonreligious individuals.

From the two studies, we would conclude that both religion and religiosity are related to ideal family size. However, these studies do not tell us anything about how the two variables *act in conjunction with one another* to exert a joint influence. Fortunately, though, the joint effects of two independent variables on a dependent variable can be studied using **factorial designs.** Table 17.1 illustrates a factorial design for the present example. There are four groups of individuals in the study: (1) religious Catholics, (2) nonreligious Catholics, (3) religious Protestants, and (4) nonreligious Protestants. The four groups are defined by combining the two levels of each of the independent variables, or **factors**, as they are commonly called in the context of two-way analysis of variance. That is, the first factor, religion, has two levels, Catholic versus Protestant. The second factor, religiosity, also has two levels, religious versus nonreligious.* If we combine the two levels of religion with the two levels of religiosity, we have a total of four groups, each of which is represented by a unique combination of variables, or **cell,** in Table 17.1.

Because it examines two independent variables, the design represented in Table 17.1 is called a *two-way* factorial design. We can also refer to this as a 2 × 2 (read "two by two") factorial design. More generally, a factorial design having two factors can be represented by the notation $a \times b$, where a refers to the number of levels of the first factor and b refers to the number of levels of the second factor. If the present study had involved three levels of religion instead of two (for instance, Catholic, Protestant, and Jewish), then the design would have been a 3 × 2 factorial design involving six groups of subjects instead of four: (1) religious Catholics, (2) nonreligious Catholics, (3) religious Protestants, (4) nonreligious Protestants, (5) religious Jews, and (6) nonreligious

* The designation of religion as the first factor and religiosity as the second factor is arbitrary. If we had wished, we could have specified religiosity as the first factor and religion as the second factor instead.

Jews. The number of groups required in a between-subjects factorial design is simply the product of the number of levels of each factor. In a 2×2 factorial design, the number of required groups is 2 multiplied by 2, or 4. In a 4×2 factorial design, the number of required groups is 4 multiplied by 2, or 8.

The present chapter is restricted to the case where the relationship between two independent variables and one dependent variable is studied. However, factorial designs can also be used to examine the relationship between a dependent variable and three or more independent variables. For example, in addition to religion and religiosity, an investigator might also study the relationship of gender to ideal family size. This would involve analyzing three factors in the context of a $2 \times 2 \times 2$ factorial design. The first factor, consisting of two levels, would be religion; the second factor, also consisting of two levels, would be religiosity; and the third factor, again consisting of two levels, would be gender. This $2 \times 2 \times 2$ factorial design would involve studying eight $[(2)(2)(2) = 8]$ groups of individuals.

17.2

Use of Two-Way Between-Subjects Analysis of Variance

The conditions for using two-way analysis of variance are similar to those required for the independent groups t test (discussed in Chapter 10) or one-way between-subjects analysis of variance (discussed in Chapter 12) except that two independent variables rather than one independent variable are studied. Thus, two-way analysis of variance is typically used when

1. the dependent variable is quantitative in nature and is measured on approximately an interval level;
2. the independent variables are both *between-subjects* in nature (they can be either qualitative or quantitative);
3. the independent variables both have two or more levels; and
4. the independent variables are combined to form a factorial design.

Consider the example involving the effects of religion and religiosity on ideal family size. Ideal family size is the dependent variable and it is quantitative in nature. Religion is one of the independent variables; it is between-subjects in nature and has two levels (Catholic versus Protestant). Religiosity is the other independent variable; it is also between-subjects in nature and has two levels (religious versus nonreligious). Finally, the two factors are combined so as to yield a 2×2 factorial design. Given these conditions, two-way between-subjects analysis of variance is the statistical technique that would typically be used to analyze the relationship between the variables.

17.3

The Concepts of Main Effects and Interactions

As noted above, two-way factorial designs allow us to study the relationship between two independent variables and a dependent variable. Specifically, they allow us to address three issues, phrased here in terms of the example relating religion and religiosity to ideal family size:

1. Is there a relationship between religion, considered alone, and ideal family size?
2. Is there a relationship between religiosity, considered alone, and ideal family size?
3. Is there a relationship between religion and religiosity, considered *in combination,* and ideal family size, independent of the effects of religion alone and religiosity alone?

In practice, we will address these issues by examining sample means and using this information to make inferences about the state of affairs in the population. However, so that we can develop the relevant theoretical concepts without regard to sampling error, the examples in this section will be phrased in terms of *population* means. The appropriate inferential procedures will then be developed in Section 17.4.

The first two of our questions are addressed in terms of **main effects.** The main effect for religion is the comparison of the mean ideal family size for Catholics with the mean ideal family size for Protestants, *collapsed across,* or disregarding, religiosity. The main effect for religiosity is the comparison of the mean ideal family size for religious individuals with the mean ideal family size for nonreligious individuals, *collapsed across,* or disregarding, religion. Each of these effects has a null hypothesis and an alternative hypothesis associated with it. In each instance, the null hypothesis states that the population means for the two groups constituting that effect are equal to one another and the alternative hypothesis states that these means are not equal to one another. Specifically, the null and alternative hypotheses for the main effect of religion are

$$H_0\colon\ \mu_C = \mu_P$$
$$H_1\colon\ \mu_C \neq \mu_P$$

where the subscript "C" denotes the Catholic group and the subscript "P" denotes the Protestant group. The null and alternative hypotheses for the main effect of religiosity are

$$H_0\colon\ \mu_R = \mu_N$$
$$H_1\colon\ \mu_R \neq \mu_N$$

TABLE 17.2 Population Means for Ideal Family Size as a Function of Religion and Religiosity

RELIGION	RELIGIOSITY		MAIN EFFECT OF RELIGION
	Religious	*Nonreligious*	
Catholic	4.00	2.00	3.00
Protestant	2.00	2.00	2.00
MAIN EFFECT OF RELIGIOSITY	3.00	2.00	

where the subscripts "R" and "N" represent the religious and the nonreligious groups, respectively.

Table 17.2 presents hypothetical population means for the four groups in question as well as the corresponding main effect means. Assuming there are equal numbers of subjects in each group, the main effect means can be calculated by taking the average of the two cells in a given row or column. In this instance, there is a relationship between religion and ideal family size, since $\mu_C \neq \mu_P$. The nature of this relationship is such that Catholics, on the average, want more children than Protestants (3.00 versus 2.00). There is also a relationship between religiosity and ideal family size as indicated by the fact that $\mu_R \neq \mu_N$. On the average, religious individuals want more children than nonreligious individuals (3.00 versus 2.00).

The third question addressed in factorial designs concerns **interaction effects.** An interaction refers to the case where the nature of the relationship between one of the independent variables and the dependent variable differs as a function of the other independent variable. Consider the relationship between religion and ideal family size. Examine this relationship just for religious individuals in Table 17.2. *Considering only religious individuals,* there is a relationship between religion and ideal family size: Catholics, on the average, want more children than Protestants (4.00 versus 2.00). Now examine the same relationship just for nonreligious individuals. For these individuals, religion is unrelated to ideal family size. On the average, Catholics want the same number of children as Protestants (2.00 in each case). This illustrates an interaction. The nature of the relationship between religion and ideal family size depends on religiosity. For religious individuals, Catholics want more children than Protestants. For nonreligious individuals, however, Catholics want the same number of children as Protestants. An important feature of factorial designs is that they allow us to test for such effects.

Although we have stated that the relationship between religion and ideal family size depends on religiosity, we could have stated the reverse instead—that the nature of the relationship between religiosity and ideal family size differs as a function of religion. Consider just Catholics. Inspection of Table 17.2 reveals that religious Catholics, on the average, want more children than nonreligious Catholics (4.00 versus 2.00). Now consider just Protestants. Table

17.2 shows that religious Protestants, on the average, want the same number of children as nonreligious Protestants (2.00 in each case). The nature of the relationship between religiosity and ideal family size depends on religion. For Catholics, religious individuals want more children than nonreligious individuals. For Protestants, religious individuals want the same number of children as nonreligious individuals.

The manner in which we choose to state an interaction is guided by the research questions being investigated. The important point for present concerns is that religion and religiosity are interacting in our example such that their joint influence uniquely affects the dependent variable.

Identifying Main Effects and Interactions

In statistical jargon, a main effect is said to be present if the null hypothesis concerning that effect is rejected. If an investigator were to conclude that there is "a main effect of religiosity" but "no main effect of religion" for a set of data, this would mean that the null hypothesis was rejected for the religiosity factor but not for the religion factor. Similarly, the "presence" of an interaction means that the null hypothesis of no interaction effect was rejected.

Figure 17.1 presents examples of various population means for the four groups defined by the two levels of religion and the two levels of religiosity. The examples have also been depicted in graphs. On the abscissa of the graphs are demarcations for the two levels of religion, Catholic and Protestant. On the ordinate of the graphs are demarcations for values of the dependent variable, ideal family size. Directly above each abscissa demarcation for Protestants is a point corresponding to the mean ideal family size for religious Protestants and directly above each abscissa demarcation for Catholics is a point corresponding to the mean ideal family size for religious Catholics.* For each graph, these two points are connected by a straight line, as are the corresponding points for nonreligious individuals.

Since we are dealing with population values, the most straightforward way to determine if there is a main effect of religion is to compare the means of the two columns. If they are the same, no main effect is present; otherwise, one is present. The presence or absence of a main effect of religiosity is similarly determined by comparing the means of the two rows. When dealing with population means, the determination of whether an interaction is present is most readily made by examining the slopes of the lines in a given graph. One of these lines represents the relationship between religion and ideal family size for religious individuals and the other line represents this relationship for nonreligious individuals. If the lines are parallel (or, as is the case in Figures 17.1a and 17.1b, overlapping), this means that the relationship is the same for both religious and nonreligious individuals. Thus, there is no interaction. If the lines are

* The representation of religion on the abscissa and religiosity in the body of the graphs is arbitrary. If we had wished, we could have represented religiosity on the abscissa and religion in the body of the graphs instead.

FIGURE 17.1 Examples of Main Effects and Interactions for a 2 × 2 Factorial Design (Adapted from Johnson & Liebert, 1977)

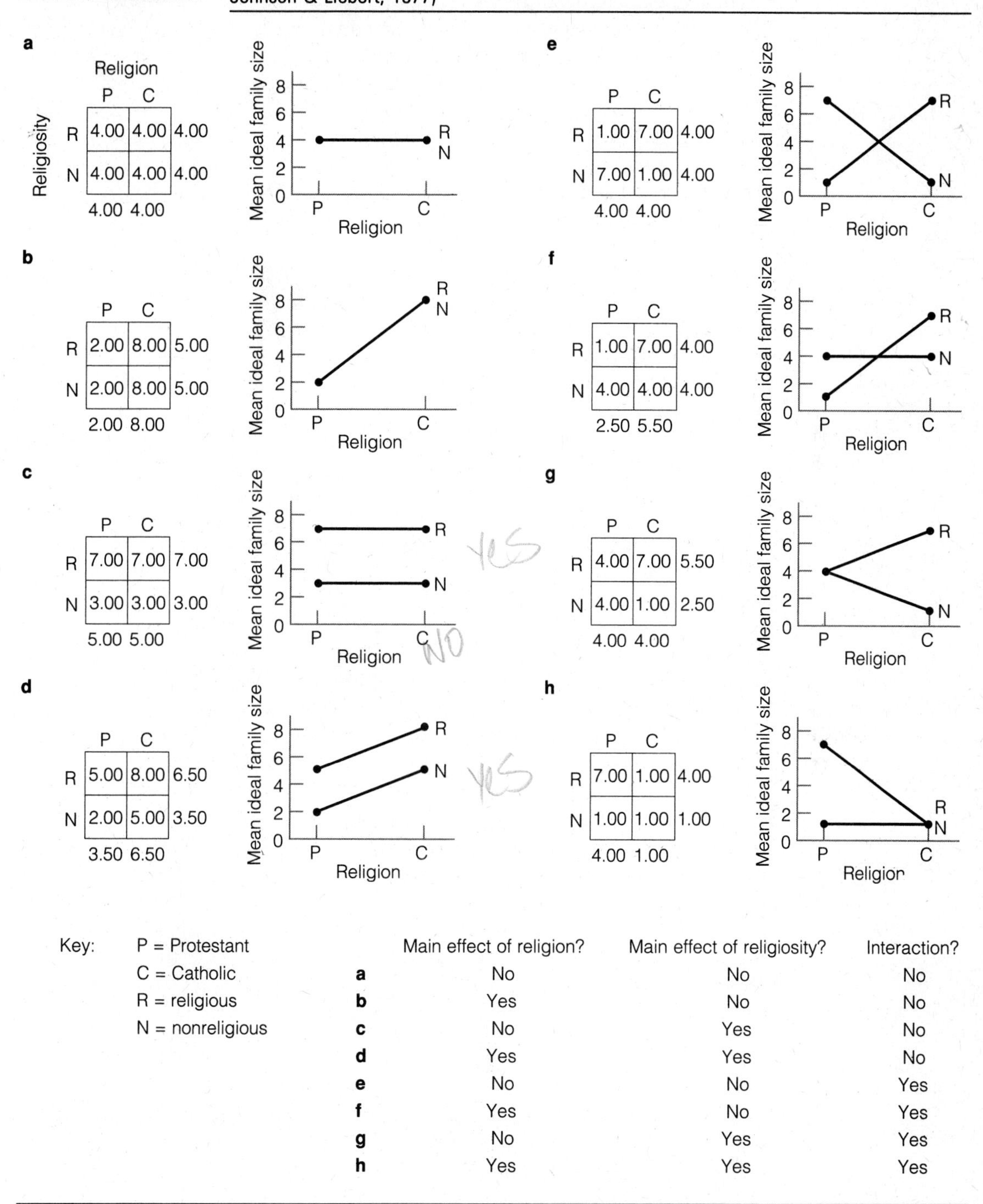

Key: P = Protestant
C = Catholic
R = religious
N = nonreligious

	Main effect of religion?	Main effect of religiosity?	Interaction?
a	No	No	No
b	Yes	No	No
c	No	Yes	No
d	Yes	Yes	No
e	No	No	Yes
f	Yes	No	Yes
g	No	Yes	Yes
h	Yes	Yes	Yes

not parallel, then the nature of the relationship between religion and ideal family size depends on religiosity, meaning that an interaction is present.

Figure 17.1a represents a case where there are neither main effects nor an interaction. In Figure 17.1b, there is a main effect of religion (Protestants, $\mu = 2.00$, want fewer children than Catholics, $\mu = 8.00$) but no main effect of religiosity and no interaction. As indicated by the parallel lines, there are also no interactions in Figures 17.1c or 17.1d. However, there is a main effect of religiosity in Figure 17.1c (religious individuals, $\mu = 7.00$, want more children than nonreligious individuals, $\mu = 3.00$), and main effects of both religion (Protestants, $\mu = 3.50$, want fewer children than Catholics, $\mu = 6.50$) and religiosity (religious individuals, $\mu = 6.50$, want more children than nonreligious individuals, $\mu = 3.50$) in Figure 17.1d.

Figures 17.1e–17.1h represent cases where interactions occur. Note that in each graph, the lines are nonparallel. This signifies that the nature of the relationship between religion and ideal family size differs as a function of religiosity. In Figure 17.1e, religious Protestants ($\mu = 1.00$) want fewer children than religious Catholics ($\mu = 7.00$) but nonreligious Protestants ($\mu = 7.00$) want more children than nonreligious Catholics ($\mu = 1.00$). In Figure 17.1f, religious Protestants ($\mu = 1.00$) again want fewer children than religious Catholics ($\mu = 7.00$). However, both nonreligious Protestants and nonreligious Catholics have an ideal family size of 4.00. There is also a mean effect of religion in Figure 17.1f such that, collapsing across religiosity, Protestants ($\mu = 2.50$) want fewer children than Catholics ($\mu = 5.50$). See if you can interpret the nature of the main effects and interactions for Figures 17.1g and 17.1h.

Main Effects and Interactions in Designs with More Than Two Levels for a Factor

The principles for detecting main effects and interactions are readily generalizable from 2×2 designs to designs with more than two levels for a factor. Figure 17.2 presents examples of a 3×2 factorial design having three levels of religion rather than two: Jewish, Protestant, and Catholic. Whenever the lines are parallel, a lack of interaction is indicated. This is the case in Figures 17.2b and 17.2d. For these examples, the nature of the relationship between one independent variable and the dependent variable is the same irrespective of the level of the other independent variable. When the lines are not parallel, the presence of an interaction is indicated. This is the case in Figures 17.2a and 17.2c. As can be seen, for these examples the nature of the relationship between religion and ideal family size differs as a function of religiosity. In all four examples, there are main effects of both religion and religiosity.

Main Effects and Interactions with Sample Data

In the following section, we will examine inferential techniques for determining the existence of main effects and interactions in populations based on sample data. At this point, we wish to emphasize that when dealing with sample data, we *cannot* make these determinations by mere visual inspection of sample means or slopes of lines in a graph. Rather, the existence of a given population main effect or interaction can be affirmed only by a significant statistical test. Thus, as with all of the previous designs that we have considered,

FIGURE 17.2 Examples of Main Effects and Interactions for a 3 × 2 Factorial Design

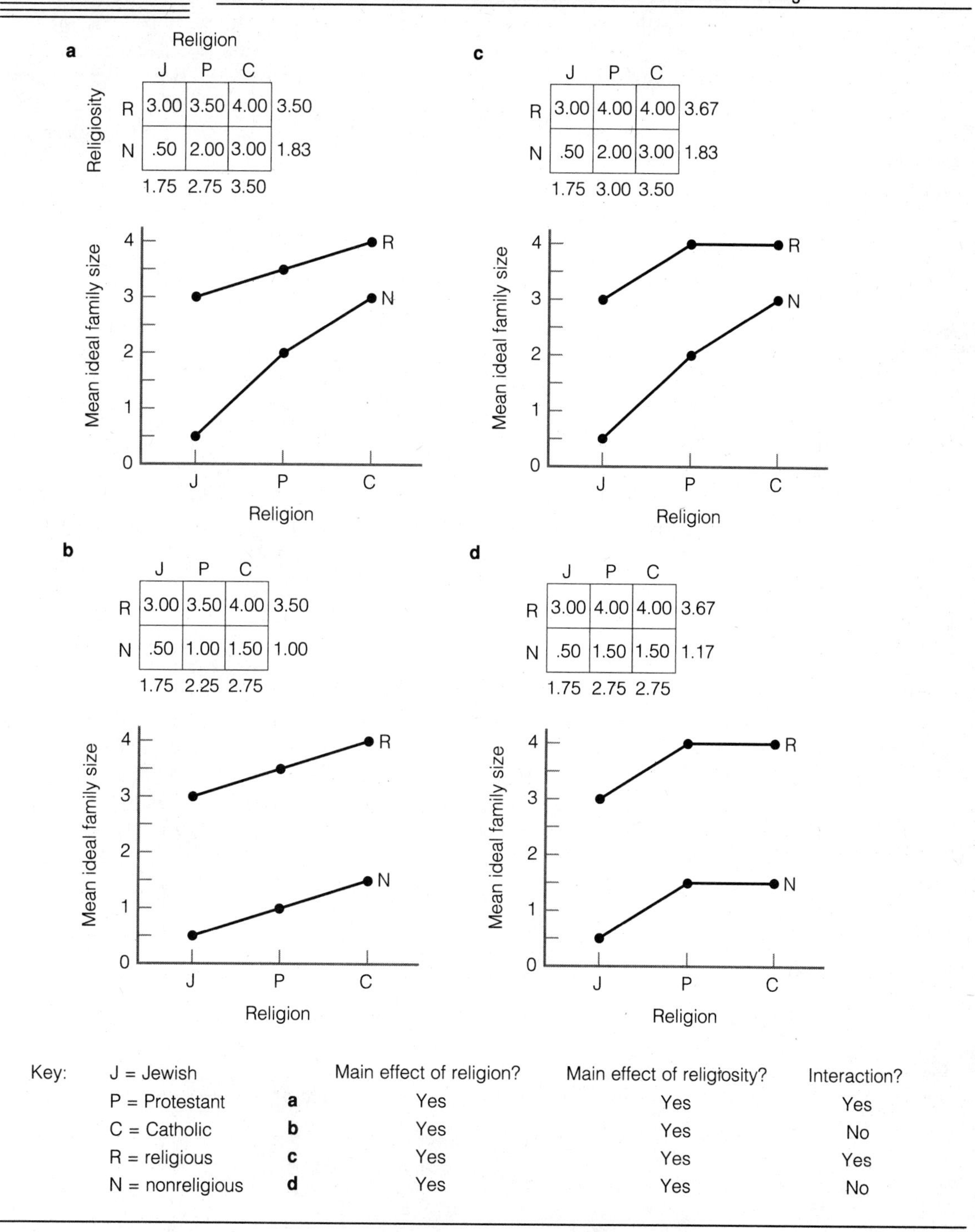

due to the role of sampling error, nonequivalent sample means do not necessarily indicate nonequivalent population means. For the same reason, we cannot assume that an interaction exists in the population just because the lines representing the sample means are nonparallel. If an interaction effect *is* statistically significant, however, these lines will indeed have different slopes and an examination of them can help to clarify the nature of the relationship between the variables.

17.4

Inference of a Relationship Using Two-Way Between-Subjects Analysis of Variance

Breakdown of $SS_{BETWEEN}$

In Chapter 12, the relationship between a between-subjects independent variable and a dependent variable was analyzed by defining the total variability in terms of two components, between-group variability and within-group variability. Recall that

$$SS_{TOTAL} = SS_{BETWEEN} + SS_{WITHIN} \quad [17.1]$$

The quantities $SS_{BETWEEN}$ and SS_{WITHIN} were each divided by their respective degrees of freedom to obtain a mean square between and a mean square within. An F ratio was then formed by dividing $MS_{BETWEEN}$ by MS_{WITHIN}. In two-way analysis of variance, similar steps are taken. However, the overall sum of squares between is further partitioned into three components: (1) variability due to the first independent variable (designated as "factor A"), (2) variability due to the second independent variable (designated as "factor B"), and (3) variability due to the interaction of factors A and B (symbolized as "$A \times B$"). Thus,

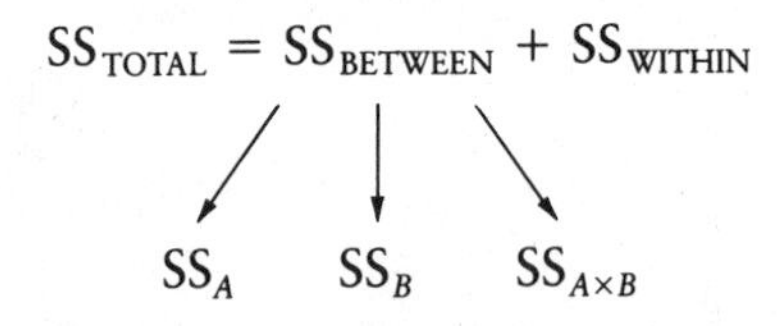

where SS_A is the sum of squares defining between-group variability for factor A, SS_B is the sum of squares defining between-group variability for factor B, and $SS_{A\times B}$ is the sum of squares defining between-group variability for the $A \times B$ interaction. Stated more formally,

$$SS_{BETWEEN} = SS_A + SS_B + SS_{A\times B} \quad [17.2]$$

The sum of squares total can thus be represented as

$$SS_{TOTAL} = SS_A + SS_B + SS_{A\times B} + SS_{WITHIN} \quad [17.3]$$

In two-way analysis of variance, each of the components of $SS_{BETWEEN}$ is divided by its respective degrees of freedom and the resulting mean squares, MS_A, MS_B, and $MS_{A \times B}$, are then divided by MS_{WITHIN} to yield F ratios. The F ratio formed by MS_A/MS_{WITHIN} is used to test the null hypothesis with regard to the main effect of factor A. The F ratio formed by MS_B/MS_{WITHIN} is used to test the null hypothesis with respect to the main effect of factor B. Finally, the F ratio formed by $MS_{A \times B}/MS_{WITHIN}$ is used to test the null hypothesis with respect to the $A \times B$ interaction.

Computation of the Sums of Squares

Table 17.3 presents data for the 2×2 factorial design from Table 17.1 that will be used as a numerical example. Unlike the situation in Section 17.3, we are now working with sample data and will continue to do so for the remainder of the chapter. The first factor, A, is religion; the second factor, B, is religiosity; and the dependent variable is ideal family size. The main effect means for the Catholic and Protestant levels of factor A are indicated by $\bar{X}_C$ and $\bar{X}_P$, respectively. The main effect means for the religious and nonreligious levels of factor B are similarly indicated by $\bar{X}_R$ and $\bar{X}_N$. The notation for the mean scores for the four groups comprising the 2×2 design requires two subscripts. The first subscript identifies the level of factor A and the second subscript identifies the level of factor B. For instance, the cell mean for nonreligious Catholics (or, consistent with the order of the subscripts, Catholic-nonreligious individuals) is indicated by $\bar{X}_{CN}$ and the cell mean for religious Protestants (that is, Protestant-religious individuals) is indicated by $\bar{X}_{PR}$.

The sum of squares total is calculated using the procedures discussed in Chapter 12 for applying the standard computational formula for a sum of squares to an entire data set. Specifically,

$$SS_{TOTAL} = \Sigma X^2 - \frac{(\Sigma X)^2}{N} \qquad [17.4]$$

where ΣX^2 is the sum of the squared scores constituting the data set, $(\Sigma X)^2$ is the square of the summed scores, and N is the total sample size. Looking at Table 17.3, we find that the sum of the 20 X^2 scores in our example is

$$\Sigma X^2 = 49 + 16 + \cdots + 4 + 4 = 246$$

and the sum of the 20 X scores is

$$\Sigma X = 7 + 4 + \cdots + 2 + 2 = 60$$

Squaring the latter value and dividing by N, we obtain

$$\frac{(\Sigma X)^2}{N} = \frac{60^2}{20} = 180$$

Thus, the sum of squares total in this instance is

$$SS_{TOTAL} = \Sigma X^2 - \frac{(\Sigma X)^2}{N}$$
$$= 246 - 180 = 66$$

TABLE 17.3 Data and Computation of Sums of Squares for Investigation on Ideal Family Size as a Function of Religion and Religiosity

RELIGION (A)	RELIGIOSITY (B): Religious X	X^2	Nonreligious X	X^2	MAIN EFFECT OF RELIGION
Catholic	7	49	5	25	$T_C = 40$
	4	16	2	4	$\bar{X}_C = 4.00$
	3	9	1	1	$T_C^2 = 1{,}600$
	5	25	3	9	
	6	36	4	16	
	$T_{CR} = 25$		$T_{CN} = 15$		
	$\bar{X}_{CR} = 5.00$		$\bar{X}_{CN} = 3.00$		
	$T_{CR}^2 = 625$		$T_{CN}^2 = 225$		
Protestant	0	0	2	4	$T_P = 20$
	4	16	4	16	$\bar{X}_P = 2.00$
	2	4	0	0	$T_P^2 = 400$
	2	4	2	4	
	2	4	2	4	
	$T_{PR} = 10$		$T_{PN} = 10$		
	$\bar{X}_{PR} = 2.00$		$\bar{X}_{PN} = 2.00$		
	$T_{PR}^2 = 100$		$T_{PN}^2 = 100$		
MAIN EFFECT OF RELIGIOSITY	$T_R = 35$		$T_N = 25$		
	$\bar{X}_R = 3.50$		$\bar{X}_N = 2.50$		
	$T_R^2 = 1{,}225$		$T_N^2 = 625$		

$$\Sigma X^2 = 49 + 16 + \cdots + 4 + 4 = 246$$
$$\Sigma X = 7 + 4 + \cdots + 2 + 2 = 60$$
$$\frac{(\Sigma X)^2}{N} = \frac{60^2}{20} = 180$$
$$\frac{\Sigma T_{A_i}^2}{nb} = \frac{1{,}600 + 400}{(5)(2)} = 200$$
$$\frac{\Sigma T_{B_j}^2}{na} = \frac{1{,}225 + 625}{(5)(2)} = 185$$
$$\frac{\Sigma T_{A_iB_j}^2}{n} = \frac{625 + 225 + 100 + 100}{5} = 210$$

The formula for calculating the sum of squares for between-group variability due to factor A (religion) is

$$\mathrm{SS}_A = \frac{\Sigma T_{A_i}^2}{nb} - \frac{(\Sigma X)^2}{N} \qquad [17.5]$$

where $T_{A_i}^2$ is the square of the sum of the scores at level i of factor A, n is the per-cell sample size (remember that we are focusing on the situation where sample sizes in the individual groups being considered are all equal), b is the number of levels of factor B, and $(\Sigma X)^2$ and N are as defined previously. In our example, $T^2_{A_i}$ equals 1,600 for Catholics and 400 for Protestants. Thus,

$$\frac{\Sigma T^2_{A_i}}{nb} = \frac{1{,}600 + 400}{(5)(2)} = 200$$

so

$$\begin{aligned} SS_A &= \frac{\Sigma T^2_{A_i}}{nb} - \frac{(\Sigma X)^2}{N} \\ &= 200 - 180 = 20 \end{aligned}$$

The formula for calculating the sum of squares for between-group variability due to factor B (religiosity) is

$$SS_B = \frac{\Sigma T^2_{B_j}}{na} - \frac{(\Sigma X)^2}{N} \qquad [17.6]$$

where $T^2_{B_j}$ is the square of the sum of the scores at level j of factor B, a is the number of levels of factor A, and all other terms are as defined above. In our example, $T^2_{B_j}$ equals 1,225 for religious individuals and 625 for nonreligious individuals. Thus,

$$\frac{\Sigma T^2_{B_j}}{na} = \frac{1{,}225 + 625}{(5)(2)} = 185$$

so

$$\begin{aligned} SS_B &= \frac{\Sigma T^2_{B_j}}{na} - \frac{(\Sigma X)^2}{N} \\ &= 185 - 180 = 5 \end{aligned}$$

The formula for calculating the sum of squares within is

$$SS_{\text{WITHIN}} = \Sigma X^2 - \frac{\Sigma T^2_{A_iB_j}}{n} \qquad [17.7]$$

where $T^2_{A_iB_j}$ is the square of the sum of the scores in cell A_iB_j and ΣX^2 and n are as defined above. In our example, $T^2_{A_iB_j}$ equals 625 for religious Catholics, 225 for nonreligious Catholics, 100 for religious Protestants, and 100 for nonreligious Protestants. Thus,

$$\frac{\Sigma T^2_{A_iB_j}}{n} = \frac{625 + 225 + 100 + 100}{5} = 210$$

so

$$\begin{aligned} SS_{\text{WITHIN}} &= \Sigma X^2 - \frac{\Sigma T^2_{A_iB_j}}{n} \\ &= 246 - 210 = 36 \end{aligned}$$

The formula for calculating the sum of squares for between-group variability due to the interaction of factor A and factor B is

$$SS_{A \times B} = \frac{(\Sigma X)^2}{N} + \frac{\Sigma T^2_{A_iB_j}}{n} - \frac{\Sigma T^2_{A_i}}{nb} - \frac{\Sigma T^2_{B_j}}{na} \qquad [17.8]$$

where all terms are as defined above. In our example,

$$SS_{A \times B} = 180 + 210 - 200 - 185 = 5$$

Of course, given that the sum of squares total is equal to the sum of squares for factor A plus the sum of squares for factor B plus the sum of squares for the $A \times B$ interaction plus the sum of squares within, if four of these quantities have already been calculated, the fifth can be determined through simple algebraic manipulation. For instance, we could have obtained the sum of squares for the $A \times B$ interaction as follows:

$$SS_{A \times B} = SS_{\text{TOTAL}} - SS_A - SS_B - SS_{\text{WITHIN}}$$
$$= 66 - 20 - 5 - 36 = 5$$

Derivation of the Summary Table

Each of the above sums of squares has a certain number of degrees of freedom associated with it. These can be represented as follows:

$$df_A = a - 1 \qquad [17.9]$$

$$df_B = b - 1 \qquad [17.10]$$

$$df_{A \times B} = (a - 1)(b - 1) \qquad [17.11]$$

$$df_{\text{WITHIN}} = (a)(b)(n - 1) \qquad [17.12]$$

As with the sums of squares, the degrees of freedom are additive:

$$df_{\text{TOTAL}} = df_A + df_B + df_{A \times B} + df_{\text{WITHIN}} \qquad [17.13]$$

Since $df_A = a - 1$, $df_B = b - 1$, $df_{A \times B} = (a - 1)(b - 1)$, and $df_{\text{WITHIN}} = (a)(b)(n - 1)$, and $(a - 1) + (b - 1) + [(a - 1)(b - 1)] + [(a)(b)(n - 1)] = a - 1 + b - 1 + (ab - a - b + 1) + (abn - ab) = abn - 1 = N - 1$, the degrees of freedom total can be calculated directly as

$$df_{\text{TOTAL}} = N - 1 \qquad [17.14]$$

In our example,

$$df_A = 2 - 1 = 1$$

$$df_B = 2 - 1 = 1$$

$$df_{A \times B} = (2 - 1)(2 - 1) = 1$$

$$df_{\text{WITHIN}} = (2)(2)(5 - 1) = 16$$

$$df_{\text{TOTAL}} = 20 - 1 = 19$$

The relevant mean squares are obtained by dividing the sum of squares for each source of between-group variability by its degrees of freedom:

$$\begin{aligned} MS_A &= \frac{SS_A}{df_A} \\ &= \frac{20}{1} = 20.00 \end{aligned} \quad [17.15]$$

$$\begin{aligned} MS_B &= \frac{SS_B}{df_B} \\ &= \frac{5}{1} = 5.00 \end{aligned} \quad [17.16]$$

$$\begin{aligned} MS_{A \times B} &= \frac{SS_{A \times B}}{df_{A \times B}} \\ &= \frac{5}{1} = 5.00 \end{aligned} \quad [17.17]$$

$$\begin{aligned} MS_{\text{WITHIN}} &= \frac{SS_{\text{WITHIN}}}{df_{\text{WITHIN}}} \\ &= \frac{36}{16} = 2.25 \end{aligned} \quad [17.18]$$

The F ratios used to test the null hypotheses for the two main effects and the $A \times B$ interaction are computed as follows:

$$\begin{aligned} F_A &= \frac{MS_A}{MS_{\text{WITHIN}}} \\ &= \frac{20.00}{2.25} = 8.89 \end{aligned} \quad [17.19]$$

$$\begin{aligned} F_B &= \frac{MS_B}{MS_{\text{WITHIN}}} \\ &= \frac{5.00}{2.25} = 2.22 \end{aligned} \quad [17.20]$$

$$\begin{aligned} F_{A \times B} &= \frac{MS_{A \times B}}{MS_{\text{WITHIN}}} \\ &= \frac{5.00}{2.25} = 2.22 \end{aligned} \quad [17.21]$$

All of the preceding calculations can be summarized in a summary table:

Source	*SS*	*df*	*MS*	*F*
A (Religion)	20.00	1	20.00	8.89
B (Religiosity)	5.00	1	5.00	2.22
$A \times B$	5.00	1	5.00	2.22
Within	36.00	16	2.25	
Total	66.00	19		

This table is similar in format to the summary tables in Chapter 12, but the sum of squares between is represented by its three components, the main effect for factor *A*, the main effect for factor *B*, and the $A \times B$ interaction. Let us now examine how we can use the information in this table to address the three questions about a relationship for each of these sources of variability.

Inference of a Relationship

The null and alternative hypotheses for the main effect of religion are

$$H_0\colon \quad \mu_C = \mu_P$$
$$H_1\colon \quad \mu_C \neq \mu_P$$

The test of the null hypothesis is made with reference to the *F* value derived from MS_A/MS_{WITHIN}. This *F* equals 8.89 with 1 and 16 degrees of freedom. Appendix F indicates that the critical value of *F* for these degrees of freedom and an alpha level of .05 is 4.49. Since the observed *F* value is greater than 4.49, we reject the null hypothesis and conclude that there is a relationship between religion and ideal family size.

The null and alternative hypotheses for the main effect of religiosity are

$$H_0\colon \quad \mu_R = \mu_N$$
$$H_1\colon \quad \mu_R \neq \mu_N$$

The test of the null hypothesis is made with reference to the *F* value derived from MS_B/MS_{WITHIN}. This *F* equals 2.22 with 1 and 16 degrees of freedom. This does not exceed the critical value of 4.49 and, hence, we fail to reject the null hypothesis. We cannot confidently conclude that there is a relationship between religiosity and ideal family size.

Depending on the research questions being investigated, the null and alternative hypotheses for the interaction effect can be phrased in either of two ways. If we are concerned with the relationship between religion and ideal family size as a function of religiosity, the null hypothesis will state that the relationship between religion and ideal family size is the same for both religious and nonreligious individuals and the alternative hypothesis will state that the relationship between religion and ideal family size is *not* the same for both religious and nonreligious individuals. If we are concerned with the relationship between religiosity and ideal family size as a function of religion, the null hypothesis will state that the relationship between religiosity and ideal family size is the same for both Catholics and Protestants and the alternative hypothesis will state that the relationship between religiosity and ideal family size is *not* the same for both Catholics and Protestants. In either instance, the test of the null hypothesis is made with reference to the F value derived from $MS_{A \times B}/MS_{WITHIN}$. This F equals 2.22 with 1 and 16 degrees of freedom. This does not exceed the critical value of 4.49 and, hence, we fail to reject the null hypothesis. We cannot confidently conclude that there is an interaction between religion and religiosity.

Assumptions of the *F* Tests

The preceding F tests are based on the same assumptions that underlie one-way between-subjects analysis of variance as discussed in Chapter 12. These are:

1. The samples are independently and randomly selected from their respective populations.
2. The scores in each population are normally distributed.
3. The scores in each population have homogeneous variances.

In addition, the dependent variable should be quantitative in nature and measured on approximately an interval level. As with one-way analysis of variance, under certain conditions the F tests are robust to violations of the normality and homogeneity of variance assumptions. This is particularly true when sample (cell) sizes are equal and relatively large. Thus, a researcher should try, when possible, to employ an equal number of subjects in each group, with sample sizes being as large as circumstances reasonably permit. Interested readers are referred to Jaccard and Becker (1988) for additional information on the robustness of two-way analysis of variance.

17.5

Strength of the Relationship

The strengths of the relationships for the three sources of between-group variability are computed using the following formulas for eta-squared:

$$\text{eta}_A^2 = \frac{SS_A}{SS_{\text{TOTAL}}} \quad [17.22]$$

$$\text{eta}_B^2 = \frac{SS_B}{SS_{\text{TOTAL}}} \quad [17.23]$$

$$\text{eta}_{A \times B}^2 = \frac{SS_{A \times B}}{SS_{\text{TOTAL}}} \quad [17.24]$$

where eta_A^2 is eta-squared for the main effect of factor A, eta_B^2 is eta-squared for the main effect of factor B, and $\text{eta}_{A \times B}^2$ is eta-squared for the $A \times B$ interaction.

For our example,

$$\text{eta}_A^2 = \frac{20}{66} = .30$$

$$\text{eta}_B^2 = \frac{5}{66} = .08$$

$$\text{eta}_{A \times B}^2 = \frac{5}{66} = .08$$

These values represent the proportion of variability in the dependent variable that is associated with the particular source of between-group variability. The proportion of variability in ideal family size that is associated with the main effect of religion is .30. This represents a strong effect. The proportion of variability in ideal family size that is associated with the main effect of religiosity and with the interaction between religion and religiosity is .08 in both instances. These represent weak effects.

17.6

Nature of the Relationship

Analysis of Main Effects

When a significant main effect has only two levels, the nature of the relationship is determined in the same fashion as for the independent groups *t* test. This involves making inferences about the two population means from examination of the two sample means. In the present example, the nature of the relationship between religion and ideal family size is such that Catholics ($\bar{X}_C$ = 4.00) want more children than Protestants ($\bar{X}_P$ = 2.00).

If a statistically significant main effect has three or more levels, then the nature of the relationship is determined using an HSD procedure conceptually identical to the one discussed in Chapter 12. This involves computing the absolute difference between all possible pairs of sample means comprising the main effect and then comparing each of these against a critical difference.

For the main effect of factor *A*, the critical difference is defined as

$$\text{CD} = q\sqrt{\frac{\text{MS}_{\text{WITHIN}}}{nb}} \qquad [17.25]$$

where CD is the abbreviation for the critical difference, *q* is a Studentized range value obtainable from Appendix G, $\text{MS}_{\text{WITHIN}}$ is the mean square within from the summary table, *n* is the per-cell sample size, and *b* is the number of levels of factor *B*. The value of *q* is determined with reference to the overall alpha level for the effect, the degrees of freedom within, and the number of levels of *factor*

A (symbolized k in Appendix G). We will follow the practice of adopting an overall alpha level of .05 for each main effect that we will analyze.

For the main effect of factor B, the critical difference is defined as

$$\text{CD} = q\sqrt{\frac{\text{MS}_{\text{WITHIN}}}{na}} \qquad [17.26]$$

where a is the number of levels of factor A and all other terms are as defined previously. In this case, however, q is determined with reference to the overall alpha level for the effect, the degrees of freedom within, and the number of levels of *factor B* (again symbolized k in Appendix G).

If the absolute difference between a given pair of sample means exceeds the critical difference, we conclude that the corresponding population means differ from one another. If the absolute difference between the sample means does not exceed the critical difference, we are unable to draw this conclusion. We will consider a numerical example involving a factor with three levels in Section 17.8.

Analysis of Interactions

When an interaction effect is statistically significant, the nature of the interaction can be determined using a statistical procedure called **simple main effects analysis.** This involves an examination of the relationship between one independent variable and the dependent variable at each level of the other independent variable. This is accomplished by phrasing and testing a series of relevant null and alternative hypotheses. For instance, in our example, the null and alternative hypotheses relating religion to ideal family size for religious individuals are

$$H_0\colon \quad \mu_{\text{CR}} = \mu_{\text{PR}}$$
$$H_1\colon \quad \mu_{\text{CR}} \neq \mu_{\text{PR}}$$

where the first subscript identifies the level of religion and the second subscript identifies the level of religiosity. Similarly, the null and alternative hypotheses relating religion to ideal family size for nonreligious individuals are

$$H_0\colon \quad \mu_{\text{CN}} = \mu_{\text{PN}}$$
$$H_1\colon \quad \mu_{\text{CN}} \neq \mu_{\text{PN}}$$

Since the interaction effect was nonsignificant, we would not actually perform a simple main effects analysis in the present instance. However, the computational procedures for this technique are provided in Appendix 17.1 for the numerical example discussed in Section 17.8.

You might wonder why we bother testing the statistical significance of the interaction effect rather than moving directly into simple main effects analysis. The reason for this is similar to the reason we conduct an F test for a main effect and then follow this up with Tukey's HSD test when the F ratio is statistically significant: Simple main effects analysis can involve a large number of statistical tests, and although the F test of the interaction effect maintains the

Type I error rate at the specified alpha level in testing for an interaction, the Type I error rate can be substantially higher when simple main effects analysis is employed. Thus, if the interaction effect is not statistically significant, simple main effects analysis should *not* be applied, as this would violate the hypothesis testing logic of maintaining the Type I error rate at the level specified by alpha.

Recently, statisticians have suggested an alternative to simple main effects analysis as a way of exploring the nature of an interaction. This technique is called **interaction comparisons** and involves identifying all possible 2 × 2 designs that are contained within the overall design. For example, in a 3 × 2 factorial design having religion (Jewish, Protestant, or Catholic) and religiosity (religious versus nonreligious) as factors, three 2 × 2 designs can be differentiated: (1) Jewish versus Protestant × religious versus nonreligious, (2) Jewish versus Catholic × religious versus nonreligious, and (3) Protestant versus Catholic × religious versus nonreligious. A separate two-way analysis of variance is performed for each 2 × 2 subtable and the *F* ratio for the interaction effect is examined. If a given *F* ratio is statistically nonsignificant, it is concluded that the comparison in question does not contribute significantly to the overall interaction. If the *F* ratio is statistically significant, then the contribution of the comparison to the overall interaction can be explored in more detail. Interested readers are referred to Keppel (1983) for a discussion of additional issues surrounding the use of interaction comparisons. For instance, given the number of comparisons involved, the researcher must take steps to control the Type I error rate.

17.7

Methodological Considerations

Several methodological considerations are worth pointing out in the context of the study relating religion and religiosity to ideal family size. First, there are the usual issues of uncontrolled disturbance variables and generalizability of results. There is also the problem of confounding associated with nonrandom assignment to groups when observational independent variables such as religion and religiosity are used. Beyond this, however, we can use the investigation to illustrate an important methodological strategy. In Chapter 12, one of the examples used to demonstrate one-way between-subjects analysis of variance was concerned with the relationship between religion and ideal family size. In that chapter, it was pointed out that religiosity was acting as a disturbance variable and thus serving to create within-group variability (that is, it was increasing the size of the sum of squares within). In the present investigation, religiosity was combined with religion to form four groups, and the within-group variability was based on the variability of scores within each of the four groups separately. As such, the disturbance effects of religiosity did not enter into the computation of within-group variability since, like religion, it was held constant within groups. This highlights an important advantage of factorial designs: Not only do they allow us to assess the interaction between two independent variables, but they also "remove" the individual and joint effects of these variables from the within-group variability. This potentially makes the tests of the main effects more sensitive than if one of the variables were left uncontrolled and took on the role of a disturbance variable. Thus, an addition to the strategies for dealing with disturbance variables discussed in Chapters 9 and 11 is to bring a variable into the research design by including it as a factor.

17.8 Numerical Example

Social psychologists have studied extensively the variables that influence the ability of a speaker to persuade an audience to take the speaker's position on an issue. One important factor influencing the amount of attitude change a speaker can generate is the discrepancy between the position advocated by the speaker relative to the position of the audience. Up to a point, the more discrepant the speaker's position, the greater the attitude change that will result. However, if the speaker's position becomes too discrepant, the speaker loses credibility and the persuasiveness of the message lessens.

It has been hypothesized that the relationship between message discrepancy and attitude change differs depending on the expertise of the speaker, formally referred to as the *source*. According to this perspective, speakers with high expertise can take much more discrepant positions than speakers with low expertise and still obtain large amounts of attitude change. As an example of how this proposition could be tested, consider the following hypothetical experiment.

College students evaluated the quality of a passage of poetry on a 21-point scale and then listened to a taped message concerning this passage that was presented as representing the opinion of either an expert (a famous poetry critic) or a nonexpert (an undergraduate student taking beginning English). The messages were identical, the only difference between them being the source they were attributed to (either the expert or the nonexpert). In addition, the messages were constructed to be either slightly discrepant, moderately discrepant, or highly discrepant from subjects' initial ratings of quality. For example, in the large-discrepancy condition, if a subject rated the passage as being relatively high in quality, the message argued that the passage was low in quality.

After listening to the message, subjects rerated the poetry. The resulting design was a 3×2 factorial design with three levels of message discrepancy (small, medium, or large) and two levels of source expertise (high versus low). The dependent variable was the amount of change in the quality ratings after listening to the message. Scores could range from -20 to $+20$, with higher scores indicating greater attitude change in the direction advocated by the source. The data for the experiment are presented in Table 17.4 along with intermediate statistics necessary for the calculation of the sums of squares.

The sum of squares for variability due to factor A (message discrepancy) is

$$SS_A = \frac{\Sigma T_{A_i}^2}{nb} - \frac{(\Sigma X)^2}{N}$$
$$= 290.00 - 270.00 = 20.00$$

The sum of squares for variability due to factor B (source expertise) is

$$SS_B = \frac{\Sigma T_{B_j}^2}{na} - \frac{(\Sigma X)^2}{N}$$
$$= 323.33 - 270.00 = 53.33$$

The sum of squares for variability due to the interaction of factor A and factor B is

$$SS_{A \times B} = \frac{(\Sigma X)^2}{N} + \frac{\Sigma T_{A_iB_j}^2}{n} - \frac{\Sigma T_{A_i}^2}{nb} - \frac{\Sigma T_{B_j}^2}{na}$$
$$= 270.00 + 350.00 - 290.00 - 323.33 = 6.67$$

The sum of squares within is

$$SS_{\text{WITHIN}} = \Sigma X^2 - \frac{\Sigma T_{A_iB_j}^2}{n}$$
$$= 362 - 350.00 = 12.00$$

Finally, the sum of squares total is

$$SS_{\text{TOTAL}} = \Sigma X^2 - \frac{(\Sigma X)^2}{N}$$
$$= 362 - 270.00 = 92.00$$

The degrees of freedom are

$$df_A = a - 1 = 3 - 1 = 2$$
$$df_B = b - 1 = 2 - 1 = 1$$
$$df_{A \times B} = (a - 1)(b - 1) = (3 - 1)(2 - 1) = 2$$
$$df_{\text{WITHIN}} = (a)(b)(n - 1) = (3)(2)(5 - 1) = 24$$
$$df_{\text{TOTAL}} = N - 1 = 30 - 1 = 29$$

TABLE 17.4 Data and Computation of Sums of Squares for Experiment on Attitude Change as a Function of Message Discrepancy and Source Expertise

MESSAGE DISCREPANCY (A)	SOURCE EXPERTISE (B): High X	High X^2	Low X	Low X^2	MAIN EFFECT OF MESSAGE DISCREPANCY
Small	3	9	1	1	$T_S = 20$
	4	16	0	0	$\bar{X}_S = 2.00$
	2	4	2	4	$T_S^2 = 400$
	3	9	1	1	
	3	9	1	1	
	$T_{SH} = 15$		$T_{SL} = 5$		
	$\bar{X}_{SH} = 3.00$		$\bar{X}_{SL} = 1.00$		
	$T_{SH}^2 = 225$		$T_{SL}^2 = 25$		
Medium	6	36	3	9	$T_M = 40$
	5	25	2	4	$\bar{X}_M = 4.00$
	5	25	3	9	$T_M^2 = 1{,}600$
	5	25	4	16	
	4	16	3	9	
	$T_{MH} = 25$		$T_{ML} = 15$		
	$\bar{X}_{MH} = 5.00$		$\bar{X}_{ML} = 3.00$		
	$T_{MH}^2 = 625$		$T_{ML}^2 = 225$		
Large	5	25	0	0	$T_L = 30$
	4	16	1	1	$\bar{X}_L = 3.00$
	6	36	1	1	$T_L^2 = 900$
	5	25	2	4	
	5	25	1	1	
	$T_{LH} = 25$		$T_{LL} = 5$		
	$\bar{X}_{LH} = 5.00$		$\bar{X}_{LL} = 1.00$		
	$T_{LH}^2 = 625$		$T_{LL}^2 = 25$		
MAIN EFFECT OF SOURCE EXPERTISE	$T_H = 65$		$T_L = 25$		
	$\bar{X}_H = 4.33$		$\bar{X}_L = 1.67$		
	$T_H^2 = 4{,}225$		$T_L^2 = 625$		

$$\Sigma X^2 = 9 + 16 + \cdots + 4 + 1 = 362$$

$$\Sigma X = 3 + 4 + \cdots + 2 + 1 = 90$$

$$\frac{(\Sigma X)^2}{N} = \frac{90^2}{30} = 270.00$$

$$\frac{\Sigma T_{A_i}^2}{nb} = \frac{400 + 1{,}600 + 900}{(5)(2)} = 290.00$$

$$\frac{\Sigma T_{B_j}^2}{na} = \frac{4{,}225 + 625}{(5)(3)} = 323.33$$

$$\frac{\Sigma T_{A_iB_j}^2}{n} = \frac{225 + 25 + 625 + 225 + 625 + 25}{5} = 350.00$$

The relevant mean squares are

$$MS_A = \frac{SS_A}{df_A} = \frac{20.00}{2} = 10.00$$

$$MS_B = \frac{SS_B}{df_B} = \frac{53.33}{1} = 53.33$$

$$MS_{A \times B} = \frac{SS_{A \times B}}{df_{A \times B}} = \frac{6.67}{2} = 3.34$$

$$MS_{\text{WITHIN}} = \frac{SS_{\text{WITHIN}}}{df_{\text{WITHIN}}} = \frac{12.00}{24} = .50$$

The F ratios are thus

$$F_A = \frac{MS_A}{MS_{\text{WITHIN}}} = \frac{10.00}{.50} = 20.00$$

$$F_B = \frac{MS_B}{MS_{\text{WITHIN}}} = \frac{53.33}{.50} = 106.66$$

$$F_{A \times B} = \frac{MS_{A \times B}}{MS_{\text{WITHIN}}} = \frac{3.34}{.50} = 6.68$$

These calculations yield the following summary table:

Source	*SS*	*df*	*MS*	*F*
A (Message discrepancy)	20.00	2	10.00	20.00
B (Source expertise)	53.33	1	53.33	106.66
A × *B*	6.67	2	3.34	6.68
Within	12.00	24	.50	
Total	92.00	29		

Inference of a Relationship

The null and alternative hypotheses for the main effect of message discrepancy are

H_0: $\mu_S = \mu_M = \mu_L$

H_1: The three population means are not all equal.

where the subscript "S" represents the small-discrepancy condition, the subscript "M" represents the medium-discrepancy condition, and the subscript "L" represents the large-discrepancy condition. The critical value of F from Appendix F for an alpha level of .05 and 2 and 24 degrees of freedom is 3.40. Since the observed F value of 20.00 for factor A is greater than 3.40, we reject the null hypothesis and conclude that there is a relationship between message discrepancy and attitude change.

The null and alternative hypotheses for the main effect of source expertise are

$$H_0: \mu_H = \mu_L$$
$$H_1: \mu_H \neq \mu_L$$

where the subscripts "H" and "L" respectively denote the high-expertise and the low-expertise conditions. The critical value of F from Appendix F for an alpha level of .05 and 1 and 24 degrees of freedom is 4.26. Since the observed F value of 106.66 for factor B is greater than 4.26, we reject the null hypothesis and conclude that there is a relationship between source expertise and attitude change.

The null hypothesis for the interaction effect states that the relationship between message discrepancy and attitude change is the same for both high-expertise and low-expertise sources. The alternative hypothesis states that the nature of the relationship between message discrepancy and attitude change depends on the expertise of the source. Since the observed F value of 6.68 for the $A \times B$ interaction is greater than the critical value of 3.40 for an alpha level of .05 and 2 and 24 degrees of freedom, we reject the null hypothesis and conclude that an interaction exists.

Strength of the Relationship

The strengths of the relationships between the dependent variable and the three sources of between-group variability are computed using Equations 17.22–17.24:

$$\text{eta}^2_A = \frac{SS_A}{SS_{TOTAL}} = \frac{20.00}{92.00} = .22$$

$$\text{eta}^2_B = \frac{SS_B}{SS_{TOTAL}} = \frac{53.33}{92.00} = .58$$

$$\text{eta}^2_{A \times B} = \frac{SS_{A \times B}}{SS_{TOTAL}} = \frac{6.67}{92.00} = .07$$

The strongest effect is for the main effect of source expertise. The proportion of variability in attitude change that is associated with this source is .58. This represents a strong effect. The proportion of variability in attitude change that is associated with message discrepancy is .22. This represents a moderate effect. Although the interaction between message discrepancy and source expertise is statistically significant, only 7% of the variability in attitude change is associated with this source. This represents a weak effect.

Nature of the Relationship

Since it has only two levels, the nature of the relationship for the main effect of source expertise is determined through examination of the sample means. Such an examination shows that the messages produced significantly more attitude change when they were attributed to a high-expertise source ($\bar{X}_H = 4.33$) than when they were attributed to a low-expertise source ($\bar{X}_L = 1.67$).

The main effect of message discrepancy has three levels and, hence, it is necessary to apply the HSD test to determine the nature of the relationship between message discrepancy and attitude change. Since message discrepancy constitutes factor A, the value of the critical difference is established using

Equation 17.25. From Appendix G, the value of q for an overall alpha level of .05, $df_{WITHIN} = 24$, and $k = a = 3$ is 3.53. Thus,

$$CD = q\sqrt{\frac{MS_{WITHIN}}{nb}}$$

$$= 3.53\sqrt{\frac{.50}{(5)(2)}} = .79$$

The identification of mean differences is accomplished using the procedures reviewed in Section 17.6. As summarized in the following table, this entails a comparison of the absolute difference between each pair of main effect means with the critical difference:

Null hypothesis tested	*Absolute difference between sample means*	*Value of CD*	*Null hypothesis rejected?*
$\mu_S = \mu_M$	$\lvert 2.00 - 4.00 \rvert = 2.00$	.79	Yes
$\mu_S = \mu_L$	$\lvert 2.00 - 3.00 \rvert = 1.00$	.79	Yes
$\mu_M = \mu_L$	$\lvert 4.00 - 3.00 \rvert = 1.00$	.79	Yes

The nature of the relationship is such that the medium-discrepancy messages ($\bar{X}_M = 4.00$) produced significantly more attitude change than either the small- ($\bar{X}_S = 2.00$) or large-discrepancy ($\bar{X}_L = 3.00$) messages. In turn, the large-discrepancy messages produced significantly more attitude change than the small-discrepancy messages.

Since the interaction effect was statistically significant, it is necessary to perform a simple main effects analysis to determine the nature of the interaction. Because of its complexity, the simple main effects analysis is presented in Appendix 17.1. The results indicate that the high-expertise source was more persuasive when message discrepancy was medium or large than when it was small, whereas the low-expertise source was most persuasive when message discrepancy was medium. This is illustrated graphically in Figure 17.3.

FIGURE 17.3 Mean Attitude Change as a Function of Message Discrepancy and Source Expertise

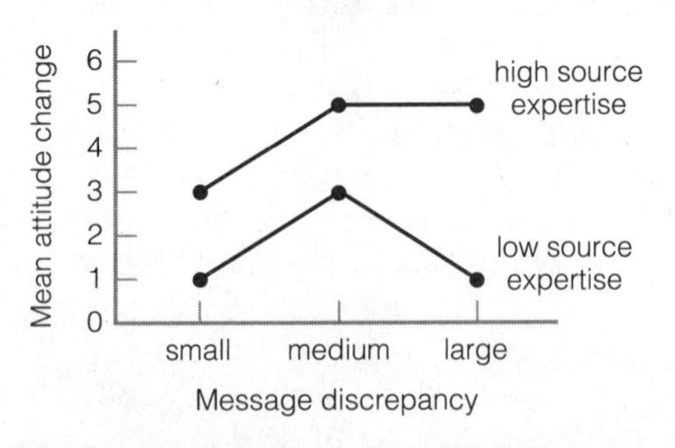

17.9 Unequal Sample Sizes

In the examples thus far, the cell sizes (n) have been equal. However, this is not always the case in behavioral science research. For example, in animal experimentation, a subject might be lost due to disease or sickness. Or a person scheduled to participate in a particular condition of an experiment might not show up. Unequal cell sizes necessitate modifications in the analysis of variance techniques discussed in this chapter.

When the sample sizes are the same in all cells, the two independent variables are unrelated to one another. Consider the following sample sizes for a 2×2 factorial design:

	RELIGION	
GENDER	*Catholic*	*Protestant*
Male	10	10
Female	10	10

In this study, there are equal sample sizes in all cells, and the two independent variables, *in the context of this investigation,* are therefore unrelated to one another. This can be thought of in terms of conditional probabilities: The probability of being a Catholic is the same as the probability of being a Protestant for both males and females. The variable of religion is independent of the variable of gender. Now consider the case of unequal sample sizes:

	RELIGION	
GENDER	*Catholic*	*Protestant*
Male	6	13
Female	13	8

In this case, a relationship exists between gender and religion. For example, if you know that an individual is a male, you know it is more likely that he is a Protestant than a Catholic. If you know that an individual is a female, you know it is more likely that she is a Catholic than a Protestant. Thus, there is a relationship between the two independent variables.*

* Another way of thinking about this is that if we were to perform a chi-square analysis on the cell frequencies (as discussed in Chapter 15), χ^2 would equal 0 in the case of equal n's (indicating no relationship in the sample) but would be greater than 0 in the case of unequal n's (indicating a relationship in the sample). Unequal sample sizes do *not* introduce a relationship between the two independent variables when the sample sizes for the groups comprising one factor are proportional across levels of the other factor (for example, when the ratio of Catholics to Protestants is 3 to 1 for both males and females).

The introduction of a relationship between the independent variables creates a number of statistical and conceptual issues for testing the two main effects and the $A \times B$ interaction. Consider factor A in a design with unequal sample sizes. On the basis of an analysis of variance, we might conclude that factor A is related to the dependent variable and accounts for 20% of the variability in it. However, because factor A is related to factor B, some of the variability in the dependent variable that we attribute to factor A may actually be due to factor B. The problem faced by statisticians is what to do about the between-group variability that is common to both factor A and factor B.

One consideration that influences how to proceed with the analysis is whether the relationship that has been introduced between factor A and factor B as a result of unequal n's is theoretically meaningful or not. Suppose one selects a random sample of 200 college students at a university and finds the following sample sizes when individuals are categorized in a 2×4 design for purpose of examining the relationship between the indicated variables and blood pressure:

	ACADEMIC MAJOR			
GENDER	*Physical sciences*	*Humanities*	*Behavioral sciences*	*Other*
Male	40	33	17	10
Female	31	42	15	12

The unequal n's produce a relationship between gender and academic major. The relationship is theoretically meaningful because it probably exists in the population of interest. That is, for the population of college students at this university, there probably *is* a relationship between a person's gender and his or her major. The observed relationship in the sample is not an artifact of the method of data collection or of subject selection procedures. Rather, it reflects a relationship that probably exists in the real world, a relationship that should be taken into consideration when analyzing the blood pressure data.

In contrast, an experiment might test the effects of gender and two different memory aids on ability to recall information. Twenty subjects are scheduled in each cell, but for some reason, four subjects do not show up, yielding the following n's:

	MEMORY AID	
GENDER	*Memory aid 1*	*Memory aid 2*
Male	18	20
Female	20	18

A relationship between the two independent variables has been introduced because of the unequal n's. However, the relationship is not theoretically meaningful. It is simply a function of four subjects not showing up, for whatever reason. In this instance, the relationship introduced by the unequal n's should not be considered in the analysis. When the unequal n's are *not* theoretically meaningful, the analysis of variance techniques described in this chapter can be altered slightly and an **unweighted means factorial analysis of variance** performed. Procedures for doing so are discussed in Winer (1971). When the unequal n's *are* theoretically meaningful, the data can be analyzed using a **least squares factorial analysis of variance.** Procedures for this technique are also discussed in Winer (1971).

17.10

Planning an Investigation Using Two-Way Between-Subjects Analysis of Variance

Appendix E.2 contains tables of the per-cell sample sizes necessary to achieve various levels of power for a given effect for two-way between-subjects analysis of variance. In two-way analysis of variance there are three relevant sample size tables to which one must refer—that for the main effect of factor A, that for the main effect of factor B, and that for the $A \times B$ interaction. Appendix E.2 presents sets of tables for various numbers of degrees of freedom for the effect of interest. Table 17.5 reproduces the portion of this appendix for $df_{EFFECT} = 1$ and an alpha level of .05 for purpose of exposition. The first column presents

TABLE 17.5 Approximate Sample Sizes Necessary to Achieve Selected Levels of Power for $df_{EFFECT} = 1$ and Alpha = .05 as a Function of Population Values of Eta-Squared

Degrees of Freedom Effect = 1
Alpha = .05

	POPULATION ETA-SQUARED										
POWER	.01	.03	.05	.07	.10	.15	.20	.25	.30	.35	.40
.10	22	8	5	4	3	2	2	2	—	—	—
.50	193	63	38	27	18	12	9	7	5	5	4
.70	310	101	60	42	29	18	13	10	8	7	6
.80	393	128	76	53	36	23	17	13	10	8	7
.90	526	171	101	71	48	31	22	17	13	11	9
.95	651	211	125	87	60	38	27	21	16	13	11
.99	920	298	176	123	84	53	38	29	22	18	15

different levels of power, and the column headings are values of eta-squared in the population.

The sample sizes given in Appendix E.2 must be adjusted to take into account the nature of the factorial design (for example, 2×3, 3×3, or 2×4). The tabled values should be adjusted as follows:

$$n' = \frac{(n_T - 1)(\mathrm{df}_{\mathrm{EFFECT}} + 1)}{ab} \qquad [17.27]$$

where n' is the adjusted per-cell sample size, n_T is the tabled per-cell sample size, a is the number of levels of factor A, b is the number of levels of factor B, and $\mathrm{df}_{\mathrm{EFFECT}}$ is the degrees of freedom for the effect in question [either $a - 1$, $b - 1$, or $(a - 1)(b - 1)$]. The value of n' is rounded up to the nearest integer. The determination of the per-cell sample size necessary to ensure that the power requirement is met for each effect requires that we calculate the necessary adjusted cell size for each effect individually and then use the *largest* sample size of the three. Possible compromises to this strategy are discussed in Cohen (1977).

As an example of these procedures, consider a 2×2 factorial design in which the investigator suspects that the population value of eta-squared is .15 for factor A, .25 for factor B, and .10 for the $A \times B$ interaction, and desires a .80 level of power for each effect for an alpha level of .05. Since $\mathrm{df}_A = a - 1 = 2 - 1 = 1$, the value of n_T for the main effect of factor A can be obtained from Table 17.5. This is equal to 23. Thus, the adjusted per-cell sample size using Equation 17.27 is

$$n' = \frac{(23 - 1)(1 + 1)}{(2)(2)} = 11$$

Since $\mathrm{df}_B = b - 1 = 2 - 1 = 1$ and $\mathrm{df}_{A \times B} = (a - 1)(b - 1) = (2 - 1)(2 - 1) = 1$, Table 17.5 can also be used to obtain the values of n_T for the main effect of factor B and for the $A \times B$ interaction. For the main effect of factor B, $n_T = 13$, so

$$n' = \frac{(13 - 1)(1 + 1)}{(2)(2)} = 6$$

For the $A \times B$ interaction, $n_T = 36$, so

$$n' = \frac{(36 - 1)(1 + 1)}{(2)(2)} = 17.50$$

or, rounded up to the nearest integer, 18. The largest adjusted cell size dictates the number of subjects required per group. In this instance, the investigator should include 18 subjects in each of the four groups. Note that this will increase the power of the F tests beyond the .80 level for the two main effects.

17.11

Method of Presentation

The results for a two-way analysis of variance are reported in much the same way as those for a one-way analysis of variance. As with the one-way case, it is not necessary to explicitly state that the design was between-subjects in nature because unless stated otherwise, it is assumed that a between-subjects analysis was performed. For each effect, the degrees of freedom, the observed value of *F*, the significance level, and the sample means should be reported. It is often more efficient to present the means for the interaction effect in a table rather than in the text itself. When this is done, the procedures described in Section 15.12 should be followed. If the interaction is statistically significant, it can sometimes be beneficial to also depict the cell means in a graph, such as that contained in Figure 17.3. In American Psychological Association format, graphs are consecutively numbered from Figure 1 and are referenced in the text in a similar manner as tables.

The results for the experiment relating message discrepancy and source expertise to attitude change discussed earlier in this chapter might be reported as follows:

Results

Attitude change scores were subjected to a two-way analysis of variance having three levels of message discrepancy (small, medium, or large) and two levels of source expertise (high versus low). All effects were found to be statistically significant. The main effect of source expertise was such that the messages from the high-expertise source (M = 4.33) produced significantly more attitude change than the messages from the low-expertise source (M = 1.67), F(1, 24) = 106.66, p < .001. The strength of the relationship, as indexed by eta², was .58. The nature of the main effect of message discrepancy, F(2, 24) = 20.00, p < .001, was determined using the Tukey HSD test. Results showed that the small (M = 2.00), medium (M = 4.00), and large (M = 3.00) discrepancy means all significantly differed from one another. As indexed by eta², the strength of the relationship was .22.

The interaction effect, $\underline{F}(2, 24) = 6.68$, $\underline{p} < .01$, was analyzed using simple main effects analysis and the HSD test. The relevant means can be found in Table 1. For

Insert Table 1 about here

subjects in the high-expertise condition, both the medium- and the large-discrepancy messages led to significantly more attitude change than the small-discrepancy message. However, the medium- and large-discrepancy means did not significantly differ. For subjects in the low-expertise condition, the medium-discrepancy message produced significantly more attitude change than either the small- or the large-discrepancy message. The means for the latter two conditions did not significantly differ. The strength of the overall interaction effect, as indexed by eta^2, was .07.

The table of means for the interaction might appear as follows:

Table 1

Mean Attitude Change Scores as a Function of Message Discrepancy and Source Expertise

	Message discrepancy		
Source expertise	Small	Medium	Large
High	3.00	5.00	5.00
Low	1.00	3.00	1.00

Note. $\underline{n} = 5$.

The note at the bottom refers to the size of the individual cells. According to American Psychological Association format, general information of this type that relates to the table as a whole is placed below the table and designated by the word "Note," which should be underlined and followed by a period.

17.12

Examples from the Literature

Attributions for Success as a Function of the Gender of the Performer and the Type of Task

When someone is successful at a task, we will sometimes attribute that success to the person's ability. Alternatively, we may simply think that the person "got lucky" and that the success had little to do with ability.

Johnson (1976) was interested in the extent to which these two attributions would be used to explain male and female task success. Subjects listened to tape recordings that indicated that either a man or a woman had succeeded at either a traditionally masculine task (identifying mechanical objects such as wrenches and screwdrivers) or a traditionally feminine task (identifying household objects such as mops and pots). Subjects were then asked to indicate on a 13-point scale the extent to which they thought performance on the task was due to luck (scores near 1) versus ability (scores near 13). The design was thus a 2×2 between-subjects factorial design with gender of the performer and type of task (masculine versus feminine) as the independent variables and ratings of luck/ability as the dependent variable.

It was hypothesized that for the masculine task, male success would be attributed more to ability than would female success, but the reverse would be true for the feminine task. Thus, Johnson predicted an interaction, with the relationship between the gender of the performer and ability/luck attributions depending on the type of task.

The main effect for the gender of the performer was statistically significant, $F(1, 96) = 4.38$, $p < .05$, and indicated that successful performance by males ($\bar{X} = 9.24$) will be attributed more to ability than will be successful performance by females ($\bar{X} = 8.40$). The strength of the relationship, as indexed by eta-squared, was only .04, however. This represents a weak effect. The main effect for the type of task was not statistically significant, $F(1, 96) = 2.46$, *ns*, the mean for the masculine task being 8.50 and the mean for the feminine task being 9.14. The strength of the relationship, as indexed by eta-squared, was .02; this again represents a weak effect.

Although a statistically significant interaction effect was observed, $F(1, 96) = 9.08$, $p < .01$, the nature of the interaction was somewhat different from what was expected. Simple main effects analysis indicated that for the masculine task, the male's performance ($\bar{X} = 9.53$) was attributed more to ability than was the female's performance on the same task ($\bar{X} = 7.48$). However, for the feminine task, the mean ratings for the male ($\bar{X} = 8.95$) and the female ($\bar{X} = 9.32$) performers were not significantly different. The strength of the overall interaction effect, as indexed by eta-squared, was .08. This represents a weak effect. What type of interpretation might you give these findings?

Restaurant Tipping as a Function of Interpersonal Touch and Diners' Gender

Crusco and Wetzel (1984) examined the effects of two types of touch on restaurant tipping. Restaurant diners were randomly assigned to conditions where, in the course of returning their change, their waitress either twice touched their palms with her fingers for .5 second (hand-touch condition), placed her hands on their shoulders for 1 to 1.5 seconds (shoulder-touch condition), or did not touch them (no-touch condition). It was predicted that the hand touch would produce positive feelings toward the waitress and thus increase the amount of her tip relative to the no-touch condition. Since a touch on the shoulder can be construed as a sign of dominance, it was felt that this might not be viewed as positively and that tipping might therefore also be greater in the hand-touch condition than in the shoulder-touch condition. So that possible gender differences in these effects could be determined, separate observations were made for male and female diners. The design was thus a 3 × 2 between-subjects factorial design with type of touch (hand, shoulder, or none) and gender of the diner as the independent variables. The dependent variable was the percentage of the bill left as a tip.

A two-way analysis of variance yielded a main effect for gender, $F(1, 108) = 3.93$, $p < .05$, such that males ($\bar{X} = 15.3\%$) tipped more than females ($\bar{X} = 12.6\%$). The main effect for the type of touch was also statistically significant, $F(2, 108) = 3.45$, $p < .05$. Analysis of the three touch means showed that tipping was greater in the hand-touch ($\bar{X} = 16.7\%$) and shoulder-touch ($\bar{X} = 14.4\%$) conditions than in the no-touch condition ($\bar{X} = 12.2\%$). The hand-touch and shoulder-touch means did not significantly differ. The interaction effect also failed to attain statistical significance, $F(2, 108) < 1$. (*Note:* It is conventional to use the indicated notation when the observed value of F is less than 1.00. It is not necessary to indicate that the F value is statistically nonsignificant because an F value of less than 1.00 can never lead to rejection of the null hypothesis. Can you think of why this is the case?) The values of eta-squared were not provided in the research report.

17.13

Summary

Two-way between-subjects analysis of variance allows for the investigation of the separate and joint effects of two independent variables on a dependent variable. This is accomplished through the analysis of main effects and interaction effects in the context of factorial designs. A factorial design is one in which the a levels of one independent variable are combined with the b levels of a second independent variable, yielding $a \times b$ groups. A test of a main effect refers to the test of the relationship between one of the independent variables and the dependent variable. A test of an interaction effect refers to the test of whether the nature of the relationship between one of the independent variables and the dependent variable differs as a function of the other independent variable. Two-way analysis of variance is typically used when (1) the dependent variable is quantitative in nature and is measured on approximately an interval level, (2) the independent variables are both between-subjects in nature, (3) the independent variables both have two or more levels, and (4) the independent variables are combined to form a factorial design.

The statistical procedures for two-way analysis of variance are similar to those for one-way analysis of variance except that the overall between-group variability is broken down into three components: between-group variability due to factor *A*, between-group variability due to factor *B*, and between-group variability due to the $A \times B$ interaction. Hypothesis testing procedures are then applied to these three sources of between-group variability. When a statistically significant main effect is obtained and there are two levels of the independent variable, the nature of the effect is determined using the same procedures as for the independent groups *t* test; when there are more than two levels of the independent variable, the HSD test is used. When a statistically significant interaction effect is obtained, the nature of the effect is determined using simple main effects analysis.

The computational procedures presented in this chapter are applicable only when there are equal numbers of subjects in each cell. When there are unequal sample sizes, a least squares factorial analysis of variance can be performed when the unequal *n*'s are theoretically meaningful, and an unweighted means factorial analysis of variance can be performed when the unequal *n*'s are not theoretically meaningful.

Appendix 17.1 Computational Procedures for Simple Main Effects Analysis

We will describe the computational procedures for simple main effects analysis using the data from the experiment on attitude change as a function of message discrepancy and source expertise for illustration. Since we are interested in whether the relationship between message discrepancy and attitude change is the same for both high-expertise and low-expertise sources, we must determine the nature of the relationship between message discrepancy and attitude change at each level of source expertise. The null and alternative hypotheses for the high-expertise source are

H_0: $\mu_{SH} = \mu_{MH} = \mu_{LH}$
H_1: The three population means are not all equal.

For the low-expertise source, the relevant hypotheses are

H_0: $\mu_{SL} = \mu_{ML} = \mu_{LL}$
H_1: The three population means are not all equal.

We will first discuss procedures for testing the **simple main effect** of message discrepancy for the high-expertise condition. This requires the calculation of a mean square reflecting between-group variability for message discrepancy for subjects in the high-expertise condition only. To do this, we must first calculate a sum of squares for between-group variability as if a one-way analysis of variance were being performed based on just these subjects. The formula for doing this is

$$SS_{A \text{ at } B_1} = \frac{\Sigma T^2_{A_i}}{n} - \frac{(\Sigma X)^2}{N} \quad [17.28]$$

where $T^2_{A_i}$ is the square of the sum of the scores at level i of factor A (message discrepancy), n is the per-cell sample size, $(\Sigma X)^2$ is the square of the summed scores, N is the total number of observations comprising the simple main effect, and *all calculations pertain only to level 1 (high expertise) of factor B (source expertise)*. In our example, $T^2_{A_i}$ equals 225 for the small-discrepancy condition, 625 for the medium-discrepancy condition, and 625 for the large-discrepancy condition, from Table 17.4. Since $n = 5$,

$$\frac{\Sigma T^2_{A_i}}{n} = \frac{225 + 625 + 625}{5} = 295.00$$

The second part of Equation 17.28 is equal to

$$\frac{(\Sigma X)^2}{N} = \frac{(3 + 4 + \cdots + 5 + 5)^2}{15} = \frac{65^2}{15} = 281.67$$

Thus,

$$SS_{A \text{ at } B_1} = 295.00 - 281.67 = 13.33$$

This sum of squares has $a - 1 = 3 - 1 = 2$ degrees of freedom associated with it. This is the same number of degrees of freedom as for the main effect of factor A because the same number of means are being compared in the two cases. The resulting mean square is

$$MS_{A \text{ at } B_1} = \frac{SS_{A \text{ at } B_1}}{df_{A \text{ at } B_1}} = \frac{13.33}{2} = 6.66 \quad [17.29]$$

An F ratio for the simple main effect is formed by dividing $MS_{A \text{ at } B_1}$ by MS_{WITHIN} from the original summary table. This mean square within is used in preference to a mean square within based only on subjects in the high-expertise condition since it encompasses more scores (30 as opposed to 15) and hence provides a better estimate of within-group variability in the population. This strategy is acceptable because an assumption of two-way analysis of variance is that the scores in each population have homogeneous variances. The mean square within based on all groups will thus be the best estimate of within-group variability. Since this was found to equal .50 in Section 17.8, the F ratio is

$$F_{A \text{ at } B_1} = \frac{MS_{A \text{ at } B_1}}{MS_{\text{WITHIN}}} = \frac{6.66}{.50} = 13.32 \quad [17.30]$$

For an alpha level of .05 and 2 and 24 degrees of freedom, the critical value of F is 3.40. Since 13.32 is greater than 3.40, the null hypothesis that $\mu_{SH} = \mu_{MH} = \mu_{LH}$ is rejected.

The nature of the relationship between message discrepancy and attitude change can be discerned using the Tukey HSD test. The critical difference is defined as

$$CD = q\sqrt{\frac{MS_{\text{WITHIN}}}{n}} \quad [17.31]$$

where MS_{WITHIN} and n are as defined above and q is a Studentized range value obtainable from Appendix G. The value of q is determined with reference to the overall alpha level for the simple main effect, the degrees of freedom within from the original summary table (in this case, 24), and the number of levels of *factor A* (symbolized k in Appendix G). We will follow the practice of adopting an overall alpha level of .05 for each simple main effect that we will analyze. In our example, $k = a = 3$, so $q = 3.53$. Thus,

$$CD = 3.53\sqrt{\frac{.50}{5}} = 1.12$$

The HSD procedure can now be applied as follows:

Null hypothesis tested	*Absolute difference between sample means*	*Value of CD*	*Null hypothesis rejected?*
$\mu_{SH} = \mu_{MH}$	$\|3.00 - 5.00\| = 2.00$	1.12	Yes
$\mu_{SH} = \mu_{LH}$	$\|3.00 - 5.00\| = 2.00$	1.12	Yes
$\mu_{MH} = \mu_{LH}$	$\|5.00 - 5.00\| = 0.00$	1.12	No

These results show that for subjects in the high-expertise condition, both the medium- ($\bar{X}_{MH} = 5.00$) and the large-discrepancy ($\bar{X}_{LH} = 5.00$) messages produced significantly more attitude change than the small-discrepancy message ($\bar{X}_{SH} = 3.00$). However, the medium-discrepancy and the large-discrepancy means did not significantly differ.

The same procedures can be applied to the analysis of message discrepancy and attitude change for subjects in the low-expertise condition. The formula for the sum of squares for between-group variability for message discrepancy for subjects in the low-expertise condition is

$$SS_{A \text{ at } B_2} = \frac{\Sigma T_{A_i}^2}{n} - \frac{(\Sigma X)^2}{N} \qquad [17.32]$$

where *all calculations pertain only to level 2 (low expertise) of factor B*. Looking at Table 17.4, we find that $T_{A_i}^2$ equals 25 for the small-discrepancy condition, 225 for the medium-discrepancy condition, and 25 for the large-discrepancy condition. Thus,

$$\frac{\Sigma T_{A_i}^2}{n} = \frac{25 + 225 + 25}{5} = 55.00$$

The second part of Equation 17.32 is equal to

$$\frac{(\Sigma X)^2}{N} = \frac{(1 + 0 + \cdots + 2 + 1)^2}{15} = \frac{25^2}{15} = 41.67$$

Thus,

$$SS_{A \text{ at } B_2} = 55.00 - 41.67 = 13.33$$

and

$$MS_{A \text{ at } B_2} = \frac{SS_{A \text{ at } B_2}}{df_{A \text{ at } B_2}} = \frac{13.33}{2} = 6.66 \qquad [17.33]$$

The F ratio for the simple main effect is

$$F_{A \text{ at } B_2} = \frac{MS_{A \text{ at } B_2}}{MS_{\text{WITHIN}}} = \frac{6.66}{.50} = 13.32 \qquad [17.34]$$

Since 13.32 is greater than the critical value of 3.40 for an alpha level of .05 and 2 and 24 degrees of freedom, the null hypothesis that $\mu_{SL} = \mu_{ML} = \mu_{LL}$ is rejected. Given that the value of the critical difference has been calculated above to be 1.12, we can directly apply the HSD test:

Null hypothesis tested	*Absolute difference between sample means*	*Value of CD*	*Null hypothesis rejected?*
$\mu_{SL} = \mu_{ML}$	$\lvert 1.00 - 3.00 \rvert = 2.00$	1.12	Yes
$\mu_{SL} = \mu_{LL}$	$\lvert 1.00 - 1.00 \rvert = 0.00$	1.12	No
$\mu_{ML} = \mu_{LL}$	$\lvert 3.00 - 1.00 \rvert = 2.00$	1.12	Yes

These results show that for subjects in the low-expertise condition, the medium-discrepancy message ($\bar{X}_{ML} = 3.00$) produced significantly more attitude change than either the small- ($\bar{X}_{SL} = 1.00$) or the large-discrepancy ($\bar{X}_{LL} = 1.00$) message. However, the means for the latter two conditions did not significantly differ.

This pattern of results differs from the pattern of results for subjects in the high-expertise condition, as reported above. The fact that the nature of the relationship between message discrepancy and attitude change differs in the observed manner as a function of the expertise of the source accounts for the significant interaction effect. The simple main effects analysis has made the nature of the interaction explicit.*

* If we had wished, we could have examined the interaction from the other perspective by performing a simple main effects analysis for the relationship between source expertise and attitude change at each level of message discrepancy.

Exercises

Answers to asterisked (*) exercises appear at the back of the book. Answers to exercises with two asterisks are also worked out step-by-step in the Study Guide.

Exercises to Review Concepts

*1. How many independent variables are there in a 3 × 3 factorial design? In a 2 × 2 × 2 factorial design?

2. How many groups of subjects are required in a 3 × 3 factorial design? In a 2 × 3 factorial design? In a 4 × 3 factorial design?

3. Give an example of an investigation that would use a 2 × 3 factorial design.

4. Under what conditions is two-way between-subjects analysis of variance typically used to analyze a relationship between variables?

***5.** What is a main effect? What is an interaction effect?

***6.** For each of the following sets of *population* means, indicate if there is a main effect of factor *A*, a main effect of factor *B*, and/or an *A* × *B* interaction:

a.

	B_1	B_2
A_1	4.00	5.00
A_2	4.00	5.00

b.

	B_1	B_2
A_1	6.00	6.00
A_2	4.00	4.00

c.

	B_1	B_2	B_3
A_1	1.00	2.00	3.00
A_2	5.00	6.00	7.00
A_3	8.00	9.00	10.00

7. For each of the following sets of *population* means, indicate if there is a main effect of factor *A*, a main effect of factor *B*, and/or an *A* × *B* interaction:

a.

	B_1	B_2
A_1	5.00	10.00
A_2	10.00	5.00

b.

	B_1	B_2	B_3
A_1	4.00	7.00	9.00
A_2	5.00	6.00	7.00
A_3	6.00	8.00	10.00

c.

	B_1	B_2	B_3
A_1	5.00	6.00	6.00
A_2	7.00	8.00	4.00
A_3	9.00	10.00	2.00

***8.** For the following 2 × 3 factorial design, generate a set of population means that would reflect a main effect of factor *A*, no main effect of factor *B*, and an *A* × *B* interaction:

	B_1	B_2	B_3
A_1			
A_2			

9. For the following 2 × 2 factorial design, generate a set of population means that would reflect no main effect of factor *A*, no main effect of factor *B*, and an *A* × *B* interaction:

	B_1	B_2
A_1		
A_2		

***10.** What do nonparallel lines indicate in a graph of population means? What do nonparallel lines indicate in a graph of sample means? What accounts for the difference in the two situations?

11. What are the three components of $SS_{BETWEEN}$ in two-way analysis of variance?

***12.** Complete the missing entries in the summary table for a 3 × 4 factorial design:

Source	SS	df	MS	F
A	—	—	—	—
B	45.00	—	—	—
A × B	60.00	—	—	—
Within	216.00	—	—	
Total	341.00	119		

13. Compute the missing entries in the summary table for a 3 × 3 factorial design:

Source	SS	df	MS	F
A	—	—	—	3.50
B	12.00	—	—	—
A × B	40.00	—	—	—
Within	—	27	2.00	
Total	120.00	—		

*14. State the critical values of F that would be used to reject the null hypothesis for the main effect of factor A, the main effect of factor B, and the $A \times B$ interaction for a two-way analysis of variance at an alpha level of .05 under each of the following conditions:

a. $a = 2, b = 3, n = 11$
b. $a = 2, b = 4, n = 7$
c. $a = 3, b = 3, n = 11$
d. $a = 4, b = 3, n = 6$

15. What are the assumptions underlying two-way between-subjects analysis of variance?

Exercises 16–21 refer to the following summary table for an experiment:

Source	SS	df	MS	F
A	50.00	1	50.00	10.00
B	40.00	2	20.00	4.00
A × B	40.00	2	20.00	4.00
Within	240.00	48	5.00	
Total	370.00	53		

*16. How many levels of factor A were there in the experiment? How many levels of factor B?

*17. What was the total number of subjects in the experiment? How many groups were there in the experiment? Assuming there were equal numbers of subjects in each group, what was the per-group sample size?

*18. What is the total amount of between-group variability represented in the table (that is, what is the value of the sum of squares between)?

*19. State the null and alternative hypotheses for the main effect of factor A, the main effect of factor B, and the $A \times B$ interaction. (The hypotheses for the interaction may be stated from either perspective.)

*20. Test the viability of the null hypotheses with respect to the main effect of factor A, the main effect of factor B, and the $A \times B$ interaction.

*21. Compute the values of eta-squared for the main effect of factor A, the main effect of factor B, and the $A \times B$ interaction. Indicate whether each value represents a weak, moderate, or strong effect.

Exercises 22–23 refer to the following means and summary table for an investigation having 11 subjects per cell:

	B_1	B_2	B_3
A_1	10.00	15.00	20.00
A_2	14.00	20.00	24.00

Source	SS	df	MS	F
A	240.00	1	240.00	24.00
B	910.00	2	455.00	45.50
A × B	10.00	2	5.00	.50
Within	600.00	60	10.00	
Total	1,760.00	65		

*22. Test the viability of the null hypothesis with respect to the main effect of factor A. If the null hypothesis is rejected, analyze the nature of the relationship between factor A and the dependent variable.

*23. Test the viability of the null hypothesis with respect to the main effect of factor B. If the null hypothesis is rejected, analyze the nature of the relationship between factor B and the dependent variable using the HSD test.

Exercises 24–25 refer to the following means and summary table for an investigation having five subjects per cell:

	B_1	B_2	B_3
A_1	10.00	12.00	14.00
A_2	12.00	8.00	4.00

Source	SS	df	MS	F
A	120.00	1	120.00	10.00
B	20.00	2	10.00	.83
A × B	380.00	2	190.00	15.83
Within	288.00	24	12.00	
Total	808.00	29		

24. Test the viability of the null hypothesis with respect to the main effect of factor *A*. If the null hypothesis is rejected, analyze the nature of the relationship between factor *A* and the dependent variable.

25. Test the viability of the null hypothesis with respect to the main effect of factor *B*. If the null hypothesis is rejected, analyze the nature of the relationship between factor *B* and the dependent variable using the HSD test.

26. An investigator was interested in the effects of gender and academic major on the number of job offers received by college seniors. Based on the following data, conduct a two-way analysis of variance to test for a relationship between each of the independent variables and their interaction with the number of job offers:

	ACADEMIC MAJOR			
GENDER	*Computer science*	*Business*	*Liberal arts*	*Behavioral sciences*
	3	4	2	2
	4	4	2	3
Male	6	5	1	3
	3	2	3	1
	3	4	2	2
	2	1	3	2
	3	2	2	3
Female	3	2	1	5
	2	3	2	3
	1	1	3	4

27. Compute the values of eta-squared for the main effect of gender, the main effect of academic major, and the interaction between gender and academic major for the data in Exercise 26. Indicate whether each value represents a weak, moderate, or strong effect.

28. Analyze the nature of the relationship between gender and the number of job offers for the data in Exercise 26.

29. Analyze the nature of the relationship between academic major and the number of job offers for the data in Exercise 26 using the HSD test.

***30.** What is the problem with unequal cell sizes in two-way analysis of variance?

31. Under what conditions is an unweighted means factorial analysis of variance appropriate? Under what conditions is a least squares factorial analysis of variance appropriate?

***32.** A researcher conducts a study involving three levels of factor *A* and four levels of factor *B*. He suspects that the strength of the relationship, as indexed by eta-squared, is .15 for factor *A*, .03 for factor *B*, and .10 for the $A \times B$ interaction. What sample size should he use for an alpha level of .05 in order to achieve power of at least .90 for each effect?

33. A researcher conducts a study involving two levels of factor *A* and five levels of factor *B*. She suspects that the strength of the relationship, as indexed by eta-squared, is .20 for factor *A*, .07 for factor *B*, and .07 for the $A \times B$ interaction. What sample size should she use for an alpha level of .05 in order to achieve power of at least .85 for each effect?

Exercises to Apply Concepts

****34.** All of us tend to classify people into social categories (for example, a "conservative businessman," a "typical housewife," and so on) and frequently these categorizations influence how we behave toward individuals. Psychologists have studied factors that influence social categorization and the effects of such categorizations on behavior. In one experiment (Rubovits & Maehr, 1973), white female undergraduates enrolled in a teacher-training course were asked to prepare a lesson for four seventh-grade students. The teachers were given information about each of the students. For one-half of the teachers, one of the students was described as being "gifted" (that is, extremely intelligent) while for the other half of the teachers, the student was described as being "nongifted" (or of average intelligence). This student was either black or white, yielding a 2×2 factorial design (gifted versus nongifted × black versus white). The teachers were then observed during a 40-minute period in which they interacted with the target student and three other students. The number of interactions directed toward the target student was noted, and this served as an index of how much attention the teacher gave to the student. Hypothetical data representative of the results of this experiment are presented below. Analyze these data as completely as possible, draw a conclusion for each effect, and report your results using the principles discussed in the Method of Presentation section.

RACE	INTELLIGENCE Gifted	Nongifted
Black	30	30
	29	29
	30	28
	30	30
	31	28
White	36	29
	36	32
	36	31
	35	30
	37	33

35. Psychologists have studied extensively the factors that contribute to weight gain in individuals. It is currently believed that individuals use at least two sources of "cues" to decide that they are hungry and should eat. One source is internal cues in the form of changes in one's physiology that create or suggest hunger. A second source is external cues that occur in the environment and suggest to the individual that he or she should be hungry. For example, when 12:00 noon approaches, this may signify that it is lunch time and that we should be hungry. Several researchers have suggested that one difference between overweight and normal-weight individuals is that normal-weight individuals attend mostly to internal rather than external cues as guidelines for eating, whereas the reverse is true for overweight individuals.

Suppose that in an investigation designed to study this issue, 15 overweight and 15 normal-weight individuals were instructed not to eat breakfast on the morning before coming to an experiment. After they arrived at the experiment and performed a number of different tasks, subjects were asked to rate how hungry they felt on a scale from 0 to 10, with higher numbers indicating greater hunger. Just prior to this, a confederate (an experimental assistant posing as a subject) asked another confederate what time it was in the presence of the subjects. For one-third of the subjects the response was 11:00, for another third the response was 1:00, and for the final third the response was 12:00, the actual time. If overweight individuals attend mostly to external cues, we would expect them to report greater hunger the later the supposed time, as this implies that they have gone longer without food. If normal-weight individuals attend mostly to internal cues, then the time manipulation should not affect their ratings of hunger. Hypothetical data for this experiment are presented below. Analyze these data as completely as possible, draw a conclusion for each effect, and report your results using the principles discussed in the Method of Presentation section.

TIME	WEIGHT Overweight	Normal
11:00	5	6
	6	6
	7	7
	6	5
	6	6
12:00	9	6
	7	6
	8	5
	9	7
	10	6
1:00	10	7
	10	6
	9	7
	8	8
	9	5

EIGHTEEN

OVERVIEW AND EXTENSION

SELECTING THE APPROPRIATE STATISTICAL TEST FOR ANALYZING BIVARIATE RELATIONSHIPS AND PROCEDURES FOR MORE COMPLEX DESIGNS

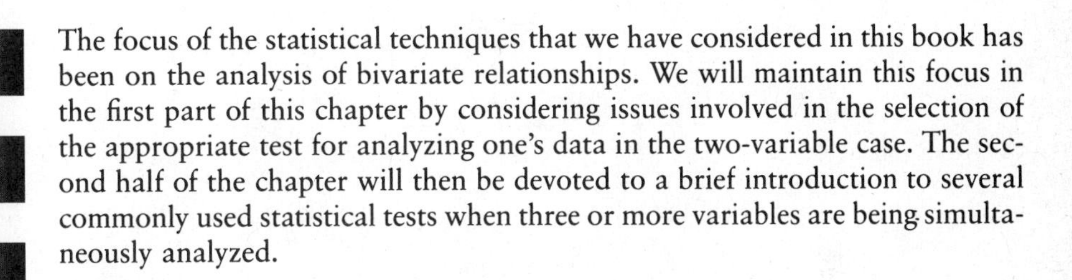

The focus of the statistical techniques that we have considered in this book has been on the analysis of bivariate relationships. We will maintain this focus in the first part of this chapter by considering issues involved in the selection of the appropriate test for analyzing one's data in the two-variable case. The second half of the chapter will then be devoted to a brief introduction to several commonly used statistical tests when three or more variables are being simultaneously analyzed.

18.1

Selecting the Appropriate Statistical Test for Analyzing Bivariate Relationships

We have considered a total of 12 statistical procedures for studying bivariate relationships.* Except for regression, three questions have guided our analysis in each instance: (1) Given sample data, can we infer that a relationship exists between two variables in the population? (2) If so, what is the strength of the relationship? (3) If so, what is the nature of the relationship? The specific statistics used to address each question are summarized for each of the relevant inferential tests in Table 18.1.

Although a large number of tests are available for analyzing bivariate relationships, the task of test selection can be simplified if one adheres to the guidelines presented below. However, it must be emphasized that these guidelines are only rules of thumb; there are potential exceptions to every one of them. Nevertheless, they should be of heuristic value in the vast majority of cases. In each instance, the recommendation of a test assumes that the relevant requirements for application of that test have been met.

* In reaching this total, the Wilcoxon rank sum test and the Mann–Whitney U test were counted as one technique.

TABLE 18.1 Statistics Used to Address the Three Questions Apropros of the Analysis of Bivariate Relationships for 11 Inferential Tests

Independent Groups t Test	
Inference of a relationship:	t statistic
Strength of the relationship:	Eta-squared
Nature of the relationship:	Inspection of group means
Correlated Groups t Test	
Inference of a relationship:	t statistic
Strength of the relationship:	Eta-squared
Nature of the relationship:	Inspection of group means
One-Way Between-Subjects Analysis of Variance	
Inference of a relationship:	F ratio
Strength of the relationship:	Eta-squared
Nature of the relationship:	Tukey HSD test
One-Way Repeated Measures Analysis of Variance	
Inference of a relationship:	F ratio
Strength of the relationship:	Eta-squared
Nature of the relationship:	Tukey HSD test
Pearson Correlation	
Inference of a relationship:	r statistic
Strength of the relationship:	r^2
Nature of the relationship:	Sign of r
Chi-square Test	
Inference of a relationship:	χ^2 statistic
Strength of the relationship:	Fourfold point correlation coefficient/Cramér's statistic
Nature of the relationship:	Goodman's simultaneous confidence interval procedure
Wilcoxon Rank Sum Test/Mann–Whitney U Test	
Inference of a relationship:	z statistic/U statistic
Strength of the relationship:	Glass rank biserial correlation coefficient
Nature of the relationship:	Inspection of mean ranks
Wilcoxon Signed-Rank Test	
Inference of a relationship:	T statistic/z statistic
Strength of the relationship:	Matched-pairs rank biserial correlation coefficient
Nature of the relationship:	Inspection of rank sums
Kruskal–Wallis Test	
Inference of a relationship:	H statistic
Strength of the relationship:	Epsilon-squared
Nature of the relationship:	Dunn procedure

(continued)

TABLE 18.1 *Continued*

Friedman Analysis of Variance by Ranks	
Inference of a relationship:	χ_r^2 statistic
Strength of the relationship:	Epsilon-squared
Nature of the relationship:	Dunn procedure

Spearman Rank-Order Correlation	
Inference of a relationship:	r_s statistic
Strength of the relationship:	Magnitude of r_s
Nature of the relationship:	Sign of r_s

The first step in determining the appropriate bivariate test is to note whether each of the two variables is quantitative or qualitative in nature. The four different situations that can arise are as follows:

Case I: both variables qualitative

Case II: independent variable qualitative and dependent variable quantitative

Case III: independent variable quantitative and dependent variable qualitative

Case IV: both variables quantitative

18.2

Case I: The Relationship Between Two Qualitative Variables

When both of the variables under study are qualitative and between-subjects in nature, the appropriate method of data analysis is the chi-square test. This situation would occur, for example, when examining the relationship between gender and political party identification, as discussed in Chapter 15. If both variables are qualitative but one or both are within-subjects in nature, none of the statistical tests that we have considered can be used. The appropriate statistical procedures in this instance are discussed in McNemar (1962) and Marascuilo and McSweeney (1977).

18.3

Case II: The Relationship Between a Qualitative Independent Variable and a Quantitative Dependent Variable

Table 18.2 presents a decision tree that can be used to guide the selection of the appropriate statistical test in situations where the independent variable is qualitative and the dependent variable is quantitative. The three decision areas represented in Table 18.2 are (1) use of a parametric versus a nonparametric test, (2) whether the independent variable is between-subjects or within-subjects in nature, and (3) the number of levels characterizing the independent variable.

Use of a Parametric Versus a Nonparametric Test

Factors to be weighed in making the decision to use a parametric versus a nonparametric test in Case II situations include the number of values of the dependent variable, the level of measurement of the dependent variable, and the extent of violation of distributional assumptions. Consider the case where the independent variable is geographical location (Northeast, South, Midwest, or West) and the dependent variable is attitude toward nuclear energy. The relevant attitude scale could conceivably have anywhere from two to a large number of response points. In situations where the dependent measure has only two or three values, most researchers will *not* use a parametric test. As noted earlier in the text, an assumption of parametric tests is that population scores are normally distributed. A dependent variable cannot be normally distributed if it has only two or three values and, hence, this assumption of parametric tests will be violated. Under this circumstance, nonparametric tests will typically be used.*

When the dependent measure has only two or three values, the rank techniques discussed in Chapter 16 might also be of questionable value. This is because with only two or three values on the dependent variable, there will be a large number of ties in ranks, and as noted in Chapter 16, numerous ties can create serious problems for rank-based approaches. The alternatives in this case are to apply the correction formulas for tied ranks or to use the chi-square test. Most researchers will choose the chi-square approach. As an example of a

* It should be noted that there is some evidence to suggest that the number of values of the dependent measure might not be a limiting factor in the application of a parametric test when sample sizes are equal and relatively large (Hsu & Feldt, 1969; Lunney, 1970). Thus, some researchers will prefer a parametric test over a nonparametric test even when the dependent variable has as few as two values.

TABLE 18.2 Decision Tree for Case II Situations

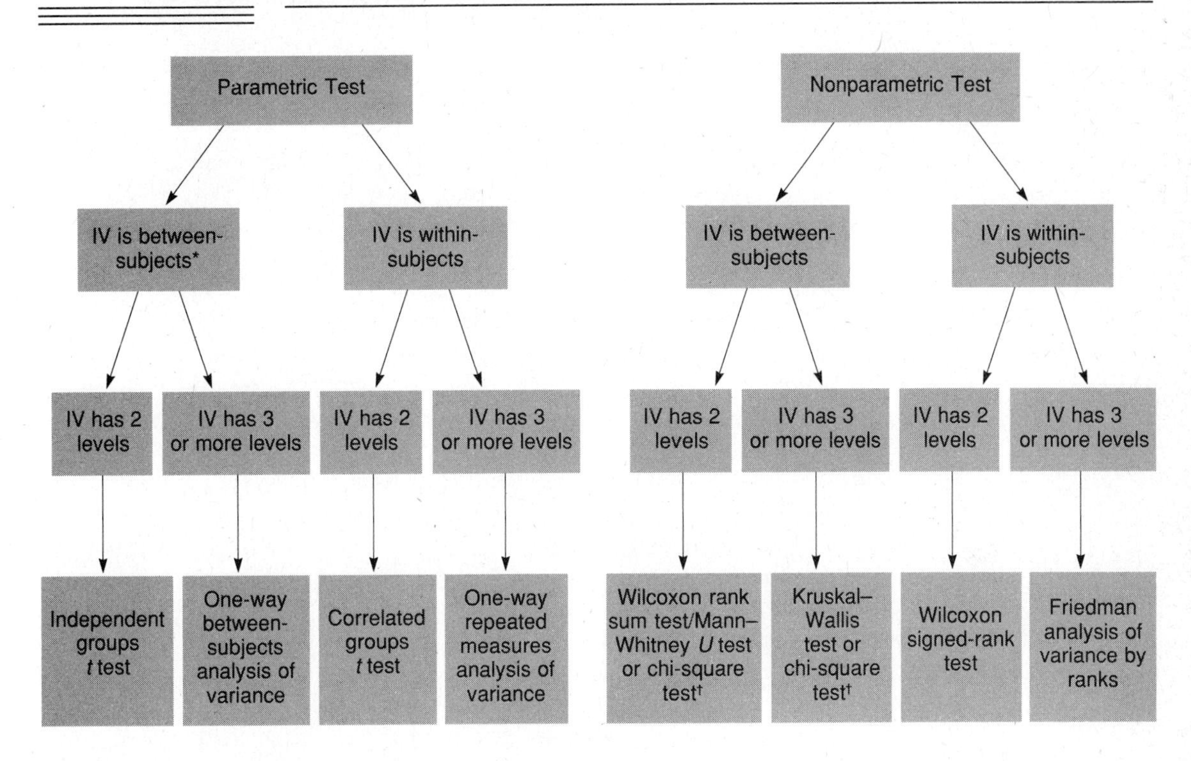

*IV = independent variable.
†The chi-square test is typically used only when the dependent variable has only two or three values.

situation where we might apply the chi-square test to a study having a qualitative independent variable and a quantitative dependent variable with only two values, consider the case where the independent variable is occupation (blue collar, white collar, or clerical) and the dependent variable is the number of hands (one or two) with which one is dexterous. A 2×3 contingency table in which manual dexterity constitutes the row variable and occupation constitutes the column variable can be formulated for this problem and the chi-square procedures developed in Chapter 15 and summarized in Table 18.1 can be applied to the observed frequencies.

Given a quantitative dependent variable with a sufficient range of values, additional characteristics of the data must be considered when deciding whether to use a parametric or a nonparametric test. If the dependent variable is measured on an ordinal level that seriously departs from interval level characteristics, a nonparametric test will typically be used. Nonparametric tests should also be used if there is reason to believe that the distributional assumptions of the corresponding parametric tests have been violated to the extent that the parametric tests are no longer robust. Otherwise, parametric tests can be used to good effect.

Using Table 18.2 to Select the Appropriate Statistical Test

As an example of the use of Table 18.2, let us consider an investigation where the researcher asked 30 single, 30 married, 30 widowed, and 30 divorced individuals to indicate their life satisfaction on a 7-point scale. In this instance, the independent variable (marital status) is qualitative and the dependent variable (life satisfaction) is quantitative. Hence, this is a Case II situation.

The first step in using Table 18.2 involves deciding whether to use a parametric test or a nonparametric test. Life satisfaction approximates an interval measure and can take on seven values. Furthermore, given the equivalence and magnitude of the sample sizes, a parametric test would probably be robust to most of the violations of distributional assumptions that might be encountered. This state of affairs would lead to the decision to use a parametric test.

The next decision in Table 18.2 requires classifying the independent variable as between-subjects or within-subjects in nature. Since each group includes 30 different individuals, the independent variable in this instance is between-subjects.

Finally, the number of levels characterizing the independent variable has to be considered. In this instance, there are four. This would dictate a one-way between-subjects analysis of variance as the data-analytic technique. Referring to Table 18.1, we would use an F ratio to infer the existence of a relationship between the independent and dependent variables, eta-squared to examine the strength of the relationship, and a Tukey HSD test to determine the nature of the relationship.

18.4

Case III: The Relationship Between a Quantitative Independent Variable and a Qualitative Dependent Variable

Our discussion of Case II situations focused on the case where the independent variable is qualitative and the dependent variable is quantitative. If both variables are between-subjects in nature, one way to handle the reverse situation of a qualitative *dependent* variable and a quantitative *independent* variable is to perform a chi-square test on the data. Depending on how many there are, the individual values of the quantitative variable will either constitute separate levels in the contingency table or be collapsed into a small number of groups. Another possibility is to apply the decision criteria outlined in Section 18.3 for Case II situations, treating the independent variable as the dependent variable and the dependent variable as the independent variable for statistical purposes. Of course, the interpretation of the selected test would maintain the original distinction between the independent variable and the dependent variable. In general, however, designs involving quantitative independent variables and qualitative dependent variables are problematic and, when possible, should be avoided.

18.5 Case IV: The Relationship Between Two Quantitative Variables

Given two variables that are measured on approximately an interval level, the most common method of analysis is Pearson correlation. As discussed in Chapter 14, Pearson correlation tests for a *linear* relationship between variables. If the expected relationship is nonlinear, procedures for nonlinear relationships discussed in Pedhazur (1982) can be applied. When one or both variables are measured on an ordinal level that seriously departs from interval level characteristics, Spearman rank-order correlation is usually the test of choice.

When one of the variables has fewer than five or so values associated with it, the Case II statistics elaborated in Table 18.2 might be applicable. As an example, a researcher might investigate the relationship between age and intelligence by administering an intelligence test to 100 5-year-old children, 100 7-year-old children, and 100 9-year-old children. In this case, both variables are quantitative in nature, but age has only a few values (5, 7, and 9). In fact, the study was explicitly structured to yield these three well-defined values on the age variable. Under these circumstances, most investigators would use Table 18.2 to decide the appropriate statistical test to use, treating age as the independent variable and intelligence as the dependent variable. Since age is between-subjects and has three levels, the resulting method of analysis would be one-way between-subjects analysis of variance if a parametric test is appropriate or the Kruskal–Wallis test if a nonparametric test is required.

Of course, based on the logic presented in Section 18.3 regarding problems caused by a quantitative variable having only two or three values, one of the tests in Table 18.2 probably *should* be applied if this situation exists. If both variables have only two or three values, the chi-square test can be used.

Given a choice between the chi-square test, the other statistics contained in Table 18.2, Pearson correlation, or Spearman correlation, the preferred alternative is usually Pearson correlation, and the least preferred alternative is usually the chi-square test. This is because Pearson correlation tends to be the most powerful and the chi-square test the least powerful of the four approaches.

18.6

Procedures for More Complex Designs

The focus of the statistical procedures that we have considered to this point has been on the analysis of bivariate relationships. However, many research problems require that more than two variables be simultaneously studied. Because of this, statistical techniques have been developed to analyze the relationship between three or more variables. Because these techniques consider the variation among multiple variables, they are referred to as **multivariate statistics.** One example of a multivariate statistic is two-way between-subjects analysis of variance considered in Chapter 17. As we have seen, this technique is a direct extension of one-way between-subjects analysis of variance for the situation where there is one dependent variable and two independent variables. In the sections that follow, we will briefly discuss several additional multivariate tests that are in common use in the behavioral sciences.

Two-Way Repeated Measures and Two-Way Mixed Design Analyses of Variance

Just as the joint influence of two between-subjects independent variables on a dependent variable can be studied using two-way between-subjects analysis of variance, it is also possible to study the joint influence of two within-subjects independent variables. For instance, returning to the wine label and perceived quality experiment discussed in Chapter 13, the researcher might decide to vary information regarding the time when the wine was supposedly produced as well as information about the supposed country of origin. In the simplest case, she might present the wine as being either an old or a new vintage. Crossing the time of production information with the country of origin information (French, Italian, or American), there are a total of six conditions (for instance, old-vintage French wine, new-vintage French wine, and so forth). Now instead of having each participant taste a total of three "different" wines, each participant will taste a total of six "different" wines. Since both independent variables are within-subjects in nature, the obtained data can be analyzed using a **two-way repeated measures analysis of variance.**

Sometimes research designs involve one between-subjects independent variable and one within-subjects independent variable. For instance, another way to conduct the study outlined above would be to use two groups of subjects, one of which tastes the supposed French, Italian, and American wines under circumstances where the labels indicate that they are all old vintage and one of which tastes the wines under circumstances where the labels indicate that they are all new vintage. In this case, the country of origin variable is within-subjects and the vintage variable is between-subjects so the appropriate means of analysis would be a **two-way mixed design analysis of variance.**

The procedures for analysis of variance with two independent variables can be extended to analyze the relationship between a dependent variable and three or more independent variables. This is true for all three types of designs—between-subjects, repeated measures, and mixed. Given the flexibility of analysis of variance procedures, it is not surprising that analysis of variance constitutes one of the most commonly used statistical approaches in the behavioral sciences.

Multivariate Analysis of Variance

A situation that sometimes arises in behavioral science research involves the analysis of two or more *dependent* variables. For instance, a researcher interested in the achievement concerns of high school, junior college, and college graduates might have members of each group complete a measure assessing concern with social achievement, a measure assessing concern with occupational achievement, and a measure assessing concern with financial achievement. In this design, there is one independent variable (education) and three dependent variables (concern with social achievement, concern with occupational achievement, and concern with financial achievement).

One approach to analyzing the data would be to conduct three one-way between-subjects analyses of variance, one for each dependent variable. The problem with this approach is that when multiple analyses of variance are performed, the probability of making at least one Type I error increases beyond the probability specified by the alpha level. This is similar to the problem discussed in Chapter 12 with conducting a series of independent groups *t* tests to determine the nature of the relationship between the independent and dependent variables following a statistically significant analysis of variance. To circumvent this problem and maintain the probability of making one or more Type I errors at the specified alpha level in the earlier case, we advocated the use of the Tukey HSD test. This problem can be similarly dealt with in the present situation by performing a **multivariate analysis of variance** before conducting any further tests.

Multivariate analysis of variance (abbreviated MANOVA) tests whether the subject groups have different population means *on the dependent variables considered jointly*. It accomplishes this by calculating a *multivariate F ratio* that can then be compared with a critical *F* value. In the present example, the multivariate analysis of variance would enable us to infer whether the three education groups differ in overall achievement concerns. If the multivariate *F* test is statistically significant, we can proceed with separate analyses of variance for each dependent variable, followed, as appropriate, by multiple comparison tests. If the multivariate F test is not statistically significant, the null hypothesis of equivalent population means on the dependent variables considered jointly cannot be rejected so no additional analyses are warranted.

Multivariate analysis of variance can be used with any number of dependent variables and any number and combination of between-subjects and within-subjects independent variables. When applied to a single independent variable having only two levels, it is referred to as the **Hotelling T^2 test.**

Multiple Regression

In Chapter 14, we saw how a regression equation can be used to predict individuals' scores on one variable (the criterion variable, symbolized Y) from knowledge of their scores on a second variable (the predictor variable, symbolized X). **Multiple regression** extends these procedures to the prediction of a criterion variable from two or more predictor variables. For instance, a common use of multiple regression is the prediction of educational achievement from previous academic performance. Admissions officers at many colleges have established regression equations for predicting grade point average at their institution from high school grade point average, SAT scores, class rank, teacher ratings, and the like. The rationale is that if several of these measures are considered simultaneously, more accurate predictions can be made than if only one predictor variable were studied. When this approach is used, predicted achievement scores are among the criteria considered when making admission decisions.

Multiple regression is a direct extension of regression with one predictor variable. For each predictor variable, a value of b, known in the context of multiple regression as a **regression coefficient,** is calculated using modifications of the procedures discussed in Chapter 5. These coefficients represent the number of units the criterion variable is predicted to change for each unit change in a given predictor variable *when the effects of the other predictor variables are held constant.* For instance, a regression coefficient of .21 for high school grade point average would mean that, holding all other predictor variables constant, college grade point average is predicted to increase by .21 unit for each unit increase in high school grade point average.

In addition to containing a regression coefficient for each predictor variable, a **multiple regression equation** contains an overall intercept symbolized, as with regression with one predictor variable, by the letter a. Thus, for instance, the general form of a regression equation involving three predictor variables is

$$\hat{Y} = a + b_1X_1 + b_2X_2 + b_3X_3 \qquad \textbf{[18.1]}$$

where $\hat{Y}$ is the predicted score on the criterion variable of interest, a is the overall intercept, b_1 is the regression coefficient for the first predictor variable, b_2 is the regression coefficient for the second predictor variable, and b_3 is the regression coefficient for the third predictor variable. In this equation, X_1, X_2, and X_3 represent a given subject's scores on the three predictor variables. For example, suppose the following equation was being used to predict college grade point average from high school grade point average (X_1), combined (verbal and quantitative) SAT scores (X_2), and teacher ratings on a 7-point scale (X_3):

$$\hat{Y} = .13 + .21X_1 + .001X_2 + .18X_3$$

Based on this equation, the predicted college grade point average for a student who had a high school grade point average of 2.78, a combined SAT score of 943, and an average teacher rating of 4.97 is

$$\hat{Y} = .13 + (.21)(2.78) + (.001)(943) + (.18)(4.97) = 2.55$$

An index of the strength of the relationship between the criterion variable and the set of predictor variables is provided by the **squared multiple correlation coefficient,** symbolized R^2. The quantity R^2 is analogous to r^2 with one predictor variable and indicates the proportion of variability in the criterion variable that is associated with the predictor variables considered simultaneously.

Factor Analysis

The goal of **factor analysis** is to determine if the correlations among a set of variables can be accounted for by one or more underlying dimensions, or **factors.** Since the calculational procedures involve the use of Pearson correlation, each variable to be included in the analysis should be measured on approximately an interval level.

As an example of the use of factor analysis, suppose a study is conducted in which nine different beliefs about abortion are measured in a sample of participants. The investigator might hypothesize that individuals' responses to the belief statements are the result of three underlying dimensions: (1) a concern with physical effects of abortion, (2) a concern with emotional effects of abortion, and (3) a concern with moral issues. Factor analysis tests hypotheses of this nature, albeit somewhat indirectly, through the calculation of a Pearson correlation coefficient between each possible pair of variables. If the investigator's hypothesis in the present study is correct, the correlations between the nine beliefs can be accounted for by the three hypothesized factors. If these three factors underlie the data, then the correlations between variables should have a predictable pattern. For example, all beliefs pertaining to physical issues should be highly correlated with one another, but relatively uncorrelated with the other beliefs. Similarly, all beliefs pertaining to emotional issues should be highly correlated with one another, but relatively uncorrelated with the other beliefs. And all beliefs pertaining to moral issues should be highly correlated with one another, but relatively uncorrelated with the other beliefs. Factor analysis formally examines the pattern of correlations among variables and provides information on the type of *factor structure* (that is, the number and makeup of factors) that might underlie the data.

Log-Linear Analysis

Sometimes a research question requires that three or more qualitative variables be simultaneously examined. For instance, we might wish to extend the altruism experiment discussed in Chapter 15 by studying how willingness to donate money to charity is influenced by one's gender in addition to the type of role model one is exposed to. Since the three variables (gender, model status, and outcome) are all qualitative and between-subjects in nature, we could form a three-way contingency table of the observed frequencies. Although the chi-square test can be extended to multidimensional tables of this type, a statistical technique known as **log-linear analysis** will usually be applied instead. While conceptually similar to chi-square analysis, log-linear analysis possesses certain statistical properties that make it more suitable for the simultaneous analysis of multiple between-subjects qualitative variables.

18.7

Summary

When selecting a test for analyzing a bivariate relationship, it is useful to distinguish between four situations: both variables qualitative (Case I), a qualitative independent variable and a quantitative dependent variable (Case II), a quantitative independent variable and a qualitative dependent variable (Case III), and both variables quantitative (Case IV). Given between-subjects variables, the appropriate method of analysis for Case I situations is the chi-square test. The selection of a statistical test in Case II situations involves decisions on the use of a parametric versus a nonparametric procedure, whether the independent variable is between-subjects or within-subjects in nature, and the number of levels characterizing the independent variable. Case III situations are generally problematic and, when possible, should be avoided. Case IV situations most commonly utilize correlational procedures.

Statistical techniques that analyze the relationship between three or more variables are known as multivariate statistics. Among the most commonly used multivariate tests are variants of analysis of variance, including between-subjects, repeated measures, mixed design, and multiple dependent variable procedures. Other multivariate statistics include multiple regression, factor analysis, and log-linear analysis.

Multiple regression is an extension of regression as discussed in Chapter 14 and allows for the prediction of a criterion variable from two or more predictor variables. The goal of factor analysis is to determine if the correlations among a set of variables can be accounted for by one or more underlying dimensions (factors). Log-linear analysis is appropriate when one wishes to simultaneously analyze three or more between-subjects qualitative variables.

Exercises

Answers to asterisked (*) exercises appear at the back of the book.

*1. What is the appropriate statistical technique for analyzing the relationship between two between-subjects qualitative variables?

2. What are the three decision areas when selecting a statistical test in situations where the independent variable is qualitative and the dependent variable is quantitative?

3. What factors should be considered when deciding to use a parametric versus a nonparametric test in situations where the independent variable is qualitative and the dependent variable is quantitative?

4. What options are there for analyzing the relationship between a quantitative independent variable and a qualitative dependent variable?

*5. What is the most common method of analysis for two quantitative variables? What other techniques are available? Identify the conditions under which each of these is used.

For each of the studies described in Exercises 6–20, indicate the appropriate statistical test for analyzing the relationship between the variables. If a parametric technique might be applicable, indicate which test you would use under conditions where (a) the underlying assumptions of the parametric technique have been satisfied, and (b) the underlying assumptions of the parametric technique have been violated to the extent that a nonparametric test is required. In each instance, state the reasons for your selection.

*6. An investigator wanted to test if changes in mood are associated with certain times of the year. Specifically, he wondered if people tend to be more depressed in the winter than in the spring. Two hundred individuals were administered a depression scale in December (winter) and again in May (spring). Scores on this scale could range from 0 to 50.

*7. A consumer psychologist was interested in the effect of the color of ice cream on taste perceptions. Four hundred individuals tasted each of three different "types" of vanilla ice cream. The ice creams were actually identical to one another but differed in the shade of yellow used for coloring. The order in which individuals tasted each of the three types of ice cream was randomized. After tasting a given type, individuals rated it in terms of the quality of taste on a scale from 1 to 10.

*8. A researcher was interested in whether the noise level of rock and roll music affects how houseplants grow. Forty seeds were randomly assigned to one of two conditions. In one condition, plants were grown with a steady background of rock and roll music playing at an average volume level. In the other condition, plants were grown under identical conditions, but with the music playing at a high volume level. After 6 months, the growth in inches was measured for each plant.

*9. A psychologist was interested in the relationship between creativity in the adult years and imagination as a child. One hundred people classified as creative and 100 people classified as noncreative were each asked whether they had had an imaginary friend during childhood. Answers to this question were in terms of a yes–no response.

*10. A researcher wanted to test the relationship between social class and how dogmatic individuals are. Dogmatism refers to closed-mindedness and the tendency to be inflexible in thought and intolerant of other viewpoints. Five hundred people were administered a dogmatism scale on which scores could range from 1 to 70. The social class of each individual was measured using an occupational index on which scores could range from 1 to 100.

*11. An investigator wanted to test the effect of social influence on drinking behavior. Sixty subjects were randomly assigned to one of three conditions, yielding 20 subjects per group. Subjects came to an experiment and were given a soft drink to consume while waiting. An experimental assistant, posing as another subject, was also given a drink. In one condition, the assistant sipped the drink at a much faster rate than the subject, in another condition the assistant sipped the drink at about the same rate as the subject, and in the last condition the assistant sipped the drink at a slower rate than the subject. The time it took for each subject to consume his or her drink was measured.

*12. An investigator wanted to test if hypnosis could influence responses on a biofeedback task. Fifty subjects were randomly assigned to one of two conditions. In one condition, subjects were hypnotized and then told to try to make one of their hands warmer by just thinking about it. The other group of subjects was given the same task but was not placed under hypnosis. Temperature was measured in centigrade units to the nearest hundredth of a degree using a special temperature gauge.

*13. A researcher was interested in the relationship between physiological arousal and time between taking a test. Fifty individuals scheduled to take an exam on a Thursday were instructed to administer the Palmer Sweat Index (a physiological measure of arousal for which scores can range from 0 to 100) to themselves just before going to bed on Tuesday (2 days before the exam), Wednesday (1 day before the exam), Thursday (day of the exam), and Friday (day after the exam). The relationship between time between the test and physiological arousal was examined.

14. In general, our bodies feel weak in the morning when we first get up. A researcher was interested in testing whether people are actually weaker when they first wake up in the morning as compared to a few hours later. Fifty subjects were instructed to squeeze a dynamometer (which measures grip strength) when they first woke up in the morning and again 3 hours later. Scores on the dynamometer could range from 0 to 30.

15. An investigator wanted to test for race discrimination in loan officers at banks. Each of 120 loan officers was given background information on an applicant and asked how much money he or she would be willing to lend the individual. Forty of the loan officers were told that the applicant was black, 40 were told that he was white, and the remaining 40 were told that he was Hispanic. Aside from this, the descriptions of the applicant were identical. The relationship between the supposed race of the applicant and the amount of money the loan officers were willing to lend was assessed.

16. An investigator wanted to test if male college students were more likely to use marijuana than female college students. Two hundred male and 200 female college students were asked whether they had ever used marijuana.

17. A researcher was interested in the relationship between individuals' performance in college and their income 5 years later. For a group of 800 people, data were obtained on grade point averages while in college and salaries after being out of college for 5 years, measured in dollars.

18. A professor wanted to test the relationship between how quickly students finish an exam and how well they perform. As part of the procedure for a scheduled course examination, the professor kept track of the order in which students turned in their tests. Specifically, she divided the class into three groups: the first third of the class to turn in the exam, the middle third of the class to turn in the exam, and the last third of the class to turn in the exam. Scores on the exam could range from 0 to 100.

19. A consumer psychologist was interested in people's impressions of individuals who buy generic foods. Two hundred individuals were given a hypothetical grocery list of a person. For half of the individuals, the list contained some generic brands whereas for the other half, the list contained only national brands. All products on the two lists were otherwise identical. After reading the list, each subject rated the shopper on a scale from 1 to 100 in terms of how discriminating he was perceived to be in his food preferences.

20. An educational psychologist was interested in the relationship between psychology departments' national reputations and the quantity of scientific articles published by members of their faculties. A national ranking of the top 100 American psychology departments was obtained (where 1 = the worst department and 100 = the best department) and for each department the number of publications generated by its faculty was tabulated. The relationship between the indices of quality and quantity was examined.

*21. What is the defining characteristic of multivariate statistics?

22. Differentiate between two-way between-subjects analysis of variance, two-way repeated measures

analysis of variance, two-way mixed design analysis of variance, multivariate analysis of variance, and the Hotelling T^2 test.

*23. What is the problem with conducting multiple analyses of variance when analyzing two or more dependent variables?

24. What is the rationale behind multiple regression?

*25. Define each of the following:

a. regression coefficient
b. multiple regression equation
c. squared multiple correlation coefficient

26. What is the goal of factor analysis? How is this accomplished?

27. Under what conditions is log-linear analysis appropriate?

APPENDIX A

Table of Random Numbers

The following table contains a listing of random numbers. Random number tables have many different uses. For example, suppose we had a list of 350 people in a population and wanted to select a random sample of 50 individuals. This could be accomplished using the following table by proceeding as follows: Arbitrarily number the 350 individuals from 1 to 350. Then, enter the random number table at any point. Suppose we start in the upper left-hand corner where the digits 19612 appear. Since we are concerned, at most, with a three-digit number, we will examine only the left-most three digits. (We could examine only the right-most three digits or only the middle three digits instead if we wished.) This yields the number 196. Therefore, the individual who is numbered 196 is included in the sample. Next, move down one row. (We could move across one column instead if we wished.) The three-digit number is 391. Since there is no individual with this number, ignore it and continue moving down rows. The next "valid" number is 035, meaning the individual who is numbered 35 is included in the sample. Continue this process until 50 individuals have been selected.

Source: Reprinted from pages 3 and 5 of *A Million Random Digits with 100,000 Normal Deviates,* by The Rand Corporation. New York: The Free Press, 1955. Copyright 1955 by The Rand Corporation. Used by permission.

19612	78430	11661	94770	77603	65669	86868	12665	30012	75989
39141	77400	28000	64238	73258	71794	31340	26256	66453	37016
64756	80457	08747	12836	03469	50678	03274	43423	66677	82556
92901	51878	56441	22998	29718	38447	06453	25311	07565	53771
03551	90070	09483	94050	45938	18135	36908	43321	11073	51803
98884	66209	06830	53656	14663	56346	71430	04909	19818	05707
27369	86882	53473	07541	53633	70863	03748	12822	19360	49088
59066	75974	63335	20483	43514	37481	58278	26967	49325	43951
91647	93783	64169	49022	98588	09495	49829	59068	38831	04838
83605	92419	39542	07772	71568	75673	35185	89759	44901	74291
24895	88530	70774	35439	46758	70472	70207	92675	91623	61275
35720	26556	95596	20094	73750	85788	34264	01703	46833	65248
14141	53410	38649	06343	57256	61342	72709	75318	90379	37562
27416	75670	92176	72535	93119	56077	06886	18244	92344	31374
82071	07429	81007	47749	40744	56974	23336	88821	53841	10536
21445	82793	24831	93241	14199	76268	70883	68002	03829	17443
72513	76400	52225	92348	62308	98481	29744	33165	33141	61020
71479	45027	76160	57411	13780	13632	52308	77762	88874	33697
83210	51466	09088	50395	26743	05306	21706	70001	99439	80767
68749	95148	94897	78636	96750	09024	94538	91143	96693	61886
05184	75763	47075	88158	05313	53439	14908	08830	60096	21551
13651	62546	96892	25240	47511	58483	87342	78818	07855	39269
00566	21220	00292	24069	25072	29519	52548	54091	21282	21296
50958	17695	58072	68990	60329	95955	71586	63417	35947	67807
57621	64547	46850	37981	38527	09037	64756	03324	04986	83666
09282	25844	79139	78435	35428	43561	69799	63314	12991	93516
23394	94206	93432	37836	94919	26846	02555	74410	94915	48199
05280	37470	93622	04345	15092	19510	18094	16613	78234	50001
95491	97976	38306	32192	82639	54624	72434	92606	23191	74693
78521	00104	18248	75583	90326	50785	54034	66251	35774	14692
96345	44579	85932	44053	75704	20840	86583	83944	52456	73766
77963	31151	32364	91691	47357	40338	23435	24065	08458	95366
07520	11294	23238	01748	41690	67328	54814	37777	10057	42332
38423	02309	70703	85736	46148	14258	29236	12152	05088	65825
02463	65533	21199	60555	33928	01817	07396	89215	30722	22102
15880	92261	17292	88190	61781	48898	92525	21283	88581	60098
71926	00819	59144	00224	30570	90194	18329	06999	26857	19238
64425	28108	16554	16016	00042	83229	10333	36168	65617	94834
79782	23924	49440	30432	81077	31543	95216	64865	13658	51081
35337	74538	44553	64672	90960	41849	93865	44608	93176	34851
05249	29329	19715	94082	14738	86667	43708	66354	93692	25527
56463	99380	38793	85774	19056	13939	46062	27647	66146	63210
96296	33121	54196	34108	75814	85986	71171	15102	28992	63165
98380	36269	60014	07201	62448	46385	42175	88350	46182	49126
52567	64350	16315	53969	80395	81114	54358	64578	47269	15747
78498	90830	25955	99236	43286	91064	99969	95144	64424	77377
49553	24241	08150	89535	08703	91041	77323	81079	45127	93686
32151	07075	83155	10252	73100	88618	23891	87418	45417	20268
11314	50363	26860	27799	49416	83534	19187	08059	76677	02110
12364	71210	87052	50241	90785	97889	81399	58130	64439	05614

03991	10461	93716	16894	66083	24653	84609	58232	88618	19161
38555	95554	32886	59780	08355	60860	29735	47762	71299	23853
17546	73704	92052	46215	55121	29281	59076	07936	27954	58909
32643	52861	95819	06831	00911	98936	76355	93779	80863	00514
69572	68777	39510	35905	14060	40619	29549	69616	33564	60780
24122	66591	27699	06494	14845	46672	61958	77100	90899	75754
61196	30231	92962	61773	41839	55382	17267	70943	78038	70267
30532	21704	10274	12202	39685	23309	10061	68829	55986	66485
03788	97599	75867	20717	74416	53166	35208	33374	87539	08823
48228	63379	85783	47619	53152	67433	35663	52972	16818	60311
60365	94653	35075	33949	42614	29297	01918	28316	98953	73231
83799	42402	56623	34442	34994	41374	70071	14736	09958	18065
32960	07405	36409	83232	99385	41600	11133	07586	15917	06253
19322	53845	57620	52606	66497	68646	78138	66559	19640	99413
11220	94747	07399	37408	48509	23929	27482	45476	85244	35159
31751	57260	68980	05339	15470	48355	88651	22596	03152	19121
88492	99382	14454	04504	20094	98977	74843	93413	22109	78508
30934	47744	07481	83828	73788	06533	28597	20405	94205	20380
22888	48893	27499	98748	60530	45128	74022	84617	82037	10268
78212	16993	35902	91386	44372	15486	65741	14014	87481	37220
41849	84547	46850	52326	34677	58300	74910	64345	19325	81549
46352	33049	69248	93460	45305	07521	61318	31855	14413	70951
11087	96294	14013	31792	59747	67277	76503	34513	39663	77544
52701	08337	56303	87315	16520	69676	11654	99893	02181	68161
57275	36898	81304	48585	68652	27376	92852	55866	88448	03584
20857	73156	70284	24326	79375	95220	01159	63267	10622	48391
15633	84924	90415	93614	33521	26665	55823	47641	86225	31704
92694	48297	39904	02115	59589	49067	66821	41575	49767	04037
77613	19019	88152	00080	20554	91409	96277	48257	50816	97616
38688	32486	45134	63545	59404	72059	43947	51680	43852	59693
25163	01889	70014	15021	41290	67312	71857	15957	68971	11403
65251	07629	37239	33295	05870	01119	92784	26340	18477	65622
36815	43625	18637	37509	82444	99005	04921	73701	14707	93997
64397	11692	05327	82162	20247	81759	45197	25332	83745	22567
04515	25624	95096	67946	48460	85558	15191	18782	16930	33361
83761	60873	43253	84145	60833	25983	01291	41349	20368	07126
14387	06345	80854	09279	43529	06318	38384	74761	41196	37480
51321	92246	80088	77074	88722	56736	66164	49431	66919	31678
72472	00008	80890	18002	94813	31900	54155	83436	35352	54131
05466	55306	93128	18464	74457	90561	72848	11834	79982	68416
39528	72484	82474	25593	48545	35247	18619	13674	18611	19241
81616	18711	53342	44276	75122	11724	74627	73707	58319	15997
07586	16120	82641	22820	92904	13141	32392	19763	61199	67940
90767	04235	13574	17200	69902	63742	78464	22501	18627	90872
40188	28193	29593	88627	94972	11598	62095	36787	00441	58997
34414	82157	86887	55087	19152	00023	12302	80783	32624	68691
63439	75363	44989	16822	36024	00867	76378	41605	65961	73488
67049	09070	93399	45547	94458	74284	05041	49807	20288	34060
79495	04146	52162	90286	54158	34243	46978	35482	59362	95938
91704	30552	04737	21031	75051	93029	47665	64382	99782	93478

APPENDIX B

Proportions of Scores in a Normal Distribution

The following table reports proportions of scores in a normal distribution that occur within selected ranges of z. A given proportion represents the probability of obtaining scores within the range of interest. Column 1 lists values of z. Column 2 indicates the probability of observing z scores greater than or equal to $-z$ and less than or equal to $+z$. Column 3 indicates the probability of obtaining z scores greater than or equal to $+z$. This column can also be used to determine the probability of obtaining z scores less than or equal to $-z$. Column 4 indicates the probability of obtaining z scores less than or equal to $-z$ or greater than or equal to $+z$. Column 5 indicates the probability of obtaining z scores between 0 and the z score of interest.

As an example, consider the z score 1.96. In column 2, we see that the probability of obtaining z scores between -1.96 and $+1.96$ is .9500. In column 3, we see that the probability of obtaining z scores greater than or equal to $+1.96$ is .0250. Since the normal distribution is symmetrical, this column also indicates the probability of obtaining z scores less than or equal to -1.96. This probability is also .0250. In column 4, we see that the probability of obtaining z scores less than or equal to -1.96 or greater than or equal to $+1.96$ is .0500. In column 5, we see that the probability of obtaining z scores between 0 and $+1.96$ is .4750. Again, because the normal distribution is symmetrical, this also represents the probability of obtaining z scores between 0 and -1.96.

Source: Adapted from R. B. McCall, *Fundamental Statistics for Psychology*. New York: Harcourt Brace Jovanovich, 1970. Used with permission.

COLUMN 1	COLUMN 2	COLUMN 3	COLUMN 4	COLUMN 5
z	$\geq -z$ and $\leq +z$	$\geq +z$ (also use to find $\leq -z$)	$\leq -z$ or $\geq +z$	≥ 0 and $\leq +z$ (also use to find $\geq -z$ and ≤ 0)
0.00	.0000	.5000	1.0000	.0000
0.01	.0080	.4960	.9920	.0040
0.02	.0160	.4920	.9840	.0080
0.03	.0240	.4880	.9768	.0170
0.04	.0320	.4840	.9680	.0160
0.05	.0398	.4801	.9602	.0199
0.06	.0478	.4761	.9522	.0239
0.07	.0558	.4721	.9442	.0279
0.08	.0638	.4681	.9362	.0319
0.09	.0718	.4641	.9282	.0359
0.10	.0796	.4602	.9204	.0398
0.11	.0876	.4562	.9124	.0438
0.12	.0956	.4522	.9044	.0478
0.13	.1034	.4483	.8966	.0517
0.14	.1114	.4443	.8886	.0557
0.15	.1192	.4404	.8808	.0596
0.16	.1272	.4364	.8728	.0636
0.17	.1350	.4325	.8650	.0675
0.18	.1428	.4286	.8572	.0714
0.19	.1506	.4247	.8494	.0753
0.20	.1586	.4207	.8414	.0793
0.21	.1664	.4168	.8336	.0832
0.22	.1742	.4129	.8258	.0871
0.23	.1820	.4090	.8180	.0910
0.24	.1896	.4052	.8104	.0948
0.25	.1974	.4013	.8026	.0987
0.26	.2052	.3974	.7948	.1026
0.27	.2128	.3936	.7872	.1064
0.28	.2206	.3897	.7794	.1103
0.29	.2282	.3859	.7718	.1141
0.30	.2358	.3821	.7642	.1179
0.31	.2434	.3783	.7566	.1217
0.32	.2510	.3745	.7490	.1255
0.33	.2586	.3707	.7414	.1293
0.34	.2662	.3669	.7338	.1331
0.35	.2736	.3632	.7264	.1368
0.36	.2812	.3594	.7188	.1406
0.37	.2886	.3557	.7114	.1443
0.38	.2960	.3520	.7040	.1480
0.39	.3034	.3483	.6966	.1517
0.40	.3108	.3446	.6892	.1554
0.41	.3182	.3409	.6818	.1591

COLUMN 1	COLUMN 2	COLUMN 3	COLUMN 4	COLUMN 5
	-z +z	+z / -z	-z +z	0 +z / -z 0
z	$\geq -z$ and $\leq +z$	$\geq +z$ (also use to find $\leq -z$)	$\leq -z$ or $\geq +z$	≥ 0 and $\leq +z$ (also use to find $\geq -z$ and ≤ 0)
0.42	.3256	.3372	.6744	.1628
0.43	.3328	.3336	.6672	.1664
0.44	.3400	.3300	.6600	.1700
0.45	.3472	.3264	.6528	.1736
0.46	.3544	.3228	.6456	.1772
0.47	.3616	.3192	.6384	.1808
0.48	.3688	.3156	.6312	.1844
0.49	.3758	.3121	.6242	.1879
0.50	.3830	.3085	.6170	.1915
0.51	.3900	.3050	.6100	.1950
0.52	.3970	.3015	.6030	.1985
0.53	.4038	.2981	.5962	.2019
0.54	.4108	.2946	.5892	.2054
0.55	.4176	.2912	.5824	.2088
0.56	.4246	.2877	.5754	.2123
0.57	.4314	.2843	.5686	.2157
0.58	.4380	.2810	.5670	.2190
0.59	.4448	.2776	.5552	.2224
0.60	.4514	.2743	.5486	.2257
0.61	.4582	.2709	.5418	.2291
0.62	.4648	.2676	.5352	.2324
0.63	.4714	.2643	.5286	.2357
0.64	.4778	.2611	.5222	.2389
0.65	.4844	.2578	.5156	.2422
0.66	.4908	.2546	.5092	.2454
0.67	.4972	.2514	.5028	.2486
0.68	.5034	.2483	.4966	.2517
0.69	.5098	.2451	.4902	.2549
0.70	.5160	.2420	.4840	.2580
0.71	.5222	.2389	.4778	.2611
0.72	.5284	.2358	.4716	.2642
0.73	.5346	.2327	.4654	.2673
0.74	.5408	.2296	.4592	.2704
0.75	.5468	.2266	.4532	.2734
0.76	.5528	.2236	.4472	.2764
0.77	.5588	.2206	.4412	.2794
0.78	.5646	.2177	.4354	.2823
0.79	.5704	.2148	.4296	.2852
0.80	.5762	.2119	.4238	.2881
0.81	.5820	.2090	.4180	.2910
0.82	.5878	.2061	.4132	.2939
0.83	.5934	.2033	.4066	.2967

COLUMN 1	COLUMN 2	COLUMN 3	COLUMN 4	COLUMN 5
z	$\geq -z$ and $\leq +z$	$\geq +z$ (also use to find $\leq -z$)	$\leq -z$ or $\geq +z$	≥ 0 and $\leq +z$ (also use to find $\geq -z$ and ≤ 0)
0.84	.5990	.2005	.4010	.2995
0.85	.6046	.1977	.3954	.3023
0.86	.6102	.1949	.3898	.3051
0.87	.6156	.1922	.3844	.3078
0.88	.6212	.1894	.3788	.3106
0.89	.6266	.1867	.3734	.3133
0.90	.6318	.1841	.3682	.3159
0.91	.6372	.1814	.3628	.3186
0.92	.6424	.1788	.3576	.3212
0.93	.6476	.1762	.3524	.3238
0.94	.6528	.1736	.3472	.3264
0.95	.6578	.1711	.3422	.3289
0.96	.6630	.1685	.3370	.3315
0.97	.6680	.1660	.3320	.3340
0.98	.6730	.1635	.3270	.3365
0.99	.6778	.1611	.3222	.3389
1.00	.6826	.1587	.3174	.3413
1.01	.6876	.1562	.3124	.3438
1.02	.6922	.1539	.3078	.3461
1.03	.6970	.1515	.3030	.3485
1.04	.7016	.1492	.2984	.3508
1.05	.7062	.1469	.2938	.3531
1.06	.7108	.1446	.2892	.3554
1.07	.7154	.1423	.2846	.3577
1.08	.7198	.1401	.2802	.3599
1.09	.7242	.1379	.2758	.3621
1.10	.7286	.1357	.2714	.3643
1.11	.7330	.1335	.2670	.3665
1.12	.7372	.1314	.2628	.3686
1.13	.7416	.1292	.2584	.3708
1.14	.7458	.1271	.2542	.3729
1.15	.7498	.1251	.2502	.3749
1.16	.7540	.1230	.2460	.3770
1.17	.7580	.1210	.2420	.3790
1.18	.7620	.1190	.2380	.3810
1.19	.7660	.1170	.2340	.3830
1.20	.7698	.1151	.2302	.3849
1.21	.7738	.1131	.2262	.3869
1.22	.7776	.1112	.2224	.3888
1.23	.7814	.1093	.2186	.3907
1.24	.7850	.1075	.2150	.3925
1.25	.7888	.1056	.2112	.3944

COLUMN 1 z	COLUMN 2 ≥ −z and ≤ +z	COLUMN 3 ≥ +z (also use to find ≤ −z)	COLUMN 4 ≤ −z or ≥ +z	COLUMN 5 ≥ 0 and ≤ +z (also use to find ≥ −z and ≤ 0)
1.26	.7924	.1038	.2076	.3962
1.27	.7960	.1020	.2040	.3980
1.28	.7994	.1003	.2006	.3997
1.29	.8030	.0985	.1970	.4015
1.30	.8064	.0968	.1936	.4032
1.31	.8098	.0951	.1902	.4049
1.32	.8132	.0934	.1868	.4066
1.33	.8164	.0918	.1836	.4082
1.34	.8198	.0901	.1802	.4099
1.35	.8230	.0885	.1770	.4115
1.36	.8262	.0869	.1738	.4131
1.37	.8294	.0853	.1706	.4147
1.38	.8324	.0838	.1676	.4162
1.39	.8354	.0823	.1646	.4177
1.40	.8384	.0808	.1616	.4192
1.41	.8414	.0793	.1586	.4207
1.42	.8444	.0778	.1556	.4222
1.43	.8472	.0764	.1528	.4236
1.44	.8502	.0749	.1498	.4251
1.45	.8530	.0735	.1470	.4265
1.46	.8558	.0721	.1442	.4279
1.47	.8584	.0708	.1416	.4292
1.48	.8612	.0694	.1388	.4306
1.49	.8638	.0681	.1362	.4319
1.50	.8664	.0668	.1336	.4332
1.51	.8690	.0655	.1310	.4345
1.52	.8714	.0643	.1286	.4357
1.53	.8740	.0630	.1260	.4370
1.54	.8764	.0618	.1236	.4382
1.55	.8788	.0606	.1212	.4394
1.56	.8812	.0594	.1188	.4406
1.57	.8836	.0582	.1164	.4418
1.58	.8858	.0571	.1142	.4429
1.59	.8882	.0559	.1118	.4441
1.60	.8904	.0548	.1096	.4452
1.61	.8926	.0537	.1074	.4463
1.62	.8948	.0526	.1052	.4474
1.63	.8968	.0516	.1032	.4484
1.64	.8990	.0505	.1010	.4495
1.65	.9010	.0495	.0990	.4505
1.66	.9030	.0485	.0970	.4515
1.67	.9050	.0475	.0950	.4525
1.68	.9070	.0465	.0930	.4535

COLUMN 1	COLUMN 2	COLUMN 3	COLUMN 4	COLUMN 5
	-z +z	+z / -z	-z +z	0 +z / -z 0
z	$\geq -z$ and $\leq +z$	$\geq +z$ (also use to find $\leq -z$)	$\leq -z$ or $\geq +z$	≥ 0 and $\leq +z$ (also use to find $\geq -z$ and ≤ 0)
1.69	.9090	.0455	.0910	.4545
1.70	.9108	.0446	.0892	.4554
1.71	.9128	.0436	.0872	.4564
1.72	.9146	.0427	.0854	.4573
1.73	.9164	.0418	.0836	.4582
1.74	.9182	.0409	.0818	.4591
1.75	.9198	.0401	.0802	.4599
1.76	.9216	.0392	.0784	.4608
1.77	.9232	.0384	.0764	.4616
1.78	.9250	.0375	.0750	.4625
1.79	.9266	.0367	.0734	.4633
1.80	.9282	.0359	.0718	.4641
1.81	.9298	.0351	.0702	.4649
1.82	.9312	.0344	.0688	.4656
1.83	.9328	.0336	.0672	.4664
1.84	.9342	.0329	.0658	.4671
1.85	.9356	.0322	.0644	.4678
1.86	.9372	.0314	.0628	.4686
1.87	.9386	.0307	.0614	.4693
1.88	.9398	.0301	.0602	.4699
1.89	.9412	.0294	.0588	.4706
1.90	.9426	.0287	.0574	.4713
1.91	.9438	.0281	.0562	.4719
1.92	.9452	.0274	.0548	.4726
1.93	.9464	.0268	.0536	.4732
1.94	.9476	.0262	.0524	.4738
1.95	.9488	.0256	.0512	.4744
1.96	.9500	.0250	.0500	.4750
1.97	.9512	.0244	.0488	.4756
1.98	.9522	.0239	.0478	.4761
1.99	.9534	.0233	.0466	.4767
2.00	.9544	.0228	.0456	.4772
2.01	.9556	.0222	.0444	.4778
2.02	.9566	.0217	.0434	.4783
2.03	.9576	.0212	.0424	.4788
2.04	.9586	.0207	.0414	.4793
2.05	.9596	.0202	.0404	.4798
2.06	.9606	.0197	.0394	.4803
2.07	.9616	.0192	.0384	.4808
2.08	.9624	.0188	.0376	.4812
2.09	.9634	.0183	.0366	.4817
2.10	.9642	.0179	.0358	.4821
2.11	.9652	.0174	.0348	.4826

COLUMN 1 z	COLUMN 2 $\geq -z$ and $\leq +z$	COLUMN 3 $\geq +z$ (also use to find $\leq -z$)	COLUMN 4 $\leq -z$ or $\geq +z$	COLUMN 5 ≥ 0 and $\leq +z$ (also use to find $\geq -z$ and ≤ 0)
2.12	.9660	.0170	.0340	.4830
2.13	.9668	.0166	.0332	.4834
2.14	.9676	.0162	.0324	.4838
2.15	.9684	.0158	.0316	.4842
2.16	.9692	.0154	.0308	.4846
2.17	.9700	.0150	.0300	.4850
2.18	.9708	.0146	.0292	.4854
2.19	.9714	.0143	.0286	.4857
2.20	.9722	.0139	.0278	.4861
2.21	.9728	.0136	.0272	.4864
2.22	.9736	.0132	.0264	.4868
2.23	.9742	.0129	.0258	.4871
2.24	.9750	.0125	.0250	.4875
2.25	.9756	.0122	.0244	.4878
2.26	.9762	.0119	.0238	.4881
2.27	.9768	.0116	.0232	.4884
2.28	.9774	.0113	.0276	.4887
2.29	.9780	.0110	.0220	.4890
2.30	.9786	.0107	.0214	.4893
2.31	.9792	.0104	.0208	.4896
2.32	.9796	.0102	.0204	.4898
2.33	.9802	.0099	.0198	.4901
2.34	.9808	.0096	.0192	.4904
2.35	.9812	.0094	.0188	.4906
2.36	.9818	.0091	.0182	.4909
2.37	.9822	.0089	.0178	.4911
2.38	.9826	.0087	.0174	.4913
2.39	.9832	.0084	.0168	.4916
2.40	.9836	.0082	.0164	.4918
2.41	.9840	.0080	.0160	.4920
2.42	.9844	.0078	.0156	.4922
2.43	.9850	.0075	.0150	.4925
2.44	.9854	.0073	.0146	.4927
2.45	.9858	.0071	.0142	.4929
2.46	.9862	.0069	.0138	.4931
2.47	.9864	.0068	.0136	.4932
2.48	.9868	.0066	.0132	.4934
2.49	.9872	.0064	.0128	.4936
2.50	.9876	.0062	.0124	.4938
2.51	.9880	.0060	.0120	.4940
2.52	.9882	.0059	.0118	.4941
2.53	.9886	.0057	.0114	.4943
2.54	.9890	.0055	.0110	.4945

COLUMN 1	COLUMN 2	COLUMN 3	COLUMN 4	COLUMN 5
z	$\geq -z$ and $\leq +z$	$\geq +z$ (also use to find $\leq -z$)	$\leq -z$ or $\geq +z$	≥ 0 and $\leq +z$ (also use to find $\geq -z$ and ≤ 0)
2.55	.9892	.0054	.0108	.4946
2.56	.9896	.0052	.0104	.4948
2.57	.9898	.0051	.0102	.4949
2.58	.9902	.0048	.0098	.4951
2.59	.9904	.0049	.0096	.4952
2.60	.9906	.0047	.0094	.4953
2.61	.9910	.0045	.0090	.4955
2.62	.9912	.0044	.0088	.4956
2.63	.9914	.0043	.0086	.4957
2.64	.9918	.0041	.0082	.4959
2.65	.9920	.0040	.0080	.4960
2.66	.9922	.0039	.0078	.4961
2.67	.9924	.0038	.0076	.4962
2.68	.9926	.0037	.0074	.4963
2.69	.9928	.0036	.0072	.4964
2.70	.9930	.0035	.0070	.4965
2.71	.9932	.0034	.0068	.4966
2.72	.9934	.0033	.0066	.4967
2.73	.9936	.0032	.0064	.4968
2.74	.9938	.0031	.0062	.4969
2.75	.9940	.0030	.0060	.4970
2.76	.9942	.0029	.0058	.4971
2.77	.9944	.0028	.0056	.4972
2.78	.9946	.0027	.0054	.4973
2.79	.9948	.0026	.0052	.4974
2.80	.9948	.0026	.0052	.4974
2.81	.9950	.0025	.0050	.4975
2.82	.9952	.0024	.0048	.4976
2.83	.9954	.0023	.0046	.4977
2.84	.9954	.0023	.0046	.4977
2.85	.9956	.0022	.0044	.4978
2.86	.9958	.0021	.0042	.4979
2.87	.9958	.0021	.0042	.4979
2.88	.9960	.0020	.0040	.4980
2.89	.9962	.0019	.0038	.4981
2.90	.9962	.0019	.0038	.4981
2.91	.9964	.0018	.0036	.4982
2.92	.9964	.0018	.0036	.4982
2.93	.9966	.0017	.0034	.4983
2.94	.9968	.0016	.0032	.4984
2.95	.9968	.0016	.0032	.4984
2.96	.9970	.0015	.0030	.4985
2.97	.9970	.0015	.0030	.4985

COLUMN 1	COLUMN 2	COLUMN 3	COLUMN 4	COLUMN 5
z	$\geq -z$ and $\leq +z$	$\geq +z$ (also use to find $\leq -z$)	$\leq -z$ or $\geq +z$	≥ 0 and $\leq +z$ (also use to find $\geq -z$ and ≤ 0)
2.98	.9972	.0014	.0028	.4986
2.99	.9972	.0014	.0028	.4986
3.00	.9974	.0013	.0026	.4987
3.01	.9974	.0013	.0026	.4987
3.02	.9974	.0013	.0026	.4987
3.03	.9976	.0012	.0024	.4988
3.04	.9976	.0012	.0024	.4988
3.05	.9978	.0011	.0022	.4989
3.06	.9978	.0011	.0022	.4989
3.07	.9978	.0011	.0022	.4989
3.08	.9980	.0010	.0020	.4990
3.09	.9980	.0010	.0020	.4990
3.10	.9980	.0010	.0020	.4990
3.11	.9982	.0009	.0018	.4991
3.12	.9982	.0009	.0018	.4991
3.13	.9982	.0009	.0018	.4991
3.14	.9984	.0008	.0016	.4992
3.15	.9984	.0008	.0016	.4992
3.16	.9984	.0008	.0016	.4992
3.17	.9984	.0008	.0016	.4992
3.18	.9986	.0007	.0014	.4993
3.19	.9986	.0007	.0014	.4993
3.20	.9986	.0007	.0014	.4993
3.21	.9986	.0007	.0014	.4993
3.22	.9988	.0006	.0012	.4994
3.23	.9988	.0006	.0012	.4994
3.24	.9988	.0006	.0012	.4994
3.25	.9988	.0006	.0012	.4994
3.30	.9990	.0005	.0010	.4995
3.35	.9992	.0004	.0008	.4996
3.40	.9994	.0003	.0006	.4997
3.45	.9994	.0003	.0006	.4997
3.50	.9996	.0002	.0004	.4998
3.60	.9996	.0002	.0004	.4998
3.70	.9998	.0001	.0002	.4999
3.80	.9998	.0001	.0002	.4999
3.90	.9999	.00005	.00010	.49995
4.00	.99994	.00003	.00006	.49997

APPENDIX C

Factorials

The following table presents factorials for the numbers 0 through 20. For example, the value of 7! is 5,040.

Number	*Factorial*
0	1
1	1
2	2
3	6
4	24
5	120
6	720
7	5,040
8	40,320
9	362,880
10	3,628,800
11	39,916,800
12	479,001,600
13	6,227,020,800
14	87,178,291,200
15	1,307,674,368,000
16	20,922,789,888,000
17	355,687,428,096,000
18	6,402,373,705,728,000
19	121,645,100,408,832,000
20	2,432,902,008,176,640,000

APPENDIX D

Critical Values for the t Distribution

The following table presents critical values of t for directional (one-tailed) and nondirectional (two-tailed) tests. The column headed "df" lists the degrees of freedom associated with the t distribution of interest. The entries in the table are the t values that define the rejection regions. The values .10, .05, .025, .01, .005, and .0005 at the top of the table under the heading "Level of Significance for Directional Test" represent alpha levels for directional tests. For instance, if we were conducting a directional test having 15 degrees of freedom at an alpha level of .05, we would locate 15 in the df column and follow across that row until it intersected the .05 column for a directional test. The entry of 1.753 is the critical value of t for a directional test using the upper tail of the t distribution. Since the t distribution is symmetrical, the critical value of t for a directional test using the lower tail of the t distribution is -1.753.

The values .20, .10, .05, .02, .01, and .001 under the heading "Level of Significance for Nondirectional Test" represent alpha levels for nondirectional tests. For instance, if we were conducting a nondirectional test having 15 degrees of freedom at an alpha level of .05, we would locate 15 in the df column and follow across that row until it intersected the .05 column for a nondirectional test. The entry of 2.131 signifies that the positive critical value of t is $+2.131$ and the negative critical value of t is -2.131. The rejection region thus consists of all values of t less than -2.131 or greater than $+2.131$.

Source: From Table III of Fisher and Yates, *Statistical Tables for Biological, Agricultural and Medical Research,* published by Longman Group Ltd., London (previously published by Oliver & Boyd Ltd., Edinburgh) and by permission of the authors and publishers.

	LEVEL OF SIGNIFICANCE FOR DIRECTIONAL TEST					
	.10	.05	.025	.01	.005	.0005
	LEVEL OF SIGNIFICANCE FOR NONDIRECTIONAL TEST					
df	.20	.10	.05	.02	.01	.001
1	3.078	6.314	12.706	31.821	63.657	636.619
2	1.886	2.920	4.303	6.965	9.925	31.598
3	1.638	2.353	3.182	4.541	5.841	12.941
4	1.533	2.132	2.776	3.747	4.604	8.610
5	1.476	2.015	2.571	3.365	4.032	6.859
6	1.440	1.943	2.447	3.143	3.707	5.959
7	1.415	1.895	2.365	2.998	3.499	5.405
8	1.397	1.860	2.306	2.896	3.355	5.041
9	1.383	1.833	2.262	2.821	3.250	4.781
10	1.372	1.812	2.228	2.764	3.169	4.587
11	1.363	1.796	2.201	2.718	3.106	4.437
12	1.356	1.782	2.179	2.681	3.055	4.318
13	1.350	1.771	2.160	2.650	3.012	4.221
14	1.345	1.761	2.145	2.624	2.977	4.140
15	1.341	1.753	2.131	2.602	2.947	4.073
16	1.337	1.746	2.120	2.583	2.921	4.015
17	1.333	1.740	2.110	2.567	2.898	3.965
18	1.330	1.734	2.101	2.552	2.878	3.922
19	1.328	1.729	2.093	2.539	2.861	3.883
20	1.325	1.725	2.086	2.528	2.845	3.850
21	1.323	1.721	2.080	2.518	2.831	3.819
22	1.321	1.717	2.074	2.508	2.819	3.792
23	1.319	1.714	2.069	2.500	2.807	3.767
24	1.318	1.711	2.064	2.492	2.797	3.745
25	1.316	1.708	2.060	2.485	2.787	3.725
26	1.315	1.706	2.056	2.479	2.779	3.707
27	1.314	1.703	2.052	2.473	2.771	3.690
28	1.313	1.701	2.048	2.467	2.763	3.674
29	1.311	1.699	2.045	2.462	2.756	3.659
30	1.310	1.697	2.042	2.457	2.750	3.646
40	1.303	1.684	2.021	2.423	2.704	3.551
60	1.296	1.671	2.000	2.390	2.660	3.460
120	1.289	1.658	1.980	2.358	2.617	3.373
∞	1.282	1.645	1.960	2.326	2.576	3.291

APPENDIX E

Power and Sample Size

The following sets of tables indicate the sample sizes necessary to achieve desired levels of power for the statistical techniques listed below. The procedure for using the tables for a given technique is described in the relevant chapter in the text.

Appendix E.1: Independent or Correlated Groups *t* Test

Directional Test, Alpha = .05

POPULATION ETA-SQUARED

POWER	.01	.03	.05	.07	.10	.15	.20	.25	.30	.35	.40	.45	.50	.55	.60	.65	.70	.75	.80
.25	48	16	10	7	5	3	3	2	2	2	—	—	—	—	—	—	—	—	—
.50	136	45	26	19	13	8	6	5	4	3	3	2	2	2	2	—	—	—	—
.60	181	59	35	25	17	11	8	6	5	4	3	3	3	2	2	2	—	—	—
.67	216	70	42	29	20	13	9	7	6	5	4	3	3	2	2	2	2	—	—
.70	236	77	45	32	22	14	10	8	6	5	4	4	3	3	2	2	2	—	—
.75	270	88	52	36	25	16	11	9	7	6	5	4	3	3	2	2	2	2	—
.80	310	101	59	42	29	18	13	10	8	6	5	4	4	3	3	2	2	2	—
.85	360	117	69	48	33	21	15	11	9	7	6	5	4	4	3	3	2	2	—
.90	429	139	82	58	39	25	18	14	11	9	7	6	5	4	4	3	3	2	2
.95	542	176	104	73	49	31	22	17	13	11	9	7	6	5	4	4	3	3	2
.99	789	256	151	105	72	45	32	24	19	15	13	10	9	7	6	5	4	3	3

Directional Test, Alpha = .01

POPULATION ETA-SQUARED

POWER	.01	.03	.05	.07	.10	.15	.20	.25	.30	.35	.40	.45	.50	.55	.60	.65	.70	.75	.80
.25	138	46	27	20	14	9	7	6	5	4	4	3	3	3	2	2	2	2	2
.50	272	89	53	37	26	17	12	10	8	7	6	5	4	4	3	3	3	2	2
.60	334	109	65	46	31	20	15	11	9	8	6	6	5	4	4	3	3	3	2
.67	382	127	75	53	36	23	17	13	11	9	7	6	5	5	4	4	3	3	2
.70	408	133	79	56	38	25	18	14	11	9	8	6	6	5	4	4	3	3	3
.75	452	147	87	61	42	27	20	15	12	10	8	7	6	5	5	4	3	3	3
.80	503	164	97	68	47	30	22	17	13	11	9	8	7	6	5	4	4	3	3
.85	567	184	109	77	52	34	24	18	15	12	10	8	7	6	5	5	4	3	3
.90	652	212	125	88	60	38	27	21	17	14	11	9	8	7	6	5	4	4	3
.95	790	257	151	106	72	46	33	25	20	16	13	11	9	8	7	6	5	4	3
.99	1,084	352	207	145	99	63	45	34	27	22	18	15	12	10	9	7	6	5	4

Nondirectional Test, Alpha = .05

POPULATION ETA-SQUARED

POWER	.01	.03	.05	.07	.10	.15	.20	.25	.30	.35	.40	.45	.50	.55	.60	.65	.70	.75	.80
.25	84	28	17	12	8	6	5	3	3	3	2	2	2	2	2	—	—	—	—
.50	193	63	38	27	18	12	9	7	5	5	4	3	3	3	2	2	2	2	—
.60	246	80	48	34	23	15	11	8	7	6	5	4	3	3	3	2	2	2	2
.67	287	93	55	39	27	17	12	10	8	6	5	4	4	3	3	3	2	2	2
.70	310	101	60	42	29	18	13	10	8	7	6	5	4	4	3	3	2	2	2
.75	348	113	67	47	32	21	15	11	9	7	6	5	4	4	3	3	2	2	2
.80	393	128	76	53	36	23	17	13	10	8	7	6	5	4	4	3	3	2	2
.85	450	146	86	61	41	26	19	14	11	9	8	6	5	5	4	3	3	2	2
.90	526	171	101	71	48	31	22	17	13	11	9	7	6	5	5	4	3	3	2
.95	651	211	125	87	60	38	27	21	16	13	11	9	8	6	5	5	4	3	3
.99	920	298	176	123	84	53	38	29	22	18	15	12	10	9	7	6	5	4	3

Nondirectional Test, Alpha = .01

POPULATION ETA-SQUARED

POWER	.01	.03	.05	.07	.10	.15	.20	.25	.30	.35	.40	.45	.50	.55	.60	.65	.70	.75	.80
.25	183	60	36	26	18	12	9	7	6	5	4	4	3	3	3	2	2	2	2
.50	333	109	65	46	31	20	15	11	9	8	6	6	5	4	4	3	3	3	2
.60	402	131	78	55	38	24	18	14	11	9	8	6	6	5	4	4	3	3	3
.67	454	148	87	62	42	27	20	15	12	10	8	7	6	5	5	4	3	3	3
.70	482	157	93	65	45	29	21	16	13	10	9	7	6	5	5	4	4	3	3
.75	528	172	102	72	49	31	23	17	14	11	9	8	7	6	5	4	4	3	3
.80	586	190	113	79	54	35	25	19	15	12	10	9	7	6	5	5	4	3	3
.85	654	213	126	88	60	38	28	21	17	14	11	9	8	7	6	5	4	4	3
.90	746	242	143	100	69	44	31	24	19	15	13	11	9	8	6	6	5	4	3
.95	892	290	171	120	82	52	37	28	22	18	15	12	10	9	7	6	5	4	4
.99	1,203	390	230	161	110	70	50	38	30	24	20	16	14	11	10	8	7	6	5

Appendix E.2: One- or Two-Way Between-Subjects Analysis of Variance

Degrees of Freedom Between or Effect = 1, Alpha = .05

POPULATION ETA-SQUARED

POWER	.01	.03	.05	.07	.10	.15	.20	.25	.30	.35	.40	.45	.50	.55	.60	.65	.70	.75	.80
.10	22	8	5	4	3	2	2	2	—	—	—	—	—	—	—	—	—	—	—
.50	193	63	38	27	18	12	9	7	5	5	4	3	3	3	2	2	2	2	—
.70	310	101	60	42	29	18	13	10	8	7	6	5	4	4	3	3	2	2	2
.80	393	128	76	53	36	23	17	13	10	8	7	6	5	4	4	3	3	2	2
.90	526	171	101	71	48	31	22	17	13	11	9	7	6	5	5	4	3	3	2
.95	651	211	125	87	60	38	27	21	16	13	11	9	8	6	5	5	4	3	3
.99	920	298	176	123	84	53	38	29	22	18	15	12	10	9	7	6	5	4	3

Degrees of Freedom Between or Effect = 2, Alpha = .05

POPULATION ETA-SQUARED

POWER	.01	.03	.05	.07	.10	.15	.20	.25	.30	.35	.40	.45	.50	.55	.60	.65	.70	.75	.80
.10	22	8	5	4	3	2	2	2	—	—	—	—	—	—	—	—	—	—	—
.50	165	55	32	23	16	10	8	6	5	4	3	3	3	2	2	2	2	2	—
.70	255	84	50	35	24	16	11	9	7	6	5	4	4	3	3	2	2	2	2
.80	319	105	62	44	30	19	14	11	9	7	6	5	4	4	3	3	2	2	2
.90	417	137	81	57	39	25	18	14	11	9	7	6	5	4	4	3	3	2	2
.95	511	168	99	69	47	30	22	16	13	11	9	7	6	5	4	4	3	3	2
.99	708	232	137	96	65	41	29	22	18	14	12	10	8	7	6	5	4	3	3

Degrees of Freedom Between or Effect = 3, Alpha = .05

POPULATION ETA-SQUARED

POWER	.01	.03	.05	.07	.10	.15	.20	.25	.30	.35	.40	.45	.50	.55	.60	.65	.70	.75	.80
.10	21	7	5	4	3	2	2	2	—	—	—	—	—	—	—	—	—	—	—
.50	144	48	28	20	14	9	7	5	4	4	3	3	2	2	2	2	2	—	—
.70	219	72	43	30	21	13	10	8	6	5	4	4	3	3	2	2	2	2	2
.80	272	90	53	37	26	17	12	9	7	6	5	4	4	3	3	2	2	2	2
.90	351	115	68	48	33	21	15	12	9	8	6	5	5	4	3	3	3	2	2
.95	426	140	83	58	40	25	18	14	11	9	7	6	5	5	4	3	3	2	2
.99	583	191	113	79	54	34	24	19	15	12	10	8	7	6	5	4	4	3	2

Degrees of Freedom Between or Effect = 4, Alpha = .05

POPULATION ETA-SQUARED

POWER	.01	.03	.05	.07	.10	.15	.20	.25	.30	.35	.40	.45	.50	.55	.60	.65	.70	.75	.80
.10	19	7	5	3	3	2	—	2	—	—	—	—	—	—	—	—	—	—	—
.50	128	43	25	18	13	8	6	5	4	3	3	3	2	2	2	2	2	—	—
.70	193	64	38	27	18	12	9	7	6	5	4	3	3	3	2	2	2	2	—
.80	238	78	46	33	23	15	10	8	7	5	5	4	3	3	3	2	2	2	2
.90	306	101	59	42	29	18	13	10	8	7	6	5	4	4	3	3	2	2	2
.95	369	121	72	50	34	22	16	12	10	8	7	6	5	4	3	3	3	2	2
.99	501	164	97	68	46	30	21	16	13	10	9	7	6	5	4	4	3	3	2

Degrees of Freedom Between or Effect = 5, Alpha = .05

POPULATION ETA-SQUARED

POWER	.01	.03	.05	.07	.10	.15	.20	.25	.30	.35	.40	.45	.50	.55	.60	.65	.70	.75	.80
.10	18	7	4	3	3	2	—	—	—	—	—	—	—	—	—	—	—	—	—
.50	117	39	23	17	12	8	6	5	4	3	3	2	2	2	2	2	2	—	—
.70	174	57	34	24	17	11	8	6	5	4	4	3	3	2	2	2	2	2	—
.80	213	70	42	29	20	13	9	7	6	5	4	4	3	3	2	2	2	2	2
.90	273	90	53	37	26	17	12	9	7	6	5	4	4	3	3	2	2	2	2
.95	328	108	64	45	31	20	14	11	9	7	6	5	4	4	3	3	2	2	2
.99	442	145	86	60	41	26	19	14	11	9	8	6	5	5	4	3	3	2	2

Degrees of Freedom Between or Effect = 6, Alpha = .05

POPULATION ETA-SQUARED

POWER	.01	.03	.05	.07	.10	.15	.20	.25	.30	.35	.40	.45	.50	.55	.60	.65	.70	.75	.80
.10	17	6	4	3	2	2	—	—	—	—	—	—	—	—	—	—	—	—	—
.50	107	36	21	15	11	7	5	4	4	3	3	2	2	2	2	2	—	—	—
.70	159	53	31	22	15	10	7	6	5	4	3	3	3	2	2	2	2	2	—
.80	194	64	38	27	19	12	9	7	6	5	4	3	3	3	2	2	2	2	—
.90	247	81	48	34	23	15	11	8	7	6	5	4	3	3	3	2	2	2	2
.95	296	97	58	41	28	18	13	10	8	7	5	5	4	3	3	3	2	2	2
.99	398	131	77	54	37	24	17	13	10	8	7	6	5	4	4	3	3	2	2

Degrees of Freedom Between or Effect = 8, Alpha = .05

POPULATION ETA-SQUARED

POWER	.01	.03	.05	.07	.10	.15	.20	.25	.30	.35	.40	.45	.50	.55	.60	.65	.70	.75	.80
.10	16	6	4	3	2	2	—	—	—	—	—	—	—	—	—	—	—	—	—
.50	94	31	19	13	9	6	5	4	3	3	2	2	2	2	2	2	—	—	—
.70	137	45	27	19	13	9	6	5	4	4	3	3	2	2	2	2	2	—	—
.80	167	55	33	23	16	10	8	6	5	4	4	3	3	2	2	2	2	2	—
.90	211	70	41	29	20	13	9	7	6	5	4	4	3	3	2	2	2	2	2
.95	251	83	49	35	24	15	11	9	7	6	5	4	4	3	3	2	2	2	2
.99	335	110	65	46	31	20	14	11	9	7	6	5	4	4	3	3	2	2	2

Degrees of Freedom Between or Effect = 10, Alpha = .05

POPULATION ETA-SQUARED

POWER	.01	.03	.05	.07	.10	.15	.20	.25	.30	.35	.40	.45	.50	.55	.60	.65	.70	.75	.80
.10	15	5	4	3	2	2	—	—	—	—	—	—	—	—	—	—	—	—	—
.50	84	28	17	12	9	6	4	4	3	3	2	2	2	2	2	—	—	—	—
.70	122	40	24	17	12	8	6	5	4	3	3	2	2	2	2	2	2	—	—
.80	147	49	29	21	14	9	7	5	4	4	3	3	2	2	2	2	2	—	—
.90	186	61	36	26	18	12	8	7	5	4	4	3	3	3	2	2	2	2	—
.95	221	73	43	30	21	14	10	8	6	5	4	4	3	3	2	2	2	2	2
.99	292	96	57	40	27	18	13	10	8	6	5	5	4	3	3	3	2	2	2

Degrees of Freedom Between or Effect = 12, Alpha = .05

POPULATION ETA-SQUARED

POWER	.01	.03	.05	.07	.10	.15	.20	.25	.30	.35	.40	.45	.50	.55	.60	.65	.70	.75	.80
.10	14	5	3	3	2	2	—	—	—	—	—	—	—	—	—	—	—	—	—
.50	77	26	16	11	8	5	4	3	3	2	2	2	2	2	2	—	—	—	—
.70	111	37	22	16	11	7	5	4	4	3	3	2	2	2	2	2	—	—	—
.80	133	44	26	19	13	9	6	5	4	3	3	3	2	2	2	2	2	—	—
.90	168	55	33	23	16	11	8	6	5	4	4	3	3	2	2	2	2	2	—
.95	198	65	39	27	19	12	9	7	6	5	4	3	3	3	2	2	2	2	—
.99	261	86	51	36	25	16	11	9	7	6	5	4	4	3	3	2	2	2	2

Degrees of Freedom Between or Effect = 15, Alpha = .05

POPULATION ETA-SQUARED

POWER	.01	.03	.05	.07	.10	.15	.20	.25	.30	.35	.40	.45	.50	.55	.60	.65	.70	.75	.80
.10	13	5	3	3	2	2	—	—	—	—	—	—	—	—	—	—	—	—	—
.50	68	23	14	10	7	5	4	3	3	2	2	2	2	2	—	—	—	—	—
.70	98	33	20	14	10	7	5	4	3	3	2	2	2	2	2	2	—	—	—
.80	118	39	23	17	12	8	6	5	4	3	3	2	2	2	2	2	2	—	—
.90	147	49	29	21	14	9	7	5	4	4	3	3	2	2	2	2	2	—	—
.95	174	57	34	24	17	11	8	6	5	4	4	3	3	2	2	2	2	2	—
.99	227	75	44	31	22	14	10	8	6	5	4	4	3	3	3	2	2	2	2

Degrees of Freedom Between or Effect = 1, Alpha = .01

POPULATION ETA-SQUARED

POWER	.01	.03	.05	.07	.10	.15	.20	.25	.30	.35	.40	.45	.50	.55	.60	.65	.70	.75	.80
.10	84	28	17	12	9	6	5	4	3	3	2	2	2	2	2	—	—	—	—
.50	333	109	65	46	31	20	15	11	9	8	6	6	5	4	4	3	3	3	2
.70	482	157	93	65	45	29	21	16	13	10	9	7	6	5	5	4	4	3	3
.80	586	190	113	79	54	35	25	19	15	12	10	9	7	6	5	5	4	3	3
.90	746	242	143	100	69	44	31	24	19	15	13	11	9	8	6	6	5	4	3
.95	892	289	171	120	82	52	37	28	22	18	15	12	10	9	7	6	5	4	4
.99	1,203	390	230	161	110	70	50	38	30	24	20	16	14	11	10	8	7	6	5

Degrees of Freedom Between or Effect = 2, Alpha = .01

POPULATION ETA-SQUARED

POWER	.01	.03	.05	.07	.10	.15	.20	.25	.30	.35	.40	.45	.50	.55	.60	.65	.70	.75	.80
.10	77	26	16	11	8	5	5	3	3	2	2	2	2	2	2	—	—	—	—
.50	272	89	53	37	26	16	13	9	7	6	5	4	4	3	3	2	2	2	2
.70	383	126	74	52	36	23	17	13	10	8	7	6	5	4	4	3	3	2	2
.80	459	151	89	62	43	27	20	15	12	10	8	7	6	5	4	3	3	3	2
.90	576	189	111	78	53	34	25	18	15	12	10	8	7	6	5	4	3	3	2
.95	683	224	132	93	63	40	29	22	17	14	11	9	8	7	6	5	4	3	3
.99	906	297	175	122	83	53	38	28	22	18	15	12	10	8	7	6	5	4	3

Degrees of Freedom Between or Effect = 3, Alpha = .01

POPULATION ETA-SQUARED

POWER	.01	.03	.05	.07	.10	.15	.20	.25	.30	.35	.40	.45	.50	.55	.60	.65	.70	.75	.80
.10	70	23	14	10	7	5	4	3	3	2	2	2	2	2	—	—	—	—	—
.50	232	76	45	32	22	14	11	8	6	5	4	4	3	3	3	2	2	2	2
.70	323	106	63	44	30	19	14	11	9	7	6	5	4	4	3	3	2	2	2
.80	384	126	75	52	36	23	17	13	10	8	7	6	5	4	4	3	3	2	2
.90	478	157	93	65	44	28	21	15	12	10	8	7	6	5	4	4	3	3	2
.95	563	184	109	76	52	33	24	18	14	12	10	8	7	6	5	4	3	3	2
.99	740	242	143	100	68	43	31	23	18	15	12	10	8	7	6	5	4	3	3

Degrees of Freedom Between or Effect = 4, Alpha = .01

POPULATION ETA-SQUARED

POWER	.01	.03	.05	.07	.10	.15	.20	.25	.30	.35	.40	.45	.50	.55	.60	.65	.70	.75	.80
.10	64	21	13	9	7	5	4	3	2	2	2	2	2	2	—	—	—	—	—
.50	204	67	40	28	19	13	10	7	6	5	4	4	3	3	2	2	2	2	2
.70	280	92	55	38	26	17	13	9	8	6	5	4	4	3	3	3	2	2	2
.80	333	109	65	46	31	20	15	11	9	7	6	5	4	4	3	3	2	2	2
.90	412	135	80	56	38	25	18	13	11	9	7	6	5	4	4	3	3	2	2
.95	483	158	94	66	45	29	21	16	12	10	8	7	6	5	4	4	3	3	2
.99	631	207	122	86	58	37	27	20	16	13	11	9	7	6	5	4	4	3	3

Degrees of Freedom Between or Effect = 5, Alpha = .01

POPULATION ETA-SQUARED

POWER	.01	.03	.05	.07	.10	.15	.20	.25	.30	.35	.40	.45	.50	.55	.60	.65	.70	.75	.80
.10	59	20	12	9	6	4	4	3	2	2	2	2	2	—	—	—	—	—	—
.50	183	61	36	25	18	11	9	7	5	4	4	3	3	3	2	2	2	2	—
.70	251	83	49	35	24	15	11	9	7	6	5	4	4	3	3	2	2	2	2
.80	296	97	58	41	28	18	13	10	8	7	5	5	4	3	3	3	2	2	2
.90	365	120	71	50	34	22	16	12	10	8	7	5	5	4	3	3	3	2	2
.95	426	140	83	58	40	25	18	14	11	9	7	6	5	5	4	3	3	2	2
.99	554	182	107	75	51	33	24	18	14	11	9	8	7	6	5	4	3	3	2

Degrees of Freedom Between or Effect = 6, Alpha = .01

POPULATION ETA-SQUARED

POWER	.01	.03	.05	.07	.10	.15	.20	.25	.30	.35	.40	.45	.50	.55	.60	.65	.70	.75	.80
.10	55	19	11	8	6	4	3	3	2	2	2	2	2	—	—	—	—	—	—
.50	168	55	33	23	16	11	8	6	5	4	4	3	3	2	2	2	2	2	—
.70	228	75	45	31	22	14	10	8	6	5	4	4	3	3	3	2	2	2	2
.80	268	88	52	37	25	16	12	9	7	6	5	4	4	3	3	2	2	2	2
.90	329	108	64	45	31	20	14	11	9	7	6	5	4	4	3	3	2	2	2
.95	384	126	74	52	36	23	17	13	10	8	7	6	5	4	4	3	3	2	2
.99	497	163	96	68	46	29	21	16	13	10	9	7	6	5	4	4	3	3	2

Degrees of Freedom Between or Effect = 8, Alpha = .01

POPULATION ETA-SQUARED

POWER	.01	.03	.05	.07	.10	.15	.20	.25	.30	.35	.40	.45	.50	.55	.60	.65	.70	.75	.80
.10	49	17	10	7	5	4	3	2	2	2	2	2	—	—	—	—	—	—	—
.50	145	48	29	20	14	9	7	5	4	4	3	3	2	2	2	2	2	—	—
.70	195	64	38	27	19	12	9	7	6	5	4	3	3	3	2	2	2	2	—
.80	228	75	45	31	22	14	10	8	6	5	4	4	3	3	3	2	2	2	2
.90	279	92	54	38	26	17	12	9	8	6	5	4	4	3	3	3	2	2	2
.95	323	106	63	44	30	19	14	11	9	7	6	5	4	4	3	3	2	2	2
.99	416	136	81	57	39	25	18	14	11	9	7	6	5	4	4	3	3	2	2

Degrees of Freedom Between or Effect = 10, Alpha = .01

POPULATION ETA-SQUARED

POWER	.01	.03	.05	.07	.10	.15	.20	.25	.30	.35	.40	.45	.50	.55	.60	.65	.70	.75	.80
.10	45	15	9	7	5	3	3	2	2	2	2	2	—	—	—	—	—	—	—
.50	128	43	25	18	13	8	6	5	4	3	3	3	2	2	2	2	2	—	—
.70	172	57	34	24	17	11	8	6	5	4	4	3	3	2	2	2	2	2	—
.80	201	66	39	28	19	12	9	7	6	5	4	3	3	3	2	2	2	2	2
.90	244	80	48	34	23	15	11	8	7	6	5	4	3	3	3	2	2	2	2
.95	283	93	55	39	27	17	12	10	8	6	5	4	4	3	3	3	2	2	2
.99	361	119	70	49	34	22	16	12	9	8	6	5	5	4	3	3	3	2	2

Degrees of Freedom Between or Effect = 12, Alpha = .01

POPULATION ETA-SQUARED

POWER	.01	.03	.05	.07	.10	.15	.20	.25	.30	.35	.40	.45	.50	.55	.60	.65	.70	.75	.80
.10	41	14	9	6	5	3	3	2	2	2	2	—	—	—	—	—	—	—	—
.50	117	39	23	17	12	8	6	5	4	3	3	2	2	2	2	2	2	—	—
.70	155	51	31	22	15	10	7	6	5	4	3	3	3	2	2	2	2	2	—
.80	181	60	35	25	17	11	8	6	5	4	4	3	3	2	2	2	2	2	—
.90	219	72	43	30	21	13	10	8	6	5	4	4	3	3	2	2	2	2	2
.95	253	83	49	35	24	15	11	9	7	6	5	4	4	3	3	2	2	2	2
.99	322	106	63	44	30	19	14	11	9	7	6	5	4	4	3	3	2	2	2

Degrees of Freedom Between or Effect = 15, Alpha = .01

POPULATION ETA-SQUARED

POWER	.01	.03	.05	.07	.10	.15	.20	.25	.30	.35	.40	.45	.50	.55	.60	.65	.70	.75	.80
.10	37	13	8	6	4	3	2	2	2	2	2	—	—	—	—	—	—	—	—
.50	103	34	21	15	10	7	5	4	3	3	3	2	2	2	2	2	—	—	—
.70	137	45	27	19	13	9	6	5	4	4	3	3	2	2	2	2	2	—	—
.80	157	52	31	22	15	10	7	6	5	4	3	3	3	2	2	2	2	2	—
.90	191	63	38	27	18	12	9	7	5	5	4	3	3	3	2	2	2	2	—
.95	220	73	43	30	21	14	10	8	6	5	4	4	3	3	2	2	2	2	2
.99	279	92	54	38	26	17	12	9	8	6	5	4	4	3	3	3	2	2	2

Appendix E.3: One-Way Repeated Measures Analysis of Variance

Degrees of Freedom Treatments = 2, Alpha = .05

POPULATION ETA-SQUARED

POWER	.01	.03	.05	.07	.10	.15	.20	.25	.30	.35	.40	.45	.50	.55	.60	.65	.70	.75	.80
.10	32	11	7	5	4	3	2	2	2	2	—	—	—	—	—	—	—	—	—
.50	247	81	48	34	23	15	11	8	7	6	5	4	3	3	3	2	2	2	2
.70	382	125	74	52	36	23	16	13	10	8	7	6	5	4	4	3	3	2	2
.80	478	157	93	65	44	28	20	15	12	10	8	7	6	5	4	4	3	3	2
.90	627	206	121	85	58	37	26	20	16	13	10	9	7	6	5	4	4	3	3
.95	765	251	148	104	70	45	32	24	19	15	13	10	9	7	6	5	4	4	3
.99	1,060	347	204	143	97	62	44	33	26	21	17	14	12	10	8	7	6	5	4

Degrees of Freedom Treatments = 3, Alpha = .05

POPULATION ETA-SQUARED

POWER	.01	.03	.05	.07	.10	.15	.20	.25	.30	.35	.40	.45	.50	.55	.60	.65	.70	.75	.80
.10	27	9	6	4	3	2	2	2	2	—	—	—	—	—	—	—	—	—	—
.50	191	63	37	27	18	12	9	7	5	5	4	3	3	3	2	2	2	2	—
.70	291	96	57	40	27	18	13	10	8	6	5	5	4	3	3	3	2	2	2
.80	361	118	70	49	34	22	16	12	9	8	6	5	5	4	3	3	3	2	2
.90	469	154	91	64	44	28	20	15	12	10	8	7	6	5	4	4	3	3	2
.95	568	186	110	77	53	33	24	18	14	12	10	8	7	6	5	4	3	3	2
.99	777	254	150	105	72	45	32	25	19	16	13	11	9	7	6	5	4	4	3

Degrees of Freedom Treatments = 4, Alpha = .05

POPULATION ETA-SQUARED

POWER	.01	.03	.05	.07	.10	.15	.20	.25	.30	.35	.40	.45	.50	.55	.60	.65	.70	.75	.80
.10	24	8	5	4	3	2	2	2	2	—	—	—	—	—	—	—	—	—	—
.50	160	53	31	22	15	10	7	6	5	4	3	3	3	2	2	2	2	2	—
.70	241	79	47	33	23	15	11	8	7	5	5	4	3	3	3	2	2	2	2
.80	297	98	58	41	28	18	13	10	8	7	5	5	4	3	3	3	2	2	2
.90	382	126	74	52	36	23	16	13	10	8	7	6	5	4	4	3	3	2	2
.95	461	151	89	63	43	27	20	15	12	10	8	7	6	5	4	3	3	3	2
.99	626	205	121	85	58	37	26	20	16	13	10	9	7	6	5	4	4	3	3

Degrees of Freedom Treatments = 5, Alpha = .05

POPULATION ETA-SQUARED

POWER	.01	.03	.05	.07	.10	.15	.20	.25	.30	.35	.40	.45	.50	.55	.60	.65	.70	.75	.80
.10	21	8	5	4	3	2	2	2	—	—	—	—	—	—	—	—	—	—	—
.50	139	46	28	20	14	9	7	5	4	4	3	3	2	2	2	2	2	—	—
.70	208	69	41	29	20	13	9	7	6	5	4	4	3	3	2	2	2	2	2
.80	225	84	50	35	24	16	11	9	7	6	5	4	4	3	3	2	2	2	2
.90	327	108	64	45	31	20	14	11	9	7	6	5	4	4	3	3	2	2	2
.95	393	129	76	54	37	23	17	13	10	8	7	6	5	4	4	3	3	2	2
.99	530	174	103	72	49	31	22	17	13	11	9	8	6	5	5	4	3	3	2

Degrees of Freedom Treatments = 6, Alpha = .05

POPULATION ETA-SQUARED

POWER	.01	.03	.05	.07	.10	.15	.20	.25	.30	.35	.40	.45	.50	.55	.60	.65	.70	.75	.80
.10	20	7	5	4	3	2	2	2	—	—	—	—	—	—	—	—	—	—	—
.50	125	41	25	18	12	8	6	5	4	3	3	3	2	2	2	2	2	—	—
.70	185	61	36	26	18	12	8	7	5	4	4	3	3	3	2	2	2	2	—
.80	226	74	44	31	21	14	10	8	6	5	4	4	3	3	3	2	2	2	2
.90	288	95	56	40	27	17	13	10	8	6	5	5	4	3	3	3	2	2	2
.95	345	113	67	47	32	21	15	11	9	7	6	5	4	4	3	3	2	2	2
.99	464	152	90	63	43	27	20	15	12	10	8	7	6	5	4	4	3	3	2

Degrees of Freedom Treatments = 2, Alpha = .01

POPULATION ETA-SQUARED

POWER	.01	.03	.05	.07	.10	.15	.20	.25	.30	.35	.40	.45	.50	.55	.60	.65	.70	.75	.80
.10	115	38	23	16	11	8	6	4	4	3	3	2	2	2	2	2	—	—	—
.50	406	133	79	55	38	24	17	13	11	9	7	6	5	4	4	3	3	2	2
.70	574	188	111	78	53	34	24	18	14	12	10	8	7	6	5	4	3	3	2
.80	688	225	133	93	63	40	29	22	17	14	11	9	8	7	6	5	4	3	3
.90	864	283	167	117	79	50	36	27	21	17	14	12	10	8	7	6	5	4	3
.95	1,023	335	197	138	94	60	42	32	25	20	16	14	11	9	8	7	5	4	4
.99	1,358	444	261	183	124	79	56	42	33	26	22	18	15	12	10	8	7	6	4

Degrees of Freedom Treatments = 3, Alpha = .01

POPULATION ETA-SQUARED

POWER	.01	.03	.05	.07	.10	.15	.20	.25	.30	.35	.40	.45	.50	.55	.60	.65	.70	.75	.80
.10	92	31	18	13	9	6	5	4	3	3	2	2	2	2	2	—	—	—	—
.50	308	101	60	42	29	19	13	10	8	7	6	5	4	4	3	3	2	2	2
.70	429	141	83	58	40	25	18	14	11	9	7	6	5	5	4	3	3	2	2
.80	511	168	99	69	47	30	22	16	13	11	9	7	6	5	4	4	3	3	2
.90	636	208	123	86	59	37	27	20	16	13	11	9	7	6	5	4	4	3	3
.95	749	245	145	101	69	44	31	24	19	15	12	10	9	7	6	5	4	4	3
.99	985	323	190	133	90	57	41	31	24	19	16	13	11	9	8	6	5	4	3

Degrees of Freedom Treatments = 4, Alpha = .01

POPULATION ETA-SQUARED

POWER	.01	.03	.05	.07	.10	.15	.20	.25	.30	.35	.40	.45	.50	.55	.60	.65	.70	.75	.80
.10	79	26	16	11	8	5	4	3	3	2	2	2	2	2	2	—	—	—	—
.50	254	84	50	35	24	15	11	9	7	6	5	4	4	3	3	2	2	2	2
.70	350	115	68	48	33	21	15	12	9	8	6	5	5	4	3	3	3	2	2
.80	416	136	81	57	39	25	18	14	11	9	7	6	5	4	4	3	3	2	2
.90	514	169	100	70	48	30	22	17	13	11	9	7	6	5	4	4	3	3	2
.95	603	198	117	82	56	35	25	19	15	12	10	8	7	6	5	4	4	3	3
.99	788	258	152	107	73	46	33	25	20	16	13	11	9	8	6	5	4	4	3

Degrees of Freedom Treatments = 5, Alpha = .01

POPULATION ETA-SQUARED

POWER	.01	.03	.05	.07	.10	.15	.20	.25	.30	.35	.40	.45	.50	.55	.60	.65	.70	.75	.80
.10	70	24	14	10	7	5	4	3	3	2	2	2	2	2	—	—	—	—	—
.50	219	72	43	30	21	14	10	8	6	5	4	4	3	3	2	2	2	2	2
.70	300	99	58	41	28	18	13	10	8	7	6	5	4	3	3	3	2	2	2
.80	355	117	69	48	33	21	15	12	9	8	6	5	5	4	3	3	3	2	2
.90	437	143	85	60	41	26	19	14	11	9	8	6	5	5	4	3	3	2	2
.95	511	168	99	69	47	30	22	16	13	11	9	7	6	5	4	4	3	3	2
.99	664	218	128	90	61	39	28	21	17	13	11	9	8	6	5	5	4	3	3

Degrees of Freedom Treatments = 6, Alpha = .01

POWER	POPULATION ETA-SQUARED .01	.03	.05	.07	.10	.15	.20	.25	.30	.35	.40	.45	.50	.55	.60	.65	.70	.75	.80
.10	64	21	13	9	7	5	4	3	2	2	2	2	2	2	—	—	—	—	—
.50	195	64	38	27	19	12	9	7	6	5	4	3	3	3	2	2	2	2	—
.70	265	87	52	36	25	16	12	9	7	6	5	4	4	3	3	2	2	2	2
.80	312	103	61	43	29	19	14	10	8	7	6	5	4	4	3	3	2	2	2
.90	383	126	74	52	36	23	16	13	10	8	7	6	5	4	4	3	3	2	2
.95	447	147	87	61	42	27	19	15	12	9	8	7	6	5	4	3	3	3	2
.99	579	190	112	79	54	34	24	19	15	12	10	8	7	6	5	4	4	3	2

Appendix E.4: Pearson Correlation

Directional Test, Alpha = .05

POWER	POPULATION CORRELATION COEFFICIENT SQUARED .01	.03	.05	.07	.10	.15	.20	.25	.30	.35	.40	.45	.50	.55	.60	.65	.70	.75	.80
.25	99	34	21	15	11	8	6	6	5	4	4	4	3	3	3	3	3	3	2
.50	277	92	56	40	28	19	14	11	9	8	7	6	6	5	5	4	4	4	3
.60	368	122	73	52	36	24	18	14	12	10	9	8	7	6	5	5	5	4	4
.67	430	142	85	61	42	28	21	16	13	11	10	9	8	7	6	5	5	4	4
.70	470	156	93	66	46	30	22	18	14	12	10	9	8	7	6	6	5	5	4
.75	537	177	106	75	52	34	25	20	16	14	12	10	9	8	7	6	6	5	5
.80	618	204	121	86	60	39	29	22	18	15	13	11	10	9	8	7	6	6	5
.85	727	239	143	101	70	46	33	26	21	18	15	13	11	10	9	8	7	6	5
.90	864	284	169	120	83	54	39	31	25	21	18	15	13	11	10	9	8	7	6
.95	1,105	363	216	152	105	68	50	39	31	26	22	19	16	14	12	11	10	8	7
.99	1,585	520	308	218	150	97	70	55	44	36	31	26	22	19	17	15	13	11	10

Directional Test, Alpha = .01

	POPULATION CORRELATION COEFFICIENT SQUARED																		
POWER	*.01*	*.03*	*.05*	*.07*	*.10*	*.15*	*.20*	*.25*	*.30*	*.35*	*.40*	*.45*	*.50*	*.55*	*.60*	*.65*	*.70*	*.75*	*.80*
.25	273	91	55	39	27	18	14	12	9	8	7	6	6	5	5	4	4	4	3
.50	540	178	106	76	52	34	25	20	16	14	12	10	9	8	7	6	6	5	5
.60	663	219	130	92	64	42	31	24	20	16	14	12	11	9	8	7	7	6	5
.67	757	249	148	105	73	47	35	28	22	18	16	13	12	10	9	8	7	6	6
.70	809	266	158	112	77	50	37	29	23	19	17	14	12	11	10	8	8	7	6
.75	897	295	175	124	86	56	41	32	26	21	18	16	14	12	10	9	8	7	6
.80	998	328	195	138	95	62	45	36	28	24	20	17	15	13	11	10	9	8	7
.85	1,126	370	220	155	107	69	51	40	32	26	22	19	16	14	13	11	10	8	7
.90	1,296	425	253	178	123	80	58	45	36	30	25	22	19	16	14	12	11	9	8
.95	1,585	520	308	218	150	97	70	55	44	36	31	26	22	19	17	15	13	11	10
.99	2,154	706	418	295	203	131	95	74	59	49	41	35	30	26	22	19	17	14	12

Nondirectional Test, Alpha = .05

	POPULATION CORRELATION COEFFICIENT SQUARED																		
POWER	*.01*	*.03*	*.05*	*.07*	*.10*	*.15*	*.20*	*.25*	*.30*	*.35*	*.40*	*.45*	*.50*	*.55*	*.60*	*.65*	*.70*	*.75*	*.80*
.25	166	56	34	25	17	12	9	8	6	6	5	5	4	4	4	3	3	3	3
.50	384	127	76	54	38	25	19	15	12	10	9	8	7	6	6	5	5	4	4
.60	489	162	97	69	48	31	23	18	15	13	11	9	8	7	7	6	5	5	4
.67	570	188	112	80	55	36	27	21	17	14	12	11	9	8	7	7	6	5	5
.70	616	203	121	86	59	39	29	23	18	15	13	11	10	9	8	7	6	6	5
.75	692	228	136	96	67	43	32	25	20	17	14	12	11	10	9	8	7	6	5
.80	783	258	153	109	75	49	36	28	23	19	16	14	12	11	9	8	7	7	6
.85	895	294	175	124	85	56	41	32	26	21	18	16	14	12	10	9	8	7	6
.90	1,046	344	204	144	100	65	47	37	30	25	21	18	15	13	12	10	9	8	7
.95	1,308	429	255	180	124	80	58	46	37	30	26	22	19	16	14	12	11	10	8
.99	1,828	599	355	251	172	111	81	63	50	42	35	30	26	22	19	17	14	13	11

Nondirectional Test, Alpha = .01

POWER	POPULATION CORRELATION COEFFICIENT SQUARED																		
	.01	*.03*	*.05*	*.07*	*.10*	*.15*	*.20*	*.25*	*.30*	*.35*	*.40*	*.45*	*.50*	*.55*	*.60*	*.65*	*.70*	*.75*	*.80*
.25	362	120	72	51	36	24	18	15	12	10	9	7	7	6	5	5	4	4	4
.50	662	218	130	92	64	42	31	24	19	16	14	12	11	9	8	7	7	6	5
.60	797	262	156	111	76	50	36	29	23	19	16	14	12	11	9	8	7	7	6
.67	901	296	176	125	86	56	41	32	26	21	18	16	14	12	10	9	8	7	6
.70	957	315	187	132	91	59	43	34	27	23	19	17	14	13	11	10	9	8	7
.75	1,052	346	205	145	100	65	47	37	30	25	21	18	16	14	12	10	9	8	7
.80	1,163	382	227	160	110	72	52	41	33	27	23	20	17	15	13	11	10	9	8
.85	1,299	426	253	179	123	80	58	45	36	30	25	22	19	16	14	12	11	9	8
.90	1,480	485	288	203	140	91	66	51	41	34	29	24	21	18	16	14	12	11	9
.95	1,790	587	348	246	169	109	79	62	49	41	34	29	25	22	19	16	14	12	11
.99	2,390	783	464	327	225	145	105	82	65	54	45	38	33	28	24	21	18	16	13

Appendix E.5: Chi-Square Test

Type of Table: 2 × 2, Alpha = .05

	POPULATION FOURFOLD POINT CORRELATION COEFFICIENT								
POWER	*.10*	*.20*	*.30*	*.40*	*.50*	*.60*	*.70*	*.80*	*.90*
.25	165	41	18	10	7	5	3	3	2
.50	384	96	43	24	15	11	8	6	5
.60	490	122	54	31	20	14	10	8	6
.70	617	154	69	39	25	17	13	10	8
.75	694	175	77	43	28	19	14	11	9
.80	785	196	87	49	31	22	16	12	10
.85	898	224	100	56	36	25	18	14	11
.90	1,051	263	117	66	42	29	21	16	13
.95	1,300	325	144	81	52	36	27	20	16
.99	1,837	459	204	115	73	51	37	29	23

Type of Table: 2 × 3, Alpha = .05

	POPULATION VALUE OF CRAMÉR'S STATISTIC								
POWER	*.10*	*.20*	*.30*	*.40*	*.50*	*.60*	*.70*	*.80*	*.90*
.25	226	56	25	14	9	6	5	4	3
.50	496	124	55	31	20	14	10	8	6
.60	621	155	69	39	25	17	13	10	8
.70	770	193	86	48	31	21	16	12	10
.75	859	215	95	54	34	24	18	13	11
.80	964	241	107	60	39	27	20	15	12
.85	1,092	273	121	68	44	30	22	17	13
.90	1,265	316	141	79	51	35	26	20	16
.95	1,544	386	172	97	62	43	32	24	19
.99	2,140	535	238	134	86	59	44	33	26

Type of Table: 2 × 4, Alpha = .05

	POPULATION VALUE OF CRAMÉR'S STATISTIC								
POWER	*.10*	*.20*	*.30*	*.40*	*.50*	*.60*	*.70*	*.80*	*.90*
.25	258	65	29	16	10	7	5	4	3
.50	576	144	64	36	23	16	12	9	7
.60	715	179	79	45	29	20	15	11	9
.70	879	220	98	55	35	24	18	14	11
.75	976	244	108	61	39	27	20	15	12
.80	1,090	273	121	68	44	30	22	17	13
.85	1,230	308	137	77	49	34	25	19	15
.90	1,417	354	157	89	57	39	29	22	17
.95	1,717	429	191	107	69	48	35	27	21
.99	2,352	588	261	147	94	65	48	37	29

Type of Table: 3 × 3, Alpha = .05

POWER	.10	.20	.30	.40	.50	.60	.70	.80	.90
	POPULATION VALUE OF CRAMÉR'S STATISTIC								
.25	154	39	17	10	6	4	3	2	2
.50	321	80	36	20	13	9	7	5	4
.60	396	99	44	25	16	11	8	6	5
.70	484	121	54	30	19	13	10	8	6
.75	536	134	60	34	21	15	11	8	7
.80	597	149	66	37	24	17	12	9	7
.85	671	168	75	42	27	19	14	10	8
.90	770	193	86	48	31	21	16	12	10
.95	929	232	103	58	37	26	19	15	11
.99	1,262	316	140	79	50	35	26	20	16

Type of Table: 3 × 4, Alpha = .05

POWER	.10	.20	.30	.40	.50	.60	.70	.80	.90
	POPULATION VALUE OF CRAMÉR'S STATISTIC								
.25	185	46	21	12	7	5	4	3	2
.50	375	94	42	23	15	10	8	6	5
.60	460	115	51	29	18	13	9	7	6
.70	557	139	62	35	22	15	11	9	7
.75	615	154	68	38	25	17	13	10	8
.80	681	170	76	43	27	19	14	11	8
.85	763	191	85	48	31	21	16	12	9
.90	871	218	97	54	35	24	18	14	11
.95	1,043	261	116	65	42	29	21	16	13
.99	1,403	351	156	88	56	39	29	22	17

Type of Table: 4 × 4, Alpha = .05

POWER	.10	.20	.30	.40	.50	.60	.70	.80	.90
	POPULATION VALUE OF CRAMÉR'S STATISTIC								
.25	148	37	16	9	6	4	3	2	2
.50	294	73	33	18	12	8	6	5	4
.60	357	89	40	22	14	10	7	6	4
.70	430	107	48	27	17	12	9	7	5
.75	472	118	52	30	19	13	10	7	6
.80	522	130	58	33	21	14	11	8	6
.85	582	145	65	36	23	16	12	9	7
.90	661	165	73	41	26	18	13	10	8
.95	786	197	87	49	31	22	16	12	10
.99	1,046	262	116	65	42	29	21	16	13

Type of Table: 2 × 2, Alpha = .01

POPULATION FOURFOLD POINT CORRELATION COEFFICIENT

POWER	*.10*	*.20*	*.30*	*.40*	*.50*	*.60*	*.70*	*.80*	*.90*
.25	362	90	40	23	14	10	7	6	4
.50	664	166	74	41	27	18	14	10	8
.60	800	200	89	50	32	22	16	13	10
.70	961	240	107	60	38	27	20	15	12
.75	1,056	264	117	66	42	29	22	17	13
.80	1,168	292	130	73	47	32	24	18	14
.85	1,305	326	145	82	52	36	27	20	16
.90	1,488	372	165	93	60	41	30	23	18
.95	1,781	445	198	111	71	49	36	28	22
.99	2,403	601	267	150	96	67	49	38	30

Type of Table: 2 × 3, Alpha = .01

POPULATION VALUE OF CRAMÉR'S STATISTIC

POWER	*.10*	*.20*	*.30*	*.40*	*.50*	*.60*	*.70*	*.80*	*.90*
.25	467	117	52	29	19	13	10	7	6
.50	819	205	91	51	33	23	17	13	10
.60	975	244	108	61	39	27	20	15	12
.70	1,157	289	129	72	46	32	24	18	14
.75	1,264	316	140	79	51	35	26	20	16
.80	1,388	347	154	87	56	39	28	22	17
.85	1,540	385	171	96	62	43	31	24	19
.90	1,743	436	194	109	70	48	36	27	22
.95	2,065	516	229	129	83	57	42	32	25
.99	2,742	685	305	171	110	76	56	43	34

Type of Table: 2 × 4, Alpha = .01

POPULATION VALUE OF CRAMÉR'S STATISTIC

POWER	*.10*	*.20*	*.30*	*.40*	*.50*	*.60*	*.70*	*.80*	*.90*
.25	544	136	60	34	22	15	11	8	7
.50	931	233	103	58	37	26	19	15	11
.60	1,101	275	122	69	44	31	22	17	14
.70	1,297	324	144	81	52	36	26	20	16
.75	1,412	353	157	88	56	39	29	22	17
.80	1,546	386	172	97	62	43	32	24	19
.85	1,709	427	190	107	68	47	35	27	21
.90	1,925	481	214	120	77	53	39	30	24
.95	2,267	567	252	142	91	63	46	35	28
.99	2,983	746	331	186	119	83	61	47	37

Type of Table: 3 × 3, Alpha = .01

POPULATION VALUE OF CRAMÉR'S STATISTIC

POWER	*.10*	*.20*	*.30*	*.40*	*.50*	*.60*	*.70*	*.80*	*.90*
.25	304	76	34	19	12	8	6	5	4
.50	512	128	57	32	20	14	10	8	6
.60	602	151	67	38	24	17	12	9	7
.70	706	177	78	44	28	20	14	11	9
.75	767	192	85	48	31	21	16	12	9
.80	824	206	92	52	33	23	17	13	10
.85	924	231	103	58	37	26	19	14	11
.90	1,037	259	115	65	41	29	21	16	13
.95	1,217	304	135	76	49	34	25	19	15
.99	1,590	398	177	99	64	44	32	25	20

Type of Table: 3 × 4, Alpha = .01

POPULATION VALUE OF CRAMÉR'S STATISTIC

POWER	*.10*	*.20*	*.30*	*.40*	*.50*	*.60*	*.70*	*.80*	*.90*
.25	357	89	40	22	14	10	7	6	4
.50	588	147	65	37	24	16	12	9	7
.60	687	172	76	43	27	19	14	11	8
.70	801	200	89	50	32	22	16	13	10
.75	867	217	96	54	35	24	18	14	11
.80	944	236	105	59	38	26	19	15	12
.85	1,037	259	115	65	41	29	21	16	13
.90	1,159	290	129	72	46	32	24	18	14
.95	1,352	338	150	85	54	38	28	21	17
.99	1,751	438	195	109	70	49	36	27	22

Type of Table: 4 × 4, Alpha = .01

POPULATION VALUE OF CRAMÉR'S STATISTIC

POWER	*.10*	*.20*	*.30*	*.40*	*.50*	*.60*	*.70*	*.80*	*.90*
.25	280	70	31	18	11	8	6	4	3
.50	453	113	50	28	18	13	9	7	6
.60	526	132	58	33	21	15	11	8	6
.70	610	153	68	38	24	17	12	10	8
.75	658	165	73	41	26	18	13	10	8
.80	714	179	79	45	29	20	15	11	9
.85	782	196	87	49	31	22	16	12	10
.90	871	218	97	54	35	24	18	14	11
.95	1,010	253	112	63	40	28	21	16	12
.99	1,296	324	144	81	52	36	26	20	16

APPENDIX F

Critical Values for the F Distribution

The following table presents critical values of F for alpha levels of .05 and .01. The .05 critical values are in roman type and the .01 critical values are in **boldface.** The values at the top of the table under the heading "Degrees of Freedom for Numerator" represent the degrees of freedom associated with the numerator of the F ratio for the analysis of interest. This will be $df_{BETWEEN}$ for one-way between-subjects analysis of variance; $df_{TREATMENTS}$ for one-way repeated measures analysis of variance; and df_A, df_B, or $df_{A \times B}$ for two-way between-subjects analysis of variance. The values at the left of the table under the heading "Degrees of Freedom for Denominator" represent the degrees of freedom associated with the denominator of the F ratio for the analysis of interest. This will be df_{WITHIN} for one- and two-way between-subjects analysis of variance, and df_{ERROR} for one-way repeated measures analysis of variance. For example, the critical value of F for a one-way between-subjects analysis of variance for $df_{BETWEEN} = 2$, $df_{WITHIN} = 11$, and an alpha level of .05 is 3.98.

Source: Reprinted by permission from *Statistical Methods* by George W. Snedecor and William G. Cochran, Seventh Edition.

Degrees of freedom for denominator	1	2	3	4	5	6	7	8	9	10	11	12	14	16	20	24	30	40	50	75	100	200	500	∞
1	161 **4,052**	200 **4,999**	216 **5,403**	225 **5,625**	230 **5,764**	234 **5,859**	237 **5,928**	239 **5,981**	241 **6,022**	242 **6,056**	243 **6,082**	244 **6,106**	245 **6,142**	246 **6,169**	248 **6,208**	249 **6,234**	250 **6,258**	251 **6,286**	252 **6,302**	253 **6,323**	253 **6,334**	254 **6,352**	254 **6,361**	254 **6,366**
2	18.51 **98.49**	19.00 **99.01**	19.16 **99.17**	19.25 **99.25**	19.30 **99.30**	19.33 **99.33**	19.36 **99.34**	19.37 **99.36**	19.38 **99.38**	19.39 **99.40**	19.40 **99.41**	19.41 **99.42**	19.42 **99.43**	19.43 **99.44**	19.44 **99.45**	19.45 **99.46**	19.46 **99.47**	19.47 **99.48**	19.47 **99.48**	19.48 **99.49**	19.49 **99.49**	19.49 **99.49**	19.50 **99.50**	19.50 **99.50**
3	10.13 **34.12**	9.55 **30.81**	9.28 **29.46**	9.12 **28.71**	9.01 **28.24**	8.94 **27.91**	8.88 **27.67**	8.84 **27.49**	8.81 **27.34**	8.78 **27.23**	8.76 **27.13**	8.74 **27.05**	8.71 **26.92**	8.69 **26.83**	8.66 **26.69**	8.64 **26.60**	8.62 **26.50**	8.60 **26.41**	8.58 **26.30**	8.57 **26.27**	8.56 **26.23**	8.54 **26.18**	8.54 **26.14**	8.53 **26.12**
4	7.71 **21.20**	6.94 **18.00**	6.59 **16.69**	6.39 **15.98**	6.26 **15.52**	6.16 **15.21**	6.09 **14.98**	6.04 **14.80**	6.00 **14.66**	5.96 **14.54**	5.93 **14.45**	5.91 **14.37**	5.87 **14.24**	5.84 **14.15**	5.80 **14.02**	5.77 **13.93**	5.74 **13.83**	5.71 **13.74**	5.70 **13.69**	5.68 **13.61**	5.66 **13.57**	5.65 **13.52**	5.64 **13.48**	5.63 **13.46**
5	6.61 **16.26**	5.79 **13.27**	5.41 **12.06**	5.19 **11.39**	5.05 **10.97**	4.95 **10.67**	4.88 **10.45**	4.82 **10.27**	4.78 **10.15**	4.74 **10.05**	4.70 **9.96**	4.68 **9.89**	4.64 **9.77**	4.60 **9.68**	4.56 **9.55**	4.53 **9.47**	4.50 **9.38**	4.46 **9.29**	4.44 **9.24**	4.42 **9.17**	4.40 **9.13**	4.38 **9.07**	4.37 **9.04**	4.36 **9.02**
6	5.99 **13.74**	5.14 **10.92**	4.76 **9.78**	4.53 **9.15**	4.39 **8.75**	4.28 **8.47**	4.21 **8.26**	4.15 **8.10**	4.10 **7.98**	4.06 **7.87**	4.03 **7.79**	4.00 **7.72**	3.96 **7.60**	3.92 **7.52**	3.87 **7.39**	3.84 **7.31**	3.81 **7.23**	3.77 **7.14**	3.75 **7.09**	3.72 **7.02**	3.71 **6.99**	3.69 **6.94**	3.68 **6.90**	3.67 **6.88**
7	5.59 **12.25**	4.74 **9.55**	4.35 **8.45**	4.12 **7.85**	3.97 **7.46**	3.87 **7.19**	3.79 **7.00**	3.73 **6.84**	3.68 **6.71**	3.63 **6.62**	3.60 **6.54**	3.57 **6.47**	3.52 **6.35**	3.49 **6.27**	3.44 **6.15**	3.41 **6.07**	3.38 **5.98**	3.34 **5.90**	3.32 **5.85**	3.29 **5.78**	3.28 **5.75**	3.25 **5.70**	3.24 **5.67**	3.23 **5.65**
8	5.32 **11.26**	4.46 **8.65**	4.07 **7.59**	3.84 **7.01**	3.69 **6.63**	3.58 **6.37**	3.50 **6.19**	3.44 **6.03**	3.39 **5.91**	3.34 **5.82**	3.31 **5.74**	3.28 **5.67**	3.23 **5.56**	3.20 **5.48**	3.15 **5.36**	3.12 **5.28**	3.08 **5.20**	3.05 **5.11**	3.03 **5.06**	3.00 **5.00**	2.98 **4.96**	2.96 **4.91**	2.94 **4.88**	2.93 **4.86**
9	5.12 **10.56**	4.26 **8.02**	3.86 **6.99**	3.63 **6.42**	3.48 **6.06**	3.37 **5.80**	3.29 **5.62**	3.23 **5.47**	3.18 **5.35**	3.13 **5.26**	3.10 **5.18**	3.07 **5.11**	3.02 **5.00**	2.98 **4.92**	2.93 **4.80**	2.90 **4.73**	2.86 **4.64**	2.82 **4.56**	2.80 **4.51**	2.77 **4.45**	2.76 **4.41**	2.73 **4.36**	2.72 **4.33**	2.71 **4.31**
10	4.96 **10.04**	4.10 **7.56**	3.71 **6.55**	3.48 **5.99**	3.33 **5.64**	3.22 **5.39**	3.14 **5.21**	3.07 **5.06**	3.02 **4.95**	2.97 **4.85**	2.94 **4.78**	2.91 **4.71**	2.86 **4.60**	2.82 **4.52**	2.77 **4.41**	2.74 **4.33**	2.70 **4.25**	2.67 **4.17**	2.64 **4.12**	2.61 **4.05**	2.59 **4.01**	2.56 **3.96**	2.55 **3.93**	2.54 **3.91**
11	4.84 **9.65**	3.98 **7.20**	3.59 **6.22**	3.36 **5.67**	3.20 **5.32**	3.09 **5.07**	3.01 **4.88**	2.95 **4.74**	2.90 **4.63**	2.86 **4.54**	2.82 **4.46**	2.79 **4.40**	2.74 **4.29**	2.70 **4.21**	2.65 **4.10**	2.61 **4.02**	2.57 **3.94**	2.53 **3.86**	2.50 **3.80**	2.47 **3.74**	2.45 **3.70**	2.42 **3.66**	2.41 **3.62**	2.40 **3.60**
12	4.75 **9.33**	3.88 **6.93**	3.49 **5.95**	3.26 **5.41**	3.11 **5.06**	3.00 **4.82**	2.92 **4.65**	2.85 **4.50**	2.80 **4.39**	2.76 **4.30**	2.72 **4.22**	2.69 **4.16**	2.64 **4.05**	2.60 **3.98**	2.54 **3.86**	2.50 **3.78**	2.46 **3.70**	2.42 **3.61**	2.40 **3.56**	2.36 **3.49**	2.35 **3.46**	2.32 **3.41**	2.31 **3.38**	2.30 **3.36**
13	4.67 **9.07**	3.80 **6.70**	3.41 **5.74**	3.18 **5.20**	3.02 **4.86**	2.92 **4.62**	2.84 **4.44**	2.77 **4.30**	2.72 **4.19**	2.67 **4.10**	2.63 **4.02**	2.60 **3.96**	2.55 **3.85**	2.51 **3.78**	2.46 **3.67**	2.42 **3.59**	2.38 **3.51**	2.34 **3.42**	2.32 **3.37**	2.28 **3.30**	2.26 **3.27**	2.24 **3.21**	2.22 **3.18**	2.21 **3.16**
14	4.60 **8.86**	3.74 **6.51**	3.34 **5.56**	3.11 **5.03**	2.96 **4.69**	2.85 **4.46**	2.77 **4.28**	2.70 **4.14**	2.65 **4.03**	2.60 **3.94**	2.56 **3.86**	2.53 **3.80**	2.48 **3.70**	2.44 **3.62**	2.39 **3.51**	2.35 **3.43**	2.31 **3.34**	2.27 **3.26**	2.24 **3.21**	2.21 **3.14**	2.19 **3.11**	2.16 **3.06**	2.14 **3.02**	2.13 **3.00**
15	4.54 **8.68**	3.68 **6.36**	3.29 **5.42**	3.06 **4.89**	2.90 **4.56**	2.79 **4.32**	2.70 **4.14**	2.64 **4.00**	2.59 **3.89**	2.55 **3.80**	2.51 **3.73**	2.48 **3.67**	2.43 **3.56**	2.39 **3.48**	2.33 **3.36**	2.29 **3.29**	2.25 **3.20**	2.21 **3.12**	2.18 **3.07**	2.15 **3.00**	2.12 **2.97**	2.10 **2.92**	2.08 **2.89**	2.07 **2.87**
16	4.49 **8.53**	3.63 **6.23**	3.24 **5.29**	3.01 **4.77**	2.85 **4.44**	2.74 **4.20**	2.66 **4.03**	2.59 **3.89**	2.54 **3.78**	2.49 **3.69**	2.45 **3.61**	2.42 **3.55**	2.37 **3.45**	2.33 **3.37**	2.28 **3.25**	2.24 **3.18**	2.20 **3.10**	2.16 **3.01**	2.13 **2.96**	2.09 **2.89**	2.07 **2.86**	2.04 **2.80**	2.02 **2.77**	2.01 **2.75**
17	4.45 **8.40**	3.59 **6.11**	3.20 **5.18**	2.96 **4.67**	2.81 **4.34**	2.70 **4.10**	2.62 **3.93**	2.55 **3.79**	2.50 **3.68**	2.45 **3.59**	2.41 **3.52**	2.38 **3.45**	2.33 **3.35**	2.29 **3.27**	2.23 **3.16**	2.19 **3.08**	2.15 **3.00**	2.11 **2.92**	2.08 **2.86**	2.04 **2.79**	2.02 **2.76**	1.99 **2.70**	1.97 **2.67**	1.96 **2.65**
18	4.41 **8.28**	3.55 **6.01**	3.16 **5.09**	2.93 **4.58**	2.77 **4.25**	2.66 **4.01**	2.58 **3.85**	2.51 **3.71**	2.46 **3.60**	2.41 **3.51**	2.37 **3.44**	2.34 **3.37**	2.29 **3.27**	2.25 **3.19**	2.19 **3.07**	2.15 **3.00**	2.11 **2.91**	2.07 **2.83**	2.04 **2.78**	2.00 **2.71**	1.98 **2.68**	1.95 **2.62**	1.93 **2.59**	1.92 **2.57**
19	4.38 **8.18**	3.52 **5.93**	3.13 **5.01**	2.90 **4.50**	2.74 **4.17**	2.63 **3.94**	2.55 **3.77**	2.48 **3.63**	2.43 **3.52**	2.38 **3.43**	2.34 **3.36**	2.31 **3.30**	2.26 **3.19**	2.21 **3.12**	2.15 **3.00**	2.11 **2.92**	2.07 **2.84**	2.02 **2.76**	2.00 **2.70**	1.96 **2.63**	1.94 **2.60**	1.91 **2.54**	1.90 **2.51**	1.88 **2.49**
20	4.35 **8.10**	3.49 **5.85**	3.10 **4.94**	2.87 **4.43**	2.71 **4.10**	2.60 **3.87**	2.52 **3.71**	2.45 **3.56**	2.40 **3.45**	2.35 **3.37**	2.31 **3.30**	2.28 **3.23**	2.23 **3.13**	2.18 **3.05**	2.12 **2.94**	2.08 **2.86**	2.04 **2.77**	1.99 **2.69**	1.96 **2.63**	1.92 **2.56**	1.90 **2.53**	1.87 **2.47**	1.85 **2.44**	1.84 **2.42**

Column headings: DEGREES OF FREEDOM FOR NUMERATOR

DEGREES OF FREEDOM FOR DENOMINATOR	DEGREES OF FREEDOM FOR NUMERATOR																							
	1	2	3	4	5	6	7	8	9	10	11	12	14	16	20	24	30	40	50	75	100	200	500	∞
21	4.32	3.47	3.07	2.84	2.68	2.57	2.49	2.42	2.37	2.32	2.28	2.25	2.20	2.15	2.09	2.05	2.00	1.96	1.93	1.80	1.87	1.84	1.82	1.81
	8.02	**5.78**	**4.87**	**4.37**	**4.04**	**3.81**	**3.65**	**3.51**	**3.40**	**3.31**	**3.24**	**3.17**	**3.07**	**2.99**	**2.88**	**2.80**	**2.72**	**2.63**	**2.58**	**2.51**	**2.47**	**2.42**	**2.38**	**2.36**
22	4.30	3.44	3.05	2.82	2.66	2.55	2.47	2.40	2.35	2.30	2.26	2.23	2.18	2.13	2.07	2.03	1.98	1.93	1.91	1.87	1.84	1.81	1.80	1.78
	7.94	**5.72**	**4.82**	**4.31**	**3.99**	**3.76**	**3.59**	**3.45**	**3.35**	**3.26**	**3.18**	**3.12**	**3.02**	**2.94**	**2.83**	**2.75**	**2.67**	**2.58**	**2.53**	**2.46**	**2.42**	**2.37**	**2.33**	**2.31**
23	4.28	3.42	3.03	2.80	2.64	2.53	2.45	2.38	2.32	2.28	2.24	2.20	2.14	2.10	2.04	2.00	1.96	1.91	1.88	1.84	1.82	1.79	1.77	1.76
	7.88	**5.66**	**4.76**	**4.26**	**3.94**	**3.71**	**3.54**	**3.41**	**3.30**	**3.21**	**3.14**	**3.07**	**2.97**	**2.89**	**2.78**	**2.70**	**2.62**	**2.53**	**2.48**	**2.41**	**2.37**	**2.32**	**2.28**	**2.26**
24	4.26	3.40	3.01	2.78	2.62	2.51	2.43	2.36	2.30	2.26	2.22	2.18	2.13	2.09	2.02	1.98	1.94	1.89	1.86	1.82	1.80	1.76	1.74	1.73
	7.82	**5.61**	**4.72**	**4.22**	**3.90**	**3.67**	**3.50**	**3.36**	**3.25**	**3.17**	**3.09**	**3.03**	**2.93**	**2.85**	**2.74**	**2.66**	**2.58**	**2.49**	**2.44**	**2.36**	**2.33**	**2.27**	**2.23**	**2.21**
25	4.24	3.38	2.99	2.76	2.60	2.49	2.41	2.34	2.28	2.24	2.20	2.16	2.11	2.06	2.00	1.96	1.92	1.87	1.84	1.80	1.77	1.74	1.72	1.71
	7.77	**5.57**	**4.68**	**4.18**	**3.86**	**3.63**	**3.46**	**3.32**	**3.21**	**3.13**	**3.05**	**2.99**	**2.89**	**2.81**	**2.70**	**2.62**	**2.54**	**2.45**	**2.40**	**2.32**	**2.29**	**2.23**	**2.19**	**2.17**
26	4.22	3.37	2.89	2.74	2.59	2.47	2.39	2.32	2.27	2.22	2.18	2.15	2.10	2.05	1.99	1.95	1.90	1.85	1.82	1.78	1.76	1.72	1.70	1.69
	7.72	**5.53**	**4.64**	**4.14**	**3.82**	**3.59**	**3.42**	**3.29**	**3.17**	**3.09**	**3.02**	**2.96**	**2.86**	**2.77**	**2.66**	**2.58**	**2.50**	**2.41**	**2.36**	**2.28**	**2.25**	**2.19**	**2.15**	**2.13**
27	4.21	3.35	2.96	2.73	2.57	2.46	2.37	2.30	2.25	2.20	2.16	2.13	2.08	2.03	1.97	1.93	1.88	1.84	1.80	1.76	1.74	1.71	1.68	1.67
	7.68	**5.49**	**4.60**	**4.11**	**3.79**	**3.56**	**3.39**	**3.26**	**3.14**	**3.06**	**2.98**	**2.93**	**2.83**	**2.74**	**2.63**	**2.55**	**2.47**	**2.38**	**2.33**	**2.25**	**2.21**	**2.16**	**2.12**	**2.10**
28	4.20	3.34	2.95	2.71	2.56	2.44	2.36	2.29	3.24	2.19	2.15	2.12	2.06	2.02	1.95	1.91	1.87	1.81	1.78	1.75	1.72	1.69	1.67	1.65
	7.64	**5.45**	**4.57**	**4.07**	**3.76**	**3.53**	**3.36**	**3.23**	**3.11**	**3.03**	**2.95**	**2.90**	**2.80**	**2.71**	**2.60**	**2.52**	**2.44**	**2.35**	**2.30**	**2.22**	**2.18**	**2.13**	**2.09**	**2.06**
29	4.18	3.33	2.93	2.70	2.54	2.43	2.35	2.28	2.22	2.18	2.14	2.10	2.05	2.00	1.94	1.90	1.85	1.80	1.77	1.73	1.71	1.68	1.65	1.64
	7.60	**5.52**	**4.54**	**4.04**	**3.73**	**3.50**	**3.32**	**3.20**	**3.08**	**3.00**	**2.92**	**2.87**	**2.77**	**2.68**	**2.57**	**2.49**	**2.41**	**2.32**	**2.27**	**2.19**	**2.15**	**2.10**	**2.06**	**2.03**
30	4.17	3.32	2.92	2.69	2.53	2.42	2.34	2.27	2.21	2.16	2.12	2.09	2.04	1.99	1.93	1.89	1.84	1.79	1.76	1.72	1.69	1.66	1.64	1.62
	7.56	**5.39**	**4.51**	**4.02**	**3.70**	**3.47**	**3.30**	**3.17**	**3.06**	**2.98**	**2.90**	**2.84**	**2.74**	**2.66**	**2.55**	**2.47**	**2.38**	**2.29**	**2.24**	**2.16**	**2.13**	**2.07**	**2.03**	**2.01**
32	4.15	3.30	2.90	2.67	2.51	2.40	2.32	2.25	2.19	2.14	2.10	2.07	2.02	1.97	1.91	1.86	1.82	1.76	1.74	1.69	1.67	1.64	1.61	1.59
	7.50	**5.34**	**4.46**	**3.97**	**3.66**	**3.42**	**3.25**	**3.12**	**3.01**	**2.94**	**2.86**	**2.80**	**2.70**	**2.62**	**2.51**	**2.42**	**2.34**	**2.25**	**2.20**	**2.12**	**2.08**	**2.02**	**1.98**	**1.96**
34	4.13	3.28	2.88	2.65	2.49	2.38	2.30	2.23	2.17	2.12	2.08	2.05	2.00	1.95	1.89	1.84	1.80	1.74	1.71	1.67	1.64	1.61	1.59	1.57
	7.44	**5.29**	**4.42**	**3.93**	**3.61**	**3.38**	**3.21**	**3.08**	**2.97**	**2.89**	**2.82**	**2.76**	**2.66**	**2.58**	**2.47**	**2.38**	**2.30**	**2.21**	**2.15**	**2.08**	**2.04**	**1.98**	**1.94**	**1.91**
36	4.11	3.26	2.86	2.63	2.48	2.36	2.28	2.21	2.15	2.10	2.06	2.03	1.89	1.93	1.87	1.82	1.78	1.72	1.69	1.65	1.62	1.59	1.56	1.55
	7.39	**5.25**	**4.38**	**3.89**	**3.58**	**3.35**	**3.18**	**3.04**	**2.94**	**2.86**	**2.78**	**2.72**	**2.62**	**2.54**	**2.43**	**2.35**	**2.26**	**2.17**	**2.12**	**2.04**	**2.00**	**1.94**	**1.90**	**1.87**
38	4.10	3.25	2.85	2.62	2.46	2.35	2.26	2.19	2.14	2.09	2.05	2.02	1.96	1.92	1.85	1.80	1.76	1.71	1.67	1.63	1.60	1.57	1.54	1.53
	7.35	**5.21**	**4.34**	**3.86**	**3.54**	**3.32**	**3.15**	**3.02**	**2.91**	**2.82**	**2.75**	**2.69**	**2.59**	**2.51**	**2.40**	**2.32**	**2.22**	**2.14**	**2.08**	**2.00**	**1.97**	**1.90**	**1.86**	**1.84**
40	4.08	3.23	2.84	2.61	2.45	2.34	2.25	2.18	2.12	2.07	2.04	2.00	1.95	1.90	1.84	1.79	1.74	1.69	1.66	1.61	1.59	1.55	1.53	1.51
	7.31	**5.18**	**4.31**	**3.83**	**3.51**	**3.29**	**3.12**	**2.99**	**2.88**	**2.80**	**2.73**	**2.66**	**2.56**	**2.49**	**2.37**	**2.29**	**2.20**	**2.11**	**2.05**	**1.97**	**1.94**	**1.88**	**1.84**	**1.81**
42	4.07	3.22	2.83	2.59	2.44	2.32	2.24	2.17	2.11	2.06	2.02	1.90	1.94	1.89	1.82	1.78	1.73	1.68	1.64	1.60	1.57	1.54	1.51	1.49
	7.27	**5.15**	**4.29**	**3.80**	**3.49**	**3.26**	**3.10**	**2.96**	**2.86**	**2.77**	**2.70**	**2.64**	**2.54**	**2.46**	**2.35**	**2.26**	**2.17**	**2.08**	**2.02**	**1.94**	**1.91**	**1.85**	**1.80**	**1.78**
44	4.06	3.21	2.82	2.58	2.43	2.31	2.23	2.16	2.10	2.05	2.01	1.98	1.92	1.88	1.81	1.76	1.72	1.66	1.63	1.58	1.56	1.52	1.50	1.48
	7.24	**5.12**	**4.26**	**3.78**	**3.46**	**3.24**	**3.07**	**2.94**	**2.84**	**2.75**	**2.68**	**2.62**	**2.52**	**2.44**	**2.32**	**2.24**	**2.15**	**2.06**	**2.09**	**1.92**	**1.88**	**1.82**	**1.78**	**1.75**
46	4.05	3.20	2.81	2.57	2.42	2.30	2.22	2.14	2.09	2.04	2.00	1.97	1.91	1.87	1.80	1.75	1.71	1.65	1.62	1.57	1.54	1.51	1.48	1.46
	7.21	**5.10**	**4.24**	**3.76**	**3.44**	**3.22**	**3.05**	**2.92**	**2.82**	**2.73**	**2.66**	**2.60**	**2.50**	**2.42**	**2.30**	**2.22**	**2.13**	**2.04**	**1.98**	**1.90**	**1.86**	**1.80**	**1.76**	**1.72**
48	4.04	3.19	2.80	2.56	2.41	2.30	2.21	2.14	2.08	2.03	1.99	1.96	1.90	1.86	1.79	1.74	1.70	1.64	1.61	1.56	1.53	1.50	1.47	1.45
	7.19	**5.08**	**4.22**	**3.74**	**3.42**	**3.20**	**3.04**	**2.90**	**2.80**	**2.71**	**2.64**	**2.58**	**2.48**	**2.40**	**2.28**	**2.20**	**2.11**	**2.02**	**1.96**	**1.88**	**1.84**	**1.78**	**1.73**	**1.70**
50	4.03	3.18	2.79	2.56	2.40	2.29	2.20	2.13	2.07	2.02	1.98	1.95	1.90	1.85	1.78	1.74	1.69	1.63	1.60	1.55	1.52	1.48	1.46	1.44
	7.17	**5.06**	**4.20**	**3.72**	**3.41**	**3.18**	**3.02**	**2.88**	**2.78**	**2.70**	**2.62**	**2.56**	**2.46**	**2.39**	**2.26**	**2.18**	**2.10**	**2.00**	**1.94**	**1.86**	**1.82**	**1.76**	**1.71**	**1.68**

DEGREES OF FREEDOM FOR DENOMINATOR	DEGREES OF FREEDOM FOR NUMERATOR																							
	1	2	3	4	5	6	7	8	9	10	11	12	14	16	20	24	30	40	50	75	100	200	500	∞
55	4.02	3.17	2.78	2.54	2.38	2.27	2.18	2.11	2.05	2.00	1.97	1.93	1.88	1.83	1.76	1.72	1.67	1.61	1.58	1.52	1.50	1.46	1.43	1.41
	7.12	**5.01**	**4.16**	**3.68**	**3.37**	**3.15**	**2.98**	**2.85**	**2.75**	**2.66**	**2.59**	**2.53**	**2.43**	**2.35**	**2.23**	**2.15**	**2.06**	**1.96**	**1.90**	**1.82**	**1.78**	**1.71**	**1.66**	**1.64**
60	4.00	3.15	2.76	2.52	2.37	2.25	2.17	2.10	2.04	1.99	1.95	1.92	1.86	1.81	1.75	1.70	1.65	1.59	1.56	1.50	1.48	1.44	1.41	1.39
	7.08	**4.98**	**4.13**	**3.65**	**3.34**	**3.12**	**2.95**	**2.82**	**2.72**	**2.63**	**2.56**	**2.50**	**2.40**	**2.32**	**2.20**	**2.12**	**2.03**	**1.93**	**1.87**	**1.79**	**1.74**	**1.68**	**1.63**	**1.60**
65	3.99	3.14	2.75	2.51	2.36	2.24	2.15	2.08	2.02	1.98	1.94	1.90	1.85	1.80	1.73	1.68	1.63	1.57	1.54	1.49	1.46	1.42	1.39	1.37
	7.04	**4.95**	**4.10**	**3.62**	**3.31**	**3.09**	**2.93**	**2.79**	**2.70**	**2.61**	**2.54**	**2.47**	**2.37**	**2.30**	**2.18**	**2.09**	**2.00**	**1.90**	**1.84**	**1.76**	**1.71**	**1.64**	**1.60**	**1.56**
70	3.98	3.13	2.74	2.50	2.35	2.32	2.14	2.07	2.01	1.97	1.93	1.89	1.84	1.79	1.72	1.67	1.62	1.56	1.53	1.47	1.45	1.40	1.37	1.35
	7.01	**4.92**	**4.08**	**3.60**	**3.29**	**3.07**	**2.91**	**2.77**	**2.67**	**2.59**	**2.51**	**2.45**	**2.35**	**2.28**	**2.15**	**2.07**	**1.98**	**1.88**	**1.82**	**1.74**	**1.69**	**1.62**	**1.56**	**1.53**
80	3.96	3.11	2.72	2.48	2.33	2.21	2.12	2.05	1.99	1.95	1.91	1.88	1.82	1.77	1.70	1.65	1.60	1.54	1.51	1.45	1.42	1.38	1.35	1.32
	6.96	**4.88**	**4.04**	**3.56**	**3.25**	**3.04**	**2.87**	**2.74**	**2.64**	**2.55**	**2.48**	**2.41**	**2.32**	**2.24**	**2.11**	**2.03**	**1.94**	**1.84**	**1.78**	**1.70**	**1.65**	**1.57**	**1.52**	**1.49**
100	3.94	3.09	2.70	2.46	2.30	2.19	2.10	2.03	1.97	1.92	1.88	1.35	1.79	1.75	1.68	1.63	1.57	1.51	1.48	1.42	1.39	1.34	1.30	1.28
	6.90	**4.82**	**3.98**	**3.51**	**3.20**	**2.99**	**2.82**	**2.69**	**2.59**	**2.51**	**2.43**	**2.36**	**2.26**	**2.19**	**2.06**	**1.98**	**1.89**	**1.79**	**1.73**	**1.64**	**1.59**	**1.51**	**1.46**	**1.43**
125	3.92	3.07	2.68	2.44	2.29	2.17	2.08	2.01	1.95	1.90	1.86	1.83	1.77	1.72	1.65	1.60	1.55	1.49	1.45	1.39	1.36	1.31	1.27	1.25
	6.84	**4.78**	**3.94**	**3.47**	**3.17**	**2.95**	**2.79**	**2.65**	**2.56**	**2.47**	**2.40**	**2.33**	**2.23**	**2.15**	**2.03**	**1.94**	**1.85**	**1.75**	**1.68**	**1.59**	**1.54**	**1.46**	**1.40**	**1.37**
150	3.91	3.06	2.67	2.43	2.27	2.16	2.07	2.00	1.94	1.89	1.85	1.82	1.76	1.71	1.64	1.59	1.54	1.47	1.44	1.37	1.34	1.29	1.25	1.22
	6.81	**4.75**	**3.91**	**3.44**	**3.13**	**2.92**	**2.76**	**2.62**	**2.53**	**2.44**	**2.37**	**2.30**	**2.20**	**2.12**	**2.00**	**1.91**	**1.83**	**1.72**	**1.66**	**1.56**	**1.51**	**1.43**	**1.37**	**1.33**
200	3.89	3.04	2.65	2.41	2.26	2.14	2.05	1.98	1.92	1.87	1.83	1.80	1.74	1.69	1.62	1.57	1.52	1.45	1.42	1.35	1.32	1.26	1.22	1.19
	6.76	**4.71**	**3.38**	**3.41**	**3.11**	**2.90**	**2.73**	**2.60**	**2.50**	**2.41**	**2.34**	**2.28**	**1.17**	**2.09**	**1.97**	**1.88**	**1.79**	**1.69**	**1.62**	**1.53**	**1.48**	**1.39**	**1.33**	**1.28**
400	3.86	3.02	2.62	2.39	2.23	2.12	2.03	1.96	1.90	1.85	1.81	1.78	1.72	1.67	1.60	1.54	1.49	1.42	1.38	1.32	1.28	1.22	1.16	1.13
	6.70	**4.66**	**3.83**	**3.36**	**3.06**	**2.85**	**2.69**	**2.55**	**2.46**	**2.37**	**2.29**	**2.23**	**2.12**	**2.04**	**1.92**	**1.84**	**1.74**	**1.64**	**1.57**	**1.47**	**1.42**	**1.32**	**1.24**	**1.19**
1,000	3.85	3.00	2.61	2.38	2.22	2.10	2.02	1.95	1.89	1.84	1.80	1.76	1.70	1.65	1.58	1.53	1.47	1.41	1.36	1.30	1.26	1.19	1.13	1.08
	6.66	**4.62**	**3.80**	**3.34**	**3.04**	**2.82**	**2.66**	**2.53**	**2.43**	**2.34**	**2.26**	**2.20**	**2.09**	**2.01**	**1.89**	**1.81**	**1.71**	**1.61**	**1.54**	**1.44**	**1.38**	**1.28**	**1.19**	**1.11**
∞	3.84	2.99	2.60	2.37	2.21	2.09	2.01	1.94	1.88	1.83	1.79	1.75	1.69	1.64	1.57	1.52	1.46	1.40	1.35	1.28	1.24	1.17	1.11	1.00
	6.64	**4.60**	**3.78**	**3.32**	**3.02**	**2.80**	**2.64**	**2.51**	**2.41**	**2.32**	**2.24**	**2.18**	**2.07**	**1.99**	**1.87**	**1.79**	**1.69**	**1.59**	**1.52**	**1.41**	**1.36**	**1.25**	**1.15**	**1.00**

APPENDIX G

Studentized Range Values (q)

The following table presents Studentized range values (q) for overall alpha levels of .05 and .01 for a set of comparisons. The values in the column headed "df for denominator" represent df_{WITHIN} when a between-subjects design is used and df_{ERROR} when a repeated measures design is used. The values in the column headed "alpha" represent overall alpha levels of .05 and .01 for each number of degrees of freedom within or degrees of freedom error. The values at the top of the table under the heading "k = number of levels of the independent variable" represent the number of levels of the independent variable of interest. For example, the Studentized range value (q) for $df_{WITHIN} = 12$, an overall alpha level of .05, and $k = 3$ (that is, an independent variable having three levels) is 3.77.

Source: R. E. Kirk, *Experimental Design: Procedures for the Behavioral Sciences*. Pacific Grove, CA: Brooks/Cole, 1967. Used by permission.

df for denominator	alpha	*k* = NUMBER OF LEVELS OF THE INDEPENDENT VARIABLE									
		2	3	4	5	6	7	8	9	10	11
5	.05	3.64	4.60	5.22	5.67	6.03	6.33	6.58	6.80	6.99	7.17
	.01	5.70	6.98	7.80	8.42	8.91	9.32	9.67	9.97	10.24	10.48
6	.05	3.46	4.34	4.90	5.30	5.63	5.90	6.12	6.32	6.49	6.65
	.01	5.24	6.33	7.03	7.56	7.97	8.32	8.61	8.87	9.10	9.30
7	.05	3.34	4.16	4.68	5.06	5.36	5.61	5.82	6.00	6.16	6.30
	.01	4.95	5.92	6.54	7.01	7.37	7.68	7.94	8.17	8.37	8.55
8	.05	3.26	4.04	4.53	4.89	5.17	5.40	5.60	5.77	5.92	6.05
	.01	4.75	5.64	6.20	6.62	6.96	7.24	7.47	7.68	7.86	8.03
9	.05	3.20	3.95	4.41	4.76	5.02	5.24	5.43	5.59	5.74	5.87
	.01	4.60	5.43	5.96	6.35	6.66	6.91	7.13	7.33	7.49	7.65
10	.05	3.15	3.88	4.33	4.65	4.91	5.12	5.30	5.46	5.60	5.72
	.01	4.48	5.27	5.77	6.14	6.43	6.67	6.87	7.05	7.21	7.36
11	.05	3.11	3.82	4.26	4.57	4.82	5.03	5.20	5.35	5.49	5.61
	.01	4.39	5.15	5.62	5.97	6.25	6.48	6.67	6.84	6.99	7.13
12	.05	3.08	3.77	4.20	4.51	4.75	4.95	5.12	5.27	5.39	5.51
	.01	4.32	5.05	5.50	5.84	6.10	6.32	6.51	6.67	6.81	6.94
13	.05	3.06	3.73	4.15	4.45	4.69	4.88	5.05	5.19	5.32	5.43
	.01	4.26	4.96	5.40	5.73	5.98	6.19	6.37	6.53	6.67	6.79
14	.05	3.03	3.70	4.11	4.41	4.64	4.83	4.99	5.13	5.25	5.36
	.01	4.21	4.89	5.32	5.63	5.88	6.08	6.26	6.41	6.54	6.66
15	.05	3.01	3.67	4.08	4.37	4.59	4.78	4.94	5.08	5.20	5.31
	.01	4.17	4.84	5.25	5.56	5.80	5.99	6.16	6.31	6.44	6.55
16	.05	3.00	3.65	4.05	4.33	4.56	4.74	4.90	5.03	5.15	5.26
	.01	4.13	4.79	5.19	5.49	5.72	5.92	6.08	6.22	6.35	6.46
17	.05	2.98	3.63	4.02	4.30	4.52	4.70	4.86	4.99	5.11	5.21
	.01	4.10	4.74	5.14	5.43	5.66	5.85	6.01	6.15	6.27	6.38
18	.05	2.97	3.61	4.00	4.28	4.49	4.67	4.82	4.96	5.07	5.17
	.01	4.07	4.70	5.09	5.38	5.60	5.79	5.94	6.08	6.20	6.31
19	.05	2.96	3.59	3.98	4.25	4.47	4.65	4.79	4.92	5.04	5.14
	.01	4.05	4.67	5.05	5.33	5.55	5.73	5.89	6.02	6.14	6.25
20	.05	2.95	3.58	3.96	4.23	4.45	4.62	4.77	4.90	5.01	5.11
	.01	4.02	4.64	5.02	5.29	5.51	5.69	5.84	5.97	6.09	6.19
24	.05	2.92	3.53	3.90	4.17	4.37	4.54	4.68	4.81	4.92	5.01
	.01	3.96	4.55	4.91	5.17	5.37	5.54	5.69	5.81	5.92	6.02
30	.05	2.89	3.49	3.85	4.10	4.30	4.46	4.60	4.72	4.82	4.92
	.01	3.89	4.45	4.80	5.05	5.24	5.40	5.54	5.65	5.76	5.85
40	.05	2.86	3.44	3.79	4.04	4.23	4.39	4.52	4.63	4.73	4.82
	.01	3.82	4.37	4.70	4.93	5.11	5.26	5.39	5.50	5.60	5.69
60	.05	2.83	3.40	3.74	3.98	4.16	4.31	4.44	4.55	4.65	4.73
	.01	3.76	4.28	4.59	4.82	4.99	5.13	5.25	5.36	5.45	5.53
120	.05	2.80	3.36	3.68	3.92	4.10	4.24	4.36	4.47	4.56	4.64
	.01	3.70	4.20	4.50	4.71	4.87	5.01	5.12	5.21	5.30	5.37
∞	.05	2.77	3.31	3.63	3.86	4.03	4.17	4.29	4.39	4.47	4.55
	.01	3.64	4.12	4.40	4.60	4.76	4.88	4.99	5.08	5.16	5.23

df for denominator	alpha	k = NUMBER OF LEVELS OF THE INDEPENDENT VARIABLE								
		12	13	14	15	16	17	18	19	20
5	.05	7.32	7.47	7.60	7.72	7.83	7.93	8.03	8.12	8.21
	.01	10.70	10.89	11.08	11.24	11.40	11.55	11.68	11.81	11.93
6	.05	6.79	6.92	7.03	7.14	7.24	7.34	7.43	7.51	7.59
	.01	9.48	9.65	9.81	9.95	10.08	10.21	10.32	10.43	10.54
7	.05	6.43	6.55	6.66	6.76	6.85	6.94	7.02	7.10	7.17
	.01	8.71	8.86	9.00	9.12	9.24	9.35	9.46	9.55	9.65
8	.05	6.18	6.29	6.39	6.48	6.57	6.65	6.73	6.80	6.87
	.01	8.18	8.31	8.44	8.55	8.66	8.76	8.85	8.94	9.03
9	.05	5.98	6.09	6.19	6.28	6.36	6.44	6.51	6.58	6.64
	.01	7.78	7.91	8.03	8.13	8.23	8.33	8.41	8.49	8.57
10	.05	5.83	5.93	6.03	6.11	6.19	6.27	6.34	6.40	6.47
	.01	7.49	7.60	7.71	7.81	7.91	7.99	8.08	8.15	8.23
11	.05	5.71	5.81	5.90	5.98	6.06	6.13	6.20	6.27	6.33
	.01	7.25	7.36	7.46	7.56	7.65	7.73	7.81	7.88	7.95
12	.05	5.61	5.71	5.80	5.88	5.95	6.02	6.09	6.15	6.21
	.01	7.06	7.17	7.26	7.36	7.44	7.52	7.59	7.66	7.73
13	.05	5.53	5.63	5.71	5.79	5.86	5.93	5.99	6.05	6.11
	.01	6.90	7.01	7.10	7.19	7.27	7.35	7.42	7.48	7.55
14	.05	5.46	5.55	5.64	5.71	5.79	5.85	5.91	5.97	6.03
	.01	6.77	6.87	6.96	7.05	7.13	7.20	7.27	7.33	7.39
15	.05	5.40	5.49	5.57	5.65	5.72	5.78	5.85	5.90	5.96
	.01	6.66	6.76	6.84	6.93	7.00	7.07	7.14	7.20	7.26
16	.05	5.35	5.44	5.52	5.59	5.66	5.73	5.79	5.84	5.90
	.01	6.56	6.66	6.74	6.82	6.90	6.97	7.03	7.09	7.15
17	.05	5.31	5.39	5.47	5.54	5.61	5.67	5.73	5.79	5.84
	.01	6.48	6.57	6.66	6.73	6.81	6.87	6.94	7.00	7.05
18	.05	5.27	5.35	5.43	5.50	5.57	5.63	5.69	5.74	5.79
	.01	6.41	6.50	6.58	6.65	6.73	6.79	6.85	6.91	6.97
19	.05	5.23	5.31	5.39	5.46	5.53	5.59	5.65	5.70	5.75
	.01	6.34	6.43	6.51	6.58	6.65	6.72	6.78	6.84	6.89
20	.05	5.20	5.28	5.36	5.43	5.49	5.55	5.61	5.66	5.71
	.01	6.28	6.37	6.45	6.52	6.59	6.65	6.71	6.77	6.82
24	.05	5.10	5.18	5.25	5.32	5.38	5.44	5.49	5.55	5.59
	.01	6.11	6.19	6.26	6.33	6.39	6.45	6.51	6.56	6.61
30	.05	5.00	5.08	5.15	5.21	5.27	5.33	5.38	5.43	5.47
	.01	5.93	6.01	6.08	6.14	6.20	6.26	6.31	6.36	6.41
40	.05	4.90	4.98	5.04	5.11	5.16	5.22	5.27	5.31	5.36
	.01	5.76	5.83	5.90	5.96	6.02	6.07	6.12	6.16	6.21
60	.05	4.81	4.88	4.94	5.00	5.06	5.11	5.15	5.20	5.24
	.01	5.60	5.67	5.73	5.78	5.84	5.89	5.93	5.97	6.01
120	.05	4.71	4.78	4.84	4.90	4.95	5.00	5.04	5.09	5.13
	.01	5.44	5.50	5.56	5.61	5.66	5.71	5.75	5.79	5.83
∞	.05	4.62	4.68	4.74	4.80	4.85	4.89	4.93	4.97	5.01
	.01	5.29	5.35	5.40	5.45	5.49	5.54	5.57	5.61	5.65

APPENDIX H

Critical Values for Pearson r

The following table presents critical values of Pearson r for directional (one-tailed) and nondirectional (two-tailed) tests of the null hypothesis $\rho = 0$. The column headed "df" lists the degrees of freedom associated with the distribution of interest. The values .05, .025, .01, and .005 at the top of the table under the heading "Level of Significance for Directional Test" represent alpha levels for directional tests. For example, the critical value of r for df = 25 and an alpha level of .05 for a directional test using the upper tail of the distribution (that is, for alternative hypotheses of the form $\rho > 0$) is +.323. Since the distribution is symmetrical, the critical value of r for a directional test using the lower tail of the distribution (that is, for null hypotheses of the form $\rho < 0$) is −.323. The values .10, .05, .02, and .01 under the heading "Level of Significance for Nondirectional Test" represent alpha levels for nondirectional tests. For example, the critical values of r for a nondirectional test for df = 25 and an alpha level of .05 are ±.381.

Source: G. A. Ferguson, *Statistical Analysis in Psychology and Education.* New York: McGraw-Hill, 1976. Used with permission.

	LEVEL OF SIGNIFICANCE FOR DIRECTIONAL TEST			
	.05	*.025*	*.01*	*.005*
	LEVEL OF SIGNIFICANCE FOR NONDIRECTIONAL TEST			
df	*.10*	*.05*	*.02*	*.01*
1	.988	.997	.9995	.9999
2	.900	.950	.980	.990
3	.805	.878	.934	.959
4	.729	.811	.882	.917
5	.669	.754	.833	.874
6	.622	.707	.789	.834
7	.582	.666	.750	.798
8	.549	.632	.716	.765
9	.521	.602	.685	.735
10	.497	.576	.658	.708
11	.476	.553	.634	.684
12	.458	.532	.612	.661
13	.441	.514	.592	.641
14	.426	.497	.574	.623
15	.412	.482	.558	.606
16	.400	.468	.542	.590
17	.389	.456	.528	.575
18	.378	.444	.516	.561
19	.369	.433	.503	.549
20	.360	.423	.492	.537
21	.352	.413	.482	.526
22	.344	.404	.472	.515
23	.337	.396	.462	.505
24	.330	.388	.453	.496
25	.323	.381	.445	.487
26	.317	.374	.437	.479
27	.311	.367	.430	.471
28	.306	.361	.423	.463
29	.301	.355	.416	.456
30	.296	.349	.409	.449
35	.275	.325	.381	.418
40	.257	.304	.358	.393
45	.243	.288	.338	.372
50	.231	.273	.322	.354
60	.211	.250	.295	.325
70	.195	.232	.274	.303
80	.183	.217	.256	.283
90	.173	.205	.242	.267
100	.164	.195	.230	.254

APPENDIX I

Fisher's Transformation of Pearson r (r')

The following table presents Fisher's transformation of Pearson r for use in testing null hypotheses other than $\rho = 0$. The columns headed "r" list values of the Pearson correlation coefficient, and the adjacent columns headed "r'" list the corresponding transformed values. For example, the transformed value (r') corresponding to an r of .200 is .203. Transformed values for negative correlations are found by inserting a negative sign before the tabled values of r and r'.

Source: G. A. Ferguson, *Statistical Analysis in Psychology and Education.* New York: McGraw-Hill, 1976. Used with permission.

r	r'	r	r'	r	r'	r	r'	r	r'
.000	.000	.200	.203	.400	.424	.600	.693	.800	1.099
.005	.005	.205	.208	.405	.430	.605	.701	.805	1.113
.010	.010	.210	.213	.410	.436	.610	.709	.810	1.127
.015	.015	.215	.218	.415	.442	.615	.717	.815	1.142
.020	.020	.220	.224	.420	.448	.620	.725	.820	1.157
.025	.025	.225	.229	.425	.454	.625	.733	.825	1.172
.030	.030	.230	.234	.430	.460	.630	.741	.830	1.188
.035	.035	.235	.239	.435	.466	.635	.750	.835	1.204
.040	.040	.240	.245	.440	.472	.640	.758	.840	1.221
.045	.045	.245	.250	.445	.478	.645	.767	.845	1.238
.050	.050	.250	.255	.450	.485	.650	.775	.850	1.256
.055	.055	.255	.261	.455	.491	.655	.784	.855	1.274
.060	.060	.260	.266	.460	.497	.660	.793	.860	1.293
.065	.065	.265	.271	.465	.504	.665	.802	.865	1.313
.070	.070	.270	.277	.470	.510	.670	.811	.870	1.333
.075	.075	.275	.282	.475	.517	.675	.820	.875	1.354
.080	.080	.280	.288	.480	.523	.680	.829	.880	1.376
.085	.085	.285	.293	.485	.530	.685	.838	.885	1.398
.090	.090	.290	.299	.490	.536	.690	.848	.890	1.422
.095	.095	.295	.304	.495	.543	.695	.858	.895	1.447
.100	.100	.300	.310	.500	.549	.700	.867	.900	1.472
.105	.105	.305	.315	.505	.556	.705	.877	.905	1.499
.110	.110	.310	.321	.510	.563	.710	.887	.910	1.528
.115	.116	.315	.326	.515	.570	.715	.897	.915	1.557
.120	.121	.320	.332	.520	.576	.720	.908	.920	1.589
.125	.126	.325	.337	.525	.583	.725	.918	.925	1.623
.130	.131	.330	.343	.530	.590	.730	.929	.930	1.658
.135	.136	.335	.348	.535	.597	.735	.940	.935	1.697
.140	.141	.340	.354	.540	.604	.740	.950	.940	1.738
.145	.146	.345	.360	.545	.611	.745	.962	.945	1.783
.150	.151	.350	.365	.550	.618	.750	.973	.950	1.832
.155	.156	.355	.371	.555	.626	.755	.984	.955	1.886
.160	.161	.360	.377	.560	.633	.760	.996	.960	1.946
.165	.167	.365	.383	.565	.640	.765	1.008	.965	2.014
.170	.172	.370	.388	.570	.648	.770	1.020	.970	2.092
.175	.177	.375	.394	.575	.655	.775	1.033	.975	2.185
.180	.182	.380	.400	.580	.662	.780	1.045	.980	2.298
.185	.187	.385	.406	.585	.670	.785	1.058	.985	2.443
.190	.192	.390	.412	.590	.678	.790	1.071	.990	2.647
.195	.198	.395	.418	.595	.685	.795	1.085	.995	2.994

APPENDIX J

Critical Values for the Chi-Square Distribution

The following table presents critical values of chi-square. The column headed "df" lists the degrees of freedom associated with the chi-square distribution of interest. The values at the top of the table represent alpha levels. For our purposes, we are only concerned with the low alpha levels appearing at the right of the table. The high alpha levels are used in certain types of advanced statistical applications. As an example of the use of this table, the critical value of chi-square for df = 4 and an alpha level of .05 is 9.488.

Source: R. P. Runyon and A. Haber, *Fundamentals of Behavioral Statistics.* Reading, MA: Addison-Wesley, 1976. Used with permission.

	ALPHA												
df	*.99*	*.98*	*.95*	*.90*	*.80*	*.70*	*.50*	*.30*	*.20*	*.10*	*.05*	*.02*	*.01*
1	.000157	.000628	.00393	.0158	.0642	.148	.455	1.074	1.642	2.706	3.841	5.412	6.635
2	.0201	.0404	.103	.211	.446	.713	1.386	2.408	3.219	4.605	5.991	7.824	9.210
3	.115	.185	.352	.584	1.005	1.424	2.366	3.665	4.642	6.251	7.815	9.837	11.341
4	.297	.429	.711	1.064	1.649	2.195	3.357	4.878	5.989	7.779	9.488	11.668	13.277
5	.554	.752	1.145	1.610	2.343	3.000	4.351	6.064	7.289	9.236	11.070	13.388	15.086
6	.872	1.134	1.635	2.204	3.070	3.828	5.348	7.231	8.558	10.645	12.592	15.033	16.812
7	1.239	1.564	2.167	2.833	3.822	4.671	6.346	8.383	9.803	12.017	14.067	16.622	18.475
8	1.646	2.032	2.733	3.490	4.594	5.527	7.344	9.524	11.030	13.362	15.507	18.168	20.090
9	2.088	2.532	3.325	4.168	5.380	6.393	8.343	10.656	12.242	14.684	16.919	19.679	21.666
10	2.558	3.059	3.940	4.865	6.179	7.267	9.342	11.781	13.442	15.987	18.307	21.161	23.209
11	3.053	3.609	4.575	5.578	6.989	8.148	10.341	12.899	14.631	17.275	19.675	22.618	24.725
12	3.571	4.178	5.226	6.304	7.807	9.034	11.340	14.011	15.812	18.549	21.026	24.054	26.217
13	4.107	4.765	5.892	7.042	8.634	9.926	12.340	15.119	16.985	19.812	22.362	25.472	27.688
14	4.660	5.368	6.571	7.790	9.467	10.821	13.339	16.222	18.151	21.064	23.685	26.873	29.141
15	5.229	5.985	7.261	8.547	10.307	11.721	14.339	17.322	19.311	22.307	24.996	28.259	30.578
16	5.812	6.614	7.962	9.312	11.152	12.624	15.338	18.418	20.465	23.542	26.296	29.633	32.000
17	6.408	7.255	8.672	10.085	12.002	13.531	16.338	19.511	21.615	24.769	27.587	30.995	33.409
18	7.015	7.906	9.390	10.865	12.857	14.440	17.338	20.601	22.760	25.989	28.869	32.346	34.805
19	7.633	8.567	10.117	11.651	13.716	15.352	18.338	21.689	23.900	27.204	30.144	33.687	36.191
20	8.260	9.237	10.851	12.443	14.578	16.266	19.337	22.775	25.038	28.412	31.410	35.020	37.566
21	8.897	9.915	11.591	13.240	15.445	17.182	20.337	23.858	26.171	29.615	32.671	36.343	38.932
22	9.542	10.600	12.338	14.041	16.314	18.101	21.337	24.939	27.301	30.813	33.924	37.659	40.289
23	10.196	11.293	13.091	14.848	17.187	19.021	22.337	26.018	28.429	32.007	35.172	38.968	41.638
24	10.856	11.992	13.848	15.659	18.062	19.943	23.337	27.096	29.553	33.196	36.415	40.270	42.980
25	11.524	12.697	14.611	16.473	18.940	20.867	24.337	28.172	30.675	34.382	37.652	41.566	44.314
26	12.198	13.409	15.379	17.292	19.820	21.792	25.336	29.246	31.795	35.563	38.885	42.856	45.642
27	12.879	14.125	16.151	18.114	20.703	22.719	26.336	30.319	32.912	36.741	40.113	44.140	46.963
28	13.565	14.847	16.928	18.939	21.588	23.647	27.336	31.391	34.027	37.916	41.337	45.419	48.278
29	14.256	15.574	17.708	19.768	22.475	24.577	28.336	32.461	35.139	39.087	42.557	46.693	49.588
30	14.953	16.306	18.493	20.599	23.364	25.508	29.336	33.530	36.250	40.256	43.773	47.962	50.892

APPENDIX K

Critical Values for the Mann–Whitney *U* Test

The following tables present critical values of the *U* statistic for the Mann–Whitney *U* test for alpha levels of .05, .025, .01, and .005 for directional (one-tailed) tests, and alpha levels of .10, .05, .02, and .01 for nondirectional (two-tailed) tests. The critical values for directional alpha levels of .01 (roman type) and .005 (**boldface**) and nondirectional alpha levels of .02 (roman type) and .01 (**boldface**) are contained in the first table. The critical values for directional alpha levels of .05 (roman type) and .025 (**boldface**) and nondirectional alpha levels of .10 (roman type) and .05 (**boldface**) are contained in the second table.

The values at the top of each table represent sample sizes (n_1) for group 1, and the values at the left of each table represent sample sizes (n_2) for group 2. The critical value of *U* is defined by the point where the sample sizes for the two groups under study intersect. For example, the critical value of *U* for a nondirectional test for an alpha level of .05 and $n_1 = 10$ and $n_2 = 12$ is 29. The observed *U* is statistically significant if it is *equal to or less than* the critical *U*.

Critical Values of U for Alpha Levels of .01 (Roman Type) and **.005 (Boldface)** for Directional Tests, and Alpha Levels of .02 (Roman Type) and **.01 (Boldface)** for Nondirectional Tests.

n_2 \ n_1	1	2	3	4	5	6	7	8	9	10	11	12	13	14	15	16	17	18	19	20
1	—	—	—	—	—	—	—	—	—	—	—	—	—	—	—	—	—	—	—	—
	—	—	—	—	—	—	—	—	—	—	—	—	—	—	—	—	—	—	—	—
2	—	—	—	—	—	—	—	—	—	—	—	—	0	0	0	0	0	0	1	1
	—	—	—	—	—	—	—	—	—	—	—	—	—	—	—	—	—	—	**0**	**0**
3	—	—	—	—	—	—	0	0	1	1	1	2	2	2	3	3	4	4	4	5
	—	—	—	—	—	—	—	—	**0**	**0**	**0**	**1**	**1**	**1**	**2**	**2**	**2**	**2**	**3**	**3**
4	—	—	—	—	0	1	1	2	3	3	4	5	5	6	7	7	8	9	9	10
	—	—	—	—	—	**0**	**0**	**1**	**1**	**2**	**2**	**3**	**3**	**4**	**5**	**5**	**6**	**6**	**7**	**8**
5	—	—	—	0	1	2	3	4	5	6	7	8	9	10	11	12	13	14	15	16
	—	—	—	—	**0**	**1**	**1**	**2**	**3**	**4**	**5**	**6**	**7**	**7**	**8**	**9**	**10**	**11**	**12**	**13**
6	—	—	—	1	2	3	4	6	7	8	9	11	12	13	15	16	18	19	20	22
	—	—	—	**0**	**1**	**2**	**3**	**4**	**5**	**6**	**7**	**9**	**10**	**11**	**12**	**13**	**15**	**16**	**17**	**18**
7	—	—	0	1	3	4	6	7	9	11	12	14	16	17	19	21	23	24	26	28
	—	—	—	**0**	**1**	**3**	**4**	**6**	**7**	**9**	**10**	**12**	**13**	**15**	**16**	**18**	**19**	**21**	**22**	**24**
8	—	—	0	2	4	6	7	9	11	13	15	17	20	22	24	26	28	30	32	34
	—	—	—	**1**	**2**	**4**	**6**	**7**	**9**	**11**	**13**	**15**	**17**	**18**	**20**	**22**	**24**	**26**	**28**	**30**
9	—	—	1	3	5	7	9	11	14	16	18	21	23	26	28	31	33	36	38	40
	—	—	**0**	**1**	**3**	**5**	**7**	**9**	**11**	**13**	**16**	**18**	**20**	**22**	**24**	**27**	**29**	**31**	**33**	**36**
10	—	—	1	3	6	8	11	13	16	19	22	24	27	30	33	36	38	41	44	47
	—	—	**0**	**2**	**4**	**6**	**9**	**11**	**13**	**16**	**18**	**21**	**24**	**26**	**29**	**31**	**34**	**37**	**39**	**42**
11	—	—	1	4	7	9	12	15	18	22	25	28	31	34	37	41	44	47	50	53
	—	—	**0**	**2**	**5**	**7**	**10**	**13**	**16**	**18**	**21**	**24**	**27**	**30**	**33**	**36**	**39**	**42**	**45**	**48**
12	—	—	2	5	8	11	14	17	21	24	28	31	35	38	42	46	49	53	56	60
	—	—	**1**	**3**	**6**	**9**	**12**	**15**	**18**	**21**	**24**	**27**	**31**	**34**	**37**	**41**	**44**	**47**	**51**	**54**
13	—	0	2	5	9	12	16	20	23	27	31	35	39	43	47	51	55	59	63	67
	—	—	**1**	**3**	**7**	**10**	**13**	**17**	**20**	**24**	**27**	**31**	**34**	**38**	**42**	**45**	**49**	**53**	**56**	**60**
14	—	0	2	6	10	13	17	22	26	30	34	38	43	47	51	56	60	65	69	73
	—	—	**1**	**4**	**7**	**11**	**15**	**18**	**22**	**26**	**30**	**34**	**38**	**42**	**46**	**50**	**54**	**58**	**63**	**67**
15	—	0	3	7	11	15	19	24	28	33	37	42	47	51	56	61	66	70	75	80
	—	—	**2**	**5**	**8**	**12**	**16**	**20**	**24**	**29**	**33**	**37**	**42**	**46**	**51**	**55**	**60**	**64**	**69**	**73**
16	—	0	3	7	12	16	21	26	31	36	41	46	51	56	61	66	71	76	82	87
	—	—	**2**	**5**	**9**	**13**	**18**	**22**	**27**	**31**	**36**	**41**	**45**	**50**	**55**	**60**	**65**	**70**	**74**	**79**
17	—	0	4	8	13	18	23	28	33	38	44	49	55	60	66	71	77	82	88	93
	—	—	**2**	**6**	**10**	**15**	**19**	**24**	**29**	**34**	**39**	**44**	**49**	**54**	**60**	**65**	**70**	**75**	**81**	**86**
18	—	0	4	9	14	19	24	30	36	41	47	53	59	65	70	76	82	88	94	100
	—	—	**2**	**6**	**11**	**16**	**21**	**26**	**31**	**37**	**42**	**47**	**53**	**58**	**64**	**70**	**75**	**81**	**87**	**92**
19	—	1	4	9	15	20	26	32	38	44	50	56	63	69	75	82	88	94	101	107
	—	**0**	**3**	**7**	**12**	**17**	**22**	**28**	**33**	**39**	**45**	**51**	**56**	**63**	**69**	**74**	**81**	**87**	**93**	**99**
20	—	1	5	10	16	22	28	34	40	47	53	60	67	73	80	87	93	100	107	114
	—	**0**	**3**	**8**	**13**	**18**	**24**	**30**	**36**	**42**	**48**	**54**	**60**	**67**	**73**	**79**	**86**	**92**	**99**	**105**

Note. To be statistically significant for any given n_1 and n_2, the observed U must be *equal to or less than* the value shown in the table. Dashes in the body of the table indicate that no decision is possible at the relevant level of alpha.

Critical Values of U for Alpha Levels of .05 (Roman Type) and **.025 (Boldface)** for Directional Tests, and Alpha Levels of .10 (Roman Type) and **.05 (Boldface)** for Nondirectional Tests.

n_2 \ n_1	1	2	3	4	5	6	7	8	9	10	11	12	13	14	15	16	17	18	19	20
1	—	—	—	—	—	—	—	—	—	—	—	—	—	—	—	—	—	—	0	0
	—	—	—	—	—	—	—	—	—	—	—	—	—	—	—	—	—	—	—	—
2	—	—	—	—	0	0	0	1	1	1	1	2	2	2	3	3	3	4	4	4
	—	—	—	—	—	—	—	**0**	**0**	**0**	**0**	**1**	**1**	**1**	**1**	**1**	**2**	**2**	**2**	**2**
3	—	—	0	0	1	2	2	3	3	4	5	5	6	7	7	8	9	9	10	11
	—	—	—	—	**0**	**1**	**1**	**2**	**2**	**3**	**3**	**4**	**4**	**5**	**5**	**6**	**6**	**7**	**7**	**8**
4	—	—	0	1	2	3	4	5	6	7	8	9	10	11	12	14	15	16	17	18
	—	—	—	**0**	**1**	**2**	**3**	**4**	**4**	**5**	**6**	**7**	**8**	**9**	**10**	**11**	**11**	**12**	**13**	**13**
5	—	0	1	2	4	5	6	8	9	11	12	13	15	16	18	19	20	22	23	25
	—	—	**0**	**1**	**2**	**3**	**5**	**6**	**7**	**8**	**9**	**11**	**12**	**13**	**14**	**15**	**17**	**18**	**19**	**20**
6	—	0	2	3	5	7	8	10	12	14	16	17	19	21	23	25	26	28	30	32
	—	—	**1**	**2**	**3**	**5**	**6**	**8**	**10**	**11**	**13**	**14**	**16**	**17**	**19**	**21**	**22**	**24**	**25**	**27**
7	—	0	2	4	6	8	11	13	15	17	19	21	24	26	28	30	33	35	37	39
	—	—	**1**	**3**	**5**	**6**	**8**	**10**	**12**	**14**	**16**	**18**	**20**	**22**	**24**	**26**	**28**	**30**	**32**	**34**
8	—	1	3	5	8	10	13	15	18	20	23	26	28	31	33	36	39	41	44	47
	—	**0**	**2**	**4**	**6**	**8**	**10**	**13**	**15**	**17**	**19**	**22**	**24**	**26**	**29**	**31**	**34**	**36**	**38**	**41**
9	—	1	3	6	9	12	15	18	21	24	27	30	33	36	39	42	45	48	51	54
	—	**0**	**2**	**4**	**7**	**10**	**12**	**15**	**17**	**20**	**23**	**26**	**28**	**31**	**34**	**37**	**39**	**42**	**45**	**48**
10	—	1	4	7	11	14	17	20	24	27	31	34	37	41	44	48	51	55	58	62
	—	**0**	**3**	**5**	**8**	**11**	**14**	**17**	**20**	**23**	**26**	**29**	**33**	**36**	**39**	**42**	**45**	**48**	**52**	**55**
11	—	1	5	8	12	16	19	23	27	31	34	38	42	46	50	54	57	61	65	69
	—	**0**	**3**	**6**	**9**	**13**	**16**	**19**	**23**	**26**	**30**	**33**	**37**	**40**	**44**	**47**	**51**	**55**	**58**	**62**
12	—	2	5	9	13	17	21	26	30	34	38	42	47	51	55	60	64	68	72	77
	—	**1**	**4**	**7**	**11**	**14**	**18**	**22**	**26**	**29**	**33**	**37**	**41**	**45**	**49**	**53**	**57**	**61**	**65**	**69**
13	—	2	6	10	15	19	24	28	33	37	42	47	51	56	61	65	70	75	80	84
	—	**1**	**4**	**8**	**12**	**16**	**20**	**24**	**28**	**33**	**37**	**41**	**45**	**50**	**54**	**59**	**63**	**67**	**72**	**76**
14	—	2	7	11	16	21	26	31	36	41	46	51	56	61	66	71	77	82	87	92
	—	**1**	**5**	**9**	**13**	**17**	**22**	**26**	**31**	**36**	**40**	**45**	**50**	**55**	**59**	**64**	**67**	**74**	**78**	**83**
15	—	3	7	12	18	23	28	33	39	44	50	55	61	66	72	77	83	88	94	100
	—	**1**	**5**	**10**	**14**	**19**	**24**	**29**	**34**	**39**	**44**	**49**	**54**	**59**	**64**	**70**	**75**	**80**	**85**	**90**
16	—	3	8	14	19	25	30	36	42	48	54	60	65	71	77	83	89	95	101	107
	—	**1**	**6**	**11**	**15**	**21**	**26**	**31**	**37**	**42**	**47**	**53**	**59**	**64**	**70**	**75**	**81**	**86**	**92**	**98**
17	—	3	9	15	20	26	33	39	45	51	57	64	70	77	83	89	96	102	109	115
	—	**2**	**6**	**11**	**17**	**22**	**28**	**34**	**39**	**45**	**51**	**57**	**63**	**67**	**75**	**81**	**87**	**93**	**99**	**105**
18	—	4	9	16	22	28	35	41	48	55	61	68	75	82	88	95	102	109	116	123
	—	**2**	**7**	**12**	**18**	**24**	**30**	**36**	**42**	**48**	**55**	**61**	**67**	**74**	**80**	**86**	**93**	**99**	**106**	**112**
19	0	4	10	17	23	30	37	44	51	58	65	72	80	87	94	101	109	116	123	130
	—	**2**	**7**	**13**	**19**	**25**	**32**	**38**	**45**	**52**	**58**	**65**	**72**	**78**	**85**	**92**	**99**	**106**	**113**	**119**
20	0	4	11	18	25	32	39	47	54	62	69	77	84	92	100	107	115	123	130	138
	—	**2**	**8**	**13**	**20**	**27**	**34**	**41**	**48**	**55**	**62**	**69**	**76**	**83**	**90**	**98**	**105**	**112**	**119**	**127**

Note. To be statistically significant for any given n_1 and n_2, the observed U must be *equal to or less than* the value shown in the table. Dashes in the body of the table indicate that no decision is possible at the relevant level of alpha.

APPENDIX L

Critical Values for the Wilcoxon Signed-Rank Test

The following table presents critical values of the *T* statistic for directional (one-tailed) and nondirectional (two-tailed) Wilcoxon signed-rank tests. The columns headed "*N*" list the sample size for the investigation of interest. The values .05, .025, .01, and .005 at the top of the table under the headings "Level of Significance for Directional Test" represent alpha levels for directional tests. The values .10, .05, .02, and .01 under the headings "Level of Significance for Nondirectional Test" represent alpha levels for nondirectional tests. For example, the critical value of *T* for a nondirectional test for $N = 35$ and an alpha level of .05 is 195. The observed *T* is statistically significant if it is *equal to or less than* the critical *T*.

Source: Wilcoxon, Katti, and Wilcox, "Critical Values and Probability Levels of the Wilcoxon Rank Sum Test and the Wilcoxon Signed Rank Test." Reproduced with the permission of American Cyanamid Company.

	Level of significance for directional test					Level of significance for directional test			
	.05	.025	.01	.005		.05	.025	.01	.005
	Level of significance for nondirectional test					Level of significance for nondirectional test			
N	.10	.05	.02	.01	*N*	.10	.05	.02	.01
5	0	—	—	—	28	130	116	101	91
6	2	0	—	—	29	140	126	110	100
7	3	2	0	—	30	151	137	120	109
8	5	3	1	0	31	163	147	130	118
9	8	5	3	1	32	175	159	140	128
10	10	8	5	3	33	187	170	151	138
11	13	10	7	5	34	200	182	162	148
12	17	13	9	7	35	213	195	173	159
13	21	17	12	9	36	227	208	185	171
14	25	21	15	12	37	241	221	198	182
15	30	25	19	15	38	256	235	211	194
16	35	29	23	19	39	271	249	224	207
17	41	34	27	23	40	286	264	238	220
18	47	40	32	27	41	302	279	252	233
19	53	46	37	32	42	319	294	266	247
20	60	52	43	37	43	336	310	281	261
21	67	58	49	42	44	353	327	296	276
22	75	65	55	48	45	371	343	312	291
23	83	73	62	54	46	389	361	328	307
24	91	81	69	61	47	407	378	345	322
25	100	89	76	68	48	426	396	362	339
26	110	98	84	75	59	446	415	379	355
27	119	107	92	83	50	466	434	397	373

Note. To be statistically significant for any given N, the observed *T* must be *equal to or less than* the value shown in the table. Slight discrepancies will be found between the critical values appearing in the table above and in Table 2 of the 1964 revision of F. Wilcoxon and R. A. Wilcox, *Some Rapid Approximate Statistical Procedures,* New York, Lederle Laboratories, 1964. The disparity reflects the latter's policy of selecting the critical value nearest a given significance level, occasionally overstepping that level. For example, for $N = 8$, the probability of a T of 3 is .0390 (nondirectional) and the probability of a T of 4 is .0546 (nondirectional). Wilcoxon and Wilcox select a T of 4 as the critical value at the .05 level of significance (nondirectional), whereas this table reflects a more conservative policy by setting a T of 3 as the critical value at this level.

APPENDIX M

Critical Values for Spearman *r*

The following table presents critical values of Spearman *r* for directional (one-tailed) and nondirectional (two-tailed) tests of the null hypothesis $\rho_s = 0$. The column headed "*N*" lists the sample size for the investigation of interest. The values .05, .025, .01, and .005 at the top of the table under the heading "Level of Significance for Directional Test" represent alpha levels for directional tests. For example, the critical value of r_s for $N = 22$ and an alpha level of .05 for a directional test using the upper tail of the distribution (that is, for alternative hypotheses of the form $\rho_s > 0$) is +.359. Since the distribution is symmetrical, the critical value of r_s for a directional test using the lower tail of the distribution (that is, for null hypotheses of the form $\rho_s < 0$) is −.359. The values .10, .05, .02, and .01 under the heading "Level of Significance for Nondirectional Test" represent alpha levels for nondirectional tests. For example, the critical values of r_s for a nondirectional test for $N = 22$ and an alpha level of .05 are ±.428.

	LEVEL OF SIGNIFICANCE FOR DIRECTIONAL TEST			
	.05	.025	.01	.005
	LEVEL OF SIGNIFICANCE FOR NONDIRECTIONAL TEST			
N	.10	.05	.02	.01
5	.900	1.000	1.000	—
6	.829	.886	.943	1.000
7	.714	.786	.893	.929
8	.643	.738	.833	.881
9	.600	.683	.783	.833
10	.564	.648	.746	.794
12	.506	.591	.712	.777
14	.456	.544	.645	.715
16	.425	.506	.601	.665
18	.399	.475	.564	.625
20	.377	.450	.534	.591
22	.359	.428	.508	.562
24	.343	.409	.485	.537
26	.329	.392	.465	.515
28	.317	.377	.448	.496
30	.306	.364	.432	.478

Source: G. A. Ferguson, *Statistical Analysis in Psychology and Education*. New York: McGraw-Hill, 1976. Used with permission.

GLOSSARY OF MAJOR SYMBOLS

Numbers in parentheses indicate the sections where the symbols are first discussed.

A Factor A (17.4)

a Intercept of a line (5.2)/Number of levels of factor A (17.4)

$A \times B$ Interaction of factor A and factor B (17.4)

B Factor B (17.4)

b Slope of a line (5.2)/Number of levels of factor B (17.4)

C Number of pairs of mean ranks or rank sums to be tested using the Dunn procedure (16.4)

c Number of levels of the column variable in a contingency table (15.4)

${}_nC_r$ Number of combinations of n things taken r at a time (6.7)

CD Critical difference for the Tukey HSD test (12.5)

cf Cumulative frequency (2.1)

CI Confidence interval (8.10)

CMF_j Column marginal frequency associated with cell j (15.4)

crf Cumulative relative frequency (2.1)

D Difference between raw (11.2) or ranked scores (16.6) for an individual

d Deviation of an individual's score from the group mean (12.2)

$\overline{D}$ Mean difference score in a sample (11.2)

df Degrees of freedom (note: if df is subscripted, the subscript indicates the source of variability to which the degrees of freedom apply) (7.4)

E Expected value of the sum of ranks for the Wilcoxon signed-rank test (16.3)

E_j Expected frequency for cell j (15.4) or category j (15.14)/Expected rank sum for group j for the Wilcoxon rank sum test (16.2)

E_R^2 Epsilon-squared (16.4)

eta^2 Eta-squared statistic (10.3)

F Variance ratio (12.2)

f Frequency of a score (2.1)

G Grand mean (10.3)

g Calculation in Goodman's simultaneous confidence interval procedure (Appendix 15.1)

H H statistic in Kruskal–Wallis test (16.4)

H_0 Null hypothesis (8.2)

Symbol	Meaning
H_1	Alternative hypothesis (8.2)
i	Size of the interval of a numerical category (3.1)
k	Number of levels of a variable (12.2)
L	Lower real limit of a numerical category (3.1)/Number of levels of the variable in a contingency table that has the fewer values (15.6)
$\log_e$	Natural logarithm of a number (Appendix 14.1)
$\underline{\text{M}}$	Sample mean in American Psychological Association format (3.7)
Mdn	Median (3.1)
MS	Mean square (note: if MS is subscripted, the subscript indicates the source of variability to which the mean square applies) (7.4)
N	Overall sample size (1.6)
n	Number of scores in a subgroup (3.7)/Number of trials in the binomial expression (6.8)
n'	Adjusted per-cell sample size necessary to achieve the desired level of power for two-way between-subjects analysis of variance (17.10)
n_L	Number of individuals with scores less than a specified value (3.1)
n_T	Tabled per-cell sample size necessary to achieve the desired level of power for two-way between-subjects analysis of variance (17.10)
n_W	Number of individuals with scores within a numerical category (3.1)
O_j	Observed frequency for cell *j* (15.4) or category *j* (15.14)
P	Percentile (4.1)
p	Probability (2.7)/Probability of a "success" in the binomial expression (6.8)/Probability or significance level (8.11)
${}_nP_r$	Number of permutations of *n* things taken *r* at a time (6.7)
$p(A)$	Probability of event *A* (1.6)
$p(A/B)$	Conditional probability of event *A*, given event *B* (6.2)
$p(A, B)$	Joint probability of *both* event *A and* event *B* (6.3)
$p(A \text{ or } B)$	Probability of *at least one* of event *A* and event *B* (6.4)
PR_X	Percentile rank of the score *X* (4.1)
q	Probability of a "failure" in the binomial expression (6.8)/Studentized range value (12.5)
r	Sample Pearson correlation coefficient (5.3)/Number of "successes" in the binomial expression (6.8)/Number of levels of the row variable in a contingency table (15.4)
R^2	Multiple correlation coefficient (18.6)
r^2	Coefficient of determination (14.3)
r' (r prime)	Fisher's logarithmic transformation of *r* (Appendix 14.1)
r_c	Matched-pairs rank biserial correlation coefficient (16.3)
r_g	Glass rank biserial correlation coefficient (16.2)
r_i	Number of mutually exclusive and exhaustive events that can occur on trial *i* (6.7)

R_j — R statistic in Wilcoxon rank sum test (16.2)/Sum of the ranks in condition j for Kruskal–Wallis test (16.4) and Friedman analysis of variance by ranks (16.5)

R_n — Sum of the ranks of the negative differences for the Wilcoxon signed-rank test (16.3)

R_p — Sum of the ranks of the positive differences for the Wilcoxon signed-rank test (16.3)

r_s — Sample Spearman rank-order correlation coefficient (16.6)

rf — Relative frequency (2.1)

rf_j — Relative frequency in the population for category j (15.14)

RMF_j — Row marginal frequency associated with cell j (15.4)

s — Sample standard deviation (3.2)

s^2 — Sample variance (3.2)

$\hat{s}$ (s-hat) — Standard deviation estimate (7.3)

$\hat{s}^2$ — Variance estimate (7.3)

$\hat{s}_D$ — Standard deviation estimate for sample difference scores (11.2)

$\hat{s}_{\overline{D}}$ — Estimated standard error of the mean of difference scores (11.2)

s_i^2 — Square of the sum of the scores of subject i (13.2)

$\hat{s}^2_{\text{pooled}}$ — Pooled variance estimate (10.2)

$\hat{s}_r$ — Estimated standard error of r (14.2)

$\hat{s}_{r_s}$ — Estimated standard error of r_s (16.6)

$\hat{s}_{\overline{X}}$ — Estimated standard error of the mean (7.5)

$\hat{s}_{\overline{X}_1 - \overline{X}_2}$ — Estimated standard error of the difference (10.2)

$\hat{s}_{\hat{y}}$ — Estimated standard error of $\hat{y}$ (Appendix 15.1)

s_{YX} — Sample standard error of estimate (5.5)

$\hat{s}_{YX}$ — Estimated standard error of estimate (14.8)

SCP — Sum of cross-products (5.3)

<u>SD</u> — Sample standard deviation in American Psychological Association format (3.7)

SS — Sum of squares (note: if SS is subscripted, the subscript indicates the source of variability to which the sum of squares applies) (3.2)

T — T score (transformed standard score in a distribution having a mean of 50.00 and a standard deviation of 10.00) (4.4)/Treatment effect (10.3)/T statistic in Wilcoxon signed-rank test (16.3)/Correction term in formulas for Wilcoxon rank sum test and Kruskal–Wallis test when a large number of ties occur (Appendix 16.1)

t — t statistic (8.9)/Number of values tied at a particular rank for Wilcoxon rank sum test and Kruskal–Wallis test (Appendix 16.1)

T_j^2 — Square of the sum of the scores in group j (12.2)

$T^2_{A_i}$ — Square of the sum of the scores at level i of factor A (17.4)

$T^2_{A_iB_j}$ — Square of the sum of the scores in cell A_iB_j (17.4)

$T^2_{B_j}$ — Square of the sum of the scores at level j of factor B (17.4)

U	Mann–Whitney U statistic (16.2)
V	Fourfold point correlation coefficient/Cramér's statistic (15.6)
X	General name for a variable (1.6)/A predictor variable (14.8)
$\bar{X}$	Sample mean for variable X (3.1)
$\bar{X}_i$	Mean X score for subject i across conditions (11.3)
X_n	Nullified score on variable X (10.3)
X_P	Score value defining the Pth percentile (4.1)
Y	A criterion variable (14.8)
$\hat{Y}$ (predicted Y)	Predicted score on variable Y (5.5)
$\hat{y}$	Sample estimate of association between two qualitative variables in the population (Appendix 15.1)
z	Standard score in a normal distribution (4.3)
z_{critical}	Critical value of z in Goodman's simultaneous confidence interval procedure (Appendix 15.1)
α (alpha)	Probability of a Type I error (8.6)
β (beta)	Probability of a Type II error (8.6)
χ^2 (chi-square)	Chi-square statistic (15.4)
χ_r^2	Test statistic for Friedman analysis of variance by ranks (16.5)
χ^2_{critical}	Critical value of chi-square associated with the original contingency table in Goodman's simultaneous confidence interval procedure (Appendix 15.1)
μ (mu)	Mean of a population (3.6) and of a sampling distribution of the mean (7.5)
$\mu_{\bar{D}}$	Mean difference score in a population and in a sampling distribution of the mean of difference scores (11.2)
ρ (rho)	Population Pearson correlation coefficient (14.2)
ρ_S	Population Spearman rank-order correlation coefficient (16.6)
Σ (sigma)	Summation sign (1.6)
σ (sigma)	Population standard deviation (3.6)
σ^2	Population variance (3.6)
σ_R	Standard deviation of the sampling distribution of R_j for the Wilcoxon rank sum test (16.2)
$\sigma_{r'}$	Standard error of r' (Appendix 14.1)
σ_T	Standard deviation of the sampling distribution of T for the Wilcoxon signed-rank test (16.3)
$\sigma_{\bar{X}}$	Population standard error of the mean (7.5)
$\sigma_{\bar{X}_1-\bar{X}_2}$	Population standard error of the difference (10.2)
! (factorial)	Factorial of a number (6.8)
$1-\beta$	Power of a statistical test (8.6)
. . .	Indication to include all relevant scores falling between the written values in the summation (12.2)

REFERENCES

Ainsworth, M. S., Blehar, M. C., Waters, E., & Wall, S. (1978). *Patterns of attachment: A psychological study of the stranger situation.* Hillsdale, NJ: Erlbaum.

American Psychological Association. (1983). *Publication manual of the American Psychological Association* (3rd ed.). Washington, DC: Author.

Anderson, N. H. (1968). Likeableness ratings of 555 personality-trait words. *Journal of Personality and Social Psychology, 9,* 272–279.

Anderson, N. H. (1970). Functional measurement and psychophysical judgment. *Psychological Review, 77,* 153–170.

Bandura, A., Ross, D., & Ross, S. A. (1963). Imitation of film-mediated aggressive models. *Journal of Abnormal and Social Psychology, 66,* 3–11.

Barcus, F. E. (1971). *Saturday children's television: A report of TV programming and advertising on Boston commercial television.* Newton, MA: Action for Children's Television.

Barron, F. (1965). The psychology of creativity. In T. Newcomb (Ed.), *New directions in psychology: II* (pp. 1–113). New York: Holt, Rinehart and Winston.

Becker, M. A., & Gaeddert, W. P. (1988). *Sex-role orientation and perceived sexual attractiveness.* Unpublished manuscript, The Pennsylvania State University at Harrisburg, Division of Behavioral Science and Education.

Bennett, E. L., Krech, D., & Rosenzweig, M. R. (1964). Reliability and regional specificity of cerebral effects of environmental complexity and training. *Journal of Comparative and Physiological Psychology, 57,* 440–441.

Bohrnstedt, G. W., & Carter, T. M. (1971). Robustness in regression analysis. *Sociological Methodology, 12,* 118–146.

Boneau, C. A. (1960). The effects of violations of assumptions underlying the t test. *Psychological Bulletin, 57,* 49–64.

Borden, R. (1978). *Environmental attitudes and beliefs in technology.* Unpublished manuscript, Purdue University, Department of Psychological Sciences, West Lafayette, IN.

Brehm, J. (1956). Post-decision changes in desirability of alternatives. *Journal of Abnormal and Social Psychology, 52,* 384–389.

Burger, J. M., & Smith, N. G. (1985). Desire for control and gambling behavior among problem gamblers. *Personality and Social Psychology Bulletin, 11,* 145–152.

Camilli, G., & Hopkins, K. D. (1978). Applicability of chi-square to 2 × 2 contingency tables with small expected cell frequencies. *Psychological Bulletin, 85,* 163–167.

Carrol, R. M., & Nordholm, L. A. (1975). Sampling characteristics of Kelley's η^2 and Hays' ω^2. *Educational and Psychological Measurement, 35,* 541–554.

Casler, L. (1964). The effects of hypnosis on ESP. *Journal of Parapsychology, 28,* 126–134.

Cohen, J. (1967). An alternative to Marascuilo's large sample multiple comparisons for proportions. *Psychological Bulletin, 67,* 199–201.

Cohen, J. (1977). *Statistical power analysis for the behavioral sciences* (2nd ed.). New York: Academic Press.

Conover, W. J. (1974a). Rejoinder. *Journal of the American Statistical Association, 69,* 382.

Conover, W. J. (1974b). Some reasons for not using the Yates continuity correction on 2 × 2 contingency tables. *Journal of the American Statistical Association, 69,* 374–376.

Cox, C. (1926). *Genetic studies of genius.* Stanford, CA: Stanford University Press.

Crusco, A. P., & Wetzel, C. G. (1984). The Midas touch: The effects of interpersonal touch on restaurant tipping. *Personality and Social Psychology Bulletin, 4,* 512–517.
Deutsch, F. M., & Mackesy, M. E. (1985). Friendship and the development of self-schemas: The effects of talking about others. *Personality and Social Psychology Bulletin, 11,* 399–408.
Dunn, O. J. (1964). Multiple comparisons using rank sums. *Technometrics, 6,* 241–252.
Eron, L. D. (1963). Relationship of TV viewing habits and aggressive behavior in children. *Journal of Abnormal and Social Psychology, 67,* 193–196.
Feldman-Summers, S., & Ashworth, C. D. (1980). *Factors related to intentions to report a rape.* Unpublished manuscript, University of Washington, Department of Psychology, Seattle.
Ferguson, G. A. (1976). *Statistical analysis in psychology and education.* New York: McGraw-Hill.
Festinger, L., & Carlsmith, J. M. (1959). Cognitive consequences of forced compliance. *Journal of Abnormal and Social Psychology, 58,* 203–210.
Fiedler, F. (1967). *A theory of leadership effectiveness.* New York: McGraw-Hill.
Fisher, R. (1950). *Statistical methods for research workers.* New York: Hafner.
Fleishman, A. I. (1980). Confidence intervals for correlation ratios. *Educational and Psychological Measurement, 40,* 659–670.
Friedman, H. (1972). *Introduction to statistics.* New York: Random House.
Frieze, I. H., Parsons, J. E., Johnson, P. B., Ruble, D. N., & Zellman, G. L. (1978). *Women and sex roles: A social psychological perspective.* New York: Norton.
Gabrielcik, A., & Fazio, R. H. (1984). Priming and frequency estimates: A strict test of the availability heuristic. *Personality and Social Psychology Bulletin, 10,* 85–89.
Gallup, G. (1976). *The sophisticated poll watcher's guide.* Princeton, NJ: Princeton Opinion Press.
Glass, G. V., & Hakstian, A. R. (1969). Measures of association in comparative experiments: Their development and interpretation. *American Educational Research Journal, 6,* 403–414.
Glass, G., & Stanley, J. C. (1970). *Statistical methods in education and psychology.* Englewood Cliffs, NJ: Prentice-Hall.
Goldberg, P. (1968). Are women prejudiced against women? *Transaction, 5,* 28–30.
Goodman, L. A. (1964). Simultaneous confidence intervals for contrasts among multinomial populations. *Annals of Mathematical Statistics, 35,* 716–725.
Greenberg, J., Williams, K. D., & O'Brien, M. K. (1986). Considering the harshest verdict first: Biasing effects on mock jury verdicts. *Personality and Social Psychology Bulletin, 12,* 41–50.
Greenwald, A. (1975). Consequences of prejudice against the null hypothesis. *Psychological Bulletin, 82,* 1–20.
Gulliksen, H. (1960). *Test theory.* New York: McGraw-Hill.
Haggard, E. A. (1958). *Intraclass correlation and analysis of variance.* New York: Dryden Press.
Harvath, J. (1943). Problem solving performance and music. *American Journal of Psychiatry, 22,* 211–212.
Hays, W. L. (1981). *Statistics* (3rd ed.). New York: Holt, Rinehart and Winston.
Heslin, R., & Boss, D. (1980). Nonverbal intimacy in airport arrival and departure. *Personality and Social Psychology Bulletin, 6,* 248–252.
Howell, D. C. (1985). *Fundamental statistics for the behavioral sciences.* Boston: Duxbury Press.
Hsu, T. C., & Feldt, L. S. (1969). The effect of limitations on the number of criterion score values on the significance level of the *F*-test. *American Educational Research Journal, 6,* 515–527.
Huck, S. W., & Sandler, H. M. (1979). *Rival hypotheses: Alternative interpretations of data based conclusions.* New York: Harper and Row.
Huff, D. (1954). *How to lie with statistics.* New York: Norton.

Hurlock, E. (1925). An evaluation of certain incentives used in schoolwork. *Journal of Educational Psychology, 16,* 145–159.

Jaccard, J. (1980). *Factors affecting the acceptance of male oral contraceptives.* Unpublished manuscript, State University of New York at Albany, Department of Psychology.

Jaccard, J., & Becker, M. A. (1988). *Selecting a statistical test for data analysis: The robustness of commonly-used statistical tests.* Unpublished manuscript, State University of New York at Albany, Department of Psychology.

Jaccard, J., Becker, M. A., & Wood, G. (1984). Pairwise multiple comparison procedures: A review. *Psychological Bulletin, 96,* 589–596.

Jensen, A. R. (1973). *Educability and group differences.* New York: Basic Books.

Johnson, M. K., & Liebert, R. M. (1977). *Statistics.* Englewood Cliffs, NJ: Prentice-Hall.

Johnson, T. (1976). *Luck versus ability: A replication.* Unpublished manuscript, Purdue University, Department of Psychological Sciences, West Lafayette, IN.

Kelman, H. C., & Hovland, C. I. (1953). Reinstatement of the communicator in delayed measurement of opinion change. *Journal of Abnormal and Social Psychology, 48,* 327–335.

Kennedy, J. J. (1970). The eta coefficient in complex ANOVA designs. *Educational and Psychological Measurement, 30,* 885–889.

Keppel, G. (1983). *Design and analysis: A researcher's handbook* (2nd ed.). Englewood Cliffs, NJ: Prentice-Hall.

Kerlinger, F. N. (1973). *Foundations of behavioral research.* New York: Holt, Rinehart and Winston.

Kesselman, H. J. (1975). A Monte Carlo investigation of three estimates of treatment magnitude: Epsilon squared, eta squared, and omega squared. *Canadian Psychological Review, 16,* 44–48.

Kirk, R. E. (1968). *Experimental design: Procedures for the behavioral sciences.* Pacific Grove, CA: Brooks/Cole.

Kirk, R. E. (1972). *Statistical issues: A reader for the behavioral sciences.* Pacific Grove, CA: Brooks/Cole.

Kirk, R. E. (1978). *Introductory statistics.* Monterey, CA: Brooks/Cole.

Kruskal, W. H., & Wallis, W. A. (1952). Use of ranks in one criterion variance analysis. *Journal of the American Statistical Association, 47,* 583–621.

Larkin, J. E. (1987). Are good teachers perceived as high self-monitors? *Personality and Social Psychology Bulletin, 13,* 64–72.

Lord, F. M. (1953). On the statistical treatment of football numbers. *American Psychologist, 8,* 750–751.

Lunney, G. H. (1970). Using analysis of variance with a dichotomous dependent variable: An empirical study. *Journal of Educational Measurement, 7,* 263–269.

Mantel, N. (1974). Comment and suggestion. *Journal of the American Statistical Association, 69,* 378–380.

Marascuilo, L. A., & McSweeney, M. (1977). *Nonparametric and distribution-free methods for the social sciences.* Pacific Grove, CA: Brooks/Cole.

McArthur, L. Z., & Eisen, S. V. (1976). Television and sex role stereotyping. *Journal of Applied Social Psychology, 6,* 329–351.

McCall, R. B. (1980). *Fundamental statistics for psychology* (3rd ed.). New York: Harcourt Brace Jovanovich.

McConnell, J. V. (1966). New evidence for the transfer of training effect in planaria. In *The biological basis of memory traces.* Symposium conducted at the International Congress of Psychology, Moscow.

McNemar, Q. (1962). *Psychological statistics.* New York: Wiley.

Miller, R. G. (1966). *Simultaneous statistical inference.* New York: McGraw-Hill.

Minium, E. (1970). *Statistical reasoning in psychology and education.* New York: Wiley.

Morrow, F., & Davidson, D. (1976). *Race and family size decisions.* Unpublished manuscript, Purdue University, Department of Psychological Sciences, West Lafayette, IN.

Nezlek, J. (1978) *Social behavior and diaries.* Unpublished manuscript, College of William and Mary, Department of Psychology, Williamsburg, VA.

Parrish, M., Lundy, R. M., & Leibowitz, H. W. (1969). Effect of hypnotic age regression on the magnitude of the Ponzo and Poggendorff illusions. *Journal of Abnormal Psychology, 74,* 693–698.

Pearson, E. S., & Please, N. W. (1975). Relations between the shape of the population distribution of four simple test statistics. *Biometrika, 62,* 223–241.

Pedhazur, E. J. (1982). *Multiple regression in behavioral research* (2nd ed.). New York: Holt, Rinehart and Winston.

Petty, R. E., Wells, G. L., Heesacker, M., Brock, T. C., & Cacioppo, J. T. (1983). The effects of recipient posture on persuasion: A cognitive response analysis. *Personality and Social Psychology Bulletin, 9,* 209–222.

Robbins, R. A. (1988). *Objective and subjective factors in estimating life expectancy.* Unpublished manuscript, The Pennsylvania State University at Harrisburg, Division of Behavioral Science and Education.

Rubovits, P. C., & Maehr, M. L. (1973). Pygmalion black and white. *Journal of Personality and Social Psychology, 25,* 210–218.

Ryan, T. A. (1959). Multiple comparisons in psychological research. *Psychological Bulletin, 56,* 26–47.

Sears, D. O. (1969). Political behavior. In G. Lindzey and E. Aronson (Eds.), *The handbook of social psychology* (pp. 315–458). Reading, MA: Addison-Wesley.

Simpson, J. A., Campbell, B., & Berscheid, E. (1986). The association between romantic love and marriage: Kephart (1967) twice revisited. *Personality and Social Psychology Bulletin, 12,* 363–372.

Smith, W. L., Phillipus, M. J., & Guard, H. L. (1968). Psychometric study of children with learning problems and 14-6 positive spike EEG patterns, treated with ethosuximide (zarontin) and placebo. *Archives of Disease in Childhood, 43,* 616–619.

Sroufe, L. A., & Waters, E. (1977). Attachment as an organizational construct. *Child Development, 48,* 1184–1199.

Steel, R. G. (1960). A rank sum test for comparing all pairs of treatments. *Technometrics, 2,* 197–207.

Steiner, I. D. (1972). *Group process and productivity.* New York: Academic Press.

Stephen, G. (1975). *Psychology and law.* New York: McGraw-Hill.

Stevens, S. S. (1951). Mathematics, measurement, and psychophysics. In S. S. Stevens (Ed.), *Handbook of experimental psychology* (pp. 28–42). New York: Wiley.

Thorndike, R. (1942). Regression fallacies in the matched groups experiment. *Psychometrika, 7,* 85–102.

Touhey, J. C. (1974). Effects of additional women professionals on ratings of occupational prestige and desirability. *Journal of Personality and Social Psychology, 29,* 86–89.

Tukey, J. W. (1953). *The problem of multiple comparisons.* Unpublished manuscript, Princeton University, Department of Statistics, Princeton, NJ.

Warkov, S., & Greeley, A. (1966). Parochial school origins and education achievement. *American Sociological Review, 31,* 406–414.

Wechsler, D. (1958). *The measurement and appraisal of adult intelligence.* Baltimore: Williams and Wilkins.

Weil, A. T., Zinberg, N. E., & Nelson, J. (1968). Clinical and psychological effects of marijuana in man. *Science, 162,* 1234–1242.

Wike, E. L. (1971). *Data analysis: A statistical primer for psychology students.* Chicago: Aldine.

Wilcoxon, F. (1949). *Some rapid approximate statistical procedures.* New York: American Cyanamid.

Willingham, W. W. (1974). Predicting success in graduate education. *Science, 183,* 275–278.

Winer, B. J. (1971). *Statistical principles in experimental design.* New York: McGraw-Hill.

Witte, R. S. (1980). *Statistics.* New York: Holt, Rinehart and Winston.

Wyer, R., & Goldberg, L. (1970). A probabilistic analysis of the relationship between beliefs and attitudes. *Psychological Review, 77,* 100–120.

Zeisel, H., & Kalven, H., Jr. (1978). Parking tickets and missing women: Statistics and the law. In J. M. Tanur, F. Mosteller, W. H. Kruskal, R. F. Link, R. S. Pieters, G. R. Rising, & E. L. Lehmann (Eds.), *Statistics: A guide to the unknown* (2nd ed.) (pp. 139–149). San Francisco: Holden-Day.

ANSWERS TO SELECTED EXERCISES

Chapter One

2. a. A constant, because there are always 24 hours in a day
   b. A variable, because different people have different attitudes toward abortion
   c. A constant, because all presidents of the United States must be born in the United States
   d. A constant, because a number divided by itself always equals 1.00
   e. A variable, because different total numbers of points are scored in different football games
   f. A variable, because different months have different numbers of days

3. a. quantitative
   b. qualitative
   c. quantitative
   d. quantitative
   e. quantitative
   f. qualitative

4. The independent variable is the preference for aggressive television shows. The dependent variable is the peer ratings of aggression. Both are quantitative in nature.

5. The independent variable is the type of occupation, which is qualitative in nature. The dependent variable is the perceived prestige, which is quantitative in nature.

10. a. discrete
    b. continuous
    c. discrete
    d. continuous

11. a. 21,384.105 and 21,384.115
    b. .6885 and .6895
    c. 12.5 and 13.5
    d. 12.95 and 13.05
    e. 12.995 and 13.005

13. a. ΣX
    b. ΣX^2
    c. $\Sigma(X - 5)$
    d. $\Sigma X^2/N$
    e. $(\Sigma Y)^2$
    f. $\sum_{i=1}^{3} Y_i$
    g. $\Sigma(X - 1)(Y - 6)$
    h. ΣXY

14. a. 36
    b. 36
    c. $\Sigma Xk = k\Sigma X$
    d. 9
    e. 9
    f. $\Sigma(X/k) = (\Sigma X)/k$

15. a. 4.893
    b. UF888.975
    c. 1.415
    d. 4.145
    e. 6.245
    f. 2.616
    g. 6.316
    h. .396
    i. 1.000
    j. 3.667
    k. 12.254
    l. 9.724
    m. 1.995
    n. 2.005

17. Calculations for the original scores:
    a. 19.96
    b. 3.99
    c. 398.34
    d. 80.06

    Calculations for the rounded scores:
    a. 19.98
    b. 4.00
    c. 399.20
    d. 80.24

    The difference between the two sets of results is due to the fact that the original scores were to four decimal places and the rounded scores were to two decimal places. The answers based on the original scores are thus more precise.

20. Your best guess as to the probability of success of the operation is .05. Your decision should depend on the danger associated with going through with versus not going through with the procedure. The information about the probability of success, in and of itself, is not sufficient to make a decision.

22. The newspaper's sample is probably not a representative sample of the community in general because it represents only those people who read the newspaper and who were willing to take the time and trouble to send in a ballot. This casts doubt on the validity of the newspaper's conclusion about the election.

25. You probably found that only a very small number of individuals were selected to participate in both samples. This indicates that every member of a population has an equal chance of being selected when random sampling is used and, thus, that random sampling will tend to yield representative samples.

Chapter Two

1–2.

Score	*f*	*rf*	*cf*	*crf*
8	2	.100	20	1.000
7	4	.200	18	.900
6	10	.500	14	.700
5	0	.000	4	.200
4	2	.100	4	.200
3	2	.100	2	.100

4. .300, .100, .900, .800

6. .100, .600, .900

7.

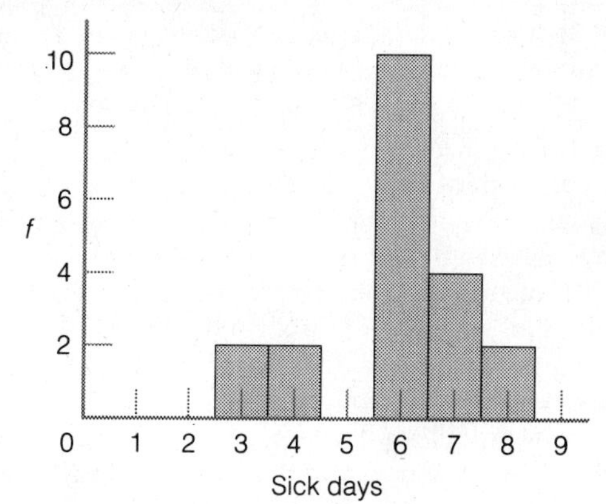

9.

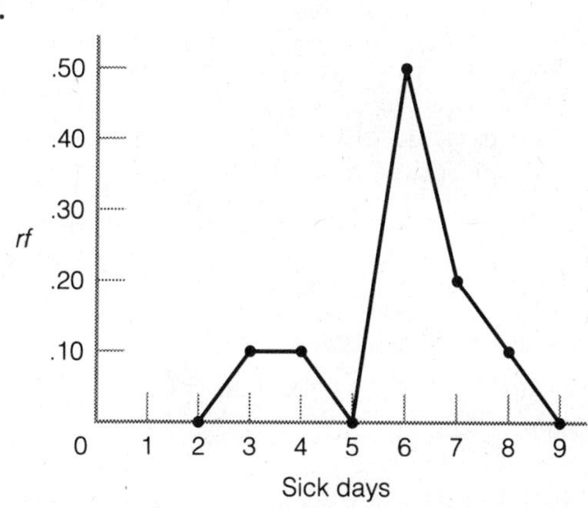

The shape of this graph is identical to the shape of the frequency polygon from Exercise 8.

11. In an ungrouped frequency distribution, each different score value is listed. In a grouped frequency distribution, scores are grouped together into intervals.

13.

Score	*f*
120–129	5
110–119	10
100–109	20
90–99	10
80–89	5

15. .70, .30

17.

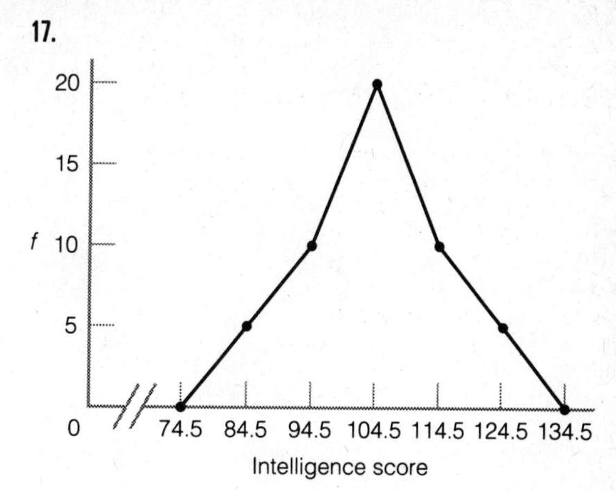

19.

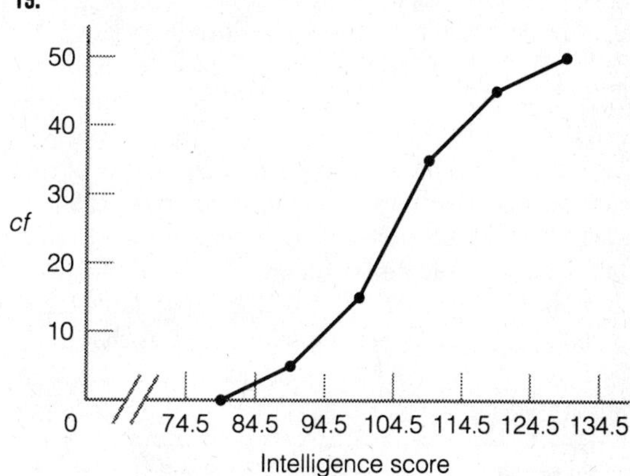

The similarity in shapes is due to the fact that the cumulative frequency line will always remain level or increase as it moves from left to right.

20. 5, 80

27. a.

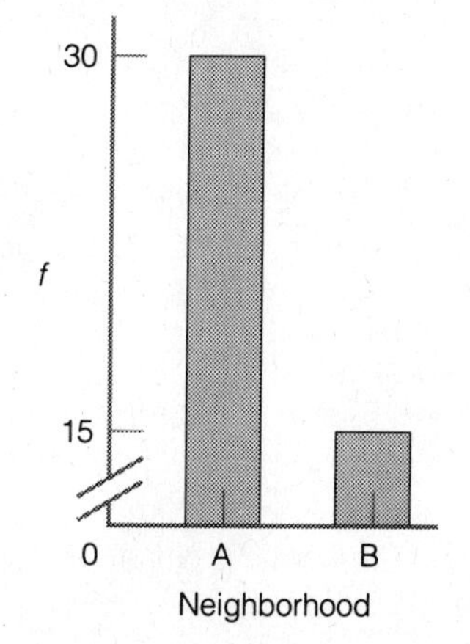

b.

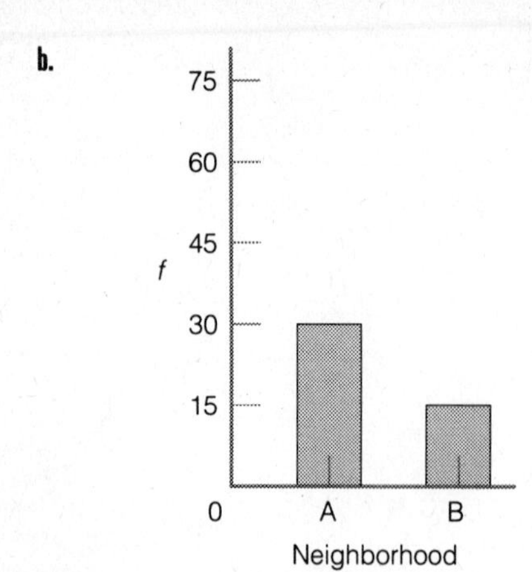

28. A probability distribution is a distribution representing the probabilities associated with all possible score values for a variable. The nature of probability distributions is different for qualitative and discrete variables versus continuous variables because in the former case it is possible to list possible values of the variable and their corresponding probabilities. Since the number of possible values a continuous variable can have is, in principle, infinite, this is not possible for continuous variables. Instead, probabilities for a continuous variable are conceptualized as the areas under corresponding intervals of the density curve.

30.

Score	*f*	*rf*	*cf*	*crf*
5	370	.200	1,850	1.000
4	555	.300	1,480	.800
3	185	.100	925	.500
2	407	.220	740	.400
1	333	.180	333	.180

Chapter Three

3. 4.33

6. Mode = 0, median = 0.00, mean = 0.00

8. The mean of the first set is 12.00; the mean of the second set is 15.00; the mean of the third set is 2.00. If a constant, k, is added to each score (X) in a set, then the mean of the new set of scores will equal $\bar{X} + k$. If k is subtracted from each score in a set, then the mean of the new set of scores will equal $\bar{X} - k$.

10. The mean of the first set is 30.00; the mean of the second set is 90.00; the mean of the third set is 3.00. If each score (X) in a set is multiplied by a constant, k, then the mean of the new set of scores will equal $k\bar{X}$. If each score in a set is divided by k, then the mean of the new set of scores will equal $\bar{X}/k$.

11. The median of the first set for the data in Exercise 8 is 12; the median of the second set is 15; the median of the third set is 2. Adding a constant to or subtracting a constant from each score in a set has the same effect on the median as on the mean.

The median of the first set for the data in Exercise 10 is 30; the median of the second set is 90; the median of the third set is 3.00. Multiplying or dividing each score in a set by a constant has the same effect on the median as on the mean.

The above manipulations have the same effects on the mode as on the mean and the median.

13. The mean is a poorer descriptor of central tendency for Set I. This is because the extreme score of 300 in Set I will substantially adjust (increase) the mean and thus lead to a distorted picture of the central tendency of the data.

15. The range is a misleading index of variability when there is an extreme score in a set of scores that are otherwise similar to one another.

17. 10.00

19. All three measures of variability must equal 0 because there is no variability since all of the scores are the same.

20. The standard deviation is more "interpretable" than the variance because it represents an average deviation from the mean *in the original unit of measurement.* In contrast, the variance is in terms of squared deviation units.

22. The sum of squares equals 104.00 using both approaches. Typically, it is more efficient to apply the computational formula than the defining formula because the computational formula requires fewer steps.

23. Range = 10, SS = 144.95, s^2 = 6.90, s = 2.63

25. The variance for each set is 2.00, and the standard deviation for each set is 1.41. Adding a constant to or subtracting a constant from each score in a set does not affect the variance or the standard deviation.

27. For the first set, the variance is 2.00 and the standard deviation is 1.41; for the second set, the variance is 18.00 and the standard deviation is 4.24; for the third set, the variance is .50 and the standard deviation is .71. If each score in a set is multiplied by a constant, k, then the variance of the new set of scores will equal the variance of the old set of scores multiplied by k^2, and the standard deviation will equal the standard deviation of the old set of scores multiplied by k. If each score in a set is divided by k, then the variance of the new set of scores will equal the variance of the old set of scores divided by k^2, and the standard deviation will equal the standard deviation of the old set of scores divided by k.

28.

Set I	*Set II*
0	9
5	10
10	10
15	10
20	11

While the mean for both sets of scores is 10.00, the standard deviation of 7.07 for Set I is substantially greater than the standard deviation of .63 for Set II.

30. The mean is 28.67, and the standard deviation is 7.74. The speed estimates considering the mean score for all subjects were quite accurate (28.67, as compared with the actual speed of 30). However, the relatively large standard deviation indicates that the *individual* estimates were not all that accurate.

32. You should conclude that the consultant should be fired—standard deviations cannot be negative.

33. The first set of scores is positively skewed, as the mean is larger than the median, which is larger than the mode. The second set of scores is negatively skewed, as the mean is smaller than the median, which is smaller than the mode. The third set of scores is not skewed, as the three measures of central tendency all take on the same value.

Chapter Four

2. a. 2.99
b. 3.75
c. 4.25
d. 5.52
e. 7.67
f. 4.00

3. a. 94.40
b. 74.90
c. 20.20
d. 5.20

6. a. .62
b. 1.00
c. −.36
d. .03
e. −1.46
f. 1.43

8. −1.23

10. A positive standard score indicates that the original score is greater than the mean. A negative standard score indicates that the original score is less than the mean.

12. John's performance on the English exam was 1.00 standard deviation above the mean, and his performance on the math exam was 6.67 standard deviations above the mean. Hence, John's performance was better (relative to his classmates') on the math exam.

14. One situation in which it would be useful to convert scores from two different distributions to standard scores before comparing them is when determining the relative ability on a given task of people of different ages. For instance, the bowling averages of a child and an adult can be compared in this manner.

18. a. .9913
b. .1210
c. .1210
d. .4798
e. .4798
f. .3550
g. .3550
h. .7850

19. a. .1587
b. .1587
c. .7865
d. .9836
e. .5000
f. .7257

21. The standard score corresponding to a galvanic skin response score of 61.40 is 4.00. Since only .003% of standard scores in a normal distribution are 4.00 or greater, the individual displayed an extreme galvanic skin response when asked the critical question. The implication is that he is lying.

23. a. 128.60
b. 75.60
c. 100.00
d. 85.00
e. 115.90
f. 107.50

24. a. 102.20
b. 108.22
c. 100.00
d. 109.80
e. 90.20

26. a. 58.7
b. 70
c. 65.6
d. 34.4
e. 50
f. 90.4

Chapter Five

2. The slope of a line indicates the number of units variable Y changes as variable X changes by 1 unit.

5. 3.00 units, 6.00 units, 21.00 units

6. The magnitude of a correlation coefficient indicates the degree to which two variables approximate a linear relationship.

8. **a.** +.37 **d.** +.26
b. −.37 **e.** −.44
c. −.76 **f.** +.61

9.

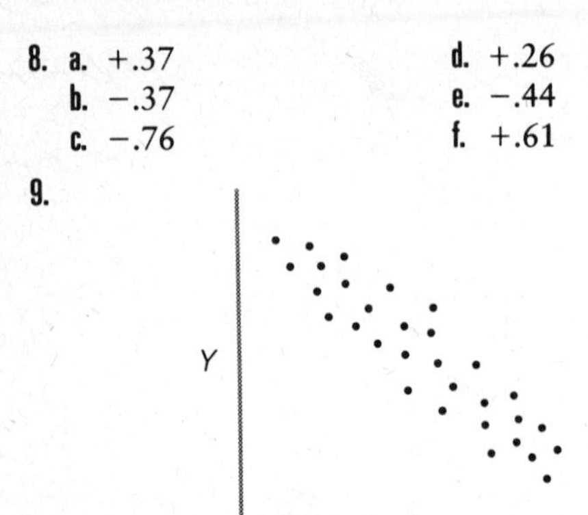

11. An example of two variables that are probably positively correlated is calorie consumption and weight gain.

13. .18

16. An example of two variables that are probably (positively) correlated with one another but that are not causally related is the number of automobiles in a household and the number of television sets in a household. The correlation between these two variables is probably attributable to such factors as the number of occupants, the ages of the occupants, and the income level of the occupants.

17. The general form of the regression equation is $\hat{Y} = a + bX$. This differs from the linear model in that the values yielded by $a + bX$ are *predicted* Y scores rather than actual Y scores.

18. The least squares criterion is a statistical criterion that defines the values of the slope and intercept of the regression line such that the sum of the squared discrepancies between individuals' actual Y scores and their predicted Y scores based on the regression equation is minimized.

19. $\hat{Y} = 5.00 + .20X$

20. 5.60, 6.20, 5.40

24. 2.41

Chapter Six

2. 180

4. 350

5. .514, .486, .500, .500

6. .343, .686, .657, .314

7. Being a male (event A) and favoring the ERA (event B) are not independent because the probability of being a male [$p(A) = .50$] is not the same as the conditional probability of being a male given that one favors the ERA [$p(A/B) = .333$].

8. .171, .157

9. Four joint probabilities are represented: (1) being a male who favors the ERA, (2) being a male who opposes the ERA, (3) being a female who favors the ERA, and (4) being a female who opposes the ERA.

10. .843, .829

17. .486

20. .516

23. **a.** 60 **d.** 120
b. 720 **e.** 24
c. 12

24. 5! equals 120, which is the same answer as for part **d** of Exercise 23. 4! equals 24, which is the same answer as for part **e** of Exercise 23. From this, we can generalize that $_nP_n = n!$.

26. **a.** 10 **d.** 1
b. 6 **e.** 1
c. 6

28. 216

30. **a.** .010 **d.** .010
b. .044 **e.** .044
c. .172

33. The correspondence between the binomial and normal distributions is influenced by the number of trials (n) and the probability of success (p). The correspondence improves as n increases and as p becomes closer to .50.

34. $\mu = 90.00$, $\sigma = 6.00$

36. A score of 30 translates into a z score of 2.50. From Appendix B, the probability of obtaining a z score of 2.50 or greater is .0062. Since .0062 is less than the criterion value of .05, the researcher should conclude that the psychotherapeutic approach is more effective than no treatment in helping clients return to normal personality.

Chapter Seven

2. Sampling error refers to the fact that values of sample statistics are likely to differ from values of their corresponding population parameters by virtue of the fact that they *are* based on only a portion of the overall population.

6. The term "degrees of freedom" refers to the number of pieces of statistical information that are independent of one another. There are $N - 1$ degrees of freedom associated with a sum of squares because given all but one deviation score for a distribution of scores, the last deviation score is determined by the other $N - 1$ deviation scores.

8. A sampling distribution of the mean is a theoretical distribution consisting of the mean scores for all possible samples of a given size that could be drawn from a population. A frequency distribution, in contrast, is concerned with the frequency with which score values occur within a set of scores.

9. The central limit theorem is a formulation that defines the mean, standard deviation, and shape of a sampling distribution of the mean.

10. The mean of a sampling distribution of the mean will always be equal to the population mean (μ). This is because when we average the means for all samples of a given size that could be drawn from a population, underestimations and overestimations of the true population mean will cancel one another, with the result being the true population mean.

12. A standard deviation of a set of raw scores represents an average deviation from the mean of the distribution. A standard error of the mean is a standard deviation of a sampling distribution of the mean and represents an average deviation of the sample means from the population mean.

13. A standard error of the mean of 0 implies that all sample means in the sampling distribution of the mean are equal to the population mean.

14. A standard error of the mean of 0 implies that there is no variability in the scores in the population, that is, that all of the scores are the same and σ therefore equals 0.

16. $\bar{X} = 5.00$, $\hat{s}_{\bar{X}} = .41$

19. The sample mean for a random sample of size 30 drawn from population A is probably a better estimator of its population mean than is the sample mean for a random sample of size 30 drawn from population B because, as indexed by its smaller standard deviation, there is less variability in population A than in population B. Consequently, the standard error of the mean for population A (.91) is smaller than the standard error of the mean for population B (1.28), thus indicating that, on the average, means for samples of a given size drawn from population A will be closer to the true population mean than will means for samples of the same size drawn from population B.

21.

Sample	*Sample mean*
2, 2	2.00
2, 4	3.00
2, 6	4.00
4, 2	3.00
4, 4	4.00
4, 6	5.00
6, 2	4.00
6, 4	5.00
6, 6	6.00
sum =	36.00

$$\text{mean} = \frac{36.00}{9} = 4.00$$

The mean across the sample means is equal to the population mean, since $\mu = (2 + 4 + 6)/3 = 4.00$. This illustrates that, as stated in the central limit theorem, the mean of a sampling distribution of the mean is always equal to the population mean.

24. The mean is usually preferred to the median and the mode as a measure of central tendency by statisticians because the sampling distribution of the mean will show less variability (that is, it will have a smaller standard error) than either the sampling distribution of the median or the sampling distribution of the mode.

Chapter Eight

2. It is necessary to assume the null hypothesis is true in the context of hypothesis testing because this allows us to determine the size of the discrepancy between an expected result and an observed result and, thus, to make inferences about population values from sample values.

4. We can never "accept" the null hypothesis in traditional statistical tests because, due to sampling error, we can never unambiguously conclude that the true population mean is equal to any one specific value based upon sample data.

5. H_0: μ = \$100; H_1: $\mu \neq$ \$100

6. z = 1.79

7. Since the observed z of 1.79 is neither less than the negative critical value of −1.96 nor greater than the positive critical value of +1.96, we fail to reject the null hypothesis. It is possible that the actual value of μ is 4.

12. The probability of a Type I error is equal to alpha because if the null hypothesis is true, the null hypothesis will be incorrectly rejected any time the observed value of the test statistic falls in the rejection region, and the probability of this occurring is equal to alpha.

13. Power is equal to 1 minus the probability of a Type II error because if the probability of making a Type II error by failing to reject a false null hypothesis is equal to β, the probability of making a correct decision by rejecting the null hypothesis under this circumstance (power) must be equal to $1 - \beta$.

14. Power decreases as alpha is set at a lower (more conservative) level because the lower the alpha level, the lower is the probability of rejecting the null hypothesis. Thus, when the null hypothesis is false, the probability of detecting an existing difference between the hypothesized and the actual population means (power) decreases as does alpha.

19. a. +1.729
 b. −1.729
 c. ±2.093
 d. +2.015
 e. −1.833
 f. ±2.262

20. a. The critical values of t for an alpha level of .05 and 9,999 degrees of freedom are ±1.960. Since the observed t value of 10.00 is greater than +1.960, we reject the null hypothesis.
 b. The critical values of t for an alpha level of .05 and 99 degrees of freedom are approximately ±1.993. Since the observed t value of 1.00 is neither less than −1.993 nor greater than +1.993, we fail to reject the null hypothesis.
 c. The observed value of t, $t(99) = 5.00$, is greater than the positive critical value of +1.993, so we reject the null hypothesis.

The null hypothesis is rejected in part **a** but not in part **b** because the t test in part **a** is based on a larger number of cases. The null hypothesis is not rejected in part **b** but is rejected in part **c** because the t test in part **b** is based on a larger standard deviation estimate.

24. The 95% confidence interval is 72.44 to 76.36. The 99% confidence interval is 71.82 to 76.98.

25. The 95% confidence interval is 72.60 to 76.20. The 99% confidence interval is 72.03 to 76.77. The effect of increasing N is to decrease the width of the intervals.

27. The 95% confidence interval is 116.98 to 125.02. The 99% confidence interval is 115.58 to 126.42.

28. A probability level or significance level is the probability of observing a statistical result as extreme as the one that was actually obtained. The alpha level, in contrast, is set by the researcher to reflect his or her decision as to how extreme the results of a statistical test should be before the null hypothesis is rejected.

29. Results

The mean number of children in the sample (2.96) was compared against a hypothesized fertility rate of 2.11 using a one-sample $\underline{t}$ test. This was found to be statistically significant, $\underline{t}(24) = 2.34$, $\underline{p} < .05$, indicating that Catholics are reproducing at an above-zero population growth rate.

Chapter Nine

2. A control group is a group in an experiment that is not exposed to the independent variable. The advantage of including a control group in the research design is that it provides a baseline for evaluating the effects of the experimental manipulation.

3. Observational research strategy; no control group

4. Observational research strategy; no control group

8. One limitation on the use of random assignment is that it is not applicable when an observational research strategy is employed. A second limitation of

random assignment is that it does not *guarantee* that the research groups will not differ beforehand on the dependent variable.

9. Sampling error can be reduced by increasing the sample sizes for the various groups, defining the populations so that the variances of scores in the groups will be relatively small, and holding potential disturbance variables constant.

12. There is not a causal relationship between individuals' height and the length of their hair. An additional variable that might account for the observed relationship is gender—males tend to be both taller than females and to have shorter hair.

13. One advantage of a within-subjects research design is that it is more economical in terms of subjects. A second advantage of a within-subjects design is that it offers better control of confounding variables related to individual differences. A disadvantage of a within-subjects design is that carry-over effects can occur.

14. The independent variable in the studies described in Exercises 3 (race), 4 (gender), and 6 (noise level) are all between-subjects in nature. The independent variable in the study described in Exercise 5 (observer status) is within-subjects in nature.

18. The major problem with existing tests of distributional assumptions is that they are insensitive to the severity of distributional violations.

19. The three general factors that influence the robustness of a statistical test are sample size, the degree of violation of distributional assumptions, and the form of violation of distributional assumptions.

Chapter Ten

2. The mean of a sampling distribution of the difference between two independent means will always be equal to the difference between the relevant population means.

3. The independent groups t test assumes that the two population variances are homogeneous. Thus, our goal is to estimate σ^2, the variance of both populations. By pooling the variance estimates from the two samples, we increase the degrees of freedom on which the estimate of σ^2 is based and thereby obtain a better estimate.

4. $\hat{s}^2_{\text{pooled}} = 5.48$, $\hat{s}_{\bar{X}_1 - \bar{X}_2} = .98$

5. **a.** .21

b. .23

c. .31

d. The estimated standard error of the difference between two independent means will always be larger than the respective estimated standard errors of the mean because there is more variability in a sampling distribution of the difference between two independent means than in the corresponding sampling distributions of the mean. For instance, if the smallest sample mean in each of two sampling distributions of the mean were 2.00 and the largest sample mean were 7.00, the range in each case would be 7.00 − 2.00 = 5.00. However, the smallest mean difference in the sampling distribution of the difference between two independent means based on the two distributions of sample means would be 2.00 − 7.00 = −5.00, and the largest value would be 7.00 − 2.00 = 5.00, a range of 5.00 − (−5.00) = 10.00. Since the estimated standard error of the difference and the estimated standard errors of the mean also reflect variability within the corresponding sampling distributions, it follows that, as with the ranges, the former measure will always be larger than the latter measures.

8. **a.** +1.734

b. ±2.101

c. ±2.048

d. −1.701

e. ±2.048

10. The critical values of t for an alpha level of .05 and 8 degrees of freedom are ±2.306. Since the observed t value of 6.71 is greater than +2.306, we reject the null hypothesis and conclude that there is a relationship between gender and discriminatory attitudes.

11. $SS_{\text{TOTAL}} = 26.50$

12. $T_M = 1.50$, $T_F = -1.50$

13.

Gender	X_n	*Gender*	X_n
Male	6.50	Female	6.50
Male	5.50	Female	5.50
Male	5.50	Female	5.50
Male	5.50	Female	5.50
Male	4.50	Female	4.50

14. $SS_{ERROR} = 4.00$, $SS_{EXPLAINED} = 22.50$

15. The value of eta-squared is .85 using both Equation 10.10 and Equation 10.11. This represents a very strong effect.

16. The nature of the relationship between gender and discriminatory attitudes toward women is that males are significantly more discriminatory than females.

17. The sum of squares total is equal to the sum of squares explained plus the sum of squares error. In other words, the total variability in the dependent variable, as represented by SS_{TOTAL}, can be partitioned into two components, one ($SS_{EXPLAINED}$) reflecting the influence of the independent variable and one (SS_{ERROR}) reflecting the influence of disturbance variables.

19. It is inappropriate to estimate the strength of the relationship between two variables in the population from the sample value of eta-squared because eta-squared is a biased estimator of the strength of the relationship in the population. Specifically, eta-squared tends to slightly overestimate the strength of the relationship in the population across random samples.

29. $n = 87$ per group

31. Results

An independent groups t test indicated that the cortex weight of rats will be greater when they are raised in an en- riched environment (M = 660.00 mg) than when they are raised in an isolated environment (M = 626.00 mg), t(12) = 2.69, p < .02. As indexed by eta², the strength of the relationship between the type of environment and cortex weight was .38.

Chapter Eleven

3. **a.** +1.833 **b.** ±2.262 **c.** −1.729 **d.** ±2.093

6. A correlated groups t test is generally more powerful than a corresponding independent groups t test because variability due to individual differences is extracted from the dependent variable as part of the correlated groups t test procedure. Because the number of degrees of freedom for the correlated groups test ($N - 1$) is smaller than the number of degrees of freedom for the independent groups test ($n_1 + n_2 - 2$), a correlated groups t test will not be more powerful than a corresponding independent groups t test when individual differences have only a minimal influence on the dependent variable. This reflects the fact that the t distribution requires more extreme values of t to reject the null hypothesis as the degrees of freedom become smaller.

7. The critical values of t for an alpha level of .05 and 4 degrees of freedom are ±2.776. Since the observed t value of −3.77 is less than −2.776, we reject the null hypothesis and conclude that there is a relationship between experimental condition and learning scores.

8. The value of eta-squared is .78. This represents a very strong effect.

9. The nature of the relationship between the experimental condition one is exposed to and learning scores is that learning scores are significantly higher in condition B than in condition A.

10. Analyzing the data as if the independent variable were between-subjects in nature, the observed value of t, $t(8) = -1.18$, is neither less than the negative critical value of −2.306 nor greater than the positive critical of +2.306, so we fail to reject the null hypothesis. The value of eta-squared in this instance is .15. This represents a moderate effect. The fact that eta-squared was .78 in the correlated groups case but only .15 in the independent groups case indicates that individual differences had a sizable effect on the dependent variable. The fact that the null hypothesis was rejected when a correlated groups t test was applied but not when an independent groups t test was applied illustrates the increased power of the statis-

Table for Exercise 11

$Subject_i$	*X for time 1*	*X for time 2*	$\bar{X}_i$	*Nullified X for time 1*	*Nullified X for time 2*
1	10	12	11.00	12.00	14.00
2	11	13	12.00	12.00	14.00
3	12	14	13.00	12.00	14.00
4	13	17	15.00	11.00	15.00
5	14	14	14.00	13.00	13.00
mean =	12.00	14.00	13.00	12.00	14.00

tical analysis when variability due to individual differences is extracted from the dependent variable.

11. See the table at the top of the next page. The respective mean values are the same because we have extracted the effects of individual differences in their role as disturbance variables. Since disturbance variables are unrelated to the independent variable, the means will not be affected.

15. The observed t using the procedures from Appendix 11.1 is -3.16. This is the same value as was obtained in Exercise 12.

16. Independent groups t test

17. Correlated groups t test

22. .60

23.

```
                 Results
      The mean desirability
 ratings for the unchosen
 alternative before and after
 making the choice between the
 two products were compared
 using a correlated groups t
 test. This showed that the al-
 ternative in question was rated
 as significantly more desirable
 on the first occasion (M = 6.00)
 than on the second (M = 4.00),
 t(9) = 3.00, p < .02. The
 strength of the relationship
 between the time of assessment
 and product desirability was
 .50, as indexed by eta².
```

Chapter Twelve

2. The alternative hypothesis for one-way analysis of variance states that the population means in question are not all equal. It cannot be summarized in a single mathematical statement because there are a number of ways in which three or more population means can pattern themselves so that they are not all equal to one another.

3. Between-group variability concerns the differences between the mean scores in the various groups under study. Within-group variability concerns the variability of scores *within* each of the groups.

6. The F ratio, over the long run, will approach 1.00 when the null hypothesis is true. The F ratio, over the long run, will be greater than 1.00 when the null hypothesis is not true.

8. The value of the sum of squares within must be 0 because all scores within a given group are the same.

9. A mean square in analysis of variance represents a sum of squares divided by its corresponding degrees of freedom—that is, it represents a measure of variance.

12. **a.** 3.55 **b.** 3.24 **c.** 3.35 **d.** 2.50

13.

Source	SS	df	MS	F
Between	30.00	2	15.00	5.62
Within	152.00	57	2.67	
Total	182.00	59		

16.

Null hypothesis tested	Absolute difference between sample means	Value of CD	Null hypothesis rejected?
$\mu_S = \mu_M$	$\lvert 6.00 - 8.00 \rvert = 2.00$	2.48	No
$\mu_S = \mu_D$	$\lvert 6.00 - 10.00 \rvert = 4.00$	2.48	Yes
$\mu_M = \mu_D$	$\lvert 8.00 - 10.00 \rvert = 2.00$	2.48	No

The nature of the relationship between marital status and attitudes toward divorce is that divorced individuals ($\bar{X}_D = 10.00$) have more positive attitudes than single individuals ($\bar{X}_S = 6.00$). However, we cannot confidently conclude that either divorced or single individuals differ in their divorce attitudes from married individuals ($\bar{X}_M = 8.00$).

17.

Source	SS	df	MS	F
Between	40.00	2	20.00	10.00
Within	24.00	12	2.00	
Total	64.00	14		

The critical value of F for an alpha level of .05 and 2 and 12 degrees of freedom is 3.88. Since the observed F value of 10.00 is greater than 3.88, we reject the null hypothesis and conclude that a relationship exists between the type of car and repair records.

18. The value of eta-squared is .62 using both Equation 12.5 and Equation 12.6. This represents a strong effect.

19.

Null hypothesis tested	Absolute difference between sample means	Value of CD	Null hypothesis rejected?
$\mu_X = \mu_Y$	$\lvert 2.00 - 4.00 \rvert = 2.00$	2.38	No
$\mu_X = \mu_Z$	$\lvert 2.00 - 6.00 \rvert = 4.00$	2.38	Yes
$\mu_Y = \mu_Z$	$\lvert 4.00 - 6.00 \rvert = 2.00$	2.38	No

The nature of the relationship between the type of car and repair records is that car X ($\bar{X}_X = 2.00$) performs better than car Z ($\bar{X}_Z = 6.00$). However, we cannot confidently conclude that either car X or car Z differs in performance from car Y ($\bar{X}_Y = 4.00$).

23.

Source	SS	df	MS	F
Between	90.00	1	90.00	36.00
Within	20.00	8	2.50	
Total	110.00	9		

The critical value of F for an alpha level of .05 and 1 and 8 degrees of freedom is 5.32. Since the observed F value of 36.00 is greater than 5.32, we reject the null hypothesis and conclude that a relationship exists between the independent and dependent variables.

24. The observed value of t is -6.00. The square of this value is 36.00, which is equal to the observed value of F from Exercise 23. The critical values of t for an alpha level of .05 and 8 degrees of freedom are ± 2.306. The square of these values is 5.32, which is equal to the critical value of F from Exercise 23. This indicates that one-way analysis of variance bears a mathematical relationship to the independent groups t test in the two-group case such that $F = t^2$.

25. One-way between-subjects analysis of variance.

28. $n = 37$ per group

30.

Source	SS	df	MS	F
Between	140.00	2	70.00	7.38
Within	256.00	27	9.48	
Total	396.00	29		

Null hypothesis tested	Absolute difference between sample means	Value of CD	Null hypothesis rejected?
$\mu_W = \mu_B$	$\lvert 3.00 - 8.00 \rvert = 5.00$	3.42	Yes
$\mu_W = \mu_M$	$\lvert 3.00 - 7.00 \rvert = 4.00$	3.42	Yes
$\mu_B = \mu_M$	$\lvert 8.00 - 7.00 \rvert = 1.00$	3.42	No

Results

A one-way analysis of variance of judgments of the probability of guilt as a function of the defendant's race (white, black, or Mexican-American) was found to be statistically significant,

F(2, 27) = 7.38, p < .01. The strength of the relationship was .35, as indexed by eta². A Tukey HSD test revealed that the mean guilt-probability judgment for the white defendant (M = 3.00) was significantly lower than the mean guilt-probability judgment for either the black (M = 8.00) or the Mexican-American (M = 7.00) defendant. The mean guilt-probability judgments for the black and the Mexican-American defendants did not significantly differ.

Chapter Thirteen

3. A repeated measures analysis of variance is a more sensitive test of the relationship between the independent and dependent variables than a between-subjects analysis of variance because variability due to individual differences is removed from the dependent variable in the form of the sum of squares subjects as part of the repeated measures analysis of variance procedure.

4.

Source	*SS*	*df*	*MS*	*F*
Treatments	20.00	4	5.00	1.67
Subjects	46.00	11		
Error	132.00	44	3.00	
Total	198.00	59		

6. $df_{TREATMENTS} = 4$, $df_{SUBJECTS} = 20$, $df_{ERROR} = 80$, $df_{TOTAL} = 104$

7. 3.25, 2.76, 3.34, 2.58

9. The Huynh–Feldt epsilon and the Greenhouse–Geisser epsilon increase the robustness of one-way repeated measures analysis of variance to violations of the sphericity assumption by adjusting the degrees of freedom for the *F* test so that they are less than or equal to the degrees of freedom that are usually used in assessing the significance of the *F* ratio. The use of these adjusted degrees of freedom serves to decrease the Type I error rate and, thus, to increase the robustness of the statistical test.

13.

Source	*SS*	*df*	*MS*	*F*
Treatments	14.53	2	7.26	4.78
Subjects	24.26	4		
Error	12.14	8	1.52	
Total	50.93	14		

The critical value of *F* for an alpha level of .05 and 2 and 8 degrees of freedom is 4.46. Since the observed *F* value of 4.78 is greater than 4.46, we reject the null hypothesis and conclude that a relationship exists between the time of assessment and anxiety scores.

14. The value of eta-squared is .54. This represents a strong effect.

15.

Null hypothesis tested	*Absolute difference between sample means*	*Value of CD*	*Null hypothesis rejected?*
$\mu_1 = \mu_2$	$\lvert 3.80 - 5.20 \rvert = 1.40$	2.23	No
$\mu_1 = \mu_3$	$\lvert 3.80 - 6.20 \rvert = 2.40$	2.23	Yes
$\mu_2 = \mu_3$	$\lvert 5.20 - 6.20 \rvert = 1.00$	2.23	No

The nature of the relationship between the time of assessment and anxiety scores is that anxiety was greater at time 3 ($\bar{X}_3 = 6.20$) than at time 1 ($\bar{X}_1 = 3.80$). However, we cannot confidently conclude that anxiety at either time 1 or time 3 differs from anxiety at time 2 ($\bar{X}_2 = 5.20$).

16.

Source	*SS*	*df*	*MS*	*F*
Between	14.53	2	7.26	2.40
Within	36.40	12	3.03	
Total	50.93	14		

Analyzing the data as if the independent variable were between-subjects in nature, the observed value of *F* does not exceed the critical value of 3.88, so we fail to reject the null hypothesis. The value of eta-squared in this instance is .29. This represents a strong effect. The fact that eta-squared was .54 in the repeated measures case but only .29 in the between-subjects case indicates that individual differences had a sizable effect on the dependent variable. The fact that the null hypothesis was rejected when a one-way repeated measures analysis of variance was applied but not when a one-way between-subjects analysis of variance was applied illustrates the increased power of the statistical analysis when variability due to individual differences is identified and removed from the dependent variable.

20. One-way between-subjects analysis of variance

21. Correlated groups t test

25. .50

26.

Source	*SS*	*df*	*MS*	*F*
Treatments	220.00	3	73.33	36.66
Subjects	16.00	4		
Error	24.00	12	2.00	
Total	260.00	19		

Null hypothesis tested	*Absolute difference between sample means*	*Value of CD*	*Null hypothesis rejected?*
$\mu_H = \mu_E$	$\|18.00 - 14.00\| = 4.00$	2.66	Yes
$\mu_H = \mu_C$	$\|18.00 - 10.00\| = 8.00$	2.66	Yes
$\mu_H = \mu_V$	$\|18.00 - 10.00\| = 8.00$	2.66	Yes
$\mu_E = \mu_C$	$\|14.00 - 10.00\| = 4.00$	2.66	Yes
$\mu_E = \mu_V$	$\|14.00 - 10.00\| = 4.00$	2.66	Yes
$\mu_C = \mu_V$	$\|10.00 - 10.00\| = 0.00$	2.66	No

Results

A one-way repeated measures analysis of variance compared the mean ratings of how important the four factors (health risks, effectiveness, cost, and convenience) were considered to be in deciding whether to use a male oral contraceptive. This was found to be statistically significant, $\underline{F}(3, 12) = 36.66$, $\underline{p} < .01$. The strength of the relationship, as indexed by eta², was .90. A Tukey HSD test indicated that the health risks factor ($\underline{M} = 18.00$) was rated as significantly more important than the other three factors, and that the effectiveness factor ($\underline{M} = 14.00$) was rated as significantly more important than either the cost ($\underline{M} = 10.00$) or the convenience ($\underline{M} = 10.00$) factor.

Chapter Fourteen

2. The t test procedure for testing the significance of a correlation coefficient using Equation 14.2 and the procedure for testing the significance of a correlation coefficient by comparing the observed value of r with critical values of r are equivalent because the critical values of r using the latter approach are the values of r that correspond to the critical values of t using the former approach.

3. **a.** $\pm.349$
b. $\pm.423$
c. $+.360$
d. $-.441$
e. $\pm.241$

6. In the context of Pearson correlation, eta-squared (r^2) represents the proportion of variability that is shared by the two variables under study. This interpretation differs from the usual interpretation of eta-squared as representing the proportion of variability in the dependent variable that is associated with the independent variable because the independent variable–dependent variable distinction is not relevant to most applications of correlational analysis. If one variable *can* be identified as the independent variable and one variable identified as the dependent variable in a given correlation problem, the more common interpretation of eta-squared will apply.

7. The critical values of r for an alpha level of .05 and 8 degrees of freedom are $\pm.632$. Since the observed r value of $+.74$ is greater than $+.632$, we reject the null hypothesis and conclude that there is a relationship between the two variables.

8. The value of eta-squared is .55. This represents a strong effect.

9. The nature of the relationship is that the two variables are positively linearly related.

13. Independent groups *t* test

14. One-way repeated measures analysis of variance

18. $N = 54$

21. The estimated standard error of estimate estimates the average error that will be made across individuals when predicting scores on *Y* from the regression equation.

23. If two variables are related in a curvilinear fashion, restricting the range of variable *X* tends to increase the magnitude of the observed correlation coefficient. If two variables are linearly related, the effect of restricting the range of one of them is often to reduce the magnitude of the observed correlation coefficient.

24. $\hat{Y} = 6.87 + .46X$; 8.25; 10.09; 11.93

25. 1.71

28. The regression equation for predicting *X* from *Y* is $\hat{X} = -4.90 + 1.17Y$. This differs from the regression equation for predicting *Y* from *X* because an imperfect relationship ($r = .74$) exists between the two variables.

29.

```
                Results

     A Pearson correlation ad-
dressed the relationship
between the leader's LPC score
(M = 14.30) and group problem-
solving performance (M = 10.90).
This was found to be statis-
tically significant, r(8) = .67,
p < .05, suggesting that these
two variables are positively
related.
```

31. The regression equation for predicting graduate grade point average from GRE-A scores is $\hat{Y} = .52 + .005X$. Based on this equation, the predicted 2-year grade point average of applicants 1, 3, 5, and 7 are all greater than 3.00, so these applicants should be admitted to the program.

Chapter Fifteen

4. Observed frequencies are the frequencies that are actually observed in an investigation. Expected frequencies are the frequencies that we would expect to observe if the two variables under study are unrelated in the population.

5. 110

8.

Cell	*E*
Romance-infrequent	10.59
Romance-moderate	15.00
Romance-frequent	19.41
Comedy-infrequent	21.18
Comedy-moderate	30.00
Comedy-frequent	38.82
Drama-infrequent	28.24
Drama-moderate	40.00
Drama-frequent	51.76

9. **a.** 3.841 **b.** 9.488 **c.** 5.991 **d.** 16.919

11. The disadvantage of using Yates' correction for continuity when analyzing 2 × 2 contingency tables is that it tends to reduce the power of the statistical test below what it would otherwise be.

13. The advantage of analyzing quantitative variables with the chi-square test is that they need be measured on only an ordinal level. The disadvantage is that the chi-square test tends to be less powerful than parametric tests.

14. The critical value of chi-square for an alpha level of .05 and 2 degrees of freedom is 5.991. Since the observed chi-square value of 5.39 does not exceed 5.991, we fail to reject the null hypothesis of no relationship between race and voting behavior.

15. .13

16. Since the null hypothesis was not rejected, we cannot conclude that race and voting behavior are related. It is neither necessary nor appropriate to further analyze the nature of the relationship.

20. The observed value of chi-square is 7.11 using both Equation 15.2 and Equation 15.4. Since this exceeds the critical value of 3.841 for an alpha level of .05 and 1 degree of freedom, we reject the null hypothesis and conclude that there is a relationship between smoking behavior and cause of death.

21. .22

22. The nature of the relationship between smoking behavior and cause of death is that smokers are more likely than nonsmokers to die of cancer as opposed to other causes.

23. Pearson correlation

24. Chi-square test

29. .85

30. The critical value of chi-square for an alpha level of .05 and 3 degrees of freedom is 7.815. Since the observed chi-square value of 11.37 is greater than 7.815, we reject the null hypothesis and conclude that the evening news program preference for the population of college students represented by this sample differs from that reflected in the national ratings. Examination of the $O - E$ and $(O - E)^2$ values indicates that college students are more likely than expected to prefer the CBS and NBC evening news programs and less likely than expected to prefer the ABC evening news program or to have no evening news program preference.

32.

Results

The relationship between gender of the central character and the role portrayed by that character was analyzed using a chi-square test. This was found to be statistically significant, $\chi^2(1, \underline{N} = 315) = 10.68$, $\underline{p} < .01$. The observed frequencies for the four cells can be found in Table 1.

Insert Table 1 about here

The strength of the relationship, as indexed by the fourfold point correlation coefficient, was .18. This reflects the fact that female central characters are more likely than male central characters to be portrayed as users as opposed to authorities.

Table 1 would be similar to the table in the exercise, but it would also include the marginal frequencies.

Chapter Sixteen

3.

Set I	*Set II*	*Set III*	*Set IV*
5	1.5	1.5	4
2	1.5	6	2
6	6	4.5	2
1	5	1.5	2
3.5	3	4.5	5
3.5	4	3	6

6. The critical values of z for an alpha level of .05 are ± 1.96. Since the observed z value of .39 is neither less than -1.96 nor greater than $+1.96$, we fail to reject the null hypothesis of no relationship between car ownership and performance in school.

7. .09

8. The critical value of U for an alpha level of .05 and $n_1 = n_2 = 8$ is 13. Since the observed U value of 24.5 is not equal to or less than 13, we fail to reject the null hypothesis of no relationship between gender and attitudes toward the pill.

9. .23

14. The critical value of χ_r^2 for an alpha level of .05 and 2 degrees of freedom is 5.991. Since the observed χ_r^2 value of 7.20 is greater than 5.991, we reject the null hypothesis and conclude that there is a relationship between brand and quality ratings. The nature of the relationship can be determined using the Dunn procedure.

15. .11

16. The critical values of r_s for an alpha level of .05 and $N = 18$ are $\pm .475$. Since the observed r_s value of $+.48$ is greater than $+.475$, we reject the null hypothesis and conclude that there is a relationship between crime rates in cities and the size of a city's police force. The strength of the relationship, as indicated by the magnitude of the correlation coefficient, is .48. The nature of the relationship, as indicated by the sign of the correlation coefficient, is that the two variables are positively related.

17. Results

A Wilcoxon signed-rank test compared individuals' self-esteem before versus after participating in the encounter session. The rank sums were found to be nonsignificantly different, $\underline{N} = 10$, $\underline{T} = 16.5$, ns.

18. Results

A Kruskal-Wallis test was applied to the ranked data relating the affective content of words to learning. The resulting value of $\underline{H}$ was found to be statistically significant, $\underline{H}(2, \underline{N} = 15) = 9.26$, $\underline{p} < .01$. The strength of the relationship was .60, as indexed by epsilon-squared. A follow-up procedure suggested by Dunn (1964) indicated that learning is better for positive words than for negative words.

Chapter Seventeen

1. 2, 3

5. The terminology "main effect" has two related meanings in the context of two-way factorial designs. First, a main effect refers to the comparison of the means for the levels of one independent variable collapsed across the second independent variable. Second, a main effect is said to be present if the null hypothesis concerning that effect is rejected. An interaction effect refers to the comparison of the cell means in terms of whether the nature of the relationship between one of the independent variables and the dependent variable differs as a function of the other independent variable. If the null hypothesis of no interaction effect is rejected, an interaction is said to be present.

6.

	Main effect of factor A?	*Main effect of factor B?*	*Interaction?*
a.	No	Yes	No
b.	Yes	No	No
c.	Yes	Yes	No

8.

	B_1	B_2	B_3
A_1	6	5	4
A_2	1	2	3

10. Nonparallel lines in a graph of population means indicate that an interaction is present. Nonparallel lines in a graph of sample means indicate that an interaction *might* be present. The difference in the two situations is attributable to the fact that, due to the role of sampling error, nonparallel lines in a graph of sample means do not necessarily indicate that an interaction exists in the population.

12.

Source	*SS*	*df*	*MS*	*F*
A	20.00	2	10.00	5.00
B	45.00	3	15.00	7.50
$A \times B$	60.00	6	10.00	5.00
Within	216.00	108	2.00	
Total	341.00	119		

14. **a.** 4.00, 3.15, 3.15
b. 4.04, 2.80, 2.80
c. 3.10, 3.10, 2.47
d. 2.76, 3.15, 2.25

16. 2, 3

17. 54, 6, 9

18. 130.00

19. Hypotheses for the main effect of factor *A*:
H_0: $\mu_{A_1} = \mu_{A_2}$
H_1: $\mu_{A_1} \neq \mu_{A_2}$

Hypotheses for the main effect of factor *B:*

H_0: $\mu_{B_1} = \mu_{B_2} = \mu_{B_3}$

H_1: The three population means are not all equal.

Hypotheses for the $A \times B$ interaction:

H_0: The relationship between factor *A* and the dependent variable is the same for all three levels of factor *B*.

H_1: The relationship between factor *A* and the dependent variable is *not* the same for all three levels of factor *B*.

20. The critical value of *F* for an alpha level of .05 and 1 and 48 degrees of freedom is 4.04. Since the observed *F* value of 10.00 for the main effect of factor *A* is greater than 4.04, we reject the null hypothesis and conclude that there is a relationship between this factor and the dependent variable.

The critical value of *F* for an alpha level of .05 and 2 and 48 degrees of freedom is 3.19. Since the observed *F* value of 4.00 for the main effect of factor *B* is greater than 3.19, we reject the null hypothesis and conclude that there is a relationship between this factor and the dependent variable.

The observed *F* value of 4.00 for the $A \times B$ interaction is greater than the critical value of 3.19, so we reject the null hypothesis and conclude that there is an interaction.

21. The value of eta-squared for the main effect of factor *A* is .14. This represents a moderate effect. The value of eta-squared for the main effect of factor *B* and for the $A \times B$ interaction is .11 in both instances. These also represent moderate effects.

22. The critical value of *F* for an alpha level of .05 and 1 and 60 degrees of freedom is 4.00. Since the observed *F* value of 24.00 for the main effect of factor *A* is greater than 4.00, we reject the null hypothesis and conclude that there is a relationship between this factor and the dependent variable. The nature of the relationship is that scores in condition A_2 ($\bar{X}_{A_2}$ = 19.33) are significantly higher than scores in condition A_1 ($\bar{X}_{A_1}$ = 15.00).

23. The critical value of *F* for an alpha level of .05 and 2 and 60 degrees of freedom is 3.15. Since the observed *F* value of 45.50 for the main effect of factor *B* is greater than 3.15, we reject the null hypothesis and conclude that there is a relationship between this factor and the dependent variable.

The HSD test can be applied as follows:

Null hypothesis tested	*Absolute difference between sample means*	*Value of CD*	*Null hypothesis rejected?*
$\mu_{B_1} = \mu_{B_2}$	$\|12.00 - 17.50\| = 5.50$	2.29	Yes
$\mu_{B_1} = \mu_{B_3}$	$\|12.00 - 22.00\| = 10.00$	2.29	Yes
$\mu_{B_2} = \mu_{B_3}$	$\|17.50 - 22.00\| = 4.50$	2.29	Yes

The nature of the relationship is that scores in condition B_3 ($\bar{X}_{B_3}$ = 22.00) are significantly higher than scores in either condition B_2 ($\bar{X}_{B_2}$ = 17.50) or condition B_1 ($\bar{X}_{B_1}$ = 12.00). In turn, scores in condition B_2 are significantly higher than scores in condition B_1.

30. The problem with unequal cell sizes in two-way analysis of variance is that they introduce a relationship between the two independent variables. The introduction of such a relationship creates a number of statistical and conceptual issues for testing the two main effects and the $A \times B$ interaction.

32. n = 38 per group

34.

Source	*SS*	*df*	*MS*	*F*
A (Race)	80.00	1	80.00	71.43
B (Intelligence)	45.00	1	45.00	40.18
$A \times B$	20.00	1	20.00	17.86
Within	18.00	16	1.12	
Total	163.00	19		

```
Results
    The number of inter-
actions directed toward the
target student were subjected
to a two-way analysis of
variance having two levels of
race (black versus white) and
two levels of supposed in-
telligence (gifted versus
nongifted). All effects were
found to be statistically
```

significant. The main effect of race was such that significantly more interactions were directed toward the white students ($\underline{M} = 33.50$) than toward the black students ($\underline{M} = 29.50$), $\underline{F}(1, 16) = 71.43$, $\underline{p} < .01$. As indexed by eta², the strength of the relationship was .49. The main effect of supposed intelligence was such that significantly more interactions were directed toward the "gifted" students ($\underline{M} = 33.00$) than toward the "nongifted" students ($\underline{M} = 30.00$), $\underline{F}(1, 16) = 40.18$, $\underline{p} < .01$. The strength of the relationship, as indexed by eta², was .28.

The interaction effect, $\underline{F}(1, 16) = 17.86$, $\underline{p} < .01$, was analyzed using simple main effects analysis. This indicated that significantly more interactions were directed toward the white students when they were described as being "gifted" ($\underline{M} = 36.00$) than when they were described as being "nongifted" ($\underline{M} = 31.00$). However, the number of interactions directed toward "gifted" ($\underline{M} = 30.00$) and "nongifted" ($\underline{M} = 29.00$) black students did not significantly differ. As indexed by eta², the strength of the overall interaction effect was .12.

Chapter Eighteen

1. The appropriate statistical technique for analyzing the relationship between two between-subjects qualitative variables is the chi-square test.

5. The most common method of analysis for two quantitative variables is Pearson correlation. This technique is appropriate when the two variables under study are measured on approximately an interval level and the researcher wants to test for a linear relationship. If the expected relationship is nonlinear, procedures for nonlinear relationships can be applied. When one or both variables are measured on an ordinal level that seriously departs from interval level characteristics, Spearman rank-order correlation is usually the test of choice. The Case II statistics elaborated in Table 18.2 might be applicable when one of the variables has fewer than five or so values associated with it and probably *should* be applied if this variable has only two or three values. If both variables have only two or three values, the chi-square test can be used.

6. The independent variable is the time of the year. This variable is qualitative in nature. The dependent variable is scores on the depression scale. This variable is quantitative in nature. Hence, this is a Case II situation. Since the independent variable is within-subjects and has two levels, the appropriate parametric and nonparametric tests for analyzing the relationship between the variables are the correlated groups *t* test and the Wilcoxon signed-rank test, respectively.

7. The independent variable is the color of the ice cream. This variable is qualitative in nature. The dependent variable is the taste ratings. This variable is quantitative in nature. Hence, this is a Case II situation. Since the independent variable is within-subjects and has three levels, the appropriate parametric and nonparametric tests for analyzing the relationship between the variables are one-way repeated measures analysis of variance and Friedman analysis of variance by ranks, respectively.

8. The independent variable is the noise level. This variable is quantitative in nature. The dependent variable is the amount of growth. This variable is also quantitative in nature. Hence, this is a Case IV situation. Since the independent variable is between-subjects and has only two levels, the appropriate parametric and nonparametric tests for analyzing the relationship between the variables are the independent groups *t* test and the Wilcoxon rank sum test/Mann–Whitney *U* test, respectively.

9. The independent variable is creativity. This variable is quantitative in nature. The dependent variable is the presence or absence of an imaginary friend. This variable is qualitative in nature. Hence, this is a Case III situation. For statistical purposes, we will treat the independent variable as the dependent variable and the dependent variable as the independent variable, and apply the decision criteria for Case II situations. Since the independent variable (for statistical purposes) is between-subjects and the dependent variable (for statistical purposes) has only two values, the appropriate test for analyzing the relationship between the variables is the chi-square test.

10. The independent variable is social class. This variable is quantitative in nature. The dependent variable is dogmatism. This variable is also quantitative in nature. Hence, this is a Case IV situation. The appropriate parametric and nonparametric tests for analyzing the relationship between the variables are Pearson correlation (assuming the expected relationship is linear) and Spearman rank-order correlation, respectively.

11. The independent variable is the sipping rate of the experimental assistant. This variable is quantitative in nature. The dependent variable is subjects' consumption time. This variable is also quantitative in nature. Hence, this is a Case IV situation. Since the independent variable is between-subjects and has only three levels, the appropriate parametric and nonparametric tests for analyzing the relationship between the variables are one-way between-subjects analysis of variance and the Kruskal–Wallis test, respectively.

12. The independent variable is the presence or absence of hypnosis. This variable is qualitative in nature. The dependent variable is hand temperature. This variable is quantitative in nature. Hence, this is a Case II situation. Since the independent variable is between-subjects and has two levels, the appropriate parametric and nonparametric tests for analyzing the relationship between the variables are the independent groups *t* test and the Wilcoxon rank sum test/Mann–Whitney *U* test, respectively.

13. The independent variable is the time between the test. This variable is quantitative in nature. The dependent variable is scores on the Palmer Sweat Index. This variable is also quantitative in nature. Hence, this is a Case IV situation. Since the independent variable is within-subjects and has four levels, the appropriate parametric and nonparametric tests for analyzing the relationship between the variables are one-way repeated measures analysis of variance and Friedman analysis of variance by ranks, respectively.

21. The defining characteristic of multivariate statistics is that they consider the variation among multiple variables.

23. The problem with conducting multiple analyses of variance when analyzing two or more dependent variables is that this would increase the probability of making at least one Type I error beyond the probability specified by the alpha level.

25. a. A regression coefficient is the name given to *b* in the context of multiple regression. It represents the number of units a criterion variable is predicted to change for each unit change in a given predictor variable when the effects of the other predictor variables are held constant.

b. A multiple regression equation is an equation used to predict individuals' scores on a criterion variable from two or more predictor variables. It contains a regression coefficient for each predictor variable and an overall intercept.

c. A squared multiple correlation coefficient is an index of the strength of the relationship between a criterion variable and a set of predictor variables. Specifically, it indicates the proportion of variability in the criterion variable that is associated with the predictor variables considered simultaneously.

INDEX